SCIENCE AND DIGITAL TECHNOLOGY FOR CULTURAL HERITAGE

PROCEEDINGS OF THE 4TH INTERNATIONAL CONGRESS SCIENCE AND TECHNOLOGY FOR THE CONSERVATION OF CULTURAL HERITAGE (TECHNOHERITAGE 2019), SEVILLE, SPAIN, 26-30 MARCH 2019

Science and Digital Technology for Cultural Heritage

Interdisciplinary Approach to Diagnosis, Vulnerability, Risk Assessment and Graphic Information Models

Editors

Pilar Ortiz Calderón

Pablo de Olavide University, Seville, Spain

Francisco Pinto Puerto

University of Seville, Seville, Spain

Philip Verhagen

Vrije Universiteit Amsterdam, Amsterdam, The Netherlands

Andrés J. Prieto

Universidad Austral de Chile, Valdivia, Chile

CRC Press is an imprint of the
Taylor & Francis Group, an **informa** business
A BALKEMA BOOK

CRC Press/Balkema is an imprint of the Taylor & Francis Group, an informa business

Typeset by Integra Software Services Pvt. Ltd., Pondicherry, India

Library of Congress Cataloging-in-Publication Data

Applied for

Published by: CRC Press/Balkema
Schipholweg 107C, 2316XC Leiden, The Netherlands

First issued in paperback 2023

ISBN: 978-1-03-257088-4 (pbk)
ISBN: 978-0-367-36368-0 (hbk)
ISBN: 978-0-429-34547-0 (ebk)

DOI: https://doi.org/10.1201/9780429345470

Publisher's Note
The publisher has gone to great lengths to ensure the quality of this reprint but points out that some imperfections in the original copies may be apparent.

Science and Digital Technology for Cultural Heritage – Ortiz Calderón et al. (Eds)
© 2020 Taylor & Francis Group, London, ISBN 978-0-367-36368-0

Table of Contents

Management and sustainability of the Cultural Heritage information. Social value, policies and applications about standardization and protocols

Science and Digital Technology for Cultural Heritage – Ortiz Calderón et al. (Eds)
© 2020 Taylor & Francis Group, London, ISBN 978-0-367-36368-0

Preface

The scientific and technological advances that influence the protection of cultural heritage are developing at an ever-increasing pace. Systems to explore, research and analyse their materiality; to control the different scopes; or to represent and model them have reached an unprecedented dimension in recent decades. Digital technologies not only improve quantitatively and qualitatively the possibilities of approaching historical legacy, but they also allow for new ways of interpreting and disseminating it. New horizons, which allow to incorporate very diverse views of heritage, coming from increasingly specialised disciplines with their own languages and algorithms, are opening up. This multidisciplinary vision enriches the processes of tutelage and joint interpretation of the problem, but it makes them more complex at the same time. On the other hand, for some years now, the difficulties of management and permanence of an increasingly specialised amount of information, which also evolves and grows almost in real time, have been made evident.

The Network of Science and Technology for the Conservation of Cultural Heritage, part of the Spanish National Research Council (CSIC, by its Spanish acronym), began operating in March 2011. It brings together sixty-seven groups, which are divided into three areas of activity: research groups from the Spanish National Research Council and from different Spanish universities; cultural institutions, foundations and museums; and companies in the sector. The Network aims to promote collaboration between the agents of the science-technology-business system, in order to facilitate the sharing of ideas and experiences that aid problem solving and to foster technology transfer, with the common goal of contributing to the conservation of Cultural Heritage.

In the context of the TechnoHeritage Network, the fourth edition of the International Congress on Science and Technology for the Conservation of Cultural Heritage was held with the collaboration of the IAPH (Andalusian Institute of Historical Heritage). This Congress was an international meeting of researchers and specialists from multiple areas, whose line of work or purpose is the knowledge and conservation of Cultural Heritage —with several sections dedicated, on this occasion, to architectural heritage —and whose strategies and working tools involve digital technologies. This publication compiles the material presented at this congress and organises it into several thematic blocks or parts: (1) New digital graphic instruments such for knowledge, analysis, protection and conservation of Cultural Heritage, (2) Management and sustainability of the Cultural Heritage information, (3) Importance, social value and policies in the management of information for the conservation of Cultural Heritage, (4) Risk Assessment and monitoring of Cultural Heritage, (5) New technologies, products and materials for conservation and maintenance of Cultural Heritage, (6) Vulnerability assessment: Agents and Mechanisms of Decay.

Among all the topics discussed, the role and impact of digital technologies for the knowledge, maintenance, management and dissemination of cultural heritage should be highlighted. Digital media not only facilitates and improves the scientific and technical processes that are traditionally used to protect heritage—it also modifies the way of understanding this heritage, of perceiving it and transmitting it, and offers a new horizon of strategies to make decision-making more sustainable over time.

Image credits: Graphic Fund IAPH-Eugenio Fernández Ruiz

Science and Digital Technology for Cultural Heritage – Ortiz Calderón et al. (Eds)
© 2020 Taylor & Francis Group, London, ISBN 978-0-367-36368-0

Organization and committees

ORGANIZING COMMITTEE

Direction

Francisco Pinto Puerto. *Universidad de Sevilla, Spain.*
Pilar Ortiz Calderón. *Universidad Pablo de Olavide, Spain.*
Manuel Castellano Román. *Universidad de Sevilla, Spain.*

Organization

Javier Becerra Luna. *Universidad Pablo de Olavide, Spain.*
Daniel Cagigas Muñiz. *Universidad de Sevilla, Spain.*
Victoria Domínguez Ruiz. *Universidad de Sevilla, Spain.*
Patricia Wanderley Ferreira Lopes. *Universidad de Sevilla, Spain.*
Marta García de Casasola Gómez. *Instituto Andaluz del Patrimonio Histórico, Junta de Andalucía, Spain.*
Auxiliadora Gómez Morón. *Instituto Andaluz del Patrimonio Histórico, Junta de Andalucía, Spain.*
Elena González Gracia. *Universidad de Sevilla, Spain.*
Jorge Moya Muñoz. *Universidad de Sevilla, Spain.*
Rocío Ortíz Calderón. *Universidad Pablo de Olavide, Spain.*
Andres J. Prieto Ibañez. *Universidad Austral de Chile, Chile.*
Reyes Rodríguez García. *Universidad de Sevilla, Spain.*
Carmen Sánchez Galiano. *Instituto Andaluz del Patrimonio Histórico, Junta de Andalucía, Spain.*
Mª Auxiliadora Vázquez González. *Universidad de Sevilla, Spain.*

COLLABORATION IN THE ORGANIZATION

Mercedes Andrades Borras. *Alumna Interna Dpto. de Construcciones Arquitectónicas I. ETSA.*
Pablo Valcárcel García. *Alumno Interno Dpto. de Construcciones Arquitectónicas I. ETSA.*
Jesús Muñoz Pérez. *Alumno Interno Dpto. de Construcciones Arquitectónicas I. ETSA.*
Sandro Massaro Jimenez. *Alumno Interno Dpto. de Construcciones Arquitectónicas I. ETSA.*
Cristina Mesa García. *Alumna Interna Dpto. de Construcciones Arquitectónicas I. ETSA.*
Victor Manuel Gonçales González. *Alumno Universidad Pablo de Olavide.*
Abián Torres González. *Alumno Interno Dpto. de Construcciones Arquitectónicas I. ETSA.*

ORGANIZED BY

TECHNOHERITAGE, Science and Technology Network for the Conservation of Cultural Heritage, Spain.
TUTSOSMOD, Universidad de Sevilla, Spain.
ART-RISK, Universidad Pablo de Olavide, Spain.
IAPH, Instituto Andaluz de Patrimonio Histórico, Spain.

INSTITUTIONAL SUPPORT

IAPH – Instituto Andaluz del patrimonio Histórico, Consejería de Cultura y Patrimonio Histórico.
Junta de Andalucía
CSIC – Consejo Superior de Investigaciones Científicas.
ETSAS – Escuela Técnica Superior de Arquitectura de Sevilla.
ETSIE – Escuela Técnica Superior de Ingeniería de Edificación.
IUACC – Instituto Universitario Arquitectura y Ciencias de la Construcción.
Casa de la Ciencia de Sevilla.

FINANCED BY

Ministerio de Ciencia, Innovación y Universidades, Gobierno de España. Project TUTSOS-MOD ("Tutela sostenible del patrimonio cultural a través de modelos digitales BIM y SIG. Contribución al conocimiento e innovación social", Ref. HAR2016-78113-R; main researcher: Francisco Pinto Puerto) and project ART-RISK ("Inteligencia artificial aplicada a la conservación preventiva de edificios patrimoniales", Ref. BIA2015-64878-R; main researcher: María del Pilar Ortiz Calderón).
VI Plan Propio de Investigación de la Universidad de Sevilla. Universidad de Sevilla, Spain.
Universidad Pablo de Olavide, Spain.
Unión Europea, Fondo Europeo de Desarrollo Regional.

COLLABORATING COMPANIES

Fundación ARQUIA, Spain.
ESRI, Spain.

HONOR COMMITTEE

Alfonso Jiménez Martín, *Académico de número, Real Academia Sevillana de Ciencias, Spain.*
José María Cabezas Méndez
Premio Nacional de Restauración y Conservación de Bienes Culturales 2014
Excmo. y Rvdmo. Sr. D. Juan José Asenjo Pelegrina, Arzobispo De Sevilla, Spain.
Antonio Tejedor Cabrera, *Director IUACC, Spain.*

ORGANIZED BY

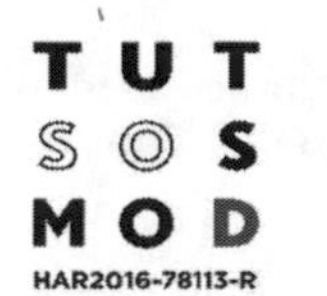

Instituto Andaluz del Patrimonio Histórico
CONSEJERÍA DE CULTURA Y PATRIMONIO HISTÓRICO

FINANCED BY

Unión Europea

Fondo Europeo
de Desarrollo Regional
"Una manera de hacer Europa"

INSTITUTIONAL SUPPORT

COLLABORATING COMPANIES

fundación **arquia**

Plenary lectures

Science and Digital Technology for Cultural Heritage – Ortiz Calderón et al. (Eds)
© 2020 Taylor & Francis Group, London, ISBN 978-0-367-36368-0

Laser remote and in situ spectroscopic diagnostics to CH surfaces

R. Fantoni, L. Caneve, F. Colao, M. Ferri De Collibus, M. Francucci, M. Guarneri & V. Spizzichino
ENEA, FSN-TECFIS, Frascati, Italy

O. Santorelli
ENEA guest

ABSTRACT: Optical and spectroscopic techniques offer unique possibilities for non-destructive or micro-destructive characterization of painted Cultural Heritage surfaces. The development of fast laser scanners in combination with sensitive CCD detector gave the chance to design and operate portable systems suitable to *in-situ* an remote spectroscopic imaging. Different prototypes have been developed and patented to collected reflectance and fluorescence images excited at different ultraviolet and visible laser wavelengths, Raman and LIBS signals. Portable integrated instruments suitable for operation at different distance from a 1.5 to 30 m, have been assembled and operated in laboratory on multilayered samples and in field campaigns on CH painted surfaces carried out within the COBRA and ADAMO regional projects. Results obtained on painted surfaces on monuments in Latium will be presented demonstrating the potentiality of combined in situ use of optical and spectroscopic tools in solving specific CH characterization problems.

1 INTRODUCTION

The preservation of CH surfaces requires suitable material diagnostics. CH surfaces have been produced on very different substrates, some of them very fragile, and often at a monumental scale (large size) with limitation of access (remote view). CH surfaces can be located in hostile environments (underwater) or must be examined in dangerous situations (after earthquakes, wars) which asks for complex interventions. There was a definite need for *in-situ* and remote surface diagnostics on CH which was satisfied with the laser break-through, i.e. the introduction of laser scanners for non-invasive or micro-destructive interrogation of the surface, which made possible to perform:

- Optical measurements consisting of collection of sets of monochromatic images by multiple visible laser scanners to reconstruct 3D model with native color information. Laser reflectance is measured, i.e. backscattered and diffused signals. Data relevant to surface appearance and morphology are acquired [1].
- Spectroscopic measurements consisting of space resolved collection of spectra containing information on surface layers. Most used laser spectroscopies are LIF [2], Raman [3], LIBS [4], with possibilities of time resolved detection. Data relevant to surface elemental and aggregate composition are collected. Possibility of subsurface analysis or stratigraphy is offered by some techniques.
- Joint application of different remote and *in-situ* diagnostics (e.g. thermography, XRF, PIXE, LIBS, Raman), with point detection or imaging capabilities.

The monochromatic laser beam interaction with a surface may cause different phenomena, with a probability depending on the incoming power for surface unit (irradiance), which determines the final energy balance. Together with partial radiation absorption, at growing irradiance, we may encounter: Back Scattering (BS) at the same wavelength as the exciting beam; Laser induced Fluorescence (LIF) at wavelengths larger than the incoming one, with shifts related to energy differences between electronic states, eventually coupled through internal relaxations processes in species at the surface; Stokes Raman Scattering (SRS) at wavelengths larger than the incoming one, with shift related to vibrational modes in species at the surface; Laser Induced Breakdown (LIBS) with atomic emission from the plasma generated at the surface, during an ablation/ionization process occurring above the threshold (~1 GW/cm^2).

Portable optical and spectroscopic instruments are nowadays available for in situ operation, specifically for multispectral/hyperspectral imaging and point diagnostics by use of local probes (e.g. XRF and Raman) operated in contact or at very short distances. Scaffolding must be used for reaching far targets.

High resolution laser scanners for imaging and analysis offer peculiar advantages first of all the remote use, other advantages include in general non-destructivity and non-invasiveness or at least only micro-destructivity (LIBS) and self-illumination, being not affected by external light sources. Furthermore additional geometrical requirements can be taken into account, automatic software handling/processing of very large data sets can be developed, and finally data can be stored as references in digital archives. In the present paper we show the novelty of integrated use of more than one single prototype (in hardware or software), their possible integration with different kinds of *in- situ* sensors, and discuss perspectives of virtual fruition.

2 LASER SCANNER PROTOTYPES

Different prototypes were realized by ENEA in the Diagnostic and Metrology Laboratory for remote characterization of CH surfaces, each addressed to obtain a digital characterization of physical and chemical properties of the outermost layers. These laser scanner prototypes can be used separately, according to the problem presented by the examined CH case study. Alternatively their combined use on the same CH target not only gives prognostic structural information indicative of potential structure damages, but it also produces multi-layered images with enhanced information contents. Results are valuable in CH characterization of surface materials, with respect to both initial constituents and restoration products. In fact the developed diagnostic methodologies with appropriate data treatments enhance colour vision, reveal material composition and document areas of early deterioration or previous restoration interventions, thus supporting both conservation and fruition.

2.1 *The RGB-ITR system*

The Red Green Blue Imaging Topological Radar (RGB-ITR) is a 3D laser scanner prototype, realized and utilized on field campaign for CH painted surface modeling. It is based on three amplitude modulated (AM) laser sources emitting continuous-wave coherent light with mW power at 440nm (blue), 517nm (green) and 660nm (red) mixed together by means of dichroic filters for obtaining a single white ray illuminating the investigated target. The system uses a motorized mirror for focusing and sweeping the laser beam onto the target. Its latest optical head, realized in order to optimize the scanning effectiveness in indoor large spaces, the actual angle of view of RGB-ITR is 90° x 360°, with a point-to-point precision of 0.002°. The data acquisition rate can currently reach the maximum value of 10000 sample points (pixels) per second. Five channels of information per each pixel are collected by three avalanche photodiodes and dedicated electronics: two distance measures (taken at high and low frequency) and three target reflectivity signals (red, green and blue channels) are acquired with typical modulation frequencies of 190MHz for red, 1–10MHz for blue and 3MHz for green. RGB-ITR is a non-invasive, AM, continuous-wave 3D laser scanner that concurrently gathers information on the surface shape and native color of the investigated target. The device is particularly suitable to reconstruct high-resolution (sub-millimetric), high-quality, dense, accurate

3D color models for a hyper-realistic rendering of colored features on a painted target and for identification of irregularities in surface morphology. Native color is recorded, after balancing intensity recorded on standard white target at each distance, and can be used as a calibrated reference. A complete description of the set-up is given elsewhere [5].

2.2 *The IR-ITR system*

The Infrared Imaging Topological Radar (IR-ITR) is a high resolution laser scanner for monochromatic imaging, operating in the near infrared @ 1.550 μm specifically realized for subsurface imaging and modeling. The principle of operation is the same as the RGB-ITR, but only two channels are recorded: near IR intensity and distance. Due to the used wavelength, subsurface penetration is possible, conversely a slightly worst space resolution is expected, current performances obtained with the first IR-ITR prototype are summarized in Table 1.

2.3 *The LIF scanner*

It is a line scanner for hyperspectral fluorescence imaging, contemporary collecting also reflectance images with the laser off option. The laser beam is shaped as a light blade, thus there is no need to vary in-plane θ angle, the scan system controls only the bending φ. The entire line profile is imaged on the monochromator slit, which spreads the entire spectrum collected at each point on the squared ICCD at 90°. Time Resolved data are obtained collecting LIF spectra at different delays, by gating the camera. On-site operation for CH surface remote characterization has been widely discussed elsewhere [6].

2.4 *The LIF imaging system FORLAB*

It is a large area fast imager, able to collect up to 8 different monochromatic fluorescence images after excitation in the UV. The apparatus is based on a KrF laser (operation at 248 nm, repetition rate of 500 Hz, pulse duration of 10 ns, maximum energy of 20 mJ) and an ICCD which collect fluorescence images behind one filter selected among 8 available on a wheel. Its main advantages are fast acquisition time (Scan time 60s/band for 10m x 10m area), reduced data processing on selected spectral bands, portable, remote operation up to 30m. Its applications in museums and restoration ateliers during COBRA project have been already reported [7].

2.5 *The ILS system*

It is an integrate laser scanner for remote surface chemical characterization at selected points. It can acquire LIF and Raman spectra upon excitation @355 nm, and perform LIBS analysis with laser ablation @1064nm, in the latter case is also capable of stratigraphy. Remote LIBS measurements with stratigraphy have been already preformed on CH ceramics [8] surfaces.

Table 1. IR-ITR performances measured in laboratory.

	@ 5m distance	@10m distance
Scanned size	418 x 452 mm	489 x 558 mm
Angular resolution	0.001 x 0.002 deg	0.001 x 0.002 deg
Pixel resolution	2400 x 2600	1400 x 1600
Space resolution	0.174 x 0.174 mm	0.349 x 0.349 mm
Acquisition time	1 hour 45 min	39 min
Spot size at the surface	1.5-2.0 mm	3.0-4.0 mm

3 APPLICATION WITHIN REGIONAL PROJECTS

3.1 *Field campaigns within COBRA project*

Most former application of ENEA laser scanner prototypes were within the regional project COBRA whose main objective was to develop and disseminate methods, technologies and advanced tools for the conservation of cultural heritage, based on the application of radiation and Enabling Technologies. More than 20 different sites were examined and main results have been already reported. A single example is here presented relevant to Priscilla catacomb in Rome, as an application in hypogea.

3.1.1 *The Greek Chapel in Priscilla Catacomb, Rome*

Priscilla Catacomb in Rome were excavated in tuff from the IInd to the IVth century a. D. forming about 13 Km of undersoil galleries in three distinct levels. In the antiquity it was known as the "Queen of catacombs" due to the large number of martyrs there built. The Greek chapel, so named after some inscriptions in Greek there found, is a rather large, structured hall with painted walls dating back to the second half of the IInd century a.D.

Two field campaigns were carried out by ENEA laser scanners at about six months distances, aimed at a 3D color reconstruction of the chapel for purposes of fruition, cataloguing, digital storage as historical memory of the artwork, diagnostics and monitoring of its preservation state (Figure 1 and 2). Specifically the fresco's were investigated by RGB-ITR, also in combination with different in-situ spectroscopic diagnostics (Raman, XRF), for discoloration and bio-degradation. The campaigns were performed in hostile environmental conditions, due to the high degree of humidity (up to 98%) coupled to a temperature constantly near 18°C, which required the use of military grade electronics for part of the data acquisition chain.

The high resolution color 3D model of the entire chapel was obtained, as Figure 1 shows, part of the fresco's had suffered significant discoloration, so a large effort in image post-processing was started to retrieve or enhance fainting features, this activity is still in progress.

Bio-degradation was investigated by LIF, the characterization of a colony found under a part of the right arcade was obtained by collecting different fluorescence images. The bio contaminant had a signature peaked in the UV and no emission in the red (Figure 2), so that photosynthetically active micro-organism could be ruled out. This spectral signature in

Figure 1. Part of the 3D color model of the Greek Chapel collected by the RGB-ITR. On the arcade the fresco representing the Eucharistic banquet "*Fractio panis*" can be recognized.

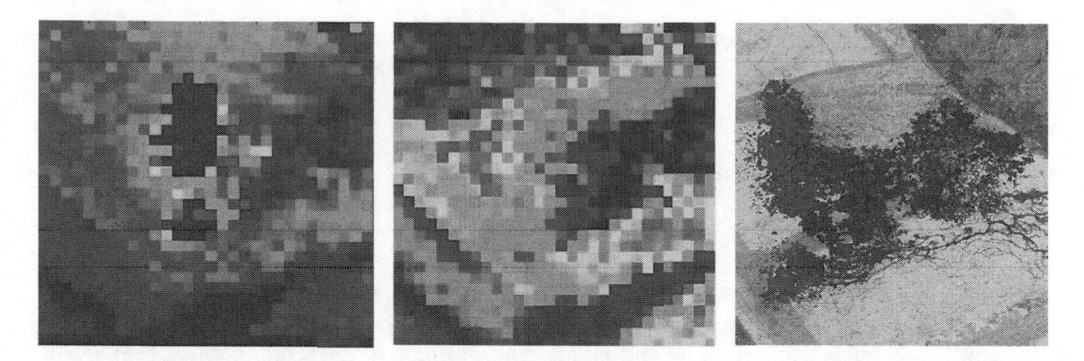

Figure 2. Characterization of bio-degradation on a fresco: color picture (right), fluorescence image at 600nm (centre), fluorescence image at 340nm (left).

combination with optical microscopy performed after sampling, resulted consistent with the presence of colonies of actinobacteria. The availability of the 3D model allowed to measure the extent of the contaminated area and to monitor its significant growth (about +25%) after six months (picture not shown).

3.2 *Field campaigns within ADAMO project*

ADAMO Technologies of Analysis, Diagnostics and Monitoring for the preservation and restoration of Cultural Heritage. It is A Research project in the Center of Excellence of the District of Technologies for Culture of Latium Region, its objectives are: technology transfer on relevant themes, services to enterprises based on facilities offered by DTC partners, demonstrations in selected cases studies, development of prototypes and test of innovative products. Several field campaigns are on-going, preliminary results on a couple of sites are reported in the following.

3.2.1 *Painted walls at Bishop Palace, Frascati*

The Hall of Landscapes in the Bishop's Palace appears as a colorful room fully decorated with tempera painted canvas covering all the walls, where painted areas and wood frames create the illusion of an architectonic structure. Interest arises on Location of current damages on canvas and other painted surfaces. The entire room was scanned by means of the RGB-ITR prototype. An example of the panted canvas on the wall is shown in Figure 3, where the red segment correspond to the line relevant to the reconstructed profile. Additionally discoloration (not shown) was observed on the painted sealing in correspondence to a former water infiltration.

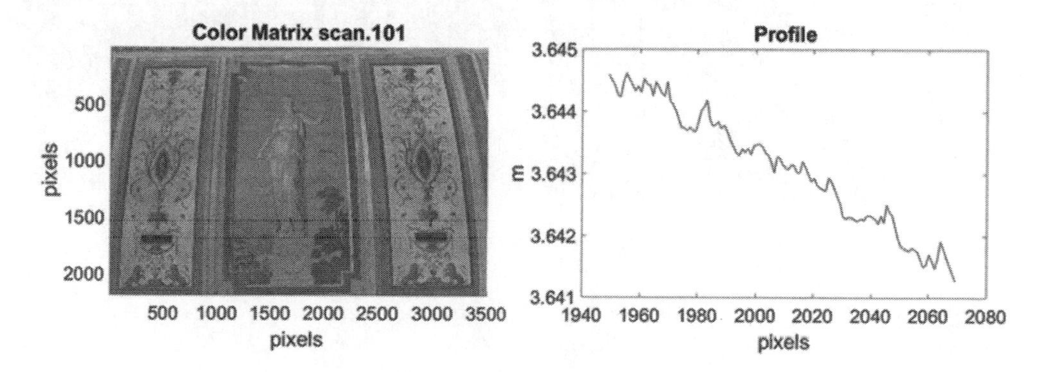

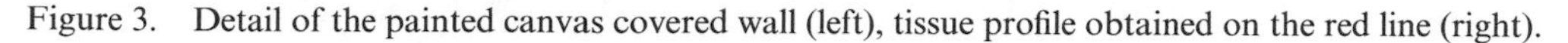

Figure 3. Detail of the painted canvas covered wall (left), tissue profile obtained on the red line (right).

A LIF investigation was performed of the wooden cover at fireplace in the same room. A carefully repainted crack is evident in the processed LIF image (Figure 4). The availability of the 3D model allowed to precisely localize the crack, after projecting the LIF image on the model.

3.2.2 *Paints at Chigi Palace, Ariccia*

Chigi Palace in Ariccia represents a unique example of baroque home remained intact over the centuries. The palace was transformed into a magnificent baroque residence by Gian Lorenzo Bernini in the XVIIth century. The picture-gallery contains paints from Mario de Fiori and other Roman baroque artists, sometimes working together on the same canvas. Interest arises to find out whether it is possible to remotely reveal changes of mind of the last author. As shown in Figure 5, in scanning the paint "The Springtime" several change-of-mind are revealed in the near infrared image collected at 1550 nm, remotely obtained by the IR-ITR at about 8 m distance.

Figure 4. Detail of a painted wood (left), evidence of restoration on the fluorescence image (right).

Figure 5. The spring time – oil on canvas: Color picture (left), image by the IR-ITR (right).

4 CONCLUSIONS

Spectroscopic techniques (LIF and TR-LIF, Raman, XRF) allowed to obtain remotely and *in situ* unique information on different CH surfaces. 3D high resolution color models are useful to precisely locate spectroscopic data collected on different points and 2D fluorescence images. The integration of different remote optical and spectroscopic techniques is often the only way to solve complex real challenges from case studies in CH samples in order to effectively answer conservators' questions.

ACKNOWLEDGEMENTS

Work supported by Latium Region (ADAMO project - District of Technology for Culture).

REFERENCES

Vacca, G., et al. "Laser scanner survey to cultural heritage conservation and restoration". 2012. International Archives of the Photogrammetry, Remote Sensing and Spatial Information Sciences, Volume XXXIX-B5, 2012 XXII ISPRS Congress, Melbourne, Australia.

Hällström, J., et al. 2009. "Documentation of soiled and biodeteriorated facades: A case study on the Coliseum, Rome, using hyperspectral imaging fluorescence lidars". *J. Cult. Heritage* 10:106–115.

A. Sodo, et al.2010."The colours of Etruscan paintings: 480 a study on the Tomba dell'Orco in the necropolis of Tarquinia", *J. Raman Spectr*. 39: 1035–1041.

I. Osticioli, et al. 2009. "A new compact instrument for Raman, laser-induced breakdown, and laser-induced fluorescence spectroscopy of works of art and their constituent materials" *Rev. Sci. Instrum*: 80, 076109.

Fantoni R., et al 2017. Laser scanners for remote diagnostic and virtual fruition of cultural heritage. *Optical and Quantum Electronics* 49, Article 120.

Fantoni R., et al. 2013. Laser-induced fluorescence study of medieval frescoes by Giusto de' Menabuoi. J. Cultural Heritage 14, S59–S65.

Spizzichino, V., et al. Origin Determination of Mediterranean Marbles by Laser Induced Fluorescence. 2018. Lecture Notes in Computer Science 11196 LNCS, pp. 212–223.

Lazic V., et al. 2016. Corrections for variable plasma parameters in laser induced breakdown spectroscopy: Application on archeological samples. Spectrochim. Acta B 122, 103–113.

Science and Digital Technology for Cultural Heritage – Ortiz Calderón et al. (Eds)
© 2020 Taylor & Francis Group, London, ISBN 978-0-367-36368-0

Architectural heritage information in 3D geospatial models: Developing opportunities and challenges

F. Noardo
Delft University of Technology (Urbanism Department, 3D geoinformation group), Delft, The Netherlands

ABSTRACT: The relationship between the cultural value of architectural heritage and its spatial features is straightforward. Digital 3D models are very effective to store such information, which can be further enhanced through the management in 3D information systems. Mainly two approaches are proposed by literature, notwithstanding the still existing challenges: the Heritage Building Information Models (HBIMs) or the archiving of the information in 3D city models. They are both extensions of systems and tools adopted in other fields (construction for BIMs and city management for 3D city models). They use different methodologies and technical solutions (e.g. different kinds of geometries; different reference standards foreseeing interoperability, etc.). However, some work is on-going for integrating them. A review of HBIM and 3D city models including heritage is described, in connection with the respective supported applications, as a starting point towards their integration for a more comprehensive 3D geospatial information for architectural heritage.

1 INTRODUCTION

Spatial values are intrinsically connected to the cultural value of architectural heritage, since, most of times, they embody the most valuable characteristics of the involved objects in many levels of detail: from the smallest objects to the widest landscape compositions.

It is essential to consider their effective representation (of multiscale shapes and spatial relationships), also in connection with non-spatial properties, because those values are critical when dealing with heritage management and preservation. Recent European projects have highlighted this need, like the Virtual Multimodal Museum project and its Manifesto, which stresses the importance of the inclusion of a richer information in documentation systems. Therefore, the spatial documentation of architectural heritage is of extreme importance, and it can be more and more accurate and complete, by means of the possibilities offered by the new advanced tools for 3D survey and spatial data management.

In this paper, the support given by the (3D) spatial documentation to the architectural heritage preservation and management is described. Moreover, a short review of recent research about some interesting solutions for the management of the heritage information using 3D spatial information systems (Heritage Building Information Models and 3D city models including heritage) will be used as base to draw some preliminary considerations towards the adoption of integration solutions, under development, for architectural heritage.

2 ARCHITECTURAL HERITAGE SPATIAL DOCUMENTATION

A great step towards a better understanding of architectural heritage and its preservation was even the 2D graphical representation, better if supported by survey measurements (Fasolo, 1954). However, many architectures have very complex shapes and require instead a 3D approach. For this reason, the 3D representation methodologies had a wide success in recent times, aided by the great advancement of survey instruments and digital representation tools (e.g. Centofanti et al., 2014;

Scopigno et al., 2011). 3D models are used effectively for communication and education (Paladini et al., 2019). As examples, Rua, Alvito (2011) employed 3D models in archaeological gaming applications; some 3D inventories, like the Cyark inventory are built as support for education aims (https://www.cyark.org); the 3D reconstruction of Pompei is used in a 3D movie reproducing the tragedy (http://www.pompeii3d.eu).

Also great support can be given by 3D models to preservation: for example, they make the accurate computation of surfaces possible for restoration projects and can be used for improved structural analysis (e.g. Bertolini-Cestari et al., 2015, Riveiro et al., 2011).

However, plain 3D data are not sufficient yet to archive the complexity of heritage: many data are connected to one single architecture (from different sources, representing different states in time, stored in various formats), and functional and spatial relationships must be considered to underline and understand complex (multi-dimensional) systems. For example, the semantic relationships (the meaning of objects) have to be considered through wide pieces of land (including the ones which build up as, or connected to intangible heritage). Moreover, the relations to risk and vulnerability elements are important to support preservation. Furthermore, a multi-scale representation is needed, since sometimes the small details are the ones that embody the proper cultural value, in elements distributed on the landscape, or as the product of traditional skills and know-hows. Finally, some administrative information, like the ownership relationship of buildings are important for their understanding and management (e.g. common properties of buildings, common managements, specific constraints, etc.). This are only examples of a very high richness and complexity characterizing heritage and related dynamics. It is therefore essential to consider this complexity in a system for improving knowledge, preservation and supporting sustainable and effective enhancement and promotion, by structuring the data in effective information. And this can be done through spatial information systems.

3 SPATIAL INFORMATION SYSTEMS AND 3D SPATIAL INFORMATION SYSTEMS

Spatial information systems can give great support to heritage understanding and preservation even when managing 2D data (e.g. Salonia, Negri, 2003; Donadio, Spanò, 2015; Noardo, Spanò, 2015; Melòn et al., 2019, Lu et al., 2019). However, even more powerful tools are given by the management of heritage in 3D information systems, namely Building Information Models (BIM) and 3D Geographical Information Systems (3D GIS).

They are usually considered as opposed, but actually they have a lot in common at a conceptual level, if considered for the architectural heritage application: they are both developed by other fields then cultural heritage: architecture engineering and construction field for BIM and cartography, geomatics and city management for 3D GIS, or 3D city models.

They both can archive 'enriched' 3D models, joining the geometric representation, in many levels of approximation, with the semantics of the models, by associating the objects to concepts, usually hierarchically organized, with attributes and reciprocal relationships. They can be georeferenced, even if it is still challenging for BIM. Finally, they are not only archives, but also include a wider framework and methods for managing processes and tools.

The key condition for information to be universally understood, used, re-used, and exchanged is interoperability. For this reason, open standards are being developed with a wide consensus among stakeholders, academics and developers, within international organizations.

The most widespread open standards in BIM and 3D city models domains are, respectively, Industry Foundation Classes (IFC) by buildingSMART and CityGML by the Open Geospatial Consortium (OGC). In the following subsections, the main characteristics of BIM and 3D city models, their standards and their use in the heritage field is summarized.

3.1 *Building information models and heritage building information models*

Building Information Modelling derives from the evolutions of Computer-Aided Drawing software, used in the field of building design, and implies not only 3D models, but more complex systems including procedures and tools for managing the project and the modelled object, being mainly buildings and infrastructures (Historic England, 2017).

Many tasks can be assisted by them: e.g. design options assessment; quantities and cost estimation; construction simulation; energy modelling; manufacture and off-site construction; project management (efficient collaboration, multi- disciplinary project team); facilities and asset management; design and construction coordination.

In building information modelling, levels of approximations are used, which are called 'Levels Of Development' (LOD), and give an indication about the progress in the design definition (from a general draft in LOD100 to the inclusion of fabrication details of each single small element in LOD400; LOD500 is for as-built BIMs) (BIMFORUM, 2019).

For BIM, IFC is the reference open standard (ISO 16739). It is part of a more complex standard including processes and further tools. It defines rules, structures and encodings for the management of interoperable, exchangeable and reusable BIM data. Moreover, the IFC data model defines a standardized semantics, which is strictly related to new constructions.

The field of architectural heritage has also understood the potential that such tool could develop for supporting preservation and management of historical constructions. Therefore, many researches are on-going about the topic: e.g. for supporting heritage interpretation (e.g. Attico et al., 2019), structural analysis and testing preservation and intervention scenarios (e.g. Malinverni et al., 2019; Chiabrando et al., 2017), as a framework for collaborative processes and sharing of coordinated datasets in multi-disciplinary environments; for architectural heritage energy-related management and retrofitting (e.g. Khodeir et al., 2016; Gigliarelli et al., 2017) and more.

The specific functionalities that can be usefully applied to heritage are also described by some official documents (e.g. Historic England 2017): decision-making support; heritage management, as work programming (conservation, repair, maintenance and reuse), visitor management, condition monitoring; project management; security, fire safety, visitor safety and health and safety planning; disaster preparedness.

However, some challenges have still to be tackled for BIM to be completely suitable for heritage. BIMs were born for reusing highly standardized components (like industry-fabricated elements), but it is not always possible to apply the same criteria with architectural heritage elements, which were often handcraft produced. As a consequence, the parametrically modelled geometries can be little consistent with the actual ones (which are often irregular) (e.g. Visintini et al., 2019). Furthermore, the available tools in software are conceived and designed for the needs of new-designed buildings, which do not always apply also to heritage, whilst others, potentially useful, can be missing.

3.2 *3D geographical information systems/3D city models including heritage*

The other very powerful 3D spatial information systems to be considered are 3D GIS, or 3D city models. They are the more natural evolution of traditional GISs, and they are usually oriented to the representation of medium levels of detail objects (e.g. the city, or portions of the city). In 3D city models, usually a boundary representation (storing coordinates) is used, georeferencing is straightforward and the semantics concern the city or land objects.

In this case the levels of detail represent the generalization from the most accurate and detailed data, taken as reference (for example the survey data). Officially 5 LoDs are used (Gröger Plümer, 2012), but some recent research gave a more granular definition of them, resulting in more effective 16 (Biljecki et al., 2016).

Use cases of 3D city models are many (Biljecki et al., 2015): e.g. energy simulations, noise modelling, shadow analysis, navigation, flood simulations, multivariate analysis, risk analysis, 3D cadaster, multitemporal analysis, flows analysis and so on.

Also heritage can have advantage of 3D city models and 3D GIS (Campanaro et al., 2016), supporting the documentation management in a unique and structured framework, permitting complex landscape analysis and cross-boundary investigations, coordinate wide preservation strategies, and, especially, they could be used as base for extracting new knowledge from the existing patterns, through using for example inferences from ontologies technologies (e.g. Lin et al., 2008, Acierno et al., 2017), data mining, machine learning (Llamas et al., 2016) and so on.

However, also in this case, some challenges are still present: for example, an LoD specification specific for heritage would be useful, much more data are needed, and some technical issues linked to software (and hardware) and data formats can make the use of such tools by current users difficult.

The OGC CityGML, is the most used open standard for 3D city models. It defines rules, structures and encodings for the management of interoperable, exchangeable and reusable 3D city models, with a semantics related to city object representations. An important feature of CityGML is the coded possibility to extend the model with application-related classes, attributes and relationships, through the CityGML Application Domain Extensions (ADEs).

In literature, CityGML ADEs have been developed also to manage the architectural heritage. For example, some studies developed a CityGML extension to manage architectural heritage in connection with energy-related information for historical centers retrofitting (Prieto el al., 2012; Egusquiza et al., 2018). Noardo (2018) developed a Cultural Heritage ADE to improve the flexibility of the model, in order to adapt it also to the detailed representation of such complex and unique buildings. In that study, the attribute 'level of specialization' was introduced, in order to represent one further dimension of the model information, that deals with the level of the semantic segmentation hierarchy even when considering one only geometric level of detail (e.g. the objects involved in part-of or subclasses relationships). Moreover, a tentative connection to external approved conceptualizations was proposed, for example reusing code-lists from existing vocabularies (e.g. the Getty Institute vocabularies). The extension also describes the entities regarding connected information, whose management is also relevant for architectural heritage, with the needed complexity (e.g. patron, usage, author, and so on). One last issue addressed in the study is the possibility to customize the segmentation criteria in order to allow different kinds of representation in an explicit way, through an attribute 'DescriptionAim' that permits the existence of different overlapping representation of the same object (according to different use cases and aims) avoiding inconsistencies.

Other studies (Fernández-Freire et al., 2013; McKeague et al., 2012) extended instead the INSPIRE data model, which is given by the European Directive for building an 'INfrastructure of Spatial Information in Europe' (INSPIRE), with the aim of a common environmental management. One of those was specifically intended to represent heritage in connection to the natural hazard, for the ResCult project (Chiabrando et al., 2018; Colucci et al., 2018).

4 GEOBIM

Both BIMs and 3D city models can offer great support to architectural heritage, and it is not possible to choose only one of the models. It is therefore necessary to integrate them. Ongoing research (e.g. Sani et al., 2018, Wang et al., 2019) about the integration of geoinformation and BIMs ('GeoBIM') can be the reference for the concepts to be applied also to heritage.

The topic 'GeoBIM' means multiple concepts, which have to be all addressed to reach an actual integration. First of all, the integration of the data (as produced) must be addressed, intended as the achievement of harmonizable and explicit characteristics, so that they can actually fit together once that more technical integration are performed. Secondly, the data must have an interoperable format, towards which the definition and use of open standard is a first step. However, it is still necessary to work for the standardized data to be suitably supported by software and correctly coded and exchanged. A third challenge is the definition of suitable procedures able to make completely consistent conversions between the two kinds of data. A study funded by the International Society for Photogrammetry and Remote Sensing

(ISPRS) and the European association for Spatial Data Research (EuroSDR) is being addressing those last two issues (GeoBIM benchmark 2019). Finally, it is also necessary to integrate and harmonize the involved procedures implied in BIM and GIS-related tools and methods, for them to be used in seamless workflows.

The advantages that can be given by the integration are many for both sides (Arroyo Ohori et al., 2018). The geo-world can obtain information for high-level-of-detail 3D cadaster, avoiding duplication of tasks related to the 3D data collection, efficiently updating 3D maps databases without additional financial costs, effective data exchange with professionals (architects, engineers, environmental scientists, etc.). Moreover, they can obtain a stronger information for lifecycle asset management and enhanced infrastructure and network management, with an enhanced level of detail, also enabling enhanced city analysis.

On the other hand, the BIM world could more easily obtain a correct, accurate and useful description of the context for design reference, perform improved analysis about the designed building into its context and exploit GIS analysis for testing the impact of the building on the city or landscape.

However, at present, the differences of the two models, explained in section 3 are not completely overcome yet. Moreover, when considering data produced by professional practice, even more challenges arise, since they are usually not fully compliant with the theoretical assumptions from academy. In addition, those technology have only recently begun to be applied, and not many data are available yet. The lack of data is moreover increased due to the implied privacy issues concerned with BIMs.

To overcome the inhomogeneity of the data in order to enable the integration, an agreement is needed between stakeholders and data producers about the requirements the data should respect (e.g. levels of details, needed entities and semantic information, need for georeferencing, specific information needed in specific use cases, etc.).

5 GEOBIM FOR ARCHITECTURAL HERITAGE

As a result of the initial literature review and initial tests on data, it is possible to envision the advantages that the use of GeoBIM, in a seamless flow of information, would allow for the heritage field, which is external to both geo and BIM.

Some of those advantages are: the archive of heritage information from dense survey and other sources in detailed semantic 3D models; the heritage analysis in connection to the context and other distant heritage; the support to heritage management and risk prevention (e.g. Kilsedar et al., 2019, Matrone et al., 2019); no duplication of data collection and representation efforts; processes optimization.

Some work will be still needed to reach those ambitious aims. However, the next steps to be addressed are already clear, which are, however, not straightforward, since an agreement among a number of interdisciplinary experts needs to be found for overcoming the present discontinuities and problems preventing a suitable integration.

Notwithstanding the integration of 3D GIS and BIM systems for heritage representation has the advantage that the source of the data for modelling is the same, that is, the existing architecture, it is anyway necessary to find agreements between data producers even within the same branch, since the existing data present differences in geometry, semantic, level of implementation of the followed data models and standards. More specific rules and guidelines are therefore needed first of all to enable a suitable harmonization between different 'geo' data, different 'BIM' data; then, to permit their integration.

With respect to geometry, guidelines are needed for overcoming the differences due to the various kinds of metric data that can be used: different kinds of survey (instruments, precision, methods, reached accuracies); different processing methods of the measurements, and so on.

Moreover, it is necessary to consider the differences due to different adopted modelling methodologies: in the data used as source (2D/3D, density, accuracy, etc.), in the modelling and interpolation methods and algorithms (manual or automatic) in the kinds of geometry storage.

Additionally, with respect to semantics, we need to agree on terms, hierarchies, semantic rules, better if using already shared and affirmed ontologies, such as the Conceptual Reference Model of the Documentation Committee (CIDOC) of the International Committee of Museums (ICOM), i.e. the 'CIDOC CRM' (Doerr, 2003) and the vocabularies published by the Getty Institute (http://vocab.getty.edu).

Requirements have to be defined for the needed results and best practices should be proposed as reference to achieve them. One fundamental need for the reuse and understanding of data is the archiving of suitable metadata, clearly explicating the data properties as listed before, preferably compliant with the existing standards (e.g. for geographic information, ISO 19115, Labetski et al., 2018).

6 CONCLUSONS

As a conclusion, it is possible to state that first of all, institutional challenges and inter-sectorial preconceptions have to be overcome in order to find agreements to fruitfully collaborate. The representation and archiving of architectural heritage in 3D city models must improve, by applying extensions systematically, to consider the heritage as special also in common city analysis and management. Moreover, the archiving of 'geo' heritage models in connection to HBIM for having advantage of both systems. As a preliminary condition, both should follow common parameters and criteria in the representation, explicated in standardized metadata, to be suitable for enabling conversions and following specific use case requirements in connected fields (e.g. history, restoration, construction, structural engineering).

Great opportunities are given by the management of architectural heritage information in HBIM and 3D city models, and their integration. Multidimensional integrated and interoperable information for architectural heritage is not yet straightforward, but initial foundations are being laid.

REFERENCES

Acierno, M., Cursi, S., Simeone, D., & Fiorani, D. 2017. Architectural heritage knowledge modelling: An ontology-based framework for conservation process. *Journal of Cultural Heritage*, 24, 124–133.

Arroyo Ohori, K., Diakité, A., Krijnen, T., Ledoux, H., & Stoter, J. 2018. Processing BIM and GIS models in practice: experiences and recommendations from a GeoBIM project in The Netherlands. *ISPRS International Journal of Geo-Information*, 7(8), 311.

Attico, D., Turrina, A., Banfi, F., Grimoldi, A., Landi, A., Condoleo, P., and Brumana, R. 2019. The HBIM analysis of the geometry to understand the constructive technique: the use of the *trompe* volume in a brick vault, *Int. Arch. Photogramm. Remote Sens. Spatial Inf. Sci.*, XLII-2/W11, 107–114, https://doi.org/10.5194/isprs-archives-XLII-2-W11-107-2019, 2019.

Bertolini-Cestari, C., Invernizzi, S., Marzi, T., & Spano, A. 2016. Numerical survey, analysis and assessment of past interventions on historical timber structures: the roof of valentino castle. *Wiadomości Konserwatorskie*.

Biljecki, F., Stoter, J., Ledoux, H., Zlatanova, S., & Çöltekin, A. 2015. Applications of 3D city models: State of the art review. *ISPRS International Journal of Geo-Information*, *4*(4), 2842–2889.

Biljecki, F., Ledoux, H., & Stoter, J. 2016. An improved LOD specification for 3D building models.*Computers, Environment and Urban Systems*, *59*, 25–37.

BIMFORUM, 2019. *Level Of Development (LOD) specification part I & commentary For Building Information Models and Data.* Accessible at: https://bimforum.org/wp-content/uploads/2019/04/LOD-Spec-2019-Part-I-and-Guide-2019-04-29.pdf Accessed on 10-05–2019.

buildingSMART Industry Foundation Classes https://www.buildingsmart.org/about/what-is-openbim/ifc-introduction/;http://www.buildingsmart-tech.org/specifications/ifc-overview Accessed 10-05-2019.

Campanaro, D. M., Landeschi, G., Dell'Unto, N., & Touati, A. M. L. 2016. 3D GIS for cultural heritage restoration: A 'white box' workflow. Journal of Cultural Heritage, 18, 321–332.

Centofanti, M., Brusaporci, S., & Lucchese, V. 2014. Architectural heritage and 3D models. In *Computational modeling of objects presented in images* (pp. 31–49). Springer, Cham.

Chiabrando, F., Lo Turco, M., & Rinaudo, F. 2017. Modeling the decay in an HBIM starting from 3d point clouds. A followed approach for cultural heritage knowledge. *International Archives of the Photogrammetry, Remote Sensing & Spatial Information Sciences, 42.*

Doerr, M. 2003. The CIDOC conceptual reference module: an ontological approach to semantic interoperability of metadata. *AI magazine, 24*(3), 75–75.

Donadio, E., & Spanò, A. 2015. Data collection and management for stratigraphic analysis of upstanding structures. In *2015 1st International Conference on Geographical Information Systems Theory, Applications and Management (GISTAM)* (pp. 1–6). IEEE.

Egusquiza, A., Prieto, I., Izkara, J. L., Béjar, R. 2018. Multi-scale urban data models for early-stage suitability assessment of energy conservation measures in historic urban areas. *Energy and Buildings, 164*, 87–98.

Fasolo, V. 1954. Guida metodica per lo studio della storia dell'architettura. Edizioni dell'Ateneo, Roma.

Fernández Freire, C., Del Bosque González, I., Vicent García, J. M., Pérez Asensio, E., Fraguas Bravo, A., Uriarte González, A., Fábrega-Álvarez, P., Parcero-Oubiña, C. 2013. A cultural heritage application schema: achieving interoperability of cultural heritage data in INSPIRE.

GeoBIM benchmark 2019 https://3d.bk.tudelft.nl/projects/geobim-benchmark/ Accessed 10-05–2019.

Getty Vocabularies http://vocab.getty.edu Accessed 11 May 2019.

Gigliarelli, E., Calcerano, F., Cessari, L. 2017. Heritage Bim, numerical simulation and decision support systems: An integrated approach for historical buildings retrofit. Energy Procedia, 133, 135–144.

Gröger, G., & Plümer, L. 2012. CityGML–Interoperable semantic 3D city models. *ISPRS Journal of Photogrammetry and Remote Sensing, 71*, 12–33.

Historic England 2017. BIM for Heritage: Developing a Historic Building Information Model. Swindon. Historic England. https://historicengland.org.uk/advice/technical-advice/recording-heritage/

Khodeir, L. M., Aly, D., & Tarek, S. 2016. Integrating HBIM (Heritage Building Information Modeling) tools in the application of sustainable retrofitting of heritage buildings in Egypt. Procedia Environmental Sciences, 34, 258–270.

Kilsedar, C. E., Fissore, F., Pirotti, F., and Brovelli, M. A. 2019. Extraction and visualization of 3D building models in urban areas for flood simulation. *Int. Arch. Photogramm. Remote Sens. Spatial Inf. Sci.*, XLII-2/W11, 669–673, https://doi.org/10.5194/isprs-archives-XLII-2-W11-669-2019, 2019.

Labetski, A., Kumar, K., Ledoux, H., Stoter, J. 2018. A metadata ADE for CityGML. *Open Geospatial Data, Software and Standards* 3(16), 2018.

Lin, C. H., Hong, J. S., Doerr, M. 2008. Issues in an inference platform for generating deductive knowledge: a case study in cultural heritage digital libraries using the CIDOC CRM. *International Journal on Digital Libraries, 8*(2), 115–132.

Llamas, J., Lerones, P. M., Zalama, E., Gómez-García-Bermejo, J. 2016. Applying deep learning techniques to cultural heritage images within the inception project. In *Euro-Mediterranean Conference* (pp. 25–32). Springer, Cham.

Lu, L., Pintossi, N., Dane, G., & Roders, A. P. 2019. The Role of ICT in Mapping Resources for Sustainable Historic Urban Regeneration: Case Studies of Amsterdam and Salerno. In *REAL CORP 2019–IS THIS THE REAL WORLD? Perfect Smart Cities vs. Real Emotional Cities. Proceedings of 24th International Conference on Urban Planning, Regional Development and Information Society* (pp. 985–991). CORP–Compentence Center of Urban and Regional Planning.

Malinverni, E. S., Mariano, F., Di Stefano, F., Petetta, L., and Onori, F. 2019. Modelling in HBIM to document materials decay by a thematic mapping to manage the cultural heritage: the case of "chiesa della pietà" in Fermo, *Int. Arch. Photogramm. Remote Sens. Spatial Inf. Sci.*, XLII-2/W11, 777–784, https://doi.org/10.5194/isprs-archives-XLII-2-W11-777-2019, 2019.

Matrone, F., Colucci, E., De Ruvo, V., Lingua, A., and Spanò, A. 2019. HBIM in a semantic 3D GIS database, *Int. Arch. Photogramm. Remote Sens. Spatial Inf. Sci.*, XLII-2/W11, 857–865, https://doi.org/10.5194/isprs-archives-XLII-2-W11-857-2019.

McKeague, P., Corns, A., Shaw, R. 2012. Developing a Spatial Data Infrastructure for Archaeological and Built Heritage. IJSDIR, 7, 38–65.

Melón, J. M. V., Miranda, Á. R., Pérez-Lorente, F., & Torices, A. 2019. The use of new web technologies for the analysis, preservation, and outreach of paleontological information and its application to La Rioja (Spain) paleontological heritage. *Palaeontologia Electronica, 22*(1), 1–10.

Noardo F. 2018. Architectural heritage semantic 3D documentation in multi-scale standard maps. *Journal of Cultural heritage*, Elsevier. ISSN: 1296–2074 DOI: 10.1016/j.culher.2018.02.009.

Noardo F., Spanó, A., 2015. Towards a spatial semantic management for the intangible cultural heritage. *International journal of heritage in the digital era* 4, 2015, pp. 133–147.

Open Geospatial Consortium CityGML www.citygmlwiki.org Accessed on 10-05–2019.

Paladini, A., Dhanda, A., Reina Ortiz, M., Weigert, A., Nofal, E., Min, A., Gyi, M., Su, S., Van Balen, K., and Santana Quintero, M. 2019. Impact of virtual reality experience on accessibility of cultural heritage, Int. *Arch. Photogramm. Remote Sens. Spatial Inf. Sci.*, XLII-2/W11, 929–936, https://doi.org/10.5194/isprs-archives-XLII-2-W11-929-2019.

Prieto, I., Izkara, J. L., & del Hoyo, F. J. D. 2012. Efficient visualization of the geometric information of CityGML: application for the documentation of built heritage. In *International Conference on Computational Science and Its Application* (pp. 529–544). Springer, Berlin, Heidelberg.

Riveiro, B., Morer, P., Arias, P., & De Arteaga, I. 2011. Terrestrial laser scanning and limit analysis of masonry arch bridges. *Construction and building materials*, *25*(4), 1726–1735.

Rua, H., & Alvito, P. 2011. Living the past: 3D models, virtual reality and game engines as tools for supporting archaeology and the reconstruction of cultural heritage–the case-study of the Roman villa of Casal de Freiria. *Journal of Archaeological Science*, *38*(12), 3296–3308.

Salonia, P., & Negri, A. 2003. Historical buildings and their decay: data recording, analysing and transferring in an ITC environment. *Int. Arch. Photogramm. Remote Sens. Spatial Inf. Sci.*, 34(5/W12), 302–306.

Sani, M. J., & Rahman, A. A. 2018. GIS and BIM integration at data level: A review. *International Archives of the Photogrammetry, Remote Sensing and Spatial Information Sciences*, *42*(4/W9).

Scopigno, R., Callieri, M., Cignoni, P., Corsini, M., Dellepiane, M., Ponchio, F., & Ranzuglia, G. 2011. 3D models for cultural heritage: beyond plain visualization. *Computer*, (7), 48–55.

Visintini, D., Marcon, E., Pantò, G., Canevese, E. P., De Gottardo, T., and Bertani, I., 2019. Advanced 3d modeling versus building information modeling: the case study of palazzo Ettoreo in Sacile (Italy). *Int. Arch. Photogramm. Remote Sens. Spatial Inf. Sci.*, XLII-2/W11, 1137–1143, https://doi.org/10.5194/isprs-archives-XLII-2-W11-1137-2019, 2019.

Wang, H., Pan, Y., & Luo, X. 2019. Integration of BIM and GIS in sustainable built environment: A review and bibliometric analysis. *Automation in Construction*, *103*, 41–52.

Development of new digital graphic instruments such as BIM, GIS and others, for knowledge, analysis, protection and conservation of Cultural Heritage

Science and Digital Technology for Cultural Heritage – Ortiz Calderón et al. (Eds)
© 2020 Taylor & Francis Group, London, ISBN 978-0-367-36368-0

Landscape indicators overview and data model proposal for a GIS-based heritage management system

M. López Sánchez, A. Tejedor Cabrera & M. Linares Gómez del Pulgar
Universidad de Sevilla, Seville, Spain

ABSTRACT: Cultural landscapes are nowadays fully recognized heritage entities whose values need to be integrated in planning processes. However, landscape assessments to that effect are still challenged by the lack of consistent policy-effective integrative methods that consider landscape both natural and cultural values and, at the same time, address its potential as a socioeconomic resource. This study aims to fill the gap by defining an indicator-oriented data model which is focused on integrating service potential estimation into landscape management systems. The model has been set by merging together Cultural Ecosystem Services (CES) indicators and focal points of landscape planning instruments through a contrasting analysis. The study reveals potential in the combination of both frameworks, pointing out towards the usefulness of CES scenario in landscape and heritage policy-making.

1 INTRODUCTION

Contemporary vision of heritage reflects a semantic expansion process that fully reached landscape dimension. The attention has moved from "monument" to notions as space, context and integration. In this line, international bodies as UNESCO, ICOMOS, IUCN and the European Council have been asserting during last decades that landscapes represent cultural identity symbols as well as recognizing their potential for socioeconomic development. The Committee of Ministers responsible for the Cultural Heritage declared in the IVth European Conference (Helsinki, 1996) awareness of the need to expand and up-date the concept of integrated conservation of heritage for the purpose of responding better to the economic and social challenges facing Europe today. In line with this, landscapes as heritage entities require an integrated approach that covers all its multidimensional benefits including the socioeconomic one. Integrative vision should be also accompanied by the development of tools with operational usefulness in decision-making processes, building together a coherent formula for landscape inclusion in spatial planning (Mrak 2013).

One of the decision-aid methods with more possibilities for offering an integrated vision are landscape indicators, as they broke down complex phenomena, which cannot be measured entirely, into its different dimensions (Cassatella & Peano 2011). Additionally, as instruments originally conceived for environmental evaluations promoted by international organizations as UN, OECD or EEA, indicators have their roots in being effective policy tools. The agricultural policies framework is also a reference field for landscape indicators, which made significant contributions by noteworthy initiatives as IRENA (EEA 2006), PAIS (LANDSIS et al. 2002) and ELISA with its subsequent ENRISK (Delbaere 2003) projects. Over the years, they have been increasingly applied in different fields as sustainability evaluations, quality assessments or CES analysis. Landscape approach in indicator´s scenario has been strengthened during last decades by integral propositional studies on the matter (Vallega 2008; Cassatella & Peano 2011).

This study aims to explore indicators potential for planning practices by analysing the possible links between CES field, where indicators have been developed for assessing landscape usefulness as a society resource (Hernández Morcillo et al. 2013), and landscape policy instruments. The result of the analysis is drawn as an indicator catalogue and subsequent data model proposal.

2 DATA MODEL PROPOSAL

During last decades, assessing the non-material benefits people can obtain from landscape (e.g. aesthetic appreciation, recreation, tourism enjoyment, inspiration, knowledge, spiritual enrichment, cultural heritage or sense of place) has received scientific attention mainly through two fields of study: CES assessments (MEA, 2005), which emerge as a contact point between ecology and economy, and landscape services (LS) assessments, rooted in landscape ecology. This text refers to CES as an integrative concept of both fields, following the point of Termorshuizen & Opdam (2009), who considers LS a specification, rather than an alternative, to ES.

The possible engagement and influence between CES indicators and landscape planning is addressed in this study by contrasting most used CES indicators, obtained by clustering three compiling-oriented indicators reviews (Egoh et al. 2012; Layke 2009; Maes et al. 2016), and focal points of two representative landscape planning instruments: Spanish Landscape Catalogues, taking as a reference pioneer Landscapes Catalogues of Catalonia (Nogue et al., 2016) and Italian Landscape Regional Planning, with Piemonte Region as the selected example (Cassatella & Paludi 2018). An indicator catalogue has been set by combining concepts from all analysed instruments (Table 1). In Table 1, each indicator is linked to the instrument/s where it appears, which shows common targets between both fields. It also denotes CES as a framework from where to complete landscape planning instruments with monetary indicators (e.g. tourism related ones) that address socioeconomic landscape dimension, which is still not fully approached in them. The indicator catalogue works as an initial step for pointing out the possibilities of combining both backgrounds and achieving the so-called integrative vision by methods with operational usefulness. It addresses multidimensional landscape features and it offers tangible quantitative results with effective communicative ability, which allows the evaluation of decision´s effect and efficacy (Botequilha Leitão et al., 2006).

The catalogue has been assumed as part of a GIS-based model, which allows the use of a single tool for characterization, evaluation and diagnosis phases. Indicator categories correspond to landscape values, which are strongly bound to landscape components and features and to their spatial and functional interrelations (Vizzari 2011). This classification system offers an alternative to traditionally service-oriented categorization schemes for CES indicators, which could present overlapping in some cases, as landscape features or dynamics could influence more than one service at the same time (Church et al. 2011). As an example, presence of water bodies could be an influential characteristic for both aesthetic value and recreation/tourism enjoyment. In order to make the structure easily adaptable to GIS, each of the features takes the shape of a GIS layer, building together a classified landscape inventory. GIS spatial analyses will be carried out on these layers as a second step in order to give response to value-based indicators. Participatory techniques will be also a necessary, in order to get a coherent conclusion for the relation between a particular landscape, the human activities that take place on it and the benefits people get from them.

The data model will be set by establishing a distinction between components/composition. Components are tangible features that are added to GIS as vector point-line-surface layers. We distinguish *physical* components related to geology, hydrology, topography and climate, *natural* components as vegetation, fauna, habitats and any relevant natural areas, *anthropogenic-natural* components as agricultural areas, and *anthropogenic* components as urban settlements, buildings, cultural heritage listed items, road network, facilities and infrastructures. Composition is related to land use/land cover layers, which will be necessary for landscape metrics analysis.

Table 1. Landscape indicators proposed catalogue.

Indicator	Egoh et al. 2012	Layke 2009	Maes et al. 2016	Nogué et al. 2016	Cassatella & Paludi 2018
Natural value					
Landscape naturalness	X			X	X
Protected natural sites and elements as monumental tress	X		X	X	X
Non-protected areas of natural interest				X	X
N° of iconic/endangered species	X		X	X	X
Biodiversity				X	X
Pollution level				X	
Historic-cultural value					
Protected cultural sites and elements	X		X	X	X
Non-protected areas and elements of cultural relevance			X	X	X
Historic paths			X	X	X
Traditional activities				X	
Time Depth				X	X
Scenic value					
Viewsheds	X			X	X
Visual landmarks	X			X	X
Structural elements for scenic composition	X			X	X
Contrasting landscapes			X	X	
Touristic/recreational value					
Number of visitors	X	X	X		
Spending on tourism	X	X	X		
Rural tourism employment		X			
N° of rural tourism enterprises			X		
Tourist average length of stay	X				
N° of tourist attractions	X			X	X
N° of outdoor recreation sites	X		X	X	X
Walking & biking trails	X		X	X	X
Viewpoints				X	X
Accessibility	X				

3 CONCLUSIONS

The study identifies common targets between CES scenario and landscape management and planning instruments, revealing that the merger of both frameworks presents potential for the development of a policy-effective landscape integral assessment method that addresses cultural and natural landscape values as well as socioeconomic potential. CES background offers a methodological corpus for assessing landscape as a resource by analyzing the services they can offer to society. The indicators catalogue represents a first approach in the process of finding common working systems. Indicators provide a common vehicle for different processes (study, documentation, planning, protection, prevention, conservation and intervention), disciplines and agents working around landscape, pointing towards integrative dynamics.

The definition of a policy-effective method for landscape assessment, which is able to stablish a relation between its characteristics and its potential as an active resource for society in an interdisciplinary framework and by means of quantitative and qualitative techniques, represents an instrument of key importance for heritage scenario, even though studies addressing the possible engagement and influence between both fields are still few (Eliasson et al., 2019; Hølleland et al., 2017; Tema Nord 2015; Tengberg et al.

2012). At the moment, as Hølleland et al. (2017) concludes, the interest in ES from the cultural heritage sector is getting increased, in particular in the Nordic countries (Tema Nord 2015), which could be a reflection of the relatively close governmental ties between environmental and cultural heritage management in these countries. The development of integrative management systems combining concepts and methods from both fields arises as a line of action for making tangible this emerging issue.

ACKNOWLEDGEMENTS

This research is funded by spanish Ministerio de Economía y Empresa through a research project entitled "Smart Architectural and Archeological Heritage. Instrumentos y Estrategias de Innovación para la Integración de la Gestión Patrimonial, Turística y Paisajística" (HAR2016-79757-R).

Main author has received a research grant by spanish Ministerio de Ciencia, Innovación y Universidades "Ayuda para la formación de profesorado universitario FPU".

REFERENCES

Botequilha Leitão, A., Miller, J., Ahern, J., & McGarigal, K. 2006. *Measuring landscapes. A planner's handbook*. London: Island Press.

Cassatella, C. & Paludi, G. (eds) 2018. *Il Piano paesaggistico del Piemonte*. Torino: Atti e Rassegna Tecnica della Societa degli Ingenieri e degli Architetti in Torino LXXII–3.

Cassatella, C. & Peano, A. (eds) 2011. *Landscape indicators. Assessing and Monitoring Landscape Quality*. Dordrecht: Springer.

Church, A., Burgess, J., Ravenscroft, N. et al. 2011. *Cultural Services: UK National Ecosystem Assessment*. Chapter 16. Cambridge: UNEP-WCMC: 633–92.

Delbaere, B. (ed) 2003. *Environmental Risk Assessment for European Agriculture: interim report*. Tilburg: European Centre for Nature Conservation.

EEA. 2006. *Integration of environment into EU agriculture policy - the IRENA indicador-based assessment report*. EEA Technical Report No 2/2006. Luxembourg: Official Publications of the European Communities.

Egoh, B., Drakou, G., Dunbar, M.B., Maes, J., Willemen, L. 2012. *Indicators for Mapping Ecosystem Services: A Review*. Luxembourg: EU Publications Office.

Eliasson, I., Knez, I. & Fredholm, S. 2019. Heritage Planning in Practice and the Role of Cultural Ecosystem Services. *Heritage & Society*, DOI: 10.1080/2159032X.2019.1576428

Hernández-Morcillo, M., Plieninger, T. & Bieling, C. 2013. An empirical review of cultural ecosystem service indicators. *Ecological Indicators* 29: 434–444.

Hølleland, H., Skrede, J. & Holmgaard, B. 2017. Cultural Heritage and Ecosystem Services: A Literature Review. *Conservation and Management of Archaeological Sites* 19 (3): 210–237.

LANDSIS g.e.i.e., Centre for Agricultural Landscape and Land Use Research, The Arkleton Centre for Rural Development Research, Scottish Agricultural College & University of Applied Science. 2002. *Proposal on Agri-enviromental indicators. PAIS-Project Summary*. Working document.

Layke, C.H. 2009. *Measuring Nature's Benefits: A Preliminary Roadmap for Improving Ecosystem Service Indicators*. WRI Working Paper. Washington: World Resources Institute.

Maes J., Liquete, C., Teller, A., Erhard, M., Paracchini, M.L. et al. 2016. An indicator framework for assessing ecosystem services in support of the EU Biodiversity Strategy to 2020. *Ecosystem Services* 17: 14–23.

M.E.A. Millenium ecosystem assessment 2005. *Ecosystem and Human Wellbeing: Synthesis*. Washington D.C.: Island Press.

Mrak, I. 2013. A Methodological Framework based on the Dynamic-Evolutionary View of Heritage. *Sustainability* 5: 3992–4023.

Nogué, J., Sala, P. & Grau, J. 2016. *The Landscapes Catalogues of Catalonia. Methodology*. Olot: Observatory of Catalonia.

Termorshuizen, J.W. & Opdam, P. 2009. Landscape services as a bridge between landscape ecology and sustainable development. *Landscape Ecology* 24: 1037–1052.

Tema Nord (2015) Kulturarv og økosystemtjenester. Sammenhenger, muligheter og begrensninger (Cultural heritage as an ecosystem service. A feasibility study). Report 2015: 540.
Tenberg, A., Fredholm, S., Eliasson, I., Knez, I., Saltzman, K. & Wettenberg, O. 2012. Cultural ecosystem services provided by landscapes. Assessment of heritage values and identity. *Ecosystem Services* 2: 14–26.
Vizzari, M. (2011) Spatial modelling of potential landscape quality. *Applied Geography* 31: 108–118.
Vallega, A. 2008. *Indicatori per il paesaggio*. Milan: FrancoAngeli.

Science and Digital Technology for Cultural Heritage – Ortiz Calderón et al. (Eds)
© 2020 Taylor & Francis Group, London, ISBN 978-0-367-36368-0

Comparison of image techniques for the detection of graffiti in the cloister of the monastery of San Millán de la Cogolla (Spain)

J.M. Valle Melón, Á. Rodríguez Miranda, G. Elorriaga Aguirre & J. Korro Bañuelos
Laboratorio de Documentación Geométrica del Patrimonio (Grupo de Investigación en Patrimonio Construido - GPAC), University of the Basque Country (UPV/EHU), Vitoria-Gasteiz, Spain

M.B. Arrúe Ugarte
University of La Rioja, Logroño, Spain

Á. Marchante Ortega
University of Las Palmas de Gran Canaria, Las Palmas de Gran Canaria, Spain

ABSTRACT: In year 2000, due to some imminent cleaning works on the walls of the cloister of the monastery of San Millán de la Cogolla, de Yuso, in La Rioja (Spain) and their foreseeable negative impact in a group of graffiti, which were known although not studied or even recorded at that time, an emergency labour was undertaken in order to trace them in plastic sheets. However, the sheets were cumbersome so no further studies were done until year 2016, when they were photographed and digitized.

On the other hand, there was no appraisal concerning the conservation status of the engravings after the cleaning, in fact, it was even thought that they were totally lost. This situation of uncertainty was tackled in 2018 with a series of explorations in different parts of the cloister, in which the visual reconnaissance was complemented with a range of image techniques, in particular: raking light photography (both with fixed and moving lighting), short distance convergent photogrammetry with image enhancement, Reflectance Transformation Imaging (RTI) and structured-light scanning.

The present work describes the techniques, their practical implementation on site and the data processing. Finally, the applicability of each one –and their possible combination- for the identification of graffiti and the determination of their state of repair are discussed.

1 INTRODUCTION

1.1 *Discovery and registration of graffiti*

The monastery of San Millán de la Cogolla was declared a World Heritage site by UNESCO in 1997. It is a renowned place, but still conceals many little know elements of great artistic and historic value. This is the case of the group of architectonic engravings on the walls of the cloister. Not even the many historians that had been written about the Monastery had notice these graffiti[1] until the year 2000, when the imminent start of some cleaning works raised the alarm of the archivist of the Monastery, who communicated their existence to the group of researchers B. Arrúe, J. Martínez and M.C. Navarro, who hurried up to copy them in large plastic sheets, just in case they were damaged during the cleaning. In total, a surface of 165 m2 was covered.

1. Here, *graffiti* is understood as "some writing or hand drawing done on the monument in the past". Note the difference with another meaning of this word as "signature or painting usually done without permission in public spaces, walls or any other durable surfaces" that is mainly circumscribed to cultural manifestations since the twentieth century.

No further studies could be done on the engravings at that moment and, eventually, the plastic sheets ended up in a box which was stored in an attic.

1.2 *Digitization of the plastic sheets*

In 2016 the sheets were digitized by the Laboratory for Geometric Documentation of Heritage of the University of Basque Country. Firstly, they were unfolded, cleaned and rolled around cardboard tubes; next, they were extended horizontally and photographed. Afterwards, the photographs were corrected from perspective and imported in a CAD software where they were scaled to their real size; then, the lines on the sheets were redrawn over the images and partial drawings joined together so as to obtain the digitized version of the ensemble (LDGP, 2016).

After this work, the engravings are available for their study in digital CAD drawings, therefore, measurements can be taken, comparison made, and so on.

2 CURRENT STATE OF THE ENGRAVINGS

Although the CAD drawings are excellent documents for studying the engravings, another concern was still pending, that is, knowing to which extent the cleaning works done on the walls of the cloister had damaged the graffiti.

To tackle this issue, we conducted an organoleptic examination of the walls and marked over the drawings which groups were still identifiable and which ones were fainted or disappeared.

Moreover, we applied some recording techniques so as to test the possibilities to both evaluate the conservation state and locate new engravings that might have been disregarded in the previous examination. The list of techniques considered were: raking light photography (both with fixed light and with moving lighting), short distance convergent photogrammetry with image enhancement, Reflectance Transformation Imaging (RTI) and structured-light scanning.

3 TESTING TECHNIQUES FOR THE GEOMETRIC DOCUMENTATION

In general, the engravings are rather shallow, no one is re-marked with charcoal and they go mostly unnoticed among the own roughness of the ashlars. Moreover, the fact of being geometric figures (arcs, squares...) without specific context and, some of them, big-sized (up to several meters) also makes difficult their identification. These characteristics are also challenging for the techniques of geometric documentation, that is why we tested some of them in order to check their pros and cons. Broadly speaking, many experiences can be referred regarding the use of enhanced photograph, close range photogrammetry, multispectral imaging and laser scanning applied to architectonic graffiti (Calvo et al., 2015; Chías & Abad, 2017, Calvo & Taín, 2018) to which we can add many others useful cases about elements which share features such as prehistoric rock art or inscriptions (Díaz-Andreu, 2005; Cortón et al., 2015; Cosentino et al., 2015; Papadaki et al, 2015; Carrero-Pazos & Espinosa-Espinosa, 2018).

3.1 *Raking light photography*

Changing the direction of the light modifies the contrast of every part of the surface. In some cases, a selected direction might enhance the complete set of engravings; however, the commonest situation is that each specific direction makes clearer some parts at the same time that hides some others.

A possibility to have multidirectional lighting pictures is to resort to long exposure photographs (e.g. 2 seconds) and to move a torch during the exposition (López-Menchero et al., 2017). Anyway, the photographs need to be taken in the dark so as to magnify the effect of the torch (Figure 1, left). For this purpose, we built a back-to-wall portable structure consisting of metallic props and covered with a black cloth that concealed a square of 3x3 meters of the wall and left another 2 meters in front in order to work with the camera.

3.2 *Structured-light scanning*

The very system for obscuring the surfaces of the wall was used to check the performance of a structured-light scanner. We used a very portable *Scan in a Box* system that was placed a meter away from the wall and captured an area of 40 x 40 cm with each scan and submillimetric resolution. The area recorded can be extended by overlapping scans, however, the procedure becomes unpractical quickly, hence, it is not recommended for areas bigger than 1 x 1 metre.

3.3 *Reflectance Transformation Imaging (RTI)*

This is another alternative of multiple lighting imaging (Mudge et al., 2006). In this case, we stack a group of images in a single file that can be examined interactively. The photographs are taken from a fixed point of view and with a movable source of light, a (hemi)spherical reflective target -that has to be present in all the images- permits the software to identify the origin of the lighting for each image and, therefore, retrieving the proper image when the user selects a specific origin for the light in a viewer for this kind of images (Figure 1, right).

3.4 *Convergent photogrammetry with image enhancement*

Short range convergent photogrammetry (less than 1 meter of distance from the camera to the wall) was also used for documenting groups of engravings. The images were processed up to the generation of orthoimages, next, the radiometric range was enhanced by equalizing the chromatic channels. In doing so, the visual appearance of the wall becomes slightly unrealistic but the engravings, tags and details are more clearly seen (Figure 2).

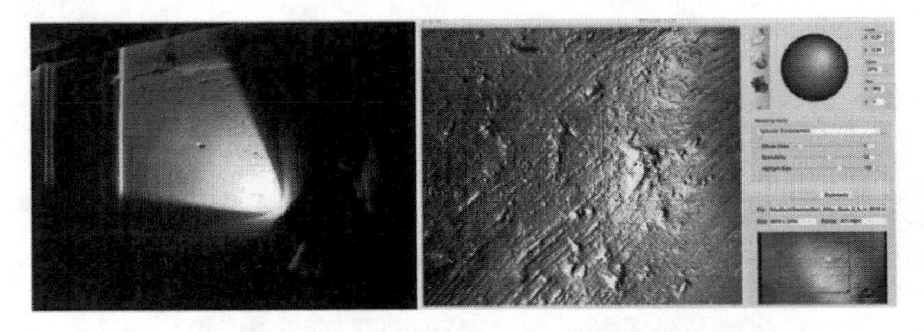

Figure 1. Example of raking light photograph (left) and a screenshot of the RTI viewer (right).

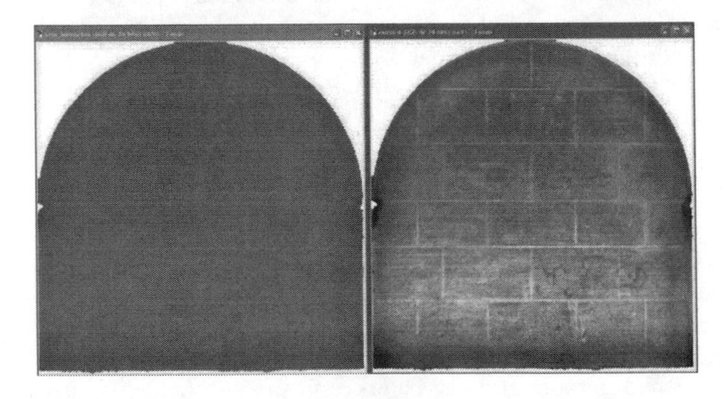

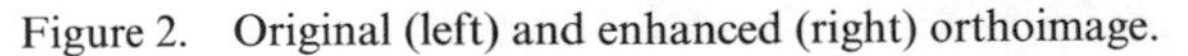

Figure 2. Original (left) and enhanced (right) orthoimage.

4 CONCLUSIONS

The copy in plastic sheets made in 2000 has proved to be essential for preserving the information about the engravings in San Millán's cloister and, also, served as a reference state for checking the effect of the cleanings works.

Concerning the current state of conservation of the graffiti, we can say that the organoleptic inspection gave a reasonable appraisal. On the other hand, we have tested several methods of geometric documentation, in general, all of them obtained significant results but there are also some considerations that we must take into account: firstly, some techniques require the use of auxiliary systems for working in the dark, which might be bulky and time consuming; secondly, the size of the areas recorded by these methods tend to be small (e.g. a square of 0.5 by 0.5 meters) so they might be hardly applicable for entire walls such as the ones of the cloister (each side of this cloister is around 35 meters long).

Finally, we would like to stress the value of this kind of monitoring works in the framework of the preventive conservation of cultural heritage and the development of plans and strategies for such a purpose.

ACKNOWLEGEMENTS

The works described in this text concerning the graffiti in the cloister were funded initially by the *Instituto de Estudios Riojanos* (Government of La Rioja) and, subsequently, by the *Fundación San Millán de la Cogolla*, the local government of San Millán de la Cogolla and the Spanish Ministry of Education Culture and Sport.

REFERENCES

Calvo, J.; Alonso, M.A.; Taín, M.; Camiruaga, I. 2015. Métodos de documentación, análisis y conservación de trazados arquitectónicos a tamaño natural. *Arqueología de la arquitectura*, 12: e026. (http://dx.doi.org/10.3989/arq.arqt.2015.024).

Calvo, J.; Taín, M. 2018. *Las monteas del convento de Santa Clara de Santiago de Compostela: un repertorio de trazados, tanteos y dibujos del Barroco español.* Biblioteca Científica del Consorcio de Santiago (ISBN: 978-84-945592-6-6).

Carrero-Pazos, M.; Espinosa-Espinosa, D. 2018. Tailoring 3D modelling techniques for epigraphic text restitution. Case studies in deteriorated roman inscriptions. *Digital Applications in Archaeology and Cultural Heritage*, 10: e00079. (https://doi.org/10.1016/j.daach.2018.e00079).

Chías, P.; Abad, T. (2017) Modelos, plantillas, trazas y monteas en los contraltos para la construcción del Monasterio del Escorial. *Informes de la Construcción*, 69 (547): e219. (http://dx.doi.org/10.3989/id55077).

Cosentino, A.; Stout, S.; Scandurra, C. 2015. Innovative imaging techniques for examination and documentation of mural paintings and historical graffiti in the catacombs of San Giovanni, Syracuse. *International Journal of Conservation Science*, 6(1): 23-34.

Díaz-Andreu, M.; Brooke, C.; Rainsbury, M.; Rosser, N. 2006. The spiral that vanished: the application of non-contact recording techniques to an elusive rock art motif at Castlerigg stone circle in Cumbria. *Journal of Archaeological Science*, 33: 1580-1587.

Cortón, N.; López, Á.; Carrera, F. 2015. Combining photogrammetry and photographic enhancement techniques for the recording of megalithic art in north-west Iberia. *Digital Applications in Archaeology and Cultural Heritage*, 2: 89-101. (http://dx.doi.org/10.1016/j.daach.2015.02.004).

LDGP. 2016. *[R_SanMillan_Yuso] Recuperación de los grafitos del claustro bajo del Monasterio de San Millán de la Cogolla, de Yuso (La Rioja): registro fotográfico y digitalización.* Laboratorio de Documentación Geométrica, University of the Basque Country (UPV/EHU). [on line] available at: http://hdl.handle.net/10810/19629 [accessed 5/4/2019].

López-Menchero, V.; Marchante, Á.; Vincent, M.L.; Cárdenas, Á.J.; Onrubia, J. 2017. Uso combinado de la fotogrametría digital nocturna y de la fotogrametría en los procesos de documentación de petroglifos: en el caso de Alcázar de San Juan (Ciudad Real, España). *Virtual Archaeology Review*, 8(17): 64-74. (http://dx.doi.org/10.4995/var.2017.6820).

Mudge, M.; Malzbender, T.; Schroer, C.; Lum. M. 2006. New Refection Transfromation Imaging methods for rock art and multiple-viewpoint display. In M. Ioannides et al. (ed.) *The 7th International Symposium on Virtual Reality, Archaeology and Cultural Heritage VAST (2006)*.

Papadaki A.I.; Agrafiotis, P.; Georgopoulos, A.; Prignitz, S. 2015. Accurate 3D scanning of damaged ancient Greek inscriptions for revealing weathered letters. *The International Archives of the Photogrammetry, Remote Sensing and Spatial Information Sciences*, Volume XL_5/W4. (http://dx.doi.org/10.5194/isprsarchives-XL-5-W4-237-2015).

Science and Digital Technology for Cultural Heritage – Ortiz Calderón et al. (Eds)
© 2020 Taylor & Francis Group, London, ISBN 978-0-367-36368-0

BIM-GIS interoperability applied to architectonic heritage: 2D and 3D digital models for the study of the ancient church of Santa Lucía in Seville (Spain).

E.J. Mascort-Albea*, A. Jaramillo-Morilla*, R. Romero-Hernández* &
F.M. Hidalgo-Sánchez*
Department of Building Structures and Soil Engineering, Higher Technical School of Architecture, Universidad de Sevilla, Seville, Spain

ABSTRACT: This contribution explores the underlying possibilities of interoperability between Building Information Modeling (BIM) and Geographic Information Systems (GIS) methodologies, applied to the management of architectural heritage. This task can be understood as a complex reality, since it must incorporate the contextualization of a specific historical period, the temporal evolution it has undergone and the interpretation based on current parameters.

The paper includes some theoretical notes on the interoperability between both methodologies based on different studied publications with the aim of establishing a reference framework for further developments and conclusions. Additionally, the theoretical basis obtained in this research has been applied to the development of a real case study model. The chosen building for this purpose was the Old Church of Santa Lucía (Seville), now used as *Centro Documental de Artes Escénicas de Andalucía (CDAEA)*. The results of this research show the need to continue working on the integration of both methodologies due to the numerous possibilities it offers.

1 INTRODUCTION

The aim of this article is to draw conclusions about the possibilities offered by the creation of BIM and GIS models for architectural heritage management purposes, taking into account the current viability offered by current interoperability strategies.

The analysis of the different integration processes used has focused on the development of a practical case: the elaboration of a digital BIM-GIS model of the Old Church of Santa Lucía (Seville), currently "*Centro de Documentación de las Artes Escénicas de Andalucía (CDAEA)*".

The interest of this building lies in its belonging to one of the medieval parish churches that were established in the city of Seville after the Christian conquest in 1248. The essential characteristics of this type of architecture were analysed through the creation of spatial databases on an architectural scale processed by GIS systems (Mascort-Albea, 2018). Through this research, a methodology was developed to establish detailed characterizations of heritage buildings using geographic tools and implementing simplified two-dimensional maps that facilitate the management of information related to their preventive conservation (Canivell, Jaramillo-Morilla, Mascort-Albea, & Romero-Hernández, 2019; Mascort-Albea & Meynier-Philip, 2017; Mascort-Albea, 2017) . Taking as a starting point the 2D GIS model of the old church of Santa Lucía, a three-dimensional BIM model of the building has

*Corresponding author: emascort@us.es, jarami@us.es, rociorome@us.es, fmhidalgos@gmail.com

been made, as well as its subsequent export in a 3D GIS environment (Hidalgo Sánchez, 2018). All this allows us to establish the reflections on interoperability that are presented below.

2 NOTIONS ON BIM-GIS INTEROPERABILITY

The concept of interoperability between BIM and GIS can be defined as the ability of the two systems to contact each other, communicate and exchange data in an attempt to achieve a common goal. However, in order to reach this point it is necessary to have a clear and total connection between the data coming from one system and another, as well as the way of interpreting them (Zhu, Wright, Wang, & Wang, 2018).

2.1 *Levels of integration*

The integration between BIM and GIS can be covered from different levels. Several groups of researchers have carried out classifications that have attempted to synthesize the complexity of this process (Amirebrahimi, Rajabifard, Mendis, & Ngo, 2015). In this work, the following classification is taken as a reference, which is structured in three levels: *data level or data interoperability, application level* and *process level* (Figure 1).

In this sense, it is essential to emphasize the *data level* or *data interoperability* as a starting point to achieve even more complex objectives, as the other levels depend on correct interoperability at this level. The *data level* in turn is divided into *geometry level* (data visualization) and *semantic level* (data visualization and analysis).

2.2 *Commonly used data formats*

The flow of information from BIM to GIS, and vice-versa, always implies a change in data structure and format. There are many formats for storing a three-dimensional geometry, such as *3D Studio (.3ds), SketchUp (.skp), COLLADA (.dae)*, etc. However, the formats that by their nature are most likely to obtain better results in this task are *Industry Foundation Classes (.IFC)* and *City Geography Markup Language (.CityGML)*. These formats are capable of storing both geometric and semantic information. An alternative to consider is the *multipatch (shapefile)* standard developed by ESRI, but it only stores data at the geometric level.

2.3 *Information exchange of data*

The search for interoperability in the *geometry level* is developed mainly for visualization purposes, focusing on the transformation of information related to the geometry of the model,

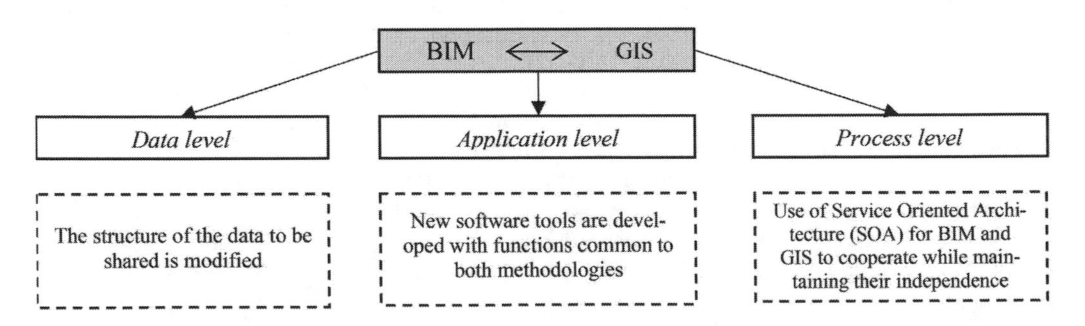

Figure 1. BIM-GIS levels of integration. Source: Prepared by the authors from Amirebrahimi et al (2015).

producing a loss of information at the semantic level. The level of difficulty to reach this objective is not too high.

The *semantic level* comprises more complex processes in order to conserve as much information as possible. By allowing the exchange of more information, it makes it possible to carry out analyses based on these, additionally to visualization. The main problem in the semantic aspect for the integration of BIM and GIS lies, on the one hand, in that each one gives different definitions for the same object, for example, a window in IFC is defined as "IfcWindow", while in CityGML is defined as "window", and on the other hand, one of them defines components or classes that the other does not. This situation is problematic for some practical applications, because when switching from the IFC format to CityGML, the relationship between objects is lost, if no procedure is carried out to fix it.

Apart from the creation of specific methodologies and processes (creation of ontologies, modification of "schemas" of the formats IFC and CityGML, etc.), there are commercial software working on the integration between BIM and GIS, such as *BIMServer, IfcExplorer, Feature Manipulation Engine (FME)* and *Data Interoperability (DI)* for ArcGIS. However, none of these tools achieves a complete transfer of information at the geometry and semantics level between BIM and GIS (Donkers, 2013).

3 STRATEGIES APPLIED TO THE CASE OF STA. LUCIA´S ANCIENT CHURCH (SEVILLE, SPAIN)

3.1 *General procedure*

In this type of practice, BIM tools are often used in the 3D modelling phase, usually using photogrammetry and laser scanners to capture data and obtain point clouds that serve as the basis for the model survey. In this case, these techniques have not been used because they suppose a high cost in relation to the pursued aims. The graphic survey has been carried out using the BIM *Graphisoft ArchiCAD* software, based on the planimetric information obtained during the research. Once the BIM (or HBIM) model has been built, an attempt has been made to include it in a 3D GIS environment.

3.2 *BIM model of the Sta. Lucía´s ancient church*

When approaching the BIM modelling of a heritage building, it is important to carry out a formal, structural and spatial preliminary study, which allows us to group elements to simplify the model and facilitates the possibility of linking heterogeneous data for subsequent analysis and management. To such ends, in this case the effort to segment the model is reduced, since it has been taken the constructive and spatial characterization of the church developed through a 2D GIS previous model (Mascort-Albea, 2018). This division groups different elements of the building into certain categories, assigning them a common coding.

Based on this structure of the architectural information of the building, the BIM model was elaborated. The construction process of the model is too extensive to be included in this article, but it can be consulted in its entirety in the Final Degree Paper that serves as the basis for the drafting of this contribution (Hidalgo Sánchez, 2018). Some images of this model are shown below (Figure 2). Finally, it is exported to IFC format using *Graphisoft's ArchiCAD* general translator. During this export process, there is a slight loss of information referring to the geometry level, mainly in those elements of the model derived from Boolean operations (subtraction, insertion, etc.). Notwithstanding, in the semantic information section, the attributes that had been linked in the original BIM model remain intact when exporting to IFC.

3.3 *BIM - GIS interoperability*

As developed in section 2.2, interoperability between BIM and GIS can be conceived from the conversion between IFC and CityGML formats. It should be remembered that

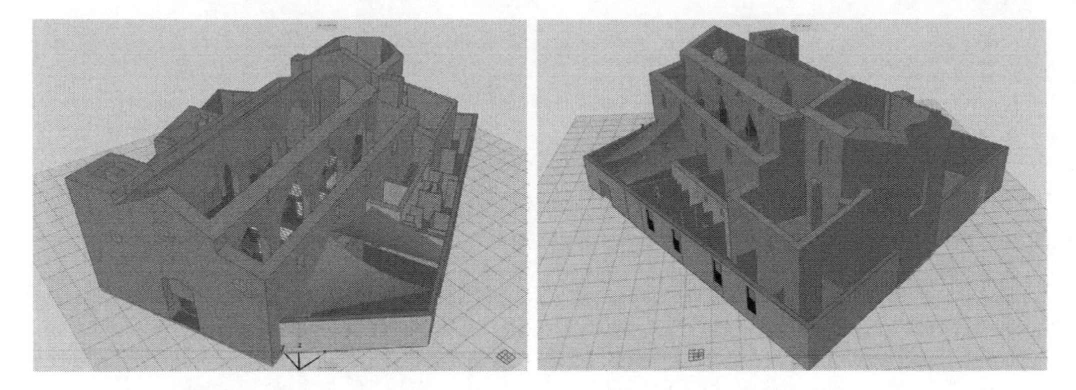

Figure 2. Santa Lucía´s BIM model images. Source: Prepared by the authors.

the main procedures for achieving interoperability between the two are basically divided into two groups. On the one hand, the use of commercial software developed especially for this task. On the other hand, the use of specific procedures, such as the creation of ADEs (Application Domain Extension), use of own ontologies, etc. Within the framework of the research carried out, both options are discarded, because of the reason previously mentioned (Hidalgo Sánchez, 2018).

In this sense, it was considered a third option, a combination of the two previous ones (Dore & Murphy, 2012). It consists of using a *Trimble SketchUp* plug-in to convert the BIM model of a heritage building to the CityGML format. This plug-in converts the objects of the BIM model taking into account the semantic classes defined by the CityGML format. It seems a good method, however, it has the disadvantage that some elements of the model, such as the pointed arches of the Church, do not present a semantic class equivalent in CityGML, which would force to create it through the development of ADEs, returning to the second of the initial options, therefore, it is also discarded.

4 CONCLUSIONS

BIM-GIS conversion currently involves tedious processes, and in spite of this, full interoperability is not guaranteed. Depending on our purpose, it would be more convenient to choose the construction of a 3D model which semantic properties are not assigned (not BIM), and then to link them in a GIS environment, or to make a BIM model with semantic starting information, being complemented and managed in the GIS environment. Which method is better will depend on the development of the standards used by BIM and GIS to communicate with each other. On the other hand, the development of both methodologies of tools enables the construction and the work with models of buildings with patrimonial characteristics, will be a very important factor, in order to facilitate and speed up this workflow.

REFERENCES

Amirebrahimi, S., Rajabifard, A., Mendis, P., & Ngo, T. 2015. A data model for integrating GIS and BIM for assessment and 3D visualisation of flood damage to building. *CEUR Workshop Proceedings*, *1323*(March), 78–89.

Canivell, J., Jaramillo-Morilla, A., Mascort-Albea, E. J., & Romero-Hernández, R. 2019. Metodología de evaluación y monitorización del patrimonio basado en la gestión cartográfica digital. La muralla de Sevilla. En M. Di Sivo & D. Ladiana (Eds.), *Le mura urbane crollano: conservazione e manutenzione programmata della cinta muraria dei centri storici.* (pp. 119–135). Pisa, Italia: Pisa University Press. https://doi.org/10.12871/97888333917559.

Donkers, S. 2013. Automatic generation of CityGML LoD3 building models from IFC models. *Zhurnal Eksperimental'noi i Teoreticheskoi Fiziki*, 127. https://doi.org/uuid:31380219-f8e8-4c66-a2dc-548c3680bb8d.
Dore, C., & Murphy, M. 2012. Integration of Historic Building Information Modeling (HBIM) and 3D GIS for recording and managing cultural heritage sites. *Proceedings of the 2012 18th International Conference on Virtual Systems and Multimedia, VSMM 2012: Virtual Systems in the Information Society*, (January 2014), 369–376. https://doi.org/10.1109/VSMM.2012.6365947.
Hidalgo Sánchez, F. M. 2018. *Interoperabilidad entre SIG y BIM aplicada al patrimonio arquitectónico. Exploración de posibilidades mediante la realización de un modelo digitalizado de la Antigua Iglesia de Santa Lucía y posterior análisis*. (Universidad de Sevilla). Retrieved from https://hdl.handle.net/11441/79394.
Mascort-Albea, E. J. 2018. *Mapas para el patrimonio: caracterización técnica de las iglesias medievales de Sevilla mediante sistemas de información geográfica (SIG)* (Universidad de Sevilla). Retrieved from https://idus.us.es/xmlui/handle/11441/70745.
Mascort-Albea, E. J., & Meynier-Philip, M. 2017. Strategies for conservation of Religious Heritage in the Metropolitan Area of Lyon/Saint-Étienne (France). Short research stay and methodological transfer. *IDA 2017. 1st International Congress on Advanced Doctoral Research in Architecture*, 675–684. Retrieved from https://idus.us.es/xmlui/handle/11441/70006
Mascort-Albea, Emilio José. 2017. Datos geográficos abiertos para la conservación preventiva del patrimonio arquitectónico. *revista PH*, (92), 228. https://doi.org/10.33349/2017.0.3948.
Zhu, J., Wright, G., Wang, J., & Wang, X. 2018. A Critical Review of the Integration of Geographic Information System and Building Information Modelling at the Data Level. *ISPRS International Journal of Geo-Information*, 7(2), 66. https://doi.org/10.3390/ijgi7020066.

Science and Digital Technology for Cultural Heritage – Ortiz Calderón et al. (Eds)
© 2020 Taylor & Francis Group, London, ISBN 978-0-367-36368-0

In search of the most efficient way to make a three-dimensional analysis of a multitude of architectural elements for a unitary project: The Moorish strip

J.F. Molina Rozalem
ETSA, Universidad de Sevilla, Seville, Spain

ABSTRACT: During the first half of the fourteenth century the council of Seville, supported by the Castilian monarchy and the local nobility is involved in a constructive program that should have required a huge logistical effort for the time and the means available in that area. This constructive program consisted in the construction of a defensive network, to the south of the territory that it controlled, in the so-called “Moorish strip”, which as a mesh, included at least 40 defensive towers and lookouts that located in strategic positions and visually communicated between they aimed to neutralize the Nasrid attacks on this border area.

Through a software to process digital images and, through the combination of digital photogrammetry and computer vision techniques, generating a 3D reconstruction of the element analysed, we will try to propose a constructive and typological analysis methodology. This work, in addition, is part of a doctoral thesis and an ID project that aims to generate strategies of value enhancement and collaborative heritage management that involve various disciplines and social actors.

1 INTRODUCTION

1.1 *Historical context*

Over the course of a historical period of two and a half centuries, a border area was formed between the southern sector of the province of Seville, together with adjacent areas of the provinces of Cádiz and Málaga.

This period will be delimited as the Late Middle Ages, between the middle of the 13th century and the end of the 15th century. Even today the toponym of “the border” in the name of many towns and villages of the area remains as a witness of that era.

In addition to this toponymic curiosity, other much more forgotten but more visible witnesses are left such as the fortifications that were erected to defend that border, which now stand as silent witnesses of time and are mostly in a dilapidated state.

The historical period to which we refer is delimited to the final stage of the Reconquest, once the last stronghold of Muslim resistance, the Nazari Sultan of Granada (García Fernández. 1980). A new border was created with the consolidation of this kingdom that remained dormant for two hundred and fifty years, and that was called “Morisca Band” in late medieval Seville (Figure 1). The fortifications studied are a consequence of this border and are based on it, in such a way that when Muslim Granada fell, the majority were abandoned and ended up in places that no longer had to be defended from any enemy, located in wild and sparsely populated areas, that will almost never be occupied except in some cases (Molina Rozalem. 2016).

In terms of the role they played, it is clear that they were part of a network of fortresses with the main objective of detecting and warning of possible Muslim attacks, as well as sheltering the very few inhabitants and their livestock, and repopulating the area (González Jiménez. 2001).

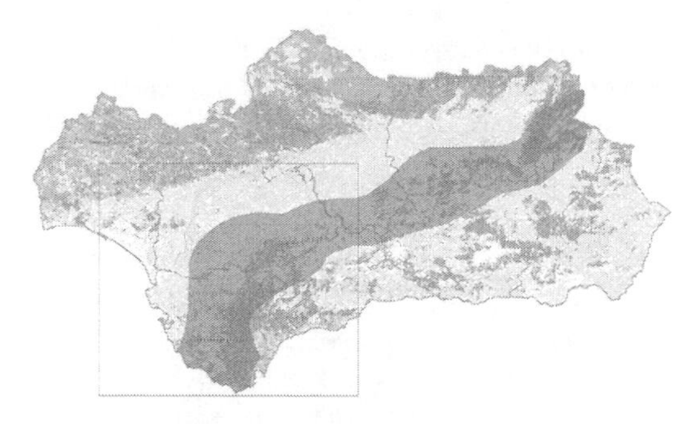

Figure 1. Location of the so-called Moorish strip.

Most of these towers are currently in a neglected state in the middle of the field, and probably no major interventions have been carried out on most of them for centuries. In our opinion, this represented a good research opportunity since the documentation on these elements is practically non-existent. In any case, it is a challenge due to the extensive field work involved and the search for historical documentation, which in most cases is reduced to the appearance of a reference or a name in certain documentary sources.

2 OBJECTIVES

The research has two objectives: the first and main will be, in view of the amount of elements that will be analyzed in the described territory, at least 40 watchtowers that we intend to analyze to find their common points, constructive and geometric characteristics, we will look for the form It is more effective to carry out this three-dimensional analysis, taking into account the use of computers, scanning or photogrammetric equipment, the benefits they offer and their value for money.

On the other hand, we want to verify with this research to what extent it is interesting to do an analysis of this type in these towers, to gauge if it is really necessary to use these technological resources to draw the conclusions we want or do not provide much more than an analysis with the conventional means.

We have chosen a fortress tower for both objectives, the Lopera tower (Figure 2-3)., a fortification that meets the condition of defensive building prototype in the Moorish strip (Rojas Gabriel. 1987), with a level of conservation that can be considered quite acceptable despite abandonment, and with geometrical and spatial elements of some complexity and interest. It consists of a keep which is about 16 meters high and divided into two levels covered by a vault of ashlars. This tower is surrounded by a walled covering that is partially preserved only

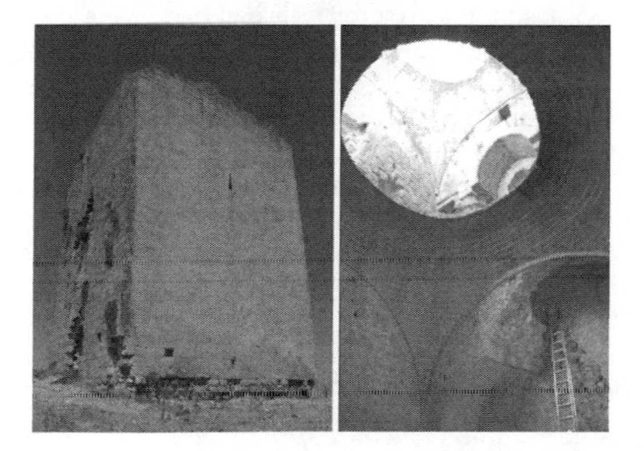

Figure 2-3. Images of the Lopera tower.

in one sector. The predominant constructive element is the stone ashlar with good carving. All of them are located on the top of a rocky hill that rises considerably on the contour. Geographically, it is located not far from the town of Montellano, in the south of the province of Seville.

3 METHODOLOGY

In the first instance, we will clarify that the photogrammetric survey is a widespread technique, and that it has been used for years by researchers at the level of architecture and archeology (Martín Talaverano, 2014). As already mentioned, what we intend is to reach the conclusion of which methodology will be most useful for a massive measurement of this type of buildings.

3.1 *Instruments used for the analysis*

The following photogrammetric measurement and scanning instruments have been used for the three-dimensional analysis of the Lopera tower:

- Leica Total Station
- "Canon EOS Digital" reflex camera.
- "DJI Phantom 2" drone, equipped with "go pro hero 3" camera.
- "3D, ZEB REVO" (GeoSLAM) mobile laser scanner.

The following programmes have been used for the processing of photos and obtaining and viewing the point cloud:

- Agisoft PhotoScan.
- MeshLab.
- Reality captur.

3.2 *Obtaining and processing the data*

Reference points and coordinates were taken with the total station to accurately position and scale the analysed element. We subsequently proceeded to take a series of photographs with the reflex camera, first sweeping the entire outer perimeter and then the interior of the keep in its two levels. In total, more than 500 photographs were taken.

Subsequently, the exterior of the tower was photographed with the drone camera, resulting in about 300 photographs.

Finally, we proceeded to scan both the interior and exterior of the building with the mobile scanner to obtain the point cloud.

Once the data was obtained, the photographs were processed through the *AgisoftPhotoScan* programme, obtaining a point cloud that was meshed and with textures.

The point cloud obtained with the scanner was visualised with the *MeshLab* programme.

Out of the point clouds processed, the external cloud produced through the drone photos and the cloud of interior points of the keep processed through photographs of the reflex camera were selected. The point cloud obtained with the 3d scanner was discarded because it did not offer points located above 15 meters.

Finally, the external and internal point cloud was combined with the *Reality captur,* programme, obtained as a result a point cloud with a lot of detail of the entire fortification.

4 ANALYSIS OF THE RESULTS

4.1 *Analysis of the data collection*

Firstly, it will be interesting to know which the most effective method is to collect data in a massive way in more than 40 heritage elements located in rural areas and that are difficult to access. We will take into account the speed of the capture, the real effectiveness of the devices, the ease of use, the reliability of the data obtained, and the price-quality ratio of the rental or purchase of the necessary equipment.

If we analyse the different options used in the data collection to obtain a cloud with good quality points, allowing us to later analyse and compare another type of data more carefully, we extract the following conclusions:

4.1.1 *Photogrammetry obtained with a Reflex camera*

Advantages: Good quality exterior photogrammetry is achieved, and also indoor if you have the right equipment (tripod, stabiliser) and there is enough light. They are affordable and are easy to use.

Disadvantages: In buildings with a certain height, photogrammetry loses quality because it does not capture many areas, or the image can be distorted if we take it with very pronounced angles.

When the building has amplitude and spatial complexity, the programme to obtain photogrammetry may not link the different images well if they are not taken with a certain general perspective of the building.

4.1.2 *Photogrammetry obtained with drone*

Advantages: Obtains a fairly accurate exterior photogrammetry. There are no problems recognising points when taking aerial images. The bird's-eye and overall views are precise and give us information that cannot be obtained otherwise. Currently the prices to acquire a drone are moderate

Disadvantages: It cannot be used indoors when the GPS signal is lost. The use of the drone is only possible in certain spaces and corresponding licenses and permits must be obtained (never populated areas). Operation requires prior learning.

4.1.3 *Point cloud obtained with mobile 3d scanner*

Advantages: Its use is easy and direct. We obtain a point cloud directly without needing processing with computer programmes. The interior point clouds are quite accurate.

Disadvantages: they usually have a height limitation, so it does not give good results in the upper and covered areas. When not taking images, it does not give us textures, so we would have to do a separate photographic studio if we want to have textures and colours and apply them to the resulting mesh. The price of rental and purchase is very high, which makes it a resource with little use if we have limited means.

4.2 *Analysis of the images*

The point cloud obtained from the combination of images of the drone and the reflex camera allows us to create a very reliable model of both the interior and the exterior of the fortification. The use of the drone gives us bird's-eye views of the building (fig. 5), with which we can decipher the level with enough accuracy, which would be very complex in another way. We also obtain a survey of annexed terrain which gives us other information for the study of the

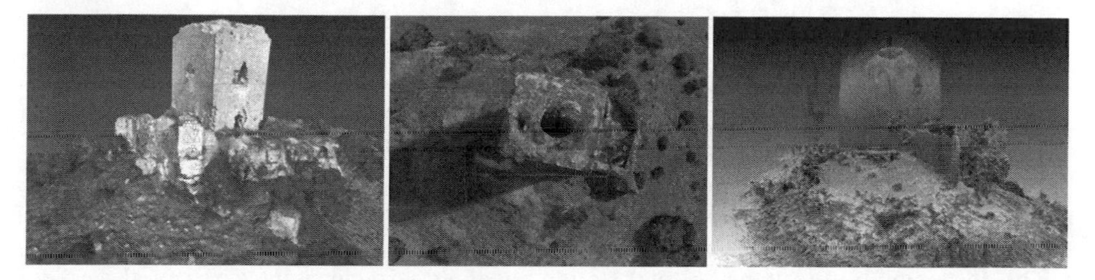

Figure 4-5. Point cloud with texture obtained from the combination of drone images and reflex camera.
Figure 6. Point cloud obtained with a mobile 3d scanner.

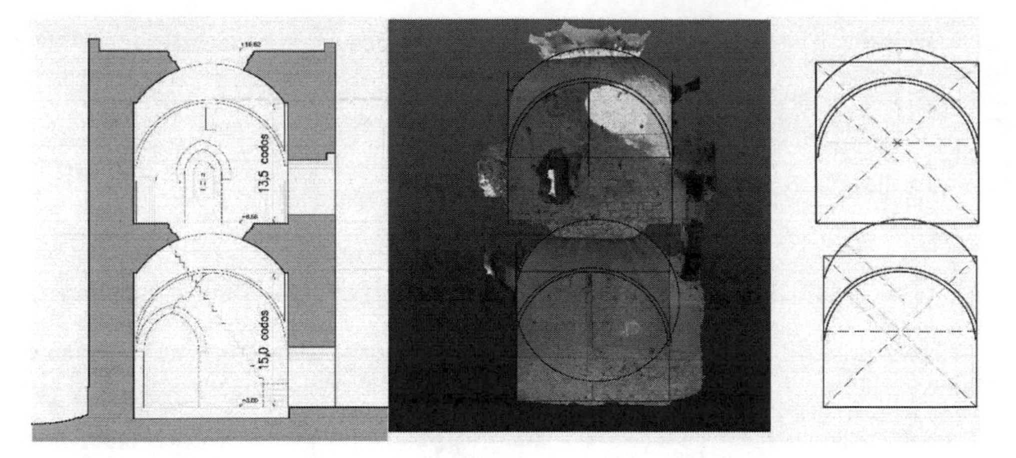

Figure 7-8-9. Lopera tower. Conventional section. Section of the point cloud. Sphericity of vaults.

fortification. As an addition, we also have the real textures of the building, so we can talk about a real scale model of the building.

On the other hand, with regard to the point cloud obtained from the 3d scanner (Figure 6), although it is true that it accurately defines the elements it captures, especially in the interior, it does not give us a defined model in the exterior where the scanner sometimes encounters its distance limitation. Unlike in the other case we do not have textures here, so in the end we are not talking about a complete model of the element.

4.3 *Geometric analysis*

One of the most interesting aspects of these fortifications are the vaults that cover the two floors of the towers. In the Lopera tower, both are preserved, albeit half-ruined. These are vaults of ashlar masonry, covering a square room of more than seven meters on each side, something very unusual to see in 13th-14th century buildings. Thanks to the three-dimensional analysis with photogrammetry we can check different previous approaches that we do, such as: if you really are vaults are hemispherical or are lowered, if the center of both spheres form an axis or if they are displaced, and if it is possible to use the same truss of wood for both vaults.

After the check in Lopera, if we can say that the vault of the ground floor is semi-spherical, however that of the upper floor is recessed and displaced with respect to the axis, mainly because the recess of the upper perimeter wall allows to inscribe a spherical higher.

Having the planes used with conventional techniques we can observe the deficiencies with respect to the planes obtained by the cloud of points, much more precise, and evaluate with this data also the suitability of using this measurement methodology in the other buildings.

5 CONCLUSIONS

With this case study, where we have chosen one of the most complex fortifications of the extensive list that we want to analyze, we have extracted that for these fortifications the most efficient way of analyzing (taking into account the number of objects that we want to analyze and the resources limited that we have) would be with the combination of the use of images obtained with a drone and the interior images and certain details obtained with the SLR camera. Both groups of images that were previously processed separately can be joined by taking the points taken with the total station, creating a single point cloud of the entire analysed element. The geometric detail that is obtained is not inferior to the 3d scan, and also gives us the texture of the elements.

On the other hand, obtaining a cloud of points of these characteristics, high definition, is considered very useful for a research project that wants to serve a set of heritage assets dispersed in rural environment (where we also have no exceptions in the use of the drone), since its access is very difficult, it will allow us to have in our computer a real model of each of these assets on our computer, being able to analyze the details and typological comparisons in a more detailed way, as well as to perform a geometric approach more precise at a later stage.

Another advantage that we obtain with this methodology is to get surveys of all its walls to be able to reproduce its historical stratigraphy, and thus be able to make a preventive conservation study as the case requires.

ACKNOWLEDGEMENT

We would like to thank the architect Jesús Rodríguez Medina for his special collaboration with the drone.

REFERENCES

Angulo Fornos, R; Pinto Puerto, F; Rodríguez, J; Palomino, A. 2017. Digital Anastylosis of the Remains of a Portal by Master Builder Hernán Ruiz: Knowledge Strategies, Methods and Modelling Results. *Digital Applications in Archaeology and Cultural Heritage*. vol. 7, pp. 32–41.

Angulo Fornos, R. 2013. La fotogrametría digital: una herramienta para la recuperación de arquitecturas perdidas. Torre del Homenaje del Castillo de Constantina. *Revista digital VAR*. Vol. 4 Número 8. 140–144.

García Fernández, M. 1980. *El Reino de Sevilla en tiempos de Alfonso XI: (1312-1350)*. Sevilla: Diputación Provincial de Sevilla.

González Jiménez, M. 2001. *La repoblación de la zona de Sevilla durante el siglo XIV*. Vol. 60. Sevilla: Secretariado de Publicaciones, Universidad de Sevilla.

Martín Talaverano, R. 2014. Graphic documentation of historical buildings: principles, applications and prospects. *Revista Arqueología de la Arquitectura, nº11*.

Molina Rozalem, J.F. 2016. *Arquitectura Defensiva en las Fronteras del Reino de Sevilla durante la Baja Edad Media.* Madrid: Subdirección General de Publicaciones y Patrimonio Cultural. Ministerio de Defensa.

Rojas Gabriel, M. 1987. La torre de Lopera, arquetipo de la Banda Morisca. *Revista Arqueología medieval española*. II Congreso: 263–271. Madrid.

Science and Digital Technology for Cultural Heritage – Ortiz Calderón et al. (Eds)
© 2020 Taylor & Francis Group, London, ISBN 978-0-367-36368-0

Augmented reality and gigapixel image technologies: Two case of study of Andalusian Barroc artifacts

M.J. González-López, B. Prado-Campos & B. Domínguez-Gómez
Departamento de Pintura, Facultad de Bellas Artes, Universidad de Sevilla, Seville, Spain

ABSTRACT: This work presents the results obtained after creating two multimedia resources made in Augmented Reality and Gigapixel Image. In the first one, the focus has been to acknowledge the main alterations of a painting, shown in a visual and intuitive way. In the second one, a further step has been taken by incorporating all the available information about the altarpiece to the interactive resource: general data, iconographic and morphological analysis, polychromy, technical information, and lastly, alterations and previous interventions.

However, each feature offers advantages and disadvantages regarding content and navigation, requiring a higher or lower participation of the user, as it will be described in the communications.

1 INTRODUCTION

This work presents the results obtained after creating two multimedia resources made in Augmented Reality and Gigapixel Image (Cabero Almenara, J, 2016). They have been made possible thanks to its selection by the Secretariat of Audiovisual Resources and New Technologies (SAV) of the University of Seville (2015 and 2016 summons).

A relevant Andalusian cultural object of study has been used in each of them. Augmented Reality technology, has been used in an Andalusian Baroque painting on canvas entitled *Inmaculada Concepción* (*Immaculate Conception*), attributed to Franscisco Meneses Osorio, disciple of Bartolomé Esteban Murillo. Owned by the City of Sanlucar la Mayor (Seville). Gigapixel Image technology, has been used an Andalusian Baroque altarpiece entitled *Descendimiento* (*Descending*), hired by Andrés de Ocampo (1603) and polychromed by Vasco de Pereira and Andrés Ramírez (Palomero, 1983). Owned by San Vicente Mártir Parish (Seville). The selection of both pieces has been made according to the following criteria: their artistic and cultural interest, their singularity, the previous information and documentation avalaible, their abordable dimensions, suitability, representativeness and easy access to carry out your study, examination and photographic documentation.

Our society is increasingly characterized by the use of Information and Communication Technologies in all areas of life, and it is difficult to imagine a world without mobile devices connected to the Internet (Pérez de Celis Herrero, 2010). Its democratization, as well as the development of technological resources based on virtual experiences, augmented reality or immersive reality is progressively modifying our way of communicating and interacting with the real world. They have quickly become an essential means for the dissemination of Cultural Heritage in all its aspects; research, intervention, dissemination and training (Kangdon, 2012). Within this context are the projects that we present IVAPTA and Gigapixel Image of the Altarpiece of San Vicente.

2 AIMS

The objectives that we intend to achieve with these two multimedia applications applied to the conservation and restoration of cultural assets are: use non invasive resources that contribute to a better knowledge of the cultural objects, contribute to increasing the multimedia resources for cultural heritage, particularly applying on paintings and altarpieces and evaluate and explore the applications of these tools.

3 OBJECT BY AUGMENTED REALITY TECHNOLOGY: IVAPTA

3.1 *The project*

This project has developed a software application for mobile devices enabling Augmented Reality to utilise the field of knowledge relating to a cultural artefact (Milgram, 1994). The aims of the application are to identify and illustrate alterations and degradation of a cultural artefact and present them to the educational community and society in general. The resource includes visual indicators of the alterations in a painting on canvas; in this case those in the work Immaculate Conception, attributed to Meneses and owned by the City of Sanlúcar la Mayor (Seville). The final application is user-friendly, affordable, visual and descriptive and allows people who are involved in the study, teaching, research and restoration of cultural artefacts to recognize, appreciate and evaluate the level of risk they pose to the integrity of the artefact.

3.2 *Organization of information*

The information is organized in a logical structure in which the descriptors considered are assigned, arranged both alphabetically and hierarchically; grouped in this last case, by families of alterations. The hierarchical structure in families, gathers groups of descriptors with common characteristics, to which they have been assigned to facilitate their identification, a distinctive color code. This classification, undoubtedly, more innovative and intuitive, constitutes the real research motor by multimedia resource. For each alteration indicator, considered, three levels of information: textual; it gathers the data of the indicator that allows its description and identification, putting it in relation, with the cause and with the development mechanisms. Visual; selected image was representative of the alteration and, easily recognizable and identifiable, by the end user of the resource. And, audio; voiceover that briefly describes the alteration. The information is organized by families of alterations: cromatic alteration, deformation, previous intervention, lost of fixation, physical lost of material and crackers.

3.3 *Information of resource in web*

You need: device: mobile, tablet, computer, etc. that contains a built-in camera to capture the image of the real world that activates digital information, and a screen where both realities, physical and virtual, are projected. With internet connection. Download of application (REPORAUS (Google Play - App Store) and marker (printed image of the painting). Download: SAV website from where you can download the resource's apps or directly from the android and appstore repositories, link: http://ra.sav.us.es/index.php?option=com_content&view=article&id=115:ivapta&catid=27:ra&Itemid=123 (Figure 1),

4 OBJECT BY OBJECT BY GIGAPIXEL IMAGE TECHNOLOGIE

4.1 *The project*

Gigapixel Image technology, has been used an Andalusian Baroque altarpiece entitled *Descendimiento* (*Descending),* hired by Andrés de Ocampo (1603) and polychromed by Vasco de

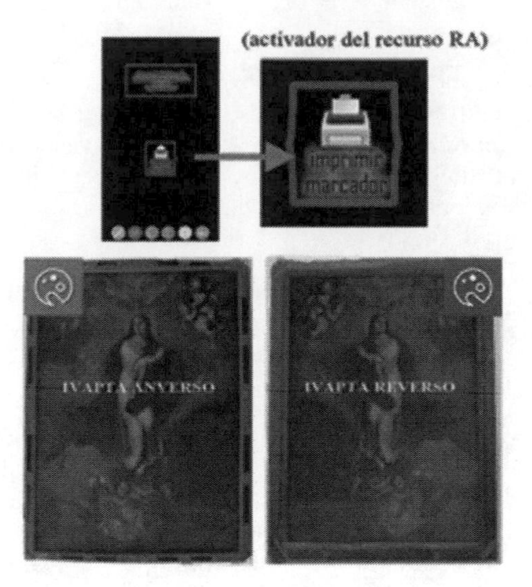

Figure 1. IVAPTA access to resource.

Pereira and Andrés Ramírez (Pérez Morales, 2012). Owned by San Vicente Mártir Parish (Seville). This project provides a novel technique such as, the super-dimensional image. More concretely, the possibilities presented by its application for the knowledge and dissemination of complex and/or large-format works, such as altarpieces, have been assessed.

Gigapixel photography allows us to obtain large images that we can zoom to see in detail areas that arouse interest of the object. From these, you can create superabled image resources developed with a special hardware-software device (SAV web); in this case the GigaPa system.

The working sequence to create a GigaPan image is simple, it consists of: selecting the scene to be captured, capturing individual images with the camera, compacting, aligning and automatically combining the individual images generating the panoramic macro-image through the specific software GigaPan Stich Software. And finally, visualize, share, embed, tag and print the images through your website.

4.2 *Organization of information*

All the information collected is included in different sections, with visual information, infographics and additional explanatory texts of certain areas of the work previously selected for their interest. The user can browse and view (general and/or in detail) the work, obtaining high quality information that allows him to study and determine aspects related to the polychrome and decorative techniques used. Or in the conservative plane, the types and the scope of the present alterations, in a comfortable way and without interfering in the physical space that shelters it. Sometimes, even getting better results than directly observing the works. Their structure is organize into diferentes sections: study, data sheet. morphology and material history, iconographic analysis, technique, previous interventions and alterations.

4.3 *Information of resource in web*

Navigation through resources can be done in two different ways: guided by the zoom (snapshot) previously selected details that evidence aspects of interest extend the information with the text the chromatic code that locates them. And free navigation arbitrated by the user with possibility to extend or decrease the image. Download: SAV website from where you can from where you can watch, through a computer or other device with internet connection, at the

Figure 2. Acces to navegation in the gigapixel imagen.

following link: http://ra.sav.us.es/realidadampliada/retablosanvicente/relieve-de-la-serpiente.html (Figure 2),

5 DISCUSSION

Both resources have allowed to achieve the established objectives. They provide and show the information with an easy and accesible language, and its adequacy has been confirmed with the students of the Conservation and Restoration of Cultural Objects Major. However, each feature offers advantages and disadvantages regarding content and navigation, requiring a higher or lower participation of the user, as it will be described.

Both of them have similar advantage, like: access to the information visual and textual including disabled people, structure the information permit evaluate and recognize the grade of alteration of painting on canvas and barroc altarpieces and direct to Conservation and Restauration specialists, and at disposal for others profesionals. And, the Altarpiece San Vicente also has: easy of navigationand Access to the information and visualization of a large-scale work in conjuntion or in detail without the need for complex Access systems. Nevertherless, Ivapta has more disadvantages than altarpieces like: some limitations of technology avalaible at this moment that difficult the navigation, heavy weight application and difficulty viewing it on the smartphones´s screen. And the main disadvantage of Altarpiece San Vicente is don't allow a big amplification factor due to the snapshot low quality. When maximize it too much, the image reflect a lost of quality and details can be also lost.

6 CONCLUSION

They are many projects focused to heritage, related to museums, archeological heritage, etc. that are developing new digital technologies. Nevertherless, there are not many samples of new digital technologies, applied to conservation and restauration, in particular to the techniques of execution of materials as well as alterations of cultural property. Because that, this proposed resources want to contribute to the study, knowledge and diffusion of Andalusian cultural heritage, as well as supply specific information about the studied cultural objects.

Finally, its dissemination has contributed to reduce the lack of specific multimedia resources available to professionals curators and restaurators.

REFERENCES

Cabero Almenara, J. & García Jiménez, F. & Barroso Osuna, J. 2016. La producción de objetos de aprendizaje en "Realidad Aumentada": la experiencia del SAV de la Universidad de Sevilla». *Revista IJERI*, 6: 110–123.

Palomero Páramo, J. 1983. *El retablo sevillano del renacimiento: análisis y evolución (1560-1629)*. Sevilla: Excma. Diputación de Sevilla.

Kangdon, L. 2012. Augmented Reality in Education and Training. *TechTrends* 56(2): 13–21.

Milgram, P. & Takemura, H. & Utsumi, A. & Kishino, F. 1994. Augmented Reality: A class of displays on the reality-virtuality continuum. *Telemanipulator and Telepresence Technologies*, 2351: 282–292.

Pérez de Celis Herrero, C. & Cossio Aguilar, G. & Lara Álvarez, J. 2010. Uso de las Tecnologías de la Información y la Comunicación (TIC) para la catalogación y difusión del patrimonio cultural. *Revista Iberoamericana de Ciencia, Tecnología y Sociedad – CTS*: 1–23.

Pérez Morales, J. C. 2012. Los límites del arte: El Descendimiento de Andrés de Ocampo. *Quinta Angustia. Boletín de la Hermandad de la Quinta Angustia*, 107: 71–72.

Salinas, J. 2010. Innovación docente y uso de las TIC en la enseñanza universitaria. *Revista. Universidad y Sociedad del Conocimiento*, 1: 1–16.

Science and Digital Technology for Cultural Heritage – Ortiz Calderón et al. (Eds)
© 2020 Taylor & Francis Group, London, ISBN 978-0-367-36368-0

Key aspects regarding BIM for historical architecture. The church of San Cebrián de Mazote

R. Martín Talaverano
Universidad Politécnica de Madrid, Madrid, Spain

J.I. Murillo-Fragero
Urbe pro Orbe Patrimonio Cultural, S.L., Madrid, Spain

M.A. Utrero-Agudo
Escuela de Estudios Árabes, CSIC, Granada, Spain

ABSTRACT: BIM (Building Information Modelling) provides an effective strategy to improve the design, construction and maintenance of buildings. Undoubtedly, historical buildings are quite different from contemporary ones. Among others, a special feature is the fact that historical architecture is not a result of a single constructive boost, but a sequence of transformations that take place throughout its history. This paper presents the results of a research focused on the adaptation of BIM methodology to the specific features of historical buildings. For this purpose, we have used as case study the 3D model of the church of San Cebrián de Mazote (Valladolid, Spain), analyzed previously by means of the methodology archaeology of the architecture. This paper addresses the creation of the model as well as the use of a specific tool designed for implementing cultural heritage objects in a BIM environment, PetroBIM, in order to create the BIM model of this church.

1 INTRODUCTION

1.1 *Research contextualization*

This research is associated with the Project HAR2016-78113-R “Sustainable guardianship of cultural heritage through digital BIM and GIS models: contribution to knowledge and social innovation”, funded by the Ministry of Economy and Competitiveness and FEDER funds. Analytical data of the case study have been obtained thanks to the projects “Archaeology of the Hispanic churches of the 10th century: the circulation of architectural and decorative models, 1st and 2nd phase (HAR2012-35222 and HAR2017-84927-P)”, both of them also funded by the Ministry of Economy and Competitiveness and FEDER funds. The survey of the building was funded by the Heritage Department (Planning Service) of the regional government of Castilla y León. The research carried out is part of a line focused on the adaptation of the BIM methodology to the specific case of historical buildings. BIM (Building Information Modelling) refers both to a methodology and the tools required for managing information on the complete life cycle of a building linked to its geometric model (García Valldecabres et al. 2016). The application of BIM within built historic heritage is increasingly being promoted, given that it offers an effective way of optimizing the works related with the analysis and diagnosis of such cultural assets. In fact, nowadays the main application of BIM in the cultural heritage frame is related with the organization, updating and management of databases which include data from several interdisciplinary fields (historical, archaeological, artistical, architectural, etc.) to obtain a global view. However, to reach an adequate use of BIM with historical buildings it is essential to understand their singularities, fundamentally those that differentiate them from contemporary works. In the same way, it is necessary to consider the particularities of the work carried out in this field and the type of information that is dealt with.

Among other singular aspects (such as their productive contexts, their specific material and constructive characteristics or their pathologies), it is important to point out the fact that historic buildings are not the product of a single constructive impulse at a specific moment in history, but the result of a sequence of transformations related to their changes in use (Utrero Agudo 2010). These changes on the initial object which occur throughout the history of the building, configure what we can call its evolutionary nature. We understand that knowing this sequence of transformations is fundamental not only for the conservation or diffusion of the cultural values of the building, but also for the adequate implementation and organization of the information associated with its BIM model.

1.2 *Objectives and research methodology*

With the intention of analysing the adequacy of the BIM methodology in the field of historical architecture, the main objective of this research has been the creation of a BIM model of a specific case study which mainly addresses two factors: on the one hand, the inclusion of its sequence of transformations in the mentioned model and, on the other, the consideration of a sufficiently high level of accuracy.

The work has been focused on a case study, the church of San Cebrián de Mazote (Valladolid), a building belonging to the set of the so-called Hispanic Mozarabic architecture, with origins dating back to the beginning of the 10th century, and which has had a rich sequence of transformations until the 20th century. The photogrammetrical survey (Figure 1) and the archaeological analysis carried out within the framework of the aforementioned projects have provided both their sequence of transformations throughout its history (Figure 2), obtained through the analysis of Archaeology of Architecture (Caballero y Escribano 1996), as well as a three-dimensional mesh model. In this way, this research has addressed both the criteria and strategies of the modelling process, as well as the implementation of the information with a specific application to obtain the BIM model of the Mazote church.

Figure 1. 3D model of the building.

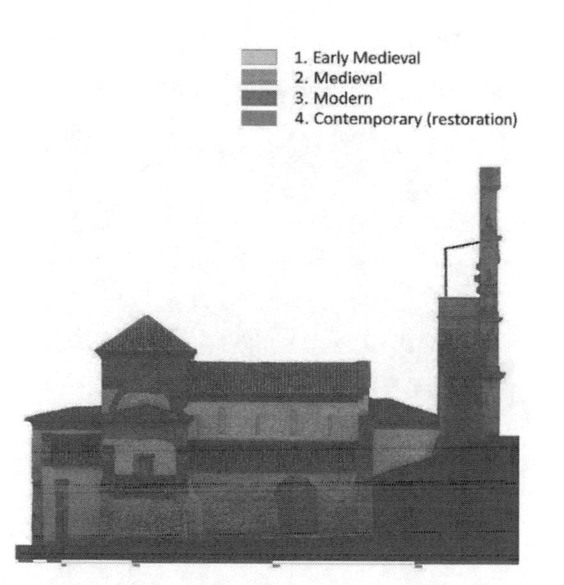

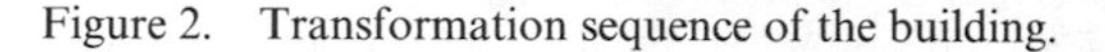

Figure 2. Transformation sequence of the building.

2 MODELLING PROCESS

2.1 *Modelling strategies*

As mentioned above, the modelling process has started from the textured triangular mesh obtained by photogrammetry, so that the level of accuracy is high, without simplifying the deformations or irregularities of the geometry of the constructive elements. With the condition of maintaining this high level of precision, the creation of a three-dimensional model has been approached by developing several strategies.

In the first place, the triangular mesh itself has been taken as a geometric support for the creation of new parametric surfaces that, as a "patch" conveniently adapted to the original geometry, allow the generation of entities which can be easily manipulated. To do this, several lines have been drawn on the original mesh that have served as geometric guides for the creation of mathematically defined NURBS surfaces. Therefore, this process has involved three steps: the drawing of guide elements (Figure 3), the creation of parametric surfaces (Figure 4)

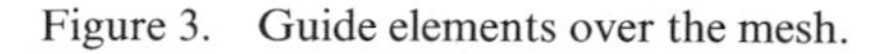

Figure 3. Guide elements over the mesh.

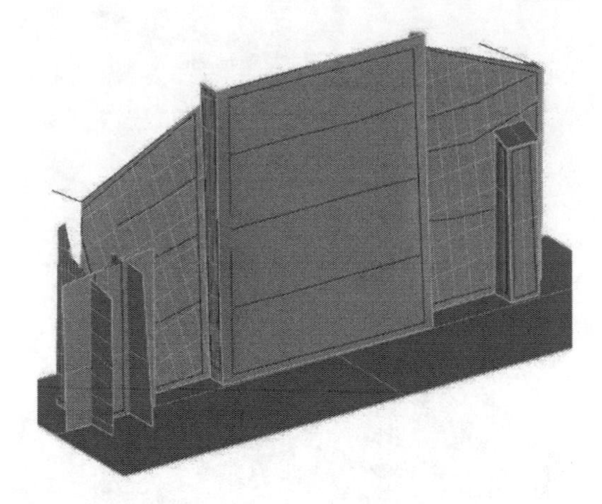

Figure 4. Generation of new NURBS surfaces.

and the reduction of such surfaces to form the real edges of the constructive elements. This strategy was analyzed with the fragment of the building corresponding to its eastern head and, although the resulting elements are simpler to handle than the triangular mesh itself, the loss of geometric precision has led us to abandon this approach and address a second strategy.

Unlike the previous case, the second line of work has been based on the use of the original triangular mesh for the creation of the building model. Since this mesh was originally unitary for the entire building, it has been necessary to approach a division into units consistent with the types of information that were to be implemented: constructive and stratigraphic. The first refers to the different elements and construction materials, while the second relates to the succession of transformations of the building, that is, with its stratigraphic sequence. Therefore, the first step has been the definition of the limits between each one of the units of the model (Figure 5), to then proceed to the division of the mesh.

Initially, such limits, conveniently transformed into surfaces by their extrusion, have been used as cutting elements with which to split the mesh. A priori, this approach seemed simple and effective, although its analysis showed that cuts were not always properly made, so the process was sometimes quite difficult, and the results were not entirely satisfactory. For this reason, a second approach was chosen for a slower, but safer and more efficient process, consisting of the removal, triangle to triangle, of the parts of the mesh that were on one of the sides of the limits of the corresponding unit of the model. Although the work was long, the

Figure 5. Definition of limits.

Figure 6. Selection of triangles of the mesh.

results were very satisfactory and, therefore, it has been used throughout the entire building (Figure 6).

2.2 *Division level*

As it has been observed, a key factor in the development of the modelling process has been the division of the original triangular mesh in the different elements in which the final model was to be composed, taking into account the type of information that was intended to be implemented later. Therefore, we understand that the model's level of division is a critical aspect in the development of BIM models of historic buildings. If the so-called "level of development" (LOD) is used in BIM models of new buildings, we believe that, in some way, the level of division, always linked to the level of knowledge of the building, must be one of the initial criteria that define the guidelines for the creation of the BIM model.

In this sense, since constructive and stratigraphic criteria have been taken to carry out the division of the model, we propose three levels in each one of them. In a constructive frame, from lower to higher level of detail, the subdivision would be: 1) large built bodies, such as the church hall, the transept or the body of the belfry, 2) constructive elements, such as walls, roofs, abutments, supports, etc., and 3) constructive materials, such as ashlars, masonry, stone walls, bricks, coatings, etc. On the other hand, in a stratigraphic frame, the methodology of the Archaeology of Architecture establishes the following categories from a lower to a higher level of definition: 1) general historical phases, 2) activities, or groups of contemporaneous stratigraphic units, and 3) stratigraphic units, or smaller homogeneous entities with precise limits (Caballero and Escribano 1996). We can therefore establish that the definition of the level of division of the model should at least address these two criteria in a complementary way, which, as mentioned, are linked to the level of knowledge of the building.

We have opted for a level of division for the transept of the Mazote temple (Figure 7) that includes the constructive elements in the constructive area (level 2) and the stratigraphic units in the stratigraphic area (level 3). However, the modelling division has been simplified for the rest of the building (Figure 8), differentiating the large building blocks (level 1), as well as the general historical phases (level 1).

3 IMPLEMENTATION OF THE INFORMATION

Once that the geometric model of the building with the corresponding divisions was created, the information of the sequence of its transformations (its stratigraphic sequence) was implemented to generate the BIM model. In this sense, the research has been focused on the analysis of a specific computer tool, PetroBIM, designed with the aim of responding to the needs of the cultural heritage and the works developed in this scope (Armisén et al. 2016).

Figure 7. Division of the model at levels 2 and 3.

Figure 8. Division of the model at level 1.

Unlike other BIM tools, PetroBIM is not a modelling application, but it focuses its functionality on the management of the information associated with the model. In this way, the model must be created externally and imported into the PetroBIM application once it has been made. Subsequently, the data is entered and organised into thematic modules (constructive elements, materials, pathology, archaeology, intervention, maintenance, etc.), creating a database with the information of the building in question. Finally, the information is linked to the model by means of a simple selection process for the geometric elements of the model and association of the desired data. The data relating to the stratigraphic sequence was concretely implemented within the framework of the present research. As mentioned above, this process was developed on two different levels: at the transept of the church, the data of each of the stratigraphic units with their relationships were introduced (maximum level of division and detail of the information), while in the rest of the temple, the data corresponding to the historical phases and their relationships were introduced (minimum level of division and detail of the information). Finally, we were able to carry out personalised consultations, establishing different search criteria combined with logical operators to formulate the specific questions to the information system and to obtain adequate answers to the research questions.

4 ANALYSIS OF RESULTS AND CONCLUSIONS

The work developed with the model of the church of San Cebrián de Mazote has allowed us to highlight some of the aspects that we understand are fundamental for the adequate adaptation of the BIM methodology to the frame of built heritage. On the one hand, in relation to the creation of the model and with the technical resources we have today, given the difficulties or the long amount of time involved in handling geometries without simplifying and with a high degree of accuracy with respect to the original object, it should be questioned to what extent such accuracy is necessary, when the final objective of the BIM model is the management of building information. In any case, the level of accuracy is something that must be defined initially and always in accordance with the objectives for which the BIM model is generated, understanding that maximum accuracy will not always be the best solution. In addition, the criterion of the division of the model is also fundamental, since the objects in which that model is composed will be the elements to which the data will be associated. Therefore, the level of division must be established coherently with both the level of knowledge about the building and the objectives of the creation of the BIM model. Finally, in the framework of the implementation of data to the BIM model, the analysis of the PetroBIM tool has allowed us to observe that one of its main advantages over other existing products in the market is to have a structuring of the information system in accordance with the reality of historic buildings, both in terms of their knowledge and their conservation or dissemination. In any case, regardless of the tool that is used, we have been able to verify that the sequence of transformations of the building (its stratigraphic sequence) offers an axis of fundamental organization for the creation of the information system, improving and facilitating the implementation, consultation, update and interrelation of the data of all the areas that are handled in the historic building frame.

REFERENCES

Armisén, A., García, B., Mateos, F. J., Valdeón L. y Rojo, A. 2016. Plataforma virtual para el diseño, planificación, control, intervención y mantenimiento en el ámbito de la conservación del patrimonio histórico 'PetroBIM'. *Actas del congreso Rehabend 2016*. Burgos.

Caballero, L. y Escribano, C. (eds.). 1996. *Arqueología de la Arquitectura. El método arqueológico aplicado al proceso de estudio y de intervención en edificios históricos*. Junta de Castilla y León, Valladolid.

García Valldecabres, J., Pellicer Armiñana, E. & Jordán Palomar, I. 2016. BIM scientific literature review for existing buildings and theoretical method: proposal for heritage data management using HBIM. *Construction Research Congress*: 2228–2238. San Juan de Puerto Rico: ASCE Library.

Martín Talaverano, R., Murillo Fragero, J.I. et al. 2018. BIM aplicado al patrimonio cultural. *Guía de usuarios BIM*. Madrid: BuildingSMART Spanish Chapter.

Utrero Agudo, María de los Ángeles. 2010. Archaeology. Archaeologia. Arqueología. Hacia el análisis de la arquitectura. *Arqueología aplicada al estudio e intervención de edificios históricos. Últimas tendencias metodológicas*: 11–23. Madrid: MCU.

Science and Digital Technology for Cultural Heritage – Ortiz Calderón et al. (Eds)
© 2020 Taylor & Francis Group, London, ISBN 978-0-367-36368-0

Protection of the architectural heritage in the agricultural landscapes. Application of GIS and BIM in oil mills of Écija

J. Moya-Muñoz
Universidad de Sevilla, Seville, Spain

A. González-Serrano & R. Rodríguez-García
Universidad de Sevilla, Seville, Spain

ABSTRACT: The conceptual consideration of the agricultural landscapes of the olive grove involves the recognition of the agrarian heritage with a patrimonial unabridged reading of the properties that conform it as a result of the history agricultural activity performed. The holistic approach to these asset protections implies their formal recognition, to avoid independent readings of each of them. This methodological proposal is raised as baseline for viable mechanisms realization to allow the enhancement of the production units associated with the agricultural landscapes of the olive grove in the municipal area of Écija - particularly, in the oil mills designed as productive units conjoined with their cultural, natural and historical aspects. The incorporation of the GIS and BIM tools are proposed for the practical interrelation between the elements of analysis, as mechanisms management of patrimonial able to generate graphic information that allows registering the real estate associated with the olive grove landscapes.

1 INTRODUCTION

The oleic activity in Écija made this municipal area of the Sevillian countryside a commercial reference of the Hispanic world in Roman times. For this reason, it was necessary to create a large territorial infrastructure where an important number of *villae* proliferated along the Astigitano land (Vega et al. 2008). After a fluctuating presence of olive activity during the Middle Ages, it was reactivated due to growing trade with America. However, the great boom in olive grove takes place in the Contemporary Age, and this remained until the second half of the 20th century. Therefore, it is considered a determining crop in the anthropisation of the Astigitano territory, where the natural and cultural conditions have shaped an agricultural landscape that can be defined from three elements: the type of crop (olive); the system for operation and ownership (consolidation of landownership); and its productive units (the oil mills with villae as antecedents (Corbacho, 1953)). However, the perception of the current landscape of Astigitano agriculture demonstrates a different view, where the quietly land dominates and the old de-contextualised oil mills stand out as testimonies of a past splendour of olive-growing activity.

The Cadastre of the Marquis of the Ensenada records a total of 258 oil mills in Écija during the middle of the 18th century. This amount has remained constant until the second half of the 20th century practically, as shown in Table 1, when a paradigm shift occurred in the countryside that affected all agricultural activity (Rodríguez, 2009). Specifically in the olive grove case, the modernisation processes and later mechanization of the tasks, the rural exodus due to the demand for labour from urban centres and the ministerial measures of the 1972-75 triennium were fundamental in worsening olive growing situation in Écija, that caused its total reconversion. As a consequence, the press and quintal oil mills ceased to be functional buildings for the production of oil. Despite the reactivation of the olive grove at the end of the twentieth century, this situation caused the oil mills to be abandoned or adapted functionally

Figure 1. Oil mill of Las Infantas and Valdecañas. Écija. Photo by Jorge Moya Muñoz 2017.

Table 1. Chronology of the number of oil mills in Écija.

Year*	1751	1824	1873	1954	2018
Mills	258	257	276	279	205

* 1751: Cadastre of the Marquis of the Ensenada. 1824 (Sánchez 1988). 1873: Cartographic Minutes. Geographic Institute. 1954: National topographic plot map. National Geographic Institute. 2018: Property Cadastre. Ministry of Finance and National Air Orthophotography Programme.

to be used as containers for implements or as garages for the new machinery. The geo-referencing from historical cartographies through GIS allows 205 locations detection where pre-existences of oil mills are identified.

These data allows us to analyse the current situation, and verify of those 205 locations where there is some trace of building (Figure 2), 65 are abandoned oil mills and have different levels of degradation. It can be verified that 21 of these 65 only have a counterweight tower. Of the 140 oil mills that maintain some type of use (linked to some agricultural activity) only 51 of them retain the original volumes of the building, although they are also in various processes of deterioration. In global terms, it can be seen that 228 of the oil mills that existed in the middle of the twentieth century have disappeared or have been modified functionally and structurally, have even been transformed into buildings that have little to do with the original structure of the oil mill.

Similarly, considering territorial historical land references, it can analyze the evolution of the olive grove in Écija; one municipalities of the largest producers of oil where the largest olive grove recess occurs in Andalusia at the end of the twentieth century. (Table 2)

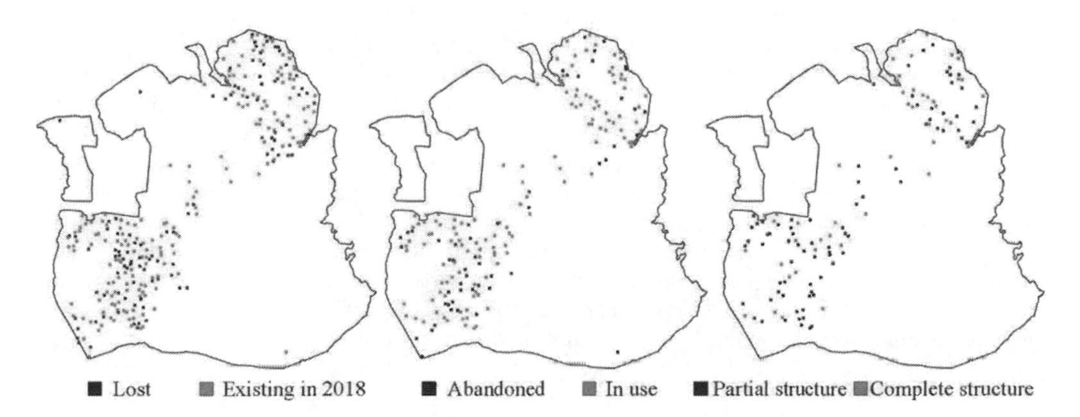

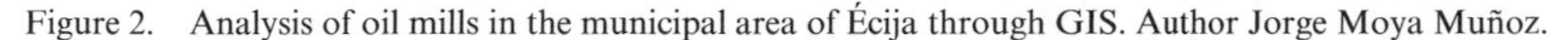

Figure 2. Analysis of oil mills in the municipal area of Écija through GIS. Author Jorge Moya Muñoz.

Table 2. Evolution of the olive groves surface in Écija (ha).

Year*	1751	1899	1922	1956	1975	1985
Hectares	17888	27258	27252	24982	8900	5380

* 1751: Cadastre of the Marquis of Ensenada. 1899/1922: Directorate-General for Agriculture. 1956/1975/1985: Ministry of Agriculture.

2 PURPOSE AND PROPOSAL

Taking into account the evident loss of heritage that derives from the changing historical processes and considering the particularities of the oil mills and the need for an exhaustive cataloguing that are included in the Écija Land-Use Plan (PGOU), the main objective of the proposed research work is to place value on the agricultural heritage of the Astigitano olive grove through knowledge. This study is included within the lines of action of the TUTSOS-MOD Project (Sustainable guardianship of cultural heritage through digital BIM and GIS models). Several specific objectives are set to achieve the main objective: study the impact of transformation dynamics as a consequence of the modernisation of the field in the olive grove since the mid-twentieth century (periods of study for the years 1956, 1977, 1988, 1998, 2009 and 2018); identify the heritage values of the historic olive grove; and create tools and mechanisms to protect their associated assets.

The proposed methodology use, as a theoretical reference, the research developed by the ASTIGIS project of the University of Seville on the historical evolution and the territorial and landscape planning of Écija in Roman times from GIS tools. In this research, the use of GIS tools is expanded with the BIM, which allow deepening in new work scales not only as tools for knowledge and analysis but also for their dissemination. A first municipal scale that includes the two areas where the historic olive grove (GIS) is located; a second scale delimited by payments as a territorial planning structure inherited from the Roman period in force until the 20th century (GIS); a third dimension related to the limit of the property linked to each oil mill (GIS); a fourth scale related to the oil mill's habitat (interoperability between GIS and BIM); and finally a fifth scale associated with the architectural object itself that represents the oil milll, its characteristics and state of building conservation over time (BIM). Conceptually considering this heritage from a holistic perspective involves recognising the oil mills in an inseparable way to their environment. To do this, we propose the creation of models defined from each oil mill, their habitat and their associated olive grove (Figure 3).

Although they can be studied individually for each of the scales proposed, each of these models must be considered together, therefore allowing the full analysis of the olive grove from the scale corresponding to both the "pagus" area and the municipal area. Olive-growing activity gives rise to each of the elements that constitute the landscape of the olive grove and corresponds to the elements of study. For the definition of each study model, it is assumed that this activity is the constituent element of Agrarian heritage (Castillo et al, 2013). At the same time, it is defined from a double perspective: the resulting of oil production and the one linked to domestic life associated with work in the olive grove. In this way, the parameters assigned to the GIS or BIM tools are derived from this double action. The data structure is grouped into five areas that are common to all productive units of the olive grove: the road system, ranked according to use; the system for water collection, transport and accumulation; the equipment and facilities that service the oil mill, annexed or dispersed in their environment; the oil grove and the oil mill itself, defined from the functionality of its spaces, its constructive systems, foundation, its morphological and stylistic aspects and its pathological state (Figure 4).

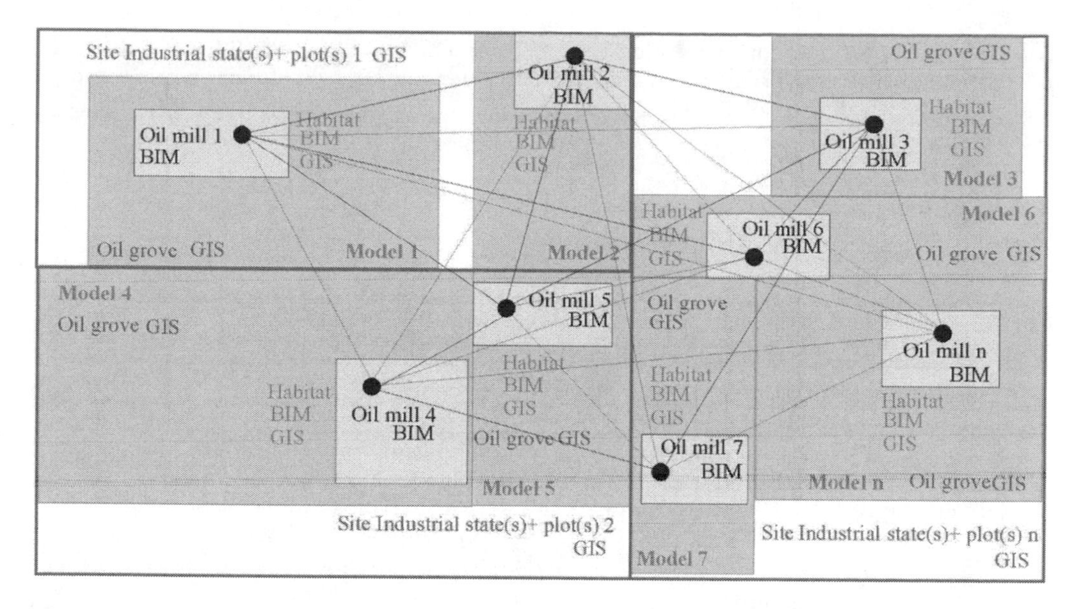

Figure 3. Methodological scheme of the models. Author Jorge Moya Muñoz.

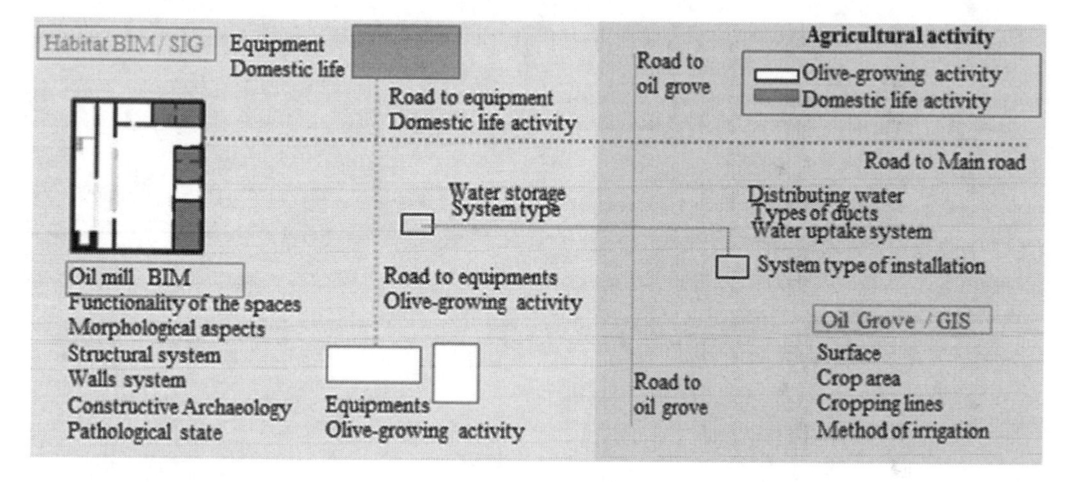

Figure 4. Scheme of analysis criteria of the model type. Author Jorge Moya Muñoz.

3 CONCLUSIONS

The use of GIS in study of this type of dispersed and complex heritage allows us to work with a large amount of georeferenced data in a systematic and organised manner. BIM-GIS interoperability allows the extrapolation of specific results of BIM at territorial analysis scales. In turn, the process of working through simple databases that can be easily translated into digital models facilitates effective knowledge in multidisciplinary teams and public institutions. On the other hand, the assimilation of the GIS as a habitual tool (SIGPAC, SIGA, etc.) would enable to identify and immediately incorporate the results obtained in the research as a new layer within these visors and to be able to transfer the patrimonial discourse to the agrarian world and also, reciprocally, to incorporate the reality of the agrarian sector in the proposals of protection of the immovable heritage. This aspect is special interest in the research, since through the obtained results, its present a methodological proposal for the cataloguing of oil mills based on its patrimonial valuation. This information is not included in the current PGOU of Écija, where the integral patrimonial valuation of the olive grove should be considered and included to adequately contextualise the oil mill and define protection measures beyond the architectural object.

REFERENCES

Castillo Ruiz, J., Anguita Cantero, R., Cañete Pérez, J. A., Cejudo García, E., Cuéllar Padilla, M. D. C., Gallar Hernández, D., & Pérez Córdoba, G. 2013. La Carta de Baeza sobre Patrimonio Agrario.

García-Dils de la Vega, S., Ordóñez Agulla, S., Sáez Fernández, P., Venegas Moreno, C., & Rodríguez.

Rodríguez, J. 2006. A preindustrial landscape in 18th-century Écija (Seville–Spain): The olive-oil production facilities as territorial markers. En *Marqueurs des paysages et systèmes socio-économiques*. Actes du colloque *COST, Le Mans* (pp. 7–9).

Rodríguez, A. M. B. 2009. Panorama histórico del campesinado andaluz en la segunda mitad del siglo XX. En *El fin del campesinado: transformaciones culturales de la sociedad rural andaluza en la segunda mitad del siglo XX* (pp. 19–32). Fundación Centro de Estudios Andaluces.

Sancho Corbacho, A. 1952. Haciendas y cortijos sevillanos. Archivo Hispalense, no 54-55, (p. 56).

Science and Digital Technology for Cultural Heritage – Ortiz Calderón et al. (Eds)
© 2020 Taylor & Francis Group, London, ISBN 978-0-367-36368-0

Building a digital platform for cultural heritage decision-making in rural areas. The case of Valverde de Burguillos

V. Domínguez Ruiz
University Institute of Architecture and Construction Sciences, Department of Architectural Construction I, School of Architecture, University of Seville, Seville, Spain

J. Rey-Pérez
University Institute of Architecture and Construction Sciences, Department of History, Theory, and Architectural Composition; School of Architecture, University of Seville, Seville, Spain

ABSTRACT: The depopulation of municipalities of the interior of Spain and the consequent deterioration of its cultural and natural resources are becoming more and more problematic. This cultural heritage is often unprotected, forgotten and increasingly neglected. However, thanks to the possibilities offered by new technologies (GIS and ICT), this delicate heritage may be understood and disseminated in such a way that it becomes a sustainable development resource. In the case of Valverde de Burguillos, the creation of an online platform for exchange, information, participation, and decision-making in relation to this cultural heritage will increase awareness, use, and dissemination. A design proposal is presented with guidelines for its implementation by the administration, which, in case of successful application, will also enable management entities–with the collaboration of citizens–to define sustainable rural development strategies supported by governance and citizen collaboration.

1 INTRODUCTION

The depopulation of municipalities of the interior of Spain and the consequent deterioration of its cultural and natural resources is becoming more and more common and problematic. This heritage is often unprotected; consequently, tangible and intangible heritage values are increasingly unknown. This means the loss of significant local knowledge that is really beneficial for the global society, which is in search of and in the process of change in terms of food, health, natural contact, and human values. Likewise, population centres require, for functionality and development, exchange among their components and between themselves and their surrounding environment. In the case of small cities and towns affected by the phenomenon of depopulation, there is a loss or reduction in the possibilities of exchange of the city as a community and of the individual inhabitants. This deterioration in the possibilities of local development limits the potential to cover the needs of its inhabitants and their communities, as regards subsistence, participation, understanding, leisure, creativity, identity, freedom, and protection of cultural heritage. Thanks to the possibilities offered by new technologies (GIS and ICT), these limitations may be overcome, enabling and amplifying exchanges among people and communities, contributing to the knowledge and development of their resources through outward projection.

From the R&D&I research project *Sustainable Protection of Cultural Heritage through BIM and GIS Digital Models. Contribution to Social Knowledge and Innovation* (ref: HAR2016-78113-R) and specifically in the line "Information systems for the knowledge of traditional architecture and reactivation of cultural heritage in depopulated areas. Study Case: Mértola (Alentejo, Portugal) and Valverde de Burguillos (Badajoz)" it is understood that reinforcing knowledge of cultural heritage in areas of low population density can improve local development. Therefore, the main objective of this applied research based on GIS technology is to value existing local resources and identify new heritage values by generating scientific and technological knowledge. Another more ambitious objective to give voice to this heritage at risk - at the same time as developing sustainable rural development strategies according to the citizenship and cultural values identified - proposes the design of a collaborative platform so that the agents involved in local development work connected and cooperate through a collaborative network. The sustainability of these results will depend on an adequate load, for which implementation guidelines are proposed.

2 STUDY CASE

The loss of population that moves inland municipalities to larger urban centres with more opportunities on the other hand, and a negative rate of natural increase, on the other, has resulted in a situation of rural depopulation which affects almost the entire Autonomous Community of Extremadura, in particular the province of Badajoz and its system of groupings of municipalities, which could be the main user of our proposal. Located 96 km from Badajoz to the southwest, the municipality of Valverde de Burguillos has 19 km2 of surface area and a population of 290 people. It belongs to the region of Zafra - Río Bodión and the judicial district of Fregenal de la Sierra. Valverde de Burguillos, which reached its maximum number of inhabitants between 1930 and 1950 and which is currently at its lowest, in addition to a negative demographic trend, having already reversed its population pyramid.

Analyzing the activities that attempt to curb this demographic trend, detected at the beginning of the collaboration with the municipality, are those carried out by the citizen platform Activa Valverde, comprised of inhabitants, emigrants, and supporters of Valverde de Burguillos. The three axes of work, under the maxim of attempting to generate involvement of the citizens, are: attracting new young population, creation of employment, and facilitating access to housing.

Very closely linked to people and lifestyle are the elements that concern us, whose integrated study reveals the important link between wisdom and knowledge of people with protected real estate, enabling their safeguard to establish measures for continuity along with a series of strategies, linked to the valuation of this type of asset. Likewise, Valverde has assets protected by the BIC declaration of "La Cultura del Agua", as well as a Chalcolithic site in its territory and, around the Bodión River, the anthropized footprint of the agroecosystems, the riverside woodland, and the constructed landscape of the Dehesa. In the urban environment, there is the Church of Nuestra Señora de la Antigua (16th century) and a hamlet with typological interest, with some traditional construction still in existence. In terms of intangible heritage, it conserves well-preserved religious, cultural, and gastronomic customs.

3 DEVELOPMENT OF RESEARCH: MATERIALS AND METHODS

In a first phase, for more than two years, the trends of improvement of local policies have been followed in strategic plans and integrated action tools as a contribution to urban and territorial planning systems. Then, taking into account the three work axes described above (new settlers, housing and employment), the following process was carried out: i) identification of existing heritage resources, but also unnoticed or undervalued so far, through an analysis of citizen perception and end-of-degree and end-of-master works with Valverde as a case study, guided by professors of this team belonging to different disciplines of the University of Seville (anthropology, computer science, construction or history of architecture); ii) contrast with the existing digital cartography, through the Territorial Information System of Extremadura (SITEX), the planning documents (Advance of the PGOU) and heritage conservation (Catalogue of Protected Assets); iii) thematic categorization of new niches of economic activity, based on the Strategic Plan Extremadura 2030 and iv) registration of the participation of local agents in forums and social networks on depopulation between 2017 and 2019, allowing to extract their requirements as potential users of the platform (Figure 1).

The methodology used for i and iii includes a series of workshops led by the authors (Domínguez & Rey, 2018; Domínguez & Rey, 2019) and attended by professionals, researchers,

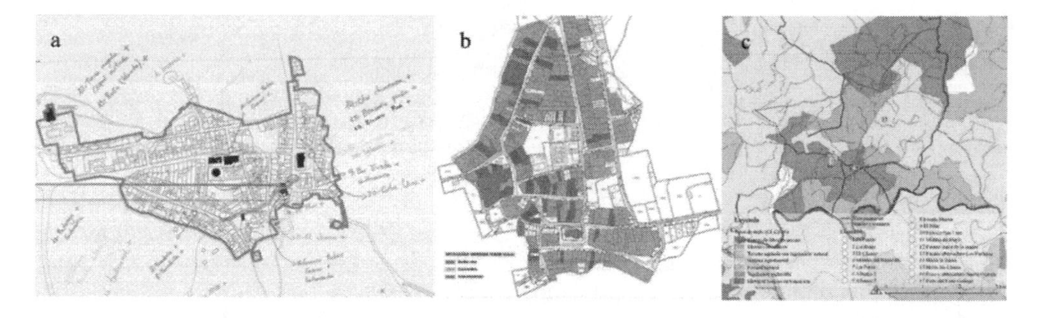

Figure 1. Phase I. Use of GIS in collaborative mapping: a) citizen perception maps; b) data analysis; c) information integration.

citizens, administration and local technicians. Their results, together with the proposals and conclusions of the academic studies mentioned in i, as well as the strategic overlap with the study of citizen perception and with urban and territorial information (supplied in ii also in GIS format) have been processed through statistics, SWOT analysis and collaborative cartographies through GIS. This has made it possible to integrate into a single georeferenced database numerous categories of graphic and thematic information established in the study, giving rise to a series of collaborative cartographies of heritage content that, after a graphic process of selection and simplification, are prepared to be disseminated on the platform. The main contribution of our research to all these methodologies, coming from a global environment and, in some cases, from overpopulated areas, is the application to the case of depopulation and the inclusion of Cultural Heritage as a structuring element, taking as a reference the parameters introduced by the New Urban Agenda (United Nations, 2015: 22).

In a second phase, the design and progressive implementation of a new online tool is considered. Through the observation of initiatives in the Extremadura environment, local needs and Phase I case studies determine the main categories or lines of action, which will give rise to the different functionalities of the application. At the same time, the necessary resources for its implementation are contemplated in brackets: Circular economy (strategic plans and individual initiatives); actual and potential inhabitants (social and human capital); environment (natural resources); sustainable habitat (quality of life and housing); local governance (participation); communications (infrastructure); development spaces (built heritage and activity niches). The main software resources used in the control panel of the prototype created have been: Bootstrap (kit for development with HTML, CSS and JavaScript); Pinegrow (multiplatform web editor such as Mac, Windows and Linux); XAMPP (distribution of free and easy to install Apache containing MariaDB, PHP and Perl); MongoDB (database NoSQL to documents, open source). It is planned to combine in the future GIS tools, social networks or participative urbanism.

In order to guarantee the sustainability of the platform, a procedure has been drawn up, partially completed for the time being, which aims to create stable structures for a management model based on governance and codecision between the different stakeholders (Figure 2).

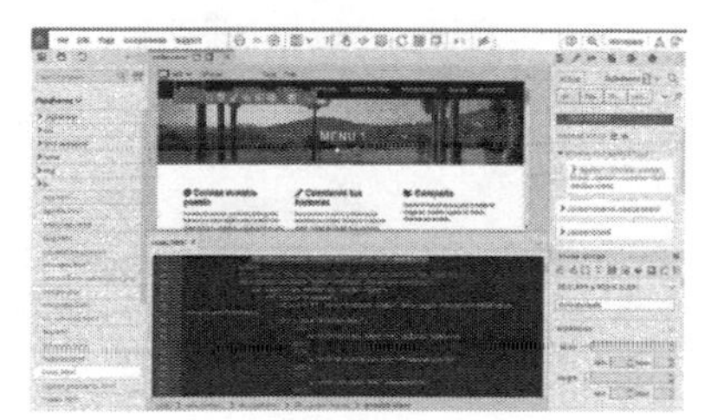

Figure 2. Prototype of online platform.

4 OUTCOMES: PARTICIPATORY PROCESS AND IMPLEMENTATION GUIDELINES

This section develops the elaboration of a heritage information system as a result of the analysis of collaborative processes with citizens (belonging to Phase I) and some guidelines prior to implementation to guarantee the sustainability of the proposal (Phase II).

Collaborative mappings enable us to decide in a participatory context what, why, and how to value neglected cultural-rural heritage and start to turn it into a sustainable development resource. In a first step the information provided by citizens is collected through a participatory workshop; in a second step the data collected by the community is contrasted with the information identified by university technicians and the corresponding Administration; and, finally, there is a third step, known as the "return of information", where a neighborhood workshop is held again to share all the information generated from the 3 stakeholders involved (citizens, university, and local government). From now on it is defined among all the stakeholders the heritage values, criteria, strategies, and action lines to implement in Valverde de Burguillos.

Step 1 was carried out in Valverde de Burguillos, with the aforementioned participatory workshop. The workshop was composed of 3 individual and group activities, with the aim of identifying heritage that distinguishes Valverde de Burguillos. Among the main results, the general landscape stands out, where nature and environment take centre stage, as well as the views from the town towards the territory, and the urban component. Issues such as gastronomy (meat, bread, sweets, asparagus, etc.), the sounds of birds, the smell of wet earth or soil, and the personality of Valverdeños are other protagonists that provide clues to the peculiarities and singularities of the heritage of this place, until now unrecognized. In step 2 is observed that the information generated from the administration at the heritage level is limited to its declaration as an Asset of Cultural Interest (BIC) for "La Cultura del Agua" in Valverde de Burguillos, Badajoz, in the category of Ethnological Interest (DOE núm. 244, 2017). However, from the University of Seville, several works focused on the commonwealth have been developed with the aim of expanding knowledge of the constructed heritage and vernacular architecture in Valverde de Burguillos. Consequently, several phases of the process have been covered: understanding the problem, contact with local reality, fieldwork, systematization of the data collected, information management, consultation with local technicians, and production of useful complex graphic information. The proposal highlights collaborative work among professors of different disciplines and students of the University of Seville, in order to reach a broader point of view as regards the problem (Figure 3).

In relation to Phase II, the following is a list of criteria taken into account for the design of the actual implementation and its sustainability:

- The design satisfies the real needs of the users and knows the limitations of the agents. To this end, it opened a participatory process of data collection and role identification. At the same time, SWOT diagnoses were carried out on these results and on the strategic plans and

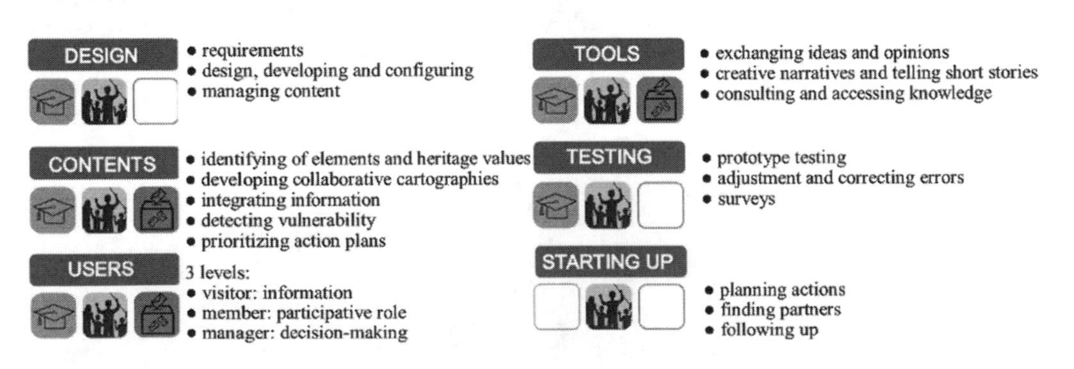

Figure 3. Methodology and roles of local development stakeholders (universities, citizens, and politicians).

tools revised. On the other hand, the integrated study plus the dissemination of results in a single platform will allow the saving of information costs, previously dispersed, as well as improving the visibility of the municipality on a global scale.

- The adequate selection of resources that feed the tool will allow its durability over time: financial resources (internal or external to the main manager: the administration), human resources and material equipment. It is also important to identify new organizations or stakeholders who must commit to making sustainable use of this service.
- The involvement and collaboration of all the stakeholders described in Figure 2 creates opportunities for future projects. The public presentation of periodic results allows for a more plural follow-up (in the initial phase we refer to the aforementioned workshops, attendance at local and national forums on depopulation, the conferences on Cultural Heritage in the Mancomunidad Río Bodión, the public presentation of the academic works carried out, etc.).
- The administration should try to activate or seek incentives to motivate individuals to present local projects aligned with regional strategic plans.
- In a future phase, the administration should exercise an effective leadership that guarantees the future of the platform: ability to raise funds in advance, be open to other quality references, enable qualified advice, clearly distribute rights and responsibilities and build trust among partners.

5 CONCLUSION

Numerous studies focused on the use of new information technologies show the social benefits of their application for heritage knowledge. An additional step would be, once the stakeholders involved–citizens, administration, technicians, and academia–have been identified, to achieve the exchange of information, participation, and decision-making among all for the reactivation of cultural resources through an online platform. The progressive and conscious implementation of this new experimental tool for work and collection of information can change not only the concept of cultural heritage but also its use and management, especially taking into account the most current dynamics in terms of economic and social development. In sum, the review of existing heritage values and identification of new elements, generation of participatory processes among citizens, politicians, and technicians, creation of a state of opinion while encouraging collective decision-making, development of an online platform as a tool for collective communication, generation of information and dissemination of knowledge, establishment of links and synergies with the resources of other surrounding municipalities, and support for the protection and sustainable management of cultural heritage, all influence the exploitation of local resources uniquely and creatively while contributing to the permanence of a population where cultural identity has been reinforced.

REFERENCES

Domínguez, V., & Rey, J., febrero 2018, I Taller de Iniciativas Vernáculas: patrimonio y sostenibilidad para una nueva ruralidad. *Crónica del Bodión* (1): 19.

Domínguez, V., & Rey, J., junio 2019, II Taller de Iniciativas Vernáculas en Valverde de Burguillos. *Crónica del Bodión* (2): 18.

Domínguez, V., & Rey, J., junio 2019, Vernacularmente: patrimonio cultural y retos de sociedad. *Revista PH* (97): 15–16. doi: https://doi.org/10.33349/2019.97.4375.

Gravagnuolo, A., & Girard, L. F. (2017). Multicriteria tools for the implementation of historic urban landscape. Quality Innovation Prosperity, 21(1): 186–201.

Paz, D., junio 2019, Valverde de Burguillos: una experiencia viva. *Crónica del Bodión* (2): 18.

Pinilla, V., & Sáez, L. A. 2016. *La despoblación rural en España: génesis de un problema y políticas innovadoras*. Zaragoza: CEDDAR.

Rey Pérez, J., & González Martínez, P. (2018). Lights and shadows over the Recommendation on the Historic Urban Landscape:'managing change'in Ballarat and Cuenca through a radical approach focused on values and authenticity. *International Journal of Heritage Studies*, *24*(1), 101–116.

Resolución: Extremadura, 28 de noviembre de 2017. Consejería de Cultura e Igualdad de la Junta de Extremadura. Publicado en DOE núm. 244 del 22 de diciembre de 2017 y en el BOE el 6 de marzo de 2018.
UN, 2012. Resoluciones aprobadas Conferencia Rio20.
UNITED NATIONS (2015) Transforming Our World: The 2030 Agenda for Sustainable Development (A/RES/70/1). https://www.un.org/ga/search/view_doc.asp?symbol=A/RES/70/1&Lang=E [Consulta: 04/04/2019].

Science and Digital Technology for Cultural Heritage – Ortiz Calderón et al. (Eds)
© 2020 Taylor & Francis Group, London, ISBN 978-0-367-36368-0

Heritage monitoring from space: Evaluation of satellite remote sensing for the development of early warning systems

S. Davies, R. Cabral & M. Correia
THEIA & CEAACP, Coimbra, Portugal

ABSTRACT: Satellite Remote Sensing (SRS) has the ability to provide a great amount of information over a large area of study. SRS technology has yet to penetrate the main scientific discourse of Archaeology. Traditional limitations of SRS in Archaeology have started to dissipate with the launch of new satellite systems such as the Sentinel constellation. Traditional methods of field monitoring are costly and labour intensive, and tend to lead to the untimely detection of the destruction of cultural heritage, as was the case in many sites in Alentejo, Portugal. In this work, we discuss the potential of Sentinel 1 Synthetic Aperture Radar and Sentinel 2 Multispectral satellites as tools for heritage monitoring, use SRS to analyse the Portuguese site of Pisões, and evaluate if it is possible to build an early warning system based on SRS, in order to solve one of the greatest problems currently affecting archaeological preservation.

1 INTRODUCTION

This study was developed by THEIA, in partnership with the Research Centre for Archaeology, Arts and Heritage Sciences of the University of Coimbra (CEAACP). The European Space Agency provided funding for this project through the Business Incubation Centre (ESA BIC), beginning in late 2017. The project resulted in the creation of a satellite monitoring and early warning system for archaeological heritage, called SENSEOS.

The area of interest for this study is the Alentejo region in the South of Portugal. It has a geographical extension of 31,551.2 km^2, corresponding to approximately one third of the country's landmass. The Alentejo region is home to thousands of archaeological findings and sites, with 1064 documented sites in the district of Beja alone. The landscape of the Alentejo region, of a rural and agricultural nature, has been changing at a rapid pace, due to the substitution of cereal based crops and dryland farming with intensive plantations of, mainly, olive and almond groves. New sources of water, supplied through the construction of dams and the Alqueva irrigation system, have enabled this process to take place. This new availability of water for the purpose of irrigation has led to the introduction of more profitable crop species that require the intervention of highly mechanized procedures. The conversion of dryland fields into intensive olive and almond groves relies on the use of these types of mechanized invasive methods for soil revolution, involving practices such as deep ploughing. Traditional farming practices, which rely on the use of animals or light mechanized procedures, rarely disturb deep soil layers located at a depth of more than half a metre. This means that most archaeological sites have remained relatively unharmed despite being located within agricultural fields and plantations. Now, with intense mechanization of agricultural practice, the continued existence of intact archaeological sites, many of which remain unexcavated, is under a serious threat. As of the time of writing (2019), the scale of the damage in the last few years is in the order of several hundreds of affected sites. The large extension of numerous agricultural fields, combined with the lack of labour to supervise landscape transformation adequately in critical areas containing archaeological features, presents a serious problem. In order to give an appropriate response to this situation, it is necessary to develop a system capable of quickly identifying events that threaten cultural heritage. The synoptic view of satellite images, combined with short revisit times of the Sentinel satellite constellation, could provide a solution to this problem, and have great potential as tools for continuous monitoring of large areas of interest.

2 METHODOLOGY

This study employed Sentinel 1 Synthetic Aperture Radar and Sentinel 2 Multispectral Data. The main advantage of radar systems like Sentinel-1 is the fact that they are capable of observing the Earth's surface in any type of weather conditions (Patel *et al*, 2010, p. 244). Passive optical systems like Sentinel-2 provide information from multiple wavelengths but are highly vulnerable to clouds, which can make it impossible to view the surface under certain weather conditions. Sentinel-1's Synthetic Aperture Radar is an active sensor that provides its own source of light. This means that it produces an image based on the light that returns to the sensor, a process called *backscatter*. A number of factors, such as surface morphology, surface characteristics, vegetation and soil humidity (Chen et al, 2018, pp. 71-72), influence the intensity of the backscatter. Because agricultural practice affects every single one of these factors, we hypothesized that radar data could serve as an indicator of agricultural activity in the proximity of archaeological sites. We would then use optical information as supporting data to extract further information when possible. The frequency of new satellite images, called *revisit time*, is 6 days for radar and approximately 5 days for optical data. For this approach, we used a method we have referred to as NDAI, or Normalized Difference Amplitude Index. The way it functions is similar to the popular NDVI, or Normalized Difference Vegetation Index used in optical images, but instead of comparing the index between Red and Infrared pixel values (Carlson & Ripley, 1997, p. 241), we compare the intensity values of radar pixels between two different acquisition dates. We normalized the data in order to allow us to work within the same ranges between different periods of observation, even if the degree of changes is different between image pairs. We filtered out a portion of the resulting data close to the value of zero (in this case between -0.2 and 0.2) in order to remove variations caused by noise or not very significant changes. This normalized data is subsequently categorized according to the severity of changes between image pairs, in order to allow the detection of different levels of changes in the landscape that could correspond to different kind of agricultural or physical events on the surface. Ploughing for example, should normally produce a different pixel value compared to events such as irrigation or germination. In order to assess if surface change may have affected the archaeological sites, the information derived from the change detection processing is crossed with geographical information of the archaeological sites in the area. A buffer is created around these sites in order to create a "protected area", and if the buffer crosses an area with identified changes, an automatic alert system is activated containing all of the affected sites. Most of these observations were confirmed with groundtruthing. We are currently working with an agricultural engineer located in Alentejo in order to classify the results obtained from radar change detection, and determine if it is possible to reliably detect the type of events, such as deep ploughing, tilling, etc. The results, so far, have confirmed that it is possible to identify these occurrences. In order to test the validity of this approach, we applied this technique to a number of sites in the Alentejo region that were destroyed in previous years. Although the damage to these sites was identified, it was not known exactly when the destructive event had taken place.

We applied this methodology to a number of sites in the Alentejo region, including Pisões, a partially excavated roman *villa*. Agricultural activity destroyed the unexcavated portion of the site, containing important features such as an aqueduct located underneath a water reservoir in a field to the Northeast, at an unknown date (Dias, 2017). In this paper, we will present the results obtained from Pisões.

3 RESULTS

Observation of Sentinel 2 time series from 2017 shows that most of the surface transformations occurred during the second trimester of the year. Evidence that supports this observation includes the disappearance of the water reservoir in the Northeast of the excavated site, between the 5th and 15th of April 2017. This water reservoir was located in very close proximity to the roman aqueduct identified at the site. Clear images are not available from Sentinel-2 until the 15th of May 2017. At this point, there seems to have been a unification of this field plot with another one adjacent to it in East-Northeast direction that had contained a small grove. These changes are indicators of possible preparation of land for future plantation. For radar analysis, we applied NDAI processing to satellite data. Pairs of images with a temporal baseline of about one month, beginning in 2014, were analysed, and we identified an event of relatively high values between the 5th of October and the 4th of November 2017. Further analysis of image pairs located between these two dates and with a temporal baseline of 6 days shows that the event occurred between the dates of 11th and the 17th of October 2017. The area corresponding to higher value NDAI pixels fits within the boundaries of the plot and shares the same shape.

Figure 1. Results of radar signal processing for Pisões using amplitude change detection. Loss of amplitude is shown as blue, while amplitude gain is shown as red.

4 CONCLUSION

THEIA is currently employing SENSEOS in a pilot programme with two public authorities, the Regional Department of Culture of Alentejo, and the Regional Department of Culture of Algarve, in order to assess the benefit of implementing an early warning system based on satellite data. The project is currently monitoring 1270 archaeological sites in the Alentejo region, and 5 in Algarve. Every 6 days, all parties receive a report, and in the event of potentially affected sites, a risk map with observations or further interpretation, if necessary. In order share access to the results of the data processing, SENSEOS includes a webGIS platform containing all the information, including archaeological site locations and their associated risk level. This work highlights one of the main advantages of satellite data applied to Archaeology, which is the ability to perform large-scale territorial analysis and monitor destructive events on thousands of sites with a weekly frequency, with a potentially global application.

REFERENCES

Carlson, T. N., & Ripley, D. A. 1997. On the relation between NDVI, fractional vegetation cover, and leaf area index. Remote sensing of Environment, 62(3), 241–252.

Chen, F., You, J., Tang, P., Zhou, W., Masini, N., & Lasaponara, R. 2018. Unique performance of spaceborne SAR remote sensing in cultural heritage applications: Overviews and perspectives. Archaeological Prospection, 25(1), 71–79. https://doi.org/10.1002/arp.1591

Dias, C. 2017, July 17. Villa romana de Pisões é agora a luxuosa casa das ervas daninhas. Público. Retrieved May 13, 2019, from https://www.publico.pt/2017/07/17/local/noticia/villa-romana-de-pisoes-foi-descoberta-ha-50-anos-e-esta-coberta-de-ervas-daninhas–1779095

Patel, V. M., Easley, G. R., Healy Jr, D. M., & Chellappa, R. 2010. Compressed synthetic aperture radar. IEEE Journal of selected topics in signal processing, 4(2), 244–254.

Science and Digital Technology for Cultural Heritage – Ortiz Calderón et al. (Eds)
© 2020 Taylor & Francis Group, London, ISBN 978-0-367-36368-0

From ink to binary: Combining GIS and 3D modelling as a method of analysis and management of an archaeological site

A.M. Ribeiro, A.V. Meireles & M.C. Lopes
Centro de Estudos em Arqueologia, Artes e Ciências do Património, Coimbra, Portugal

ABSTRACT: One of the greatest problems archaeologists face when their field work is the quantity of information that is yet to organize within the archaeological site being studied. Traditionally, most of this information would be registered by hand on paper. However, the last thirty years have allowed for the democratisation of computer systems and Geographic Information System (GIS) software, assuring a much more efficient and automated approach to information registration, querying and extraction. Therefore, we have employed GIS as a method of organising and formatting a great amount of work that has been executed in the archaeological project of "Arqueologia das Cidades de Beja". The result of this work is a platform in which data that was previously indirectly correlated becomes conjugated in a digital singularity that allows for a simplified, fast and straightforward access and analysis of information without requiring a very large investment in training.

1 INTRODUCTION

Seeing that the heritage digital record is defined as being the activity that produces and stores digital information in a computer, such as plans/drawings, photographs, photogrammetric records and other electronic data that make the heritage record, and are projected for use in a personal computer (Letellier, 2007, p. 39). The use of new technologies in Archaeology already has a lot of followers, and from the available software, the Geographic Information Systems has been the method used for the longest time relative to other techniques, such as photogrammetry or modelling.

Until last century, the data collected by the total station, such as spatial documentation of artefacts, distance measuring or altimetry of archaeological units have their own limitations: these include the necessity of a computer to visualize all data, and the fact that the data is, by itself, individualised, meaning there is no direct relation between all of its parts. The solution found to relate this different data was GIS (Kimball, 2016, p. 8).

As said before, the implications that GIS bring to Archaeology are very profound. Traditionally, field archaeology divided itself in two phases: the field work and the office work (Conolly and Lake, 2006, p. 36 *in* Kimbal, 2016, p. 9). The datasets collected in the field needed a further post processing with a level of detail that was not possible during excavation work. With technological progress, computers became more compact and with contained larger storage capacities, allowing for the possibility of doing office work in parallel with excavation, meaning that this process - storage and relation between archaeological data - can be conciliated with their acquisition, leading to increased efficiency (Conolly and Lake, 2006, p. 36-37 *in* Kimbal, 2016, p. 9).

For this study, our aim was to create a workflow that could allow us a fast and direct access to the information of the project "Arqueologia das Cidades de Beja", with a series of steps that would ultimately result in a platform that would merge data from the excavation, and would allow a full view of the area while simultaneously allowing access to information contained in descriptive databases.

Such task would have to go through many steps, starting with digital surveys, gathering all data, creation of 3D models through photogrammetry, image processing, introduction of data in a GIS platform, vectorisation of the plan of the excavation and database linking in the same system.

I emphasize the work of Justin Kimbal (2016) who, for the archaeological site of Uppakra, in Sweden, developed a line of work that consisted on the creation of 3D models through the use of photography, in order to later on insert into a GIS platform, vectorizing the tridimensional surfaces and supply the respective attributes in a database. This allowed for an agregation of multiple types of archaeological information in a single platform.

The work of Sjoerd Van Riel (2016) also presents a similar methodology, where for the same archaeological site, in Sweden, he integrated diferente types of tridimensional data from the creation of 3D models to analytical operations and vector drawing in GIS.

2 "ARQUEOLOGIA DAS CIDADES DE BEJA"

The project "Arqueologia das Cidades de Beja" carries an archaeological excavation in the city of Beja, more specifically *Pax Iulia* in the south of Portugal, and aims to safeguard heritage through an approach that links the study of the past with the actual city, and conjugates this with a perspective of participation and development of the community.

The first archaeological evidences date to the decade of 1940, when the archaeologist Abel Viana came to the conclusion that we were witnessing the building foundations of a roman temple, similar to the one in Évora (Lopes, 2000, p. 140).

The site is divided in four sectors: A, B, C and D, with B being subdivided into three subareas. Each sector has a numbering system for the structures that have been excavated.

2.1 *Methodology*

The workflow consisted in converting paper information into digital data, assembling it in a single platform - in this case GIS - and allowing the user to consult and manage that same data.

The data acquired, although was uniformed, had derived from different registration sources, each containing different levels of information. For example, in the beginning of the excavation, the record was made in notebooks and later swapped to recording sheets and registers. In addition to data of an architectural nature, another type of information was also analysed, such as stratigraphic, that we cannot see in here. It was necessary to have an image of the site that was correctly scaled, and for that we used the point cloud from a laser scanning performed in 2011.

For the inside of the buildings, there was a computer assisted drawing to help with the assembly and vectorisation, and it was also necessary to have a plan that contained all of the altimetric dimensions to enable modelling of the area of the excavation. Similarly, a database was required of all the structures, that had to be updated so we could connect it to the GIS.

There were some 3D models of the structures and the excavation site, but it was still necessary to create some more. The rest of the 3D models were created and assembled using photogrammetry.

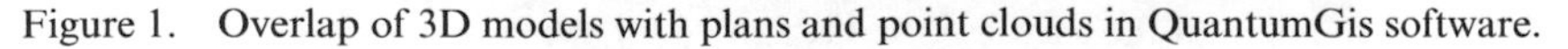

Figure 1. Overlap of 3D models with plans and point clouds in QuantumGis software.

3 RESULTS

The results obtained made it much easier to survey the archaeological structures along with the descriptive database. This can be very useful when there is a need for a more intensive search in a short amount of time. What was obtained with this line of work was the simultaneous visualisation of 3D models and graphic interpretations and the total area of the excavation in a single platform, since several structures are inside the buildings.

The vectorisation resulted in a schematic plan, and once it was complete, the database of the structures was linked so we could directly and quickly access information related to the description of the features. With the support of a plugin, the querying of the structures by different fields, like chronology or even date became much easier because we could visualize the structures we were looking for. The advantage in this system is that it is possible to create scenarios that can assist the field investigation, meaning this that the GIS can be used to select and show tridimensional artificial environments, excavated in different periods of time, allowing a comparison with data that no longer exist in the field, and without the need of being there.

Later on a 3D model was created in the software Blender with the altimetric points in order to have a spatial and tridimensional perception of the area of excavation and the different chronological levels and construction phases. The altimetric dimensions used were from one of the plans of the excavation.

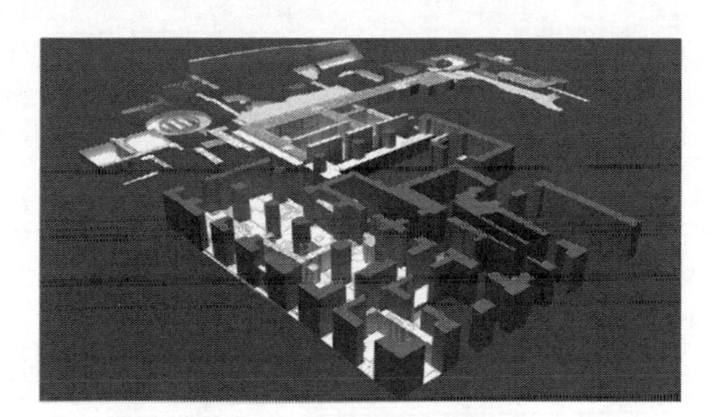

Figure 2. Tridimensional model made in the software blender.

4 CONCLUSIONS

The result of this workflow is a single platform (GIS) which enables visual interpretation of graphics and tridimensional models with a much more simplified and easy search, and a complete visualisation of the whole area of the excavation, allowing a good visual panorama for the researcher.

It also enables the researcher to create objective drawings, with a lower probability of making mistakes when vectorising, because the 3D models and point clouds have the right measurements, instead of the traditional drawing in the field, which leads to a more subjective result. Besides, it is important to refer that a large amount of stratigraphical, architectural and material data, without the support of an integrated system such as the one here presented, would make it almost impossible to manage this data efficiently.

This platform allows for the association of data from the field, to archival materials held in inventory. In this sense, it is a multi featured instrument that makes it possible to process the data as a whole. On the other hand, as Beja is a historical site, with different periods of reutilisation and recycling, this platform brings together many different variables, allowing us to study the city in any of the historical moments that define it.

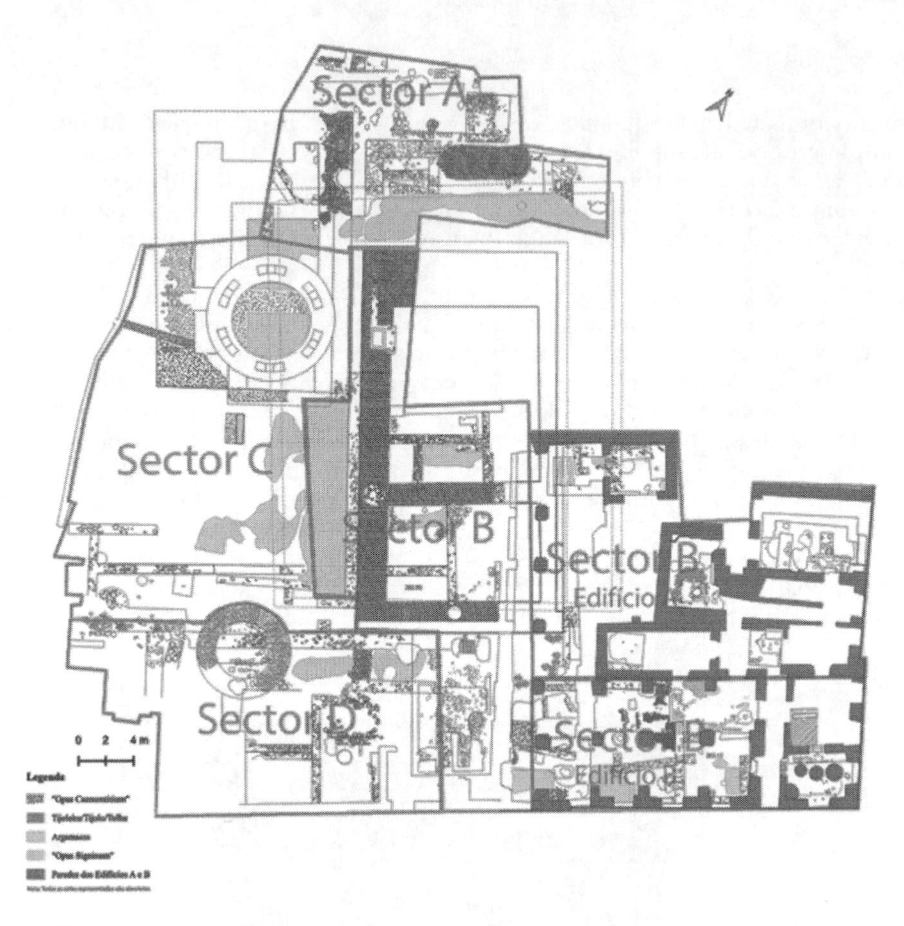

Figure 3. Final vectorisation of the area of the excavation.

REFERENCES

Conolly, J.; Lake, M. 2006. Geographical Information Systems in Archaeology. Cambridge, University Press: United Kingdom.

Kimball, J. J. L. 2016.3D Delineation: A modernisation of drawing method for Field Archaeology. Oxford: Archaeopress.

Letellier, R. 2007. Documentation and Information Management for the Conservation of Heritage Places. Guiding Principles. The Getty Conservation Institute: Los Angeles.

Lopes, M. C. 2000. A Cidade Romana de Beja - Percursos e Debates em torno de PAX IVLIA. PhD Dissertation in Archaeology presented to Faculdade de Letras da Universidade de Coimbra.

Van Riel, S. 2016. *Exploring the use of 3D GIS as an analytical tool in arcaheological excavation practice.* Dissertação de Mestrado apresentada ao Departamento de Arqueologia e História Antiga da Universidade de Lund.

Science and Digital Technology for Cultural Heritage – Ortiz Calderón et al. (Eds)
 ISBN 978-0-367-36368-0

BIM procedure for heritage construction site simulation on complex buildings: Coordination and safety management

C. Cioni*, M. Di Sivo & M. Martino
DESTEC, University Pisa, Pisa, Italy

Mª R. Chaza Chimeno
Departamento Expresión Gráfica e Ingeniería en la Edificación, University of Seville, Seville, Spain

J.M. Macías-Bernal
Departamento de Construcciones Arquitectónicas II, University of Seville, Seville, Spain

ABSTRACT: The aim of this study is to apply the BIM tools in the execution phase of a restoration project, focusing on safety management and coordination. The case study is a hypothetical construction site project on a complex façade, as it happens in cultural heritage. The software used for this research are Autodesk Revit, STRvisionCPM, and Autodesk Navisworks Simulate. We carried out a simulation of the execution work's phase, to optimize the coordination of the tasks in order to obtain a high-quality result together with a greater capacity in safety management. It is possible to simulate different types of intervention or conditions and that allows the designer to choose the more efficient solution in advance. The most important point for this kind of approach is the possibility of foreseeing clashes and problems. The prevention is the most efficient way to reduce the risk in safety management and avoid unforeseen events that could negatively affect the budget.

1 INTRODUCTION

Nowadays, the BIM (Building Information Modeling) methodology is widely used in building domain; it is now a real necessity to improve the research in this field. In practice, the use of this method is limited to the design phase. Indeed, the potential of BIM is the opportunity to support the development of a building from its conception to the end of its life cycle, including maintenance or restauration. Thinking about the application of IT in building domain we realize that we are now working in an incredibly fast-growing environment and there is no common and concerted knowledge base in this field. Analyzing the available literature on the subject it is clear that the research studies in this area are punctual and isolated. Even the legislative framework in this regard is decidedly immature. Nevertheless, there are interesting initiatives that touch the heritage sphere. For example, in Europe an interesting project for the monitoring and preventive conservation of cultural heritage uses BIM technologies. It is called HeritageCare and includes many different big partners (Talon, 2017). On the contrary, in the field of safety management and coordination of the works the literature is really weak. The possibilities of application are infinite and dispersive. It is essential to study and select a methodological approach to be suitable for the practical applications we need.

*Corresponding author: costanzacioni@gmail.com

2 METHODOLOGY

2.1 *Approach*

This work is divided into two phases: the design and the simulation. The design phase of the existing building is referred to the methodology elaborated at the University of Seville based on photogrammetry and 3D scanning techniques. The objects are represented as 3D-based vector entities that allow a real work of auscultation and analysis of the building besides a great representation "as build" of the geometry (Nieto, 2013).

Afterwards the entire organizational process is broken down into a branched structure of three levels, which the Project Management Institute of Philadelphia called Work Breakdown Structure (Project Management Institute, 2017). This operation allows having a general index of the works that contains the codes used by the software. The aim is to obtain a 3D model, which not only contains information about the geometry but also about the complete management of the project. The main objectives and the level of detail (LoD) must be fixed at the first setup phase. The final model must contain physical, functional and management characteristics of the project. In this case, the choice is to focus on this procedure in the construction phase and in the scaffolding simulation. On Autodesk Navisworks we can elaborate a 4D simulation of how our hypothesis will be executed, so it will be possible to detect clashes in advance.

2.2 *Scaffolding design*

To design the scaffolding on Autodesk Revit (Figure 1), it is necessary to consider an articulated geometry that imposes a rigorous and careful design methodology. The details of all the working phases are provided below.

The definition of the construction site plan and the layout of the fences must be defined in accordance with the project's needs, paying attention to the critical issues of the site organization.

1. First, the geometry of the scaffolding base is drawn, marking the supports in plan at ground floor level. It is necessary to adapt the design to the complex geometry of the building and respect the limitations according to safety regulations. To build the 3D Revit model, a virtual catalogue of Revit families of generic models has to be created. This catalogue contains all the pieces necessaries to compose the scaffolding design and the models are based on a real catalogue of a scaffolding constructor.
2. After the creation of the scaffold geometry framework at the ground floor, all the structural layers are superimposed. The basic elements are repeated, in vertical alignment, for the number of levels required.
3. Now, the clash detection starts. It highlights the points where the scaffold structure interferes with the facade elements. Once the clashes have been identified, they must be solved

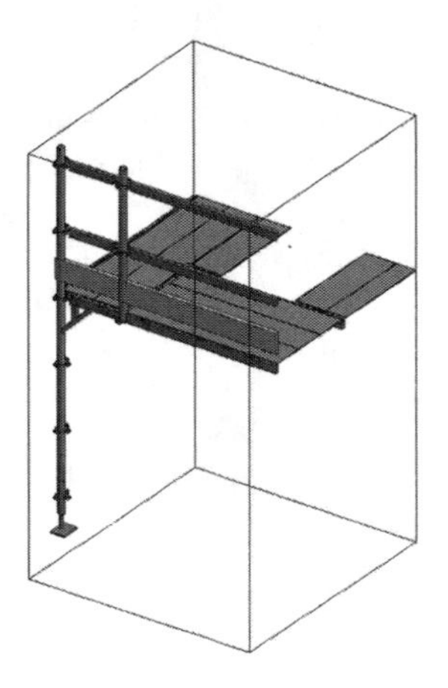

Figure 1. Scaffolding detail modeled on autodesk Revit.

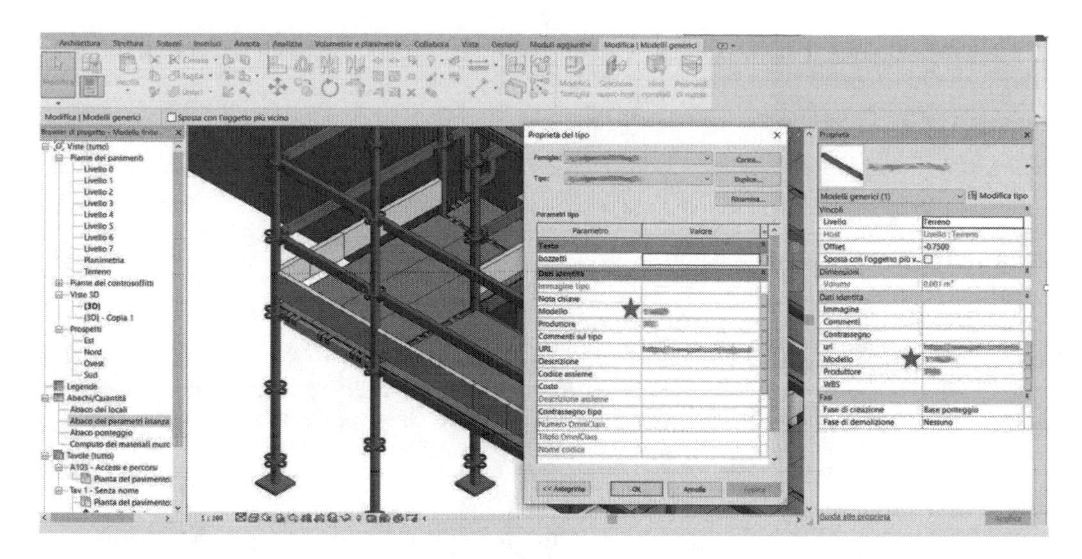

Figure 2. Workspace in Revit, focus on type and instances properties.

by the designer using the "subtraction" method. He eliminates the critical elements and he studies, case by case, an alternative design to solve the structure.

4. It is now necessary to fill all the gaps between the internal edge of the scaffolding and the façade. The goal is to ensure that at no point there are empty spaces with a depth greater than 20cm. Analyzing the plans of each level and manually measuring the distances of the edges from the façade the designer can find the critical points.
5. Using a certified modular structure, it is necessary to verify that the load limits of the catalogue are respected. As for stability and resistance to the wind, the anchors must be placed and each facade has to be stabilized with diagonals on the front.
6. One of the most important aspects of modelling is the creation of an abacus that helps understanding and elaborating the model. Two key abacuses were studied for this project. One that lists the parameters of the "type" with temporal phases and one with the properties of the "instances", in parallel with the type parameters (Figure 2).

2.3 *IT simulation*

The whole organizational process is broken down into a Work Breakdown Structure; this allows us to have a general index of the works, containing the codes used by the software to define each work phase. To do this we focused on the survey chapter dedicated to the construction site organization.

For the simulation phase two comparing software were used:

– STR Vision CPM is an Italian software of quantity surveying that contains a BIM platform for the IFC model vision. It allows linking the model, the measurements and the GANTT diagram to directly execute the simulation. There are two possibilities to do the 3D linked quantity survey: one is to manually execute the quantity take off from each model family and for each WBS work package. This option is relatively simple although quite cumbersome. The alternative is to create a parametric calculation of the quantity survey that contains the detection rules set for each item, repeatable for all the different levels of WBS. Once this procedure has been performed, by updating the model the measurements and the calculations are automatically updated at the same time. Now, each work package can be associated with one or more elements in the model. In this way, it is possible to create visualization styles to represent temporal phases and the works progress. The elements can be differentiated between computed and non-computed, or work to be performed, executed or in execution. This allows the development of a realistic simulation of the works throughout the duration of the construction site. At the end, the simulation can be recorded as a video.

– Navisworks Simulate belongs to the Autodesk group and it can be used to perform simulations starting from a previous time schedule. It is set up to work in agreement with other Autodesk software and with Project MS. The interface is very intuitive and the importing procedure of the model, directly in .rvt format, is optimal. Unlike STR, the model is imported as a link, so when the base file changes, the simulation is automatically updated as well. Simulations can also be carried out by manually reporting the temporal phases on the timeline.

3 RESULTS IN SAFETY MANAGEMENT

It is very important to be able to predict and anticipate the interferences and the unforeseen events that may occur on the construction site because they represent the greatest risk in safety field. This is the reason why safety is not to be considered a "phase" of the project, but a "constant" that accompanies all the project timeline and every work package. Specifically, BIM can help safety management by visualization, communication, and education improvement. In this case the clash detection had the power to identify the geometrical clashes between the elements of a complex façade and the scaffolding and to detect the areas in the construction site where risks were higher. The results of this analysis were a key instrument for the evaluation of the risk on simulation. The same methodology could be used to perform a lot of different types of clash detections in this domain or in others.

4 CONCLUSIONS

Nowadays the potential of the technology is impressive, for sure the studies to be pursued in this field should refer essentially to the method.

It is important to understand that new technologies need new researches and frameworks to be applied. The legislative, economic and professional apparatus is not now properly prepared to work with this philosophy. It is necessary to catch up with the times.

The study on the method performed in this research could also be applied in other areas of design; and it represents a line of thought to share with different disciplines, to facilitate the integration of technology with a concrete organizational process.

ACKNOWLEDGMENTS

The case study was developed within an Erasmus program by the post-graduate student in Building Engineering – Architecture Costanza Cioni, and reviewed by professors Michele Di Sivo, Massimiliano Martino, María Rosario Chaza Chimeno and Juan Manuel Macías Bernal, in collaboration with the University of Pisa and the University of Seville.

REFERENCES

Biagini C., Capone P., Donato V. & Facchini N. 2015. *IT Procedures for Simulation of Historical Building Restoration Site*. Firenze.

Feng C.W. & Lu S.W. 2017 *Using BIM to Automate Scaffolding Planning for Risk Analysis at Construction Sites*. Taiwan.

Gökgür A. 2015. *Current and future use of BIM in renovation projects*. Chalmers.

Nieto J.E., Moyano J., Rico F. & Anton D. 2013 *The need for information model applied to architectural heritage*, Seville.

Osello A. 2012. *Il futuro del disegno con il BIM per ingegneri e architetti*. Turin.

Project Management Institute. 2017. *A Guide to the Project Management Body of Knowledge*. Phyadelphya.

Raineri A. A. C. & Scoglio M. 2014. *BIM management e fase di esecuzione – Modelli operativi per la caratterizzazione del cantiere e delle azioni di controllo*. Milan.

Sadeghia H., Mohandesb S. R.& Abdul A. R 2016. *Reviewing the usefulness of BIM adoption in improving safety environment of construction projects*. Jurnal Teknologi.

Talon A., Cauvin C. & Chateauneuf A. 2017 *State of the Art of HBIM to develop the HBIM of the HeritageCare Project*, Clermont Auvergne.

Trani M., Cassano M., Todaro D., & Bossi B. 2016. *BIM Level of Detail for Construction Site Design*.

Zou Y., Kiviniemi A. & Jones S.W. 2015. *BIM and Knowledge Based Risk Management System: A Conceptual Model*. ResearchGate.

Science and Digital Technology for Cultural Heritage – Ortiz Calderón et al. (Eds)
© 2020 Taylor & Francis Group, London, ISBN 978-0-367-36368-0

Urban vulnerability assessment as a monitoring tool for cultural heritage preservation. Medium-sized cities in Andalusia

D. Navas-Carrillo
Departamento de Urbanística y Ordenación del Territorio, Universidad de Sevilla, Seville, Spain

B. Del Espino Hidalgo
Centro de Inmuebles, Obras e Infraestructuras, Instituto Andaluz de Patrimonio Histórico, Seville, Spain

J.A. Rodríguez-Lora & M.T. Pérez Cano
Departamento de Urbanística y Ordenación del Territorio, Universidad de Sevilla, Seville, Spain

ABSTRACT: The paper focuses on medium-sized cities, a system of urban settlements which dates back to more than two thousand years. Specifically, the case study is defined in Andalusia, the most populous and the second largest region in Spain, considering the Valley of Guadalquivir River as a geographical framework. Twenty-six out of these cities have been declared as cultural heritage sites due to their spatial and landscape configuration, as well as to the relevant examples of civil, military or religious architecture they hold. However, the conservation of these assets is conditioned by high exposure to certain risks and uncertainties, concerning socio-demographic, socioeconomic or residential aspects. Particularly, the research proposes to adopt the methodology implemented by the Spanish Ministry of Development in the Atlas of Urban Vulnerability to detect conditions of the social and structural disadvantage of the population of the abovementioned cities for effective cultural heritage tutelage.

1 INTRODUCTION

Andalusia has historically been a mainly urban territory (Instituto de Cartografía de Andalucía, 2009). According to the regional land plan, namely POTA-Plan de Ordenación del Territorio de Andalucía- (Junta de Andalucía, 2006), the Andalusian urban system is structured into three categories of urban networks: regional centres, networks of medium-sized cities - coastal and inner ones- and, finally, rural areas (Figure 1). Furthermore, this document recognises that 60% of the Andalusian territorial system is under the influence of a medium-sized city, which places them in the focus of attention from a quantitative and representative point of view.

It is also relevant to the fact that the clear majority of them have a historical origin. Indeed, these settlements have served as intermediate cities (namely they have played a role as secondary territorial centres) for centuries (Merinero Rodríguez & Lara de Vicente, 2010). The historical relevance of these cities has been shown by historical cartography, both at national and international scales, such as the case of *Civitates Orbis Terrarum* (Braun and Hogenberg, 1572). It is also essential to consider that the founding cores of many of these cities have been declared as cultural heritage (Navas-Carrillo, et al. 2018). Thus, together they must be regarded as a valuable historical heritage, as they represent the result of the superposition of cultures, ways of life and productive systems that have given rise to landscapes, tangible and intangible assets of unquestionable value both individually and as a set (Díaz Quidiello, 2007).

However, these cities may be affected by certain fragile urban conditions which, when applied to a social space, may put at risk the transmission of the historic urban fabric to future generations. In this sense, the research proposes analyse these weaknesses thought the concept of Urban Vulnerability and the methodology that have been developed by the

Spanish Ministry of Development to study it. This analysis would help local governments to develop specific policies to solve those detected weaknesses. Therefore, acting in reducing the vulnerability would result in a better and more efficient organisational capacity to manage the conservation of the cultural heritage of our cities.

2 METHODOLOGY

The vulnerability can be defined as a state of high exposure to certain risks and uncertainties, in combination with a reduced ability to protect or defend oneself against those risks and uncertainties and cope with their negative consequences (United Nations, 2013). Vulnerability applied to urban status is inexorably linked to the application of preventive actions (Ministerio de Fomento, 2012). In Spain, the Ministry of Development has been the promoter of a methodology that tries to objectively analyse Urban Vulnerability through twenty urban indicators and from four perspectives (socio-demographic, socioeconomic, residential and subjective). The resulting product is a web application: the Atlas of Urban Vulnerability that was firstly published in 2011. The methodology was reviewed to become part of the Urban Vulnerability Observatory in 2015, thanks to the approval of the Land and Urban Rehabilitation Law (Ministerio de Fomento, 2015).

This work adopts the methodology provided by the Atlas using the data supplied by the Spanish National Statistics Institute (INE) in its most recent Population and Housing Census. The authors have not been involved in its development. Given that the information needed to Subjective indicators is not available, the application has been limited to the first fifteen indicators:

- Socio-demographic Vulnerability: Percentage of seniors aged 75 or more (I1), Percentage of families with only a person aged 64 or more (I2), Percentage of families with only an adult and a child (I3), Percentage of foreign population (i4), Percentage of foreign children (i5).
- Socio-economic Vulnerability: Percentage of unemployed population (I6), Percentage of unemployed young population (I7), Percentage of contingent workers (I8), Percentage of workers without qualification (I9), Percentage of people without primary education (I10).
- Residential Vulnerability: Percentage of dwellings with less than 30 m² (I11), Average adequate living area by inhabitant (I12), Percentage of population in dwellings without toilet or WC (I13), Percentage of dwellings in ruined or deficient buildings (I14), Percentage of dwellings in buildings built before 1940 (I15).

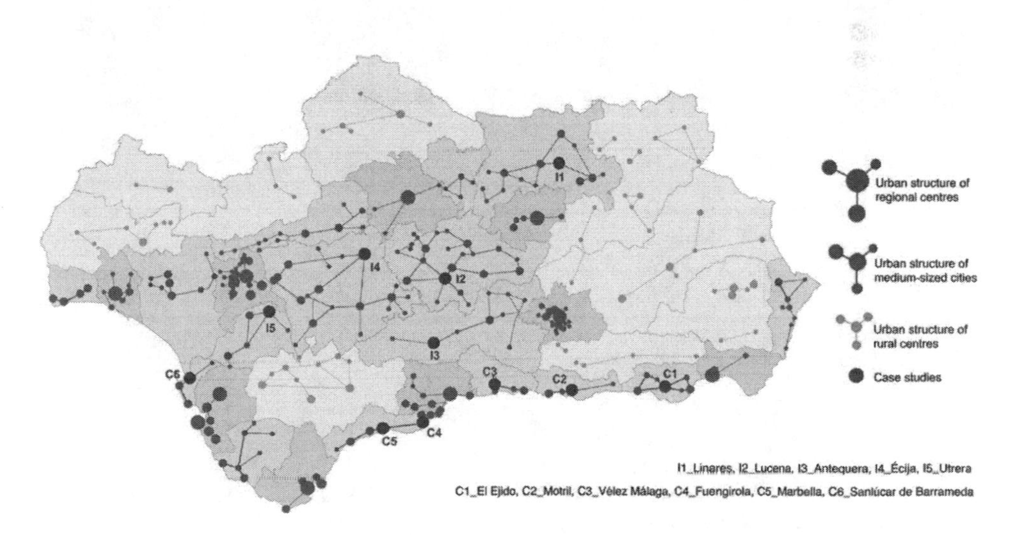

Figure 1. Andalusian urban system. Source: Compiled by the authors based on POTA (2006).

This methodology has been validated through its application in different urban contexts of the Spanish geography (De Santiago Rodríguez, 2018), proposing here two vectors of innovation. The first one is its application to the Andalusian system of medium-sized cities. The vulnerability of most of them has not been analysed to date. Furthermore, the results have not been confronted to provide an overview linked to territorial organisation factors. The second novelty is to propose the inclusion of urban vulnerability assessment as an additional tool to be taken into account in the cultural heritage preservation policies. This hypothesis is based on the fact that, just as other external aggressors are taken into account in preventive conservation, the social, economic and residential urban weaknesses can reduce the capacity as collective that cities have for the preservation of their heritage. This vision utterly implies an approach to the urban fabric as a whole, beyond the individual assessment of its cultural assets.

The indicators have been applied to the eleven cities that are considered by the POTA the primary functional centres in their respective territorial regions: El Ejido in Almería; Sanlúcar de Barrameda in Cádiz; Lucena in Córdoba; Motril in Granada; Linares in Jaén; Fuengirola, Antequera, Vélez Málaga and Marbella in Málaga; Utrera and Écija in Seville. All the indicators have been obtained in their numerical and percentage value from the data available in the INE. Thus, the results have been grouped by getting quintiles for each of them, been represented in graphs, as shown in Figure 2. The current territorial discourse defends a polycentric

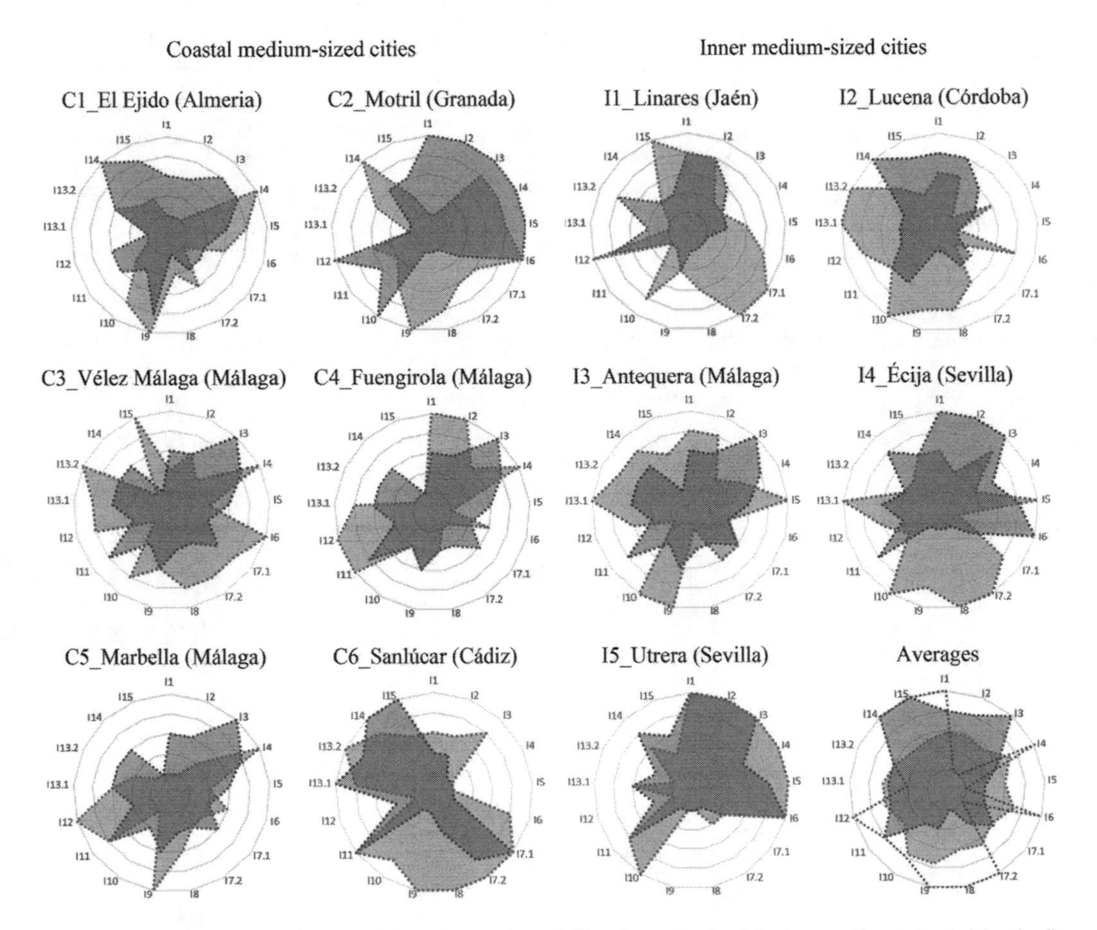

Figure 2. Comparative graphs of the urban vulnerability the main Andalusian medium-sized cities (red) and their corresponding regional centres (grey), in addition to the comparison between regional and national average (green). Source: Compiled by the authors based on data from INE (2011).

model, whose potential as sustainable and balanced stands out. In this sense, it has been particularly interesting to analyse the response of these cities compared to their regional centres. Besides, a comparative graph has been added that includes the arithmetic mean obtained from the eleven representative study cases and their respective provincial capitals. In addition, the national average has been included. That allows us to contextualise the situation of the Andalusian region.

3 RESULTS AND CONCLUSIONS

The analysis of the obtained results reveals two different situations between some cities and others. However, specific behaviour patterns are observed that are given generically and allow obtaining conclusions to the study. Although most medium-sized cities stand out for their vulnerability in indicators six to ten, relating to economic aspects, the weaknesses of regional centres are significantly higher considering social (1 to 5) or residential (11 to 15) issues. In this sense, it seems logical that the proximity of the natural environment of intermediate settlements favours a situation of positive social equity concerning the large capitals, which also affects the state of housing, as there are not sizeable residential overcrowding situations more typical of large urban agglomerations. These data are positively interpreted since they could increase urban features such as social cohesion and even the sense of belonging to a place, immaterial factors that are essential for effective heritage tutelage.

On the one hand, we have also observed that medium-sized coastal cities present the highest levels of economic vulnerability, as well as the high levels of foreigners. While the economic weakness is clearly seen as a risk factor for heritage preservation, the number of foreigners requires further analysis. It would be necessary to determine the degree of interaction with the daily urban functioning and their possible contribution to the conservation of built urban heritage. In this equation, we should also consider the positive and negative effects of tourism (Navas-Carrillo, et al. 2019). These cities have traditionally been characterized by coastal tourism, not drawing attention to their heritage legacy. In recent years, they have been promoting cultural tourism as a means to overcome seasonality. Tourism begins to play an essential role in the urban regeneration of many of these cities. However, it begins to pose serious problems when it becomes the predominant activity in heritage ensembles as vulnerable as the historic centres.

On the other hand, most inner cities have high levels of uneducated population, probably caused by their traditional characterization as agrarian cities. In general, low educational levels also entail limited knowledge on heritage matters. This fact would require the development of formative and informative actions that would lead the citizens to know, value and respect their legacy.

In general, according to the research results, we can affirm that Andalusian medium-sized cities have a level of vulnerability that is not clearly inferior to the regional centres, not reducing their capacity to manage the conservation of the cultural heritage compared to them. Nevertheless, critical urban weaknesses have been detected that should be addressed by future urban policies to increase the efficiency of the current tutelage mechanisms.

ACKNOWLEDGEMENTS

This paper has been developed under the R&D project entitled Urban Heritage Characterization and Cultural Tourism Model in Medium-Sized Cities. Potentialities and Challenges for its Internationalization: Inner Baetica, funded by the competitive call of the State Plan 2013-2016 Excellence - R&D Projects of the Ministry of Economy and Competitiveness of the Government of Spain (HAR2016-79788-P).

REFERENCES

Braun, G. & Hogenberg, F. 1572. *Civitates Orbis Terrarum*. Colonia.

De Santiago Rodríguez, E. 2018. Herramientas de diagnóstico para las intervenciones de regeneración urbana integrada en la ciudad consolidada: ejemplos de España. *Limaq. Revista de Arquitectura de la Universidad de Lima*, 4: 219–246.

Díaz Quidiello, J. L. 2007. Las ciudades medias interiores en el Plan de Ordenación del Territorio de Andalucía. *PH Boletín del Instituto Andaluz del Patrimonio Histórico*, 63: 42–91.

Instituto de Cartografía de Andalucía. 2009. *Atlas de la historia del territorio de Andalucía*. Sevilla: Consejería de Vivienda y Ordenación del Territorio de la Junta de Andalucía.

Junta de Andalucía. 2006. *Plan de Ordenación del Territorio de Andalucía*. Sevilla: Consejería de Medio Ambiente y Ordenación del Territorio, Sevilla, Spain.

Ministerio de Fomento. 2012. *Análisis urbanístico de Barrios Vulnerables en España Sobre la Vulnerabilidad Urbana*. Madrid: Ministerio de Fomento. Gobierno de España.

Ministerio de Fomento. 2015. *Atlas de la Vulnerabilidad Urbana en España 2001 y 2011. Metodología, contenidos y créditos*. Madrid: Ministerio de Fomento. Gobierno de España.

Merinero Rodríguez, R. & Lara de Vicente, F. 2010. *Las Ciudades Medias del interior de Andalucía. Caracterización y retos para el desarrollo turístico en un nuevo entorno*. Córdoba: Aecit.

Navas-Carrillo, D.; Del Espino Hidalgo, B.; Navarro de Pablos, J. & Pérez Cano, M. T. 2018. The Urban Heritage Characterization using 3D Geographic Information Systems. The system of medium-sized cities in Andalusia. *International Archives of the Photogrammetry, Remote Sensing and Spatial Information Sciences*. XLII-4 (W10): 127–134.

Navas-Carrillo, D.; Pérez Cano, M. T.; Del Espino Hidalgo, B. & Royo Naranjo, L. 2019. Dualidad Turística Residencial de las ciudades medias del litoral andaluz. In: *TOURISCAPE, Turismo y Paisaje*, pp. 247–260. Madrid: Tirant Lo Blanch.

United Nations. 2013. Report on the world. Social situation. Social Vulnerability: Sources and Challenges. New York: United Nations. Department of Economic and Social Affairs.

Science and Digital Technology for Cultural Heritage – Ortiz Calderón et al. (Eds)
© 2020 Taylor & Francis Group, London, ISBN 978-0-367-36368-0

Methodological framework to assess military rammed-earth walls. The case of Seville city ramparts

J. Canivell*
Department of Architectural Construction 2, Higher Technical School of Building Engineering, Universidad de Sevilla, Seville, Spain

A. Jaramillo-Morilla*, E.J. Mascort-Albea* & R. Romero-Hernández*
Department of Building Structures and Soil Engineering, Higher Technical School of Architecture, Universidad de Sevilla, Seville, Spain

ABSTRACT: The high density of historic rammed earth military samples in the Iberian Peninsula is mainly due to Almoravid and Almohad presence during the 11th and 13rd centuries.

The aim of this paper is to provide a methodological framework that enables the assessment of these sites, using the city walls of Seville as a case study. Hence, multi-scale and multidisciplinary approaches have been developed employing methodologies based on the use of CAD-GIS-BIM digital models, conceived as a Digital Cartographic Management (DCM) protocol.

This research will contribute to a better knowledge of this medieval heritage that will enable the development of future intervention criteria and the creation of preventive conservation strategies. The scientific knowledge achieved will also contribute to the regulation and standardization of the restoration of monuments built with rammed earth. Conservation and repair of historic rammed earth sites should only be undertaken if there is a good understanding of the consequences of any intervention technique.

1 INTRODUCTION

The historic wall of Seville is one of the most important urban symbols of the Muslim presence in the city. Its Islamic origin design and its great area are the result of the extension of the previous Roman enclosure, which has no visible remains today. Thus, the medieval fortification of Seville consisted of an urban fortress occupying a 17 hectares site and a walled enclosure, which during the period of Almohad domination reached an occupation of 273 hectares (Valor Piechotta, 2014). Additionally, this monument also stands out for its material nature, based on the use of soils by means of the *tapial* technique. This consist of a reusable formwork within which earth is manually compacted. While the walls of residential or civil buildings are around 40-60 cm, in the case of fortifications, the thickness is usually well over one meter.

Once built, the urban rampart of Seville has suffered partial transformations, often related to the conservation and repair of walls and the renovation of its gates. However, from the second half of the 19th century, it began a continuous process of demolitions which affected several parts of the monument, as a consequence of the growing that the city experienced (Suárez Garmendia, 1986; Morales, 2013). Consequently, of the more than 6,000 meters from the original perimeter, only about 2,000 meters remain (Valor Piechotta, 2014). Within the preserved parts, a significant percentage corresponds to the sector located in the northern area of the old historical center, which is placed between the old gates of Cordoba and Macarena. This sector is known by the name of *Muralla de la Macarena* and was declared National

*Corresponding author: jacanivell@us.es, jarami@us.es, emascort@us.es, rociorome@us.es

Historical Monument in January 1908. Precisely, this consideration was requested in order to avoid their demolition. As a result, a 500 meter long perimeter stands for one outstanding monument consisting of walls, barbicans, towers, gates portholes (Suárez Garmendia, 1986).

In this context, it is necessary to point out that the interventions carried out during the 19th and 20th centuries have revealed conservation work as a punctual and unplanned task. This question, as well as the weak response of certain restorations that have been performed with restoring techniques not totally compatible to the original and has therefore resulted in a deficient state of conservation. This issue can be checked through studies about interventions undertaken from the 1980s to the most recent (García-Tapial, 2006; Canivell and Graciani, 2014; Graciani and Canivell, 2014). Nowadays, erosion, loss of material, sandblasting, dirt, loss of cohesion or partial detachments of blocks and merlons are common. Even though these affections do not structurally compromise this thick masonry wall, they do totally undermine this valuable and protected heritage asset, and may even accidentally affect the integrity of visitors due to the proximity of the public pedestrian road.

The abovementioned situation exemplifies the need of an integral look for the conservation, maintenance and dissemination of the values of the patrimonial architecture built on earth. This question should be channeled through a common plan applied to the Macarena sector, through a strategy of collaboration proposed by the City Planning Department and the University of Seville. The objective of the work is to propose a methodological procedure that facilitates a progressive follow-up of the results that will be obtained during the future intervention work on the building (Canivell, Jaramillo-Morilla, Mascort-Albea, & Romero-Hernández, 2019).

2 METHODOLOGICAL APPROACH AND OBJECTIVES

Nowadays, the use of digital models for the management of information related to cultural heritage constitutes a very relevant research line. Additionally, the interoperability between different tools such as *Computer Aided Design* (CAD), and more recently, *Building Information Modelling* (BIM) and *Geographic Informatic Systems* (GIS) can provide a very complete graphic, parametric and spatial control of any cultural assets. Based on this approach, the authors of this document define *Digital Cartographic Management* (DMC) as the use of geographical tools for the creation of spatial and architectonic databases related to build heritage. This is achieved through a multi-scale conception of the use of GIS that ranges from territorial and urban to an architectural scale that allows defining through geographical entities the main spatial and architectural components of a building, as well as the movable assets contained therein.

Precisely, in the use of geographical techniques at a level of architectural detail resides the novelty of the current approach. Thus, it is possible to fill a gap of information existing in most spatial geo-viewers and generate in a flexible and affordable way new digital and simplified models that analyze the particularities of buildings (Mascort-Albea, 2018), constituting a solid basis for further 3D developments on BIM models or CityGML formats (Hidalgo Sánchez, 2018). Through the developed experiences in various case studies, it has been shown that this strategy contributes to the preventive conservation of historic buildings, from a transversal perspective. In this sense, the use of GCD stands out for the development of tourist applications for heritage interpretation (Mascort-Albea et al., 2016), cataloguing of risk of abandonment buildings (Mascort-Albea and Meynier-Philip, 2017) or the multidisciplinary analysis of historical typologies (Mascort-Albea, 2017).

Based on the above considerations, a predictive monitoring of the historic wall of Seville is proposed and based on the following methodological approach:

- A common methodological framework for evaluating the state of the study case, as a basis for (1) improving the management of the knowledge acquired about this monument, (2) knowing and diagnosing its current state of conservation and (3) establishing more effective and unitary criteria for action.

- Assess the risks and vulnerabilities of the wall with in order to detect the critical sectors and prioritize the interventions.
- A technical protocol to manage the maintenance of the wall, based on a procedure for monitoring physical-mechanical parameters and on the same cartographic-digital models.
- A tool that, based on the use of interactive cartographic-digital models and institutional web platforms, serves to spread the heritage values of the historic Seville city wall.

3 WORKPLAN AND GDC SYSTEM STRUCTURE

The proposed methodological procedure will be developed through a work plan, which contains the following phases:

A. Documentary phase. Documentation on archaeological interventions will be compiled and analysed. Additionally, the historical documentation about the evolution of the monument, as well as the technical restoration will be studied.
B. Digitalization phase. Starting from the available graphic documentation, mainly plans and elevations, a detailed cartography will be generated. Especially significant are the elevations of the walls from the archaeological reports, where specific aspects of the wall are detailed.

As a complementary resource, a digital scan could be employed to make a digital 3D model. As a result, not only the volume, but also rectified images of each elevation with real textures will be obtained. Starting from the digital cartography designed by the Urban Development Department of Seville (GUS), the detailed elements not already considered will be completed.

C. Structure of the Digital Cartographic Management (GCD) system. Based on the characterization and evaluation procedure carried out by Canivell (Canivell and Graciani, 2015; Canivell, 2012) and the cartographic representation at architectural scale proposed by Mascort-Albea (Mascort-Albea, 2017, 2018), a geo-referenced spatial database will be structured. The database will include descriptive, constructive and material parameters. At the same time, a standard data collection form will be created. Likewise, a sectorisation of the walls and towers will be designed in logical units at different levels building a related hierarchical tree, where the data collected during the field work and documentation will be dumped (Figure 1).
D. Data collection and characterization. In successive in situ surveys, the available plans will be checked and updated. Afterwards, the characterization studies will be carried out using

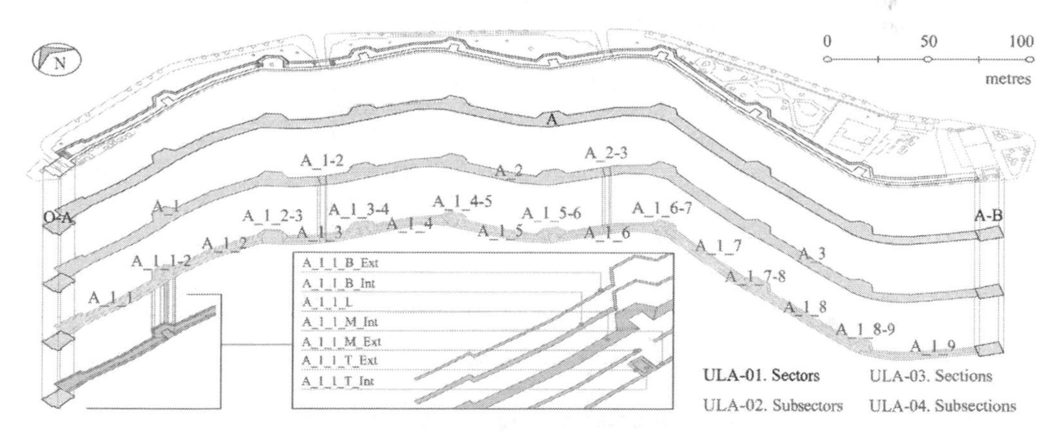

Figure 1. Macarena sector main GDC units. Codes and numbers are related to a hierarchical increase of the architectonic level of detail. Source: Prepared by the authors from IDE Sevilla plans.

non-destructive inspection methods, such as thermography and ground-penetrating radar (GPR), despite these data will be analysed at a later stage. Finally, once the vulnerable sectors have been determined, samples will be extracted to be characterised and analysed in the laboratory.

E. Diagnosis and sensorization. A first level preliminary assessment of the current state of conservation and vulnerabilities will be prepared in order to establish the most critical sectors. Next, and starting with the samples from the previous phase, the characterization tests of the samples will be run. At the same time, a prototype of sensors based on Open Access technologies will be developed to monitor the physical-mechanical parameters of the walls (pressure, relative humidity and movements of the masonry, as well as climatological data of the environment).

F. Data uploading, update and analysis. The information upload into the developed GCD system is validated, both the data corresponding to the inspection and those from laboratory tests and inspection methods. The parameters and factors of the GCD system will then be updated in order to carry out a more precise second level risk assessment, in order to establish the critical sectors, action strategies and planning of a conservation and preventive maintenance protocol.

G. Transfer and dissemination of the achieved results. On the basis of the GCD system and by selecting the contents, an interactive guide will be prepared through an institutional website. In this way, the results will be reverted, both to citizens and to those specialized technicians related to the conservation of the built heritage on earth. The institutional tools and open data platforms that GUS has set up for this purpose will be used for this transfer work.

4 CONCLUSIONS

Given the current need to define simple procedures for the conservation and predictive maintenance of cultural assets, strategies are proposed for the widespread use of GIS on an architectural scale. The proposed GDC provides a methodological approach that encourages the use of geographical tools for the detailed analysis of heritage architecture. In this way, the historic wall of Seville due to its extension and urban influence as a great monumental infrastructure, constitutes an ideal case study for the application of the proposed strategies. Through this work, a predictive model has been designed that adapts to the needs of the project, the existing resources and the particular nature of the case study. In this sense, the development of a methodological strategy that seeks to integrate different scales of resolution and disciplinary components, it is required that the work achieve a high degree of graphic, thematic and terminological coherence.

The information will be structured through different thematic components, which due to their transversal nature will serve as an internal working tool for the control and monitoring of the material state of the property, and additionally, employed as an instrument for widespread of the monument through the subsequent technological development of digital applications. Finally, it is observed how the case study adapts to the limits of a work that intends to be a pilot episode. In this way, a preliminary selection of representative elements was made for each work unit, so that they could be significant to extend the model created to the whole city's historic wall.

REFERENCES

Canivell, J. and Graciani, A. 2014. 'Muralla de Sevilla (1984–2008)', La restauración de la tapia en la Península Ibérica : criterios, técnicas, resultados y perspectivas, pp. 218–221.

Canivell, J. 2012. 'Characterization methodology to efficiently manage the conservation of historical rammed-earth buildings', in C Mileto et al. (eds.) Rammed Earth Conservation. Rammed Earth Conservation. London: Taylor & Francis Group. pp. 283–288.

Canivell, J. and Graciani, A. 2015. Constructive characterization of historical rammed-earth walls in Almohad fortresses in the ancient Reign of Seville. Arqueologia de la Arquitectura (12) Retrieved from http://arqarqt.revistas.csic.es/index.php/arqarqt/article/view/178.

Canivell, J., Jaramillo-Morilla, A., Mascort-Albea, E. J., & Romero-Hernández, R. 2019. Metodología de evaluación y monitorización del patrimonio basado en la gestión cartográfica digital. La muralla de Sevilla. En M. Di Sivo & D. Ladiana (Eds.), *Le mura urbane crollano: conservazione e manutenzione programmata della cinta muraria dei centri storici.* (pp. 119–135). Pisa, Italia: Pisa University Press. https://doi.org/10.12871/97888333917559.

García-Tapial, J. 2006. 'Proyecto de rehabilitación de la Muralla de la Macarena. Fase III'. Sevilla.

Graciani, A. and Canivell, J. 2014. 'Revisión de las Intervenciones en Fábricas de Tapia en Andalucía Occidental', in Mileto, C., Vegas, F. (ed.) Restauración de la tapia en la península ibérica. Criterios, técnicas, resultados, perspectivas. General de Ediciones de Arquitectura, pp. 30–41.

Hidalgo Sánchez, F. M. 2018. Interoperabilidad entre SIG y BIM aplicada al patrimonio arquitectónico. Exploración de posibilidades mediante la realización de un modelo digitalizado de la Antigua Iglesia de Santa Lucía y posterior análisis. (Universidad de Sevilla). Retrieved from https://hdl.handle.net/11441/79394

Mascort-Albea, E. J. 2018. Mapas para el patrimonio: caracterización técnica de las iglesias medievales de Sevilla mediante sistemas de información geográfica (SIG) (Universidad de Sevilla). Retrieved from https://idus.us.es/xmlui/handle/11441/70745

Mascort-Albea, E. J. and Meynier-Philip, M. 2017. Strategies for conservation of Religious Heritage in the Metropolitan Area of Lyon/Saint-Étienne (France). Short research stay and methodological transfer. *IDA 2017. 1st International Congress on Advanced Doctoral Research in Architecture*, 675–684. Retrieved from https://idus.us.es/xmlui/handle/11441/70006

Mascort-Albea, E. J. 2017. Datos geográficos abiertos para la conservación preventiva del patrimonio arquitectónico. *revista PH*, (92), 228. Retrieved from https://doi.org/10.33349/2017.0.3948

Mascort-Albea, E. J. et al. 2016 Sevilla, Patrimonio Mundial: guía cultural interactiva para dispositivos móviles. revista PH. (90), 152. Retrieved from https://doi.org/10.33349/2016.0.3778

Morales, A. J. 2013. 'Un episodio en el derribo de las murallas de Sevilla', Lab.Arte, 25(2), pp.689–689700.

Suárez Garmendia, J. M. 1986. Arquitectura y urbanismo en la Sevilla del siglo XIX. Sevilla, España: Diputación Provincial de Sevilla.

Valor Piechotta, M. 2014. 'La muralla medieval de Sevilla. Otra interpretación', in Sevilla Arqueológica. La ciudad en época protohistórica, antigua y andalusí. Sevilla, España: Ayuntamiento de Sevilla.

Science and Digital Technology for Cultural Heritage – Ortiz Calderón et al. (Eds)
 ISBN 978-0-367-36368-0

Testing digital photogrammetry and terrestrial laser scanners in surveying dry stone vernacular constructions

C. Mallafrè
Fundació el Solà, Universitat Rovira i Virgili, Tarragona, Spain

A. Costa & S. Coll
Universitat Rovira i Virgili, Tarragona, Spain

ABSTRACT: The dry stone construction technique is the oldest method used by humanity. Moreover, it is the type used most frequently worldwide. In the Iberian Peninsula, this technique was used regularly between from the end of the eighteenth century and the beginning of the twentieth century. Because of the mechanization of fieldwork and the lack of generational relief, dry stone heritage constructions are in severe danger. Currently, it is impossible to conserve all the constructions. Though it might seem unimportant, new surveying technologies make it possible to register these structures with unprecedented accuracy. Thus, they may be catalogued before they are ruined.

Digital photogrammetry and terrestrial laser scanning are two massive data capture techniques. Both are commonly used in architectural heritage surveys. These safe, contactless, and non-invasive procedures allow registration of millions of topographical points quickly, and with great accuracy. In the case of dry stone vernacular heritage, photogrammetry offers great advantages, as the surfaces of these constructions have very marked textures. Additionally, these constructions are often built in irregular locations that are difficult to access. This causes difficulties in the use of heavy equipment.

Despite its positive features, the appearance of smaller and lighter laser devices leads us to reconsider photogrammetry's use in the field. This article presents a comparative analysis between digital photogrammetry (using the Structure from Motion (SfM) technique) and a recently released laser scanner in performing a 3D topographical survey of dry stone constructions.

1 INTRODUCTION

In the field of built heritage, massive data capture techniques have had a great impact on topographic registration processes. Several techniques have been developed during the last decade (Pavlidis et al., 2007). Digital photogrammetry and terrestrial laser scanning may be the most widespread. Both techniques have been tested extensively in architectural heritage surveys, with highly positive results (Kadobayashi et al., 2004; Grussenmeyer et al., 2008). Additionally, they have proven compatible with (and complementary to) one another. In recent years, the use of Structure From Motion (SFM) algorithms for digital photogrammetry has become an increasingly important tool. It is capable of obtaining high-precision, 3D built heritage surveys quickly, safely, and easily (Koutsoudis, 2014).

Because of the abandonment of rural areas and agricultural activities, dry stone construction is a built heritage in danger. Nevertheless, dry stone construction techniques can have great impacts on methods of surveying, cataloguing, and conserving. Traditional surveying techniques, together with drawings and sketches, have been used in the twentieth century in Spain to catalogue this type of construction (Rubió, 1914) (Lison & Zaragozà, 2000).

In recent years, numerous investigations have used massive data capture techniques to approach the registration and study of dry stone constructions. To name a few, Restruccia et al.

(2012) used the terrestrial laser scanner to study dry stone constructions in the Ibleos Mountains of Sicily, Italy. Rossi & Leserri (2013) delved into the difficulties of carrying out accurate surveys of dry stone constructions. Consequently (using massive data capture techniques to analyse Italy's varying typologies), they have attempted different graphical approaches. Barroso et al. (2018) proposed the use of terrestrial laser scanners to study dry stone constructions in the area of Geres-Xurés (between Spain and Portugal). Because of the great irregularity in the constructions, Geres-Xurés provides very useful terrain for employing this technique.

Both techniques are suitable for surveying this type of construction. However, there are some specific limitations that differ from methods used for monumental heritage sites. The primary issues regard the constructions' locations. Typically, they are found in abandoned rural areas, where vegetation has grown out of control. This overgrowth usually causes occlusions. Additionally, locations are usually far from paved roads, making access to buildings possible only on foot. Photogrammetry was once considered a satisfactory technique, because of its ease of use (and the minimal technical equipment required). Yet, the arrival of smaller, lighter laser devices on the market leads us to reconsider photogrammetry's usefulness.

Thus, the aim of the present investigation was to test the use of laser scanning and digital photogrammetry to survey dry stone heritage constructions. The methods' reliability has been proven. Thus, the test is focused on the comparison of the techniques in the field. To do so, a new, lightweight terrestrial laser scanner is used, making it possible to access difficult locations. The test aims to determine the most suitable techniques, both in terms of accuracy and the time needed to employ each.

Moreover, this investigation presents the results of the study on a singular typology, known as *cocó*. This typology is found in a rural area of the township of Tivenys, a province of Tarragon located northeast of the Iberian Peninsula. Despite proximity to Ebre River, the zone is dry and its dominant crop is the olive. Accordingly, the function of the *cocó* is collecting and storing rainwater for *domestic* use (for both animals and humans) (Lluch Martin & Giral Quintana, 1967). The *cocó* is formed by a rounded rounded wall and a covering dome, which protects the water from the sun. The bottom part use to be excavated. It is here that the water is preserved. These are located on inclined rock surfaces that conduct rainwater towards each *cocó*. The stones are very irregular, and the construction lacks any formwork. According to oral fonts, there are about 250 *cocons* along the municipal boundary (Figure 1).

2 METHODOLOGY

The investigation's methodology contained four main steps: 1) data collection; 2) generation of point clouds and the comparison between surveying techniques; 3) the resultant point clouds; and 4) determination of which technique offers the greatest advantage in reaching the objectives.

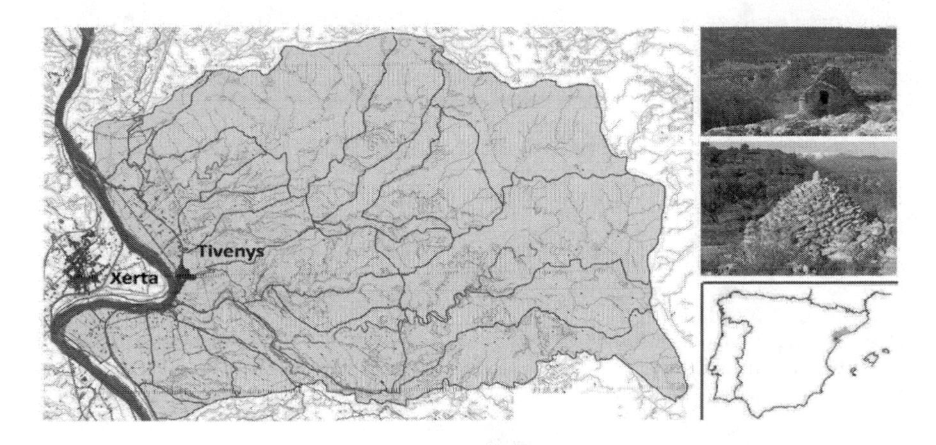

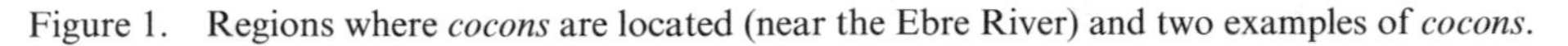
Figure 1. Regions where *cocons* are located (near the Ebre River) and two examples of *cocons*.

Figure 2. Point clouds obtained using photogrammetry (L) and scan laser (R).

Data collection took place simultaneously for both surveying techniques, so that the ambient conditions and light were the same. The procedures used were the Terrestrial Laser Scanner (TLS) and the SFM photogrammetric technique. Both methodologies allowed us to obtain a complete point cloud of the object. Four black-and-white targets were used to enter both surveys into the same coordinate system.

The scanner laser used was a Leica BLK360, a small (165 x 100 mm) and light (1kg) device. We used the Time of Flight (ToF) technique, with a measuring range of 0.6–60m. Our measuring tax was 360000 points per second, with an error of 4 mm at 10 m. Ten positions were required to construct the complete point cloud, all of which were processed by the Cyclone Register 360 software. The error of joining of all the point clouds was 1.50 centimetres.

For photogrammetry, a total of 120 photographs were taken using a Nikon D7000 camera and TAMRON LD XR DI AF 17-50mm 67 a16 if The commercial Agisoft Photoscan package was used for processing of the point cloud.

Once we obtained both point clouds, they were entered into the same coordinate system using the 3DReshaper program, via the control points. The maximum error of the alignment was 1.25 cm. Subsequently, the command measure was used to compare and obtain the range of distances between points. The compared assessment of both surveys considers 1) the time consumed, 2) the number of points obtained, and 3) a comparison of the range of distances between the two point clouds (Figure 2).

Although comparison of the techniques has been applied to dry stone constructions, applying them to the typology of the *cocons* is problematic. Complicating factors include 1) difficulty of access, 2) the fact that most are full of water, 3) state regulations limiting drone use, and 4) the amount of vegetation surrounding construction areas.

3 RESULTS AND DISCUSSION

The study allowed analysis of the point clouds obtained by means of the two massive data capture techniques. Table 1 summarizes the main parameters we compared: time data collection, time data processing, quantity of point clouds, points for each photo, points for each position and points per second inverted.

Table 1. Average of parameters obtained by photogrammetry and laser scanning.

Parameters	Photogrammetry	Scan laser
Time data collection (seconds)	540	2.700
Time data processing (seconds)	11.700	4.200
Quantity of points (points)	40.374.796	8.638.819
Points for each photo (points)	340.000	-
Points for each position (points)	-	863.881
Points per second inverted (points)	3.298,59	1.252

In addition, we made a dimensional comparison between the two point clouds. From the comparison, we obtained a graphic showing the range of distances between corresponding points of the two point clouds (Table 2) (Figure 3):

Table 2. Average range of distances between the two point clouds.

Interval (centimetres)	Percent
0,00 - 1,25 (blue)	58,9 %
1,25- 2,50	20,3 %
2,50 - 3,75	8,2 %
3,75 - 5,00	5,06 %
5,00 - 6,25	3,07 %
6,25 - 7,50	1,94 %
7,50 - 8,75	1,37 %
8,75 - 10,00 (red)	1,15 %

Figure 3. Comparison between the two point clouds.

A total of 79.2 % points are separated by a distance between 0 and 2.5 cm. The remaining 20.8 % of points are separated by a distance between 2.5 and 10 cm. These points are those that take considerable distortions among two point clouds. As can be observed in Figure 3, these appear at the outside of the *cocó* (next to the floor), and also inside the *cocó*. Regarding the *cocó* outside inferior zones, the differences among the two point clouds are due to the noise that supposes the vegetation. On the other hand, the distance between the points increases inside the *cocó*. This is due to lighting conditions, which affect the generation of the point cloud by photogrammetry. The lack of light causes a loss of quality and definition in the photos, and creates distortions in the generation of the indoor point cloud.

It was possible to obtain a complete point cloud of the object by means of the two techniques. By analysing the inverted time, data collection was faster using photogrammetry, but processing time was better for the laser scanner. Although the point cloud obtained by photogrammetry has 2.5 times more points than the scanner data, the subsequent mesh will incorporate the textures of the object with photographical quality. The higher density of the point cloud influences the resultant mesh, which can more accurately reproduce the object's irregularities. Notably, a denser mesh does not necessarily indicate better quality. It will be necessary to establish a correlation between density and the survey's needs in every case.

On the other hand, distances between the two point clouds reveal accuracy differences of about 2.5 cm. This is significant for a small construction such as the studied case. Several issues can affect the accuracy of photogrammetry: 1) internal lighting conditions; 2) high-contrast; irregular surfaces; 3) short ranges for taking interior (and sometimes exterior) photographs; 4) the existence of abundant vegetation outside; and 5) irregular shapes. These conditions can have a negative impact on the accuracy of the point cloud. Otherwise, the laser scanner rarely presents distortions in the measurements (although the amount of data collected is lower).

Once the point cloud and the mesh are obtained using either of the two techniques, it is possible to process the data to obtain different dimensional information2D dimensions, volumes, etc.)

There are differences regarding the drawings that can be produced by each method. On one hand, laser scanning requires use of another application to mesh and texturize the object. The quality of the model's texture is conditioned by the incorporated camera's device, and is usually lower than in photogrammetry. On the other hand, photogrammetry allows texturization of the model using one application, and with higher quality. However, this is also dependent upon the camera's function. Thus, ortho-images can be obtained using the two procedures.

In addition, the models we obtained open the door to new means of diffusion, developed in novel ways. Free web platforms such as Sketchlab allow users to handle the modelled object, move it, visualize its interior, and even experience it through augmented reality.

4 CONCLUSIONS

The investigation tested the use of two different massive data capture techniques to survey dry stone constructions: motion photogrammetry, and terrestrial laser scanning. Both techniques are well known in architectural heritage. However, their use has been negligible in comparing dry stone constructions. Thus, the present investigation has not focused on testing them exhaustively. Rather, priority has been given to obtaining results which provide information on dry stone constructions. Special consideration was paid to ease of use in the field, and to processing time.

We made use of a light laser device (BLK) that has recently appeared on the market. While it is much lighter than other devices, structure from motion photogrammetry still offers greater advantages compared to laser scanning. Of particular note is its ease of use in the field, and the dissemination possibilities of the models obtained.

In terms of the surveyed objects, placing the laser device inside the structure was problematic. The cocons are small and used to contain water, thus making it impossible to perform a complete survey of the interior. Photogrammetry allows image generation of the complete structure, but the lighting conditions and close vegetation can cause the appearance of noise and false points in the point cloud. Moreover, it must be considered that the surveying time in the field is faster with photogrammetry, but processing time is faster using the laser scanner.

The registration of dry stone elements has progressed over the last few years. The application of modern techniques and the possibilities of dissemination may prove to have a great impact on conservation, study, and cataloguing.

REFERENCES

Barroso, C E; Barros, F C; Riveiro, B y Oliveira, D V. 2018. The construction of the transhumance territory of the Gerês-Xurés: Vernacular heritage identification, analysis and charecterization. *Rehabend 2018 construction pathology, rehabilitation technology and heritage management*.

Grussenmeyer, P.; Landes, T.; Voegtle, T. and Ringle, K. 2008. Comparison methods of terrestrial laser scanning, photogrammetry and tacheometry data for recording of cultural heritage buildings. *Int. Arch. Photogramm. Remote Sens. Spat. Inf. Sci.*, vol. 37 (B5), pp. 213–218.

Kadobayashi, R.; Kochi, N.; Otani, H. and Furukawa, R. 2004. Comparison and evaluation of laser scanning and photogrammetry and their combined use for digital recording of cultural heritage,' *Int. Arch. Photogramm.*

Koutsoudis, A., Vidmar, B., Ioannakis, G., Arnaoutoglou, F., Pavlidis, G., & Chamzas, C. 2014. Multi-image 3D reconstruction data evaluation. *Journal of Cultural Heritage*, *15*(1), 73–79. https://doi.org/10.1016/j.culher.2012.12.003.

García, M. & Zaragozá, A., 2000. Arquitectura rural primitiva en secà.

Lluch Martin, E., & Giral Quintana, E. (1967). L'econolia de la regió del Baix Ebre. Barcelona: Servei d'Estudis de la Banca Catalana.
Pavlidis, G.; Koutsoudis, A. Fotis, A., Vassilios, T. and Christodoulos, C. 2007. Methods for 3D digitization of Cultural Heritage. *J. Cult. Herit.*, vol. 8, no. 1, pp. 93–98, Jan.
Restuccia, F.; Fianchino, C.; Galizia, M. y Santagati, C. 2012. The stone landscape of the Hyblaen Mountains: the geometry, structure, shape and natureof the muragghio.
Rossi, G. y Leserri, M. 2013. Arquitecturas de piedra seca, un levantamiento problemático. *Expresión gráfica arquitectónica*, 184–195.
Rubió, J. (1914). Construccions de pedra en sec. Barcelona: Col·legi d'arquitectes de Barcelona.

Science and Digital Technology for Cultural Heritage – Ortiz Calderón et al. (Eds)
© 2020 Taylor & Francis Group, London, ISBN 978-0-367-36368-0

GIS as a tool for effective preventive conservation: Its utility in studies of conventual outdoor spaces

M. Molina-Liñán & C. López-Bravo
Department of Architectural History, Theory and Composition, University of Seville, Seville, Spain

F.J. Ostos-Prieto
Department of Urbanistic and Land Management, University of Seville, Seville, Spain

ABSTRACT: The aim of this research is to update conservation methods currently employed for the study of the conventual heritage in Seville. Within these buildings, there are several typologies of open spaces -compasses, courtyards, gardens, orchards and cloisters-, which add complexity to a historical and particularly dense urban layout. Some of these constructions, which undergo the contemporary lack of vocations, have the maximum Spanish heritage recognition as Cultural Assets. A situation that enhances the coexistence of cultural and religious functions in some cases. For these reasons, it is necessary to have a better knowledge about these spaces conservation conditions, specially since they are, sometimes, closed to the public. Consequently, this research proposes the use of Geographic Information Systems as a new way of storage and georeference the pathologies and needs of this buildings on an urban scale. On this matter, priority is given to the visual exploration of the data, reinforcing their spatiality. Then, they can be quantitatively analysed on a territorial scale, not just as isolated buildings. As a result, the aim is to obtain a certain priority order for future interventions and heritage protection of this building typology.

1 INTRODUCTION. THE PROCESSING, ON AN URBAN SCALE, OF DATA RELATED TO CONVENT CONSERVATION

Seville owes its urban configuration to the establishment of most of its convents and monasteries. This architectural typology configured almost two thirds of the city center area during the 18th century (Pérez de Lama, 1996). However, their conservation condition is, in many cases, unknown, due to the restricted access to them.

Many of them have had numerous transformations, interventions and works over the centuries. Sometimes these works are difficult to reference in chronological or systematic ways through traditional data storage methods, oftenly used by architects, conservators and restorers in past times.

That is the reason why this research suggests employing Geographic Information Systems as a means of storing and geolocating the pathologies and needs of this buildings, on an urban and systematic scale. This makes possible to prioritise the data visual exploration, which can be quantitatively analysed on a territorial scale (Mascort-Albea, 2017). As a result, an order of priority in the interventions is obtained, contributing to the building heritage protection.

1.1 *Precedents. What is GIS being used for on an urban level?*

The conservation of the Sevillian convent structure has been studied on a small scale for decades. Related to these buildings, the consultation of several intervention projects has been carried out. It has been verified how the diagnoses of them have been executed, normally, as isolated interventions, even within the different parts of the same convent (Consejería de Cultura. Delegación Provincial de Sevilla, 2004).

Traditionally, intervention reports are complex to read and interpret. In most cases, especially in interventions before to the second half of the 20th century, it is very difficult to reconstruct a history of the actions carried out. However, it should be mentioned that some efforts have been made trying to combine this information, facilitating the research and consultation for future interventions (Database of Interventions in buildings declared to be of Cultural Interest, 2003).

In last decades, the use of Geographic Information Systems has increased considerably in the municipality and the Andalusian Government. In this way, there have been developed catalogues of Cultural Interest Buildings (Gerencia de Urbanismo, 2017), World Heritage Sites (Consejería de Economía, Conocimiento, Empresas y Universidad, 2019) or Public Green Areas (Consejería de Agricultura, Ganadería, Pesca y Desarrollo Sostenible, 2005). In this situation, this research has seen the need to improve them introducing data about the conservation conditions of these buildings. The idea is to use the processing of these data through GIS to obtain not only visual or interpretative results, but also comparative and quantitative results.

2 RESEARCH OBJECTIVES AND METHODOLOGY

2.1 *Employing GIS for conventual buildings preventive conservation: Main and specific objectives*

Convent buildings are built by complex additions which suffer from numerous pathologies, both in their indoor and, above all, in their outdoor spaces. These can be due to the many different causes, such as the multiple uses they have accommodated over the years (Molina, López and Mosquera, 2018), the materials and complex structures used during their primitive construction, the plots on which they settle, or directly, to the action of the time. Their diagnosis requires a methodical and analytical control for their preservation and the determination of the causes. Then the main objective of this work is the chronological storage of specific data, through georeferenced layers.

Among the most common damages that convents of Seville present, are those related to the absence of conservative patterns or weak preventive and corrective measures: cracks and fissures, humidity, biological colonization, degradation of whitewash or plaster, among others. All these directly affect their current conditions and physical integrity

As specific objectives derived from this, are the recognition of positioning, extension, shape, temporality and distribution of the damages, as well as possible spatial interactions between specific pathologies.

The intention is to obtain results related to appearance patterns and to address a priority map of conservation actions on an urban scale.

2.2 *Methodology, the use of GIS as a support for spatial conservation*

As we have already outlined along with the approach and objectives of this research, it evident that the use of GIS allows multiple and diverse options for the work of conservation and heritage intervention (López and Marquez, 2013). Chasing the objectives set out above, the software employed for this study case is QGIS, an open-source Geographic Information System that allows handling raster and vector formats.

Thanks to its application, this research is organized along three stages: taking, georeferencing, and processing the data.

To complete this method, a specific case that qualifies the study is determined. The one chosen is Santa Inés Convent, a female cloister located in the center of the city, which has been accessible for taking data and photographs. This convent also gathers specific typological characteristics that make its case comparably to the rest of convents in the city.

The first step is the technical visit, in which data have been taken about the state of conservation of the outdoor rooms (cloister and courtyards), taking photographs of them and identifying the location, shape, extent, and distribution of such pathologies.

Subsequently, this data is uploaded into the QGIS software. There, there have previously been arranged different layers, which determine modifications in the extension and

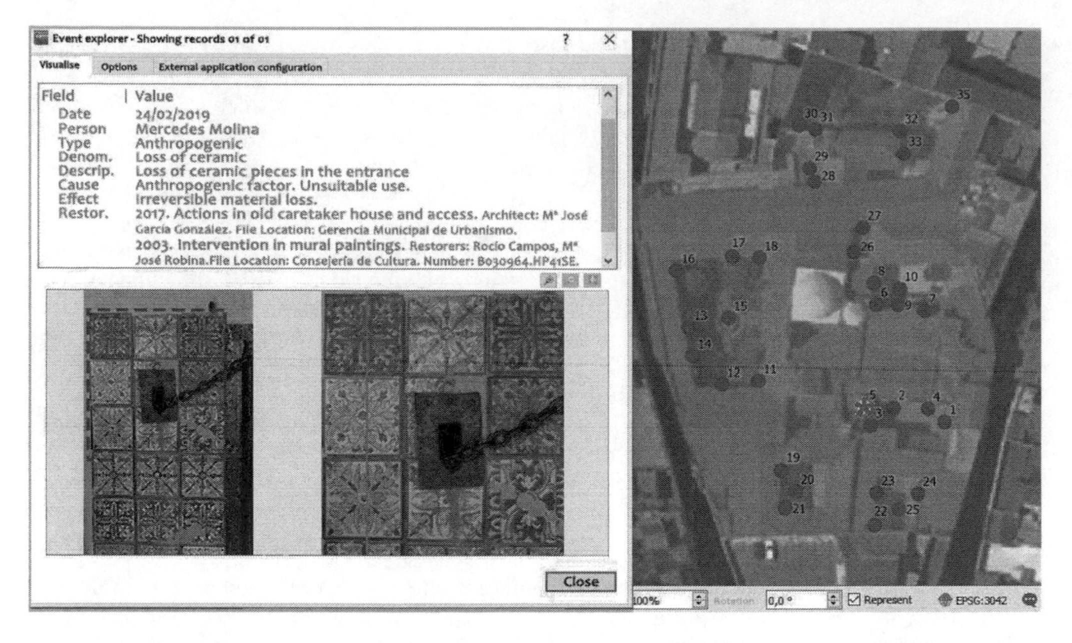

Figure 1. QGIS applied to the study case of the santa inés convent, Seville. The authors.

architecture of the convent. By entering the data in the layer that corresponds to the current situation, it gets referenced.

Graphically, the current convent space is highlighted with a shadow. As we can see in Figure 1, red dots are introduced, thus marking the position of the damages detected. Each point contains attributes of its own. Among the information described in each point, fields on the table are shown. The main advantage of using attributes is that they allow us to select any damage according to its interest or characteristics. It is also possible to add photographs to their description. However, this software is not yet intended as a photography administrator so, for its classification, is used another application, in this case, the software of Adobe Lightroom, given its usefulness in creating image maps in combination with Google Maps.

As the last step, the processing of the data takes place. All information incorporated into the program is represented graphically by point planimetry or tables with photographs. At the same time, we can make use of different queries and thus, instantly, get the selection of points according to the chosen parameter, year, damage, extension, etcetera. At each point, the user can open all the related data. The program allows an accurate, simple, and effective way of seeing the state of conservation of as many points as were added.

3 RESULTS

There are many advantages related to the use of Geographic Information Systems in historical heritage researches. This is due to the fact that these types of tools allows the collaborative and interdisciplinary work in current interventions. In this research, we have seen the interest in using this type of software, in this case free code, in the conservation discipline and cultural heritage restoration. The main reason is because it extracts important data from technical visits: injuries recognition, deterioration factors identification, establishment of possible causes, etcetera. The technical information obtained is materialized in this type of software in a fast and accessible way, facilitating the work of specialist technicians. In this way, it is possible to georeference any type of damage or injury to the building, being able to add complementary data and images that support the diagnosis issued.

The information obtained in the technical inspection must be complemented with an exhaustive knowledge of the previous interventions that have taken place. These data respond to pathologies that may affect the current state of the convent conservation. In this sense, the Geographic Information Systems can also be used as great data managers. They detail the right location of the previous intervention project and the architect or conservator-restorer who has executed it.

The main results of the use of GIS are focused on the diagnosis of the conservation in Sevillian convents. In this case, in the Convent of Santa Inés. These results have been able to georeference each one of the lesions in all the spaces, as well as to add the technical data obtained in the technical visits and archives.

On an urban scale, gathering this information in the software facilitates the knowledge of the conservation urgency that exists in Sevillian convents. This allows the management of the conservation of the buildings in the face of the measures type be adopted, the prioritization of the intervention and even the planning before possible investments.

REFERENCES

Consejería de Agricultura, Ganadería, Pesca y Desarrollo Sostenible. Junta de Andalucía. 2005. Zonas verdes públicas de las capitales de provincia de Andalucía y los municipios mayores de 100.000 habitantes a escala 1:5.000. Sevilla: Red de Información Ambiental de Andalucía. Retrieved from http://www.juntadeandalucia.es/medioambiente/mapwms/REDIAM_Zonas_Verdes_Publicas?

Consejería de Cultura. Junta de Andalucía. Delegación Provincial de Sevilla. 2004. Monasterio de Santa Inés. Proyecto básico y de ejecución. Restauración del zócalo de azulejos del claustro principal. CPPH 01/12/04. 557.

Consejería de Economía, Conocimiento, Empresas y Universidad. 2019. Infraestructura de Datos Espaciales de Andalucía (IDEA). Retrieved from http://www.ideandalucia.es/portal/herramientas/descargas;jsessionid=3671F77DB6685FBFE6F77FE88D436C09.

Gerencia de Urbanismo. Ayuntamiento de Sevilla. 2017. Catálogo General del Patrimonio Histórico Andaluz en Sevilla. Seville: IDE. Sevilla. Retrieved from http://sig.urbanismosevilla.org/jsapi/ideS/SocialMediaViewer/index_BIC.html?webmap=e18ed1ee72694170bd0788245c70f57a&showAboutDialogOnLoad=true.

López Heredia, M., & Márquez Pérez, J. 2013. El papel de la tecnología SIG en la investigación sobre Patrimonio Cultural. In *Compartiendo el patrimonio: Paisajes culturales y modelos de gestión en Andalucía y Piura*: 91–104. Retrieved from https://dialnet.unirioja.es/servlet/extart?codigo=4421973.

Mascort-Albea, E. J. 2017. Datos geográficos abiertos para la conservación preventiva del patrimonio arquitectónico. *PH: Boletín Del Instituto Andaluz Del Patrimonio Histórico*, *25*(92): 228–229. Retrieved from https://dialnet.unirioja.es/servlet/articulo?codigo=6205497&orden=0&info=link.

Molina-Liñán, M., López-Bravo, C. & Mosquera Adell, E. 2018. El Convento en una nueva sociedad: Transformaciones y nuevos usos. El Caso De Los Conventos Sevillanos. In *XI Congreso Internacional AR&PA: El papel del Patrimonio Cultural en la construcción de la Europa de los Ciudadanos*. Valladolid, 8-10 November 2018.

Pérez de Lama Halcón, J. L. 1996. Biografía del patio mediterráneo. Permanencia y cambio en la historia del mediterráneo. Seville: Escuela Técnica Superior de Arquitectura.

Science and Digital Technology for Cultural Heritage – Ortiz Calderón et al. (Eds)
© 2020 Taylor & Francis Group, London, ISBN 978-0-367-36368-0

Graphic instruments for visualizing hypotheses

D. Lengyel
Brandenburg University of Technology, Cottbus-Senftenberg, Germany

C. Toulouse
Lengyel Toulouse Architects, Berlin, Germany

ABSTRACT: Within the last decade we have developed a new digital graphic instrument for knowledge mediation and knowledge development of archaeological hypotheses concerning architectural topoi. The key feature of archaeological hypotheses is the varying degree of certainty ranging from undoubtful findings over most probably deductions to vague yet scientifically based assumptions. For illustrations, the discipline of archaeology has set up a number of analogue, two dimensional drawing techniques, that represent the range of certainty, but also three dimensional diagrams preferably with color-shaded volumetric elements. Drawn perspectives provide spatial impression. Our method consists of the two complementary parts Virtual Modeling and Virtual Photography. Other than usual, we do not consider the spatial model as the decisive key in mediating archaeology but only as an integral part of the visual mediation. Instead, we consider its counter part, the virtual photography, the re-translation of three dimensional data into visually perceivable information, as at least equally important.

1 INTRODUCTION

The mediation of archaeological knowledge towards the scientific community as well as towards the public is usually based on scientific, written text accompanied by imagery that – on its turn – tends to be perceived even preferentially.

Drawn perspectives by archaeologists traditionally provide spatial impressions, but neglect the advantages of digital models, whereas three dimensional diagrams usually serve as data base access without providing an architectural impact by their spatiality. The usual contemporary imagery, that is based on three dimensional models, typically used in public mediation of archaeology, and that at the same time tries to provide an immersive spatial impression, usually appears as pseudo-realistic simulations, that cause the high risk of misunderstandings due to their overwhelming amount of additions based on pure phantasy.

2 VISUALIZING HYPOTHESES

2.1 *Scientific content and spatial impression*

To solve this ostensive dilemma, for architectural topoi we have developed a method that explicitly respects the scientific content and still offers an immersive spatial impression. Contrasting the geometric abstraction of the model, that is strictly based on the verbal hypotheses, our way of depicting the hypotheses uses, transforms and re-arranges traditional methods of realistic architectural photography. By this referring on visual traditions our imagery enables its recipients to immediately engage with the content – as both archaeology and architecture have always dealt with and visually mediated spatial structures – instead of adapting an uncommon visual language.

2.2 *Design of abstraction and virtual photograhpy*

The major task is architectonic design in two steps. The first step is the creation of abstract representations of the hypothetical architectonic entities in different scales from buildings viabuilding parts to ornamental details. Every single element undergoes an individual design process until the cooperating scientist attest at least a state that we call free from contradiction. The second, complementary step is the projection of the model, the depiction, or the "virtual photography" as we call it according to its intention and importance. It means, that the abstract geometry is treated as if the architecture as well as the photographer were real, and it underlines the importance of the photographic design in our visualisation of hypotheses.

Both tasks are not trivial as they cannot be fully formalized. Their intention can be described, and there are a number of targets that can be achieved accordingly. But there is still a critical need for design, the search for a formal composition, a synthesis of diverging necessaries.

2.3 *Proximity to verbal hypotheses*

The proximity of a visualisation to its verbal hypothesis is indispensable for letting them act as mediators of science. This influences, among others, the decision of black and white or coloured images. As colours always win over geometry, in cases where the certainty in knowledge concerning the geometry outstands the knowledge about the singles colours, the scientists insisted on monochromatic representations. If the knowledge suffices, colours are applied (Hardering 2016).

The proximity to the hypotheses also concerns the need for a photography that despite its design in composition aims at a projection that does not distort the geometry but makes it unambiguous and reliable to ensure its appropriate interpretation. Unambiguity in eye level means a clear distinction of human eye level or bird's eye view. Unambiguity in tilt means either a horizonal view direction – leading to a vertical image plane leading to a vertical depiction of vertical building edges – or a clear tilt view used for example as a view into a steep spiral stair case.

3 PROJECTS

All depicted projects were developed in close cooperation with archaeologists in public scientific cultural institution, Cologne Cathedral additionally accompanied by art historians:

3.1 *Metropolis of Pergamon*

The project of the metropolis of Pergamon (Figure 1 and 2, by Lengyel and Toulouse, Laufer et al. 2011) including 3D Digital image acquisition for the re-contextualisation of 3D scanned sculptures in their hypothetical original architectonic context was funded by and elaborated in cooperation with the German Research Fund Excellence Cluster TOPOI. Its last monographic exhibition took place in Leipzig in 2018 as part of Sharing Heritage, the European Cultural Heritage Year 2018. Our visualisations are published in major scienfitic architectural works about Pergamon as well as in the main museum catalogue of the antique collection of the Pergamon Museum Berlin and in a survey of research in museums by the German Ministry of Culture and Science (Hauser et al. 2012).

Since 2008 scientists in archaeology and building archaeology work with the geometric accuracy of this first three-dimensional overall model of the metropolis. The model does not only evaluate spatial relations, it also reveals unforeseen visual phenomena. The re-contextualisation of antique sculptures furthermore allowed for the first time to visually present and prove the positioning of a number of antique sculptures that were exclusively scanned three-dimensionally for this project.

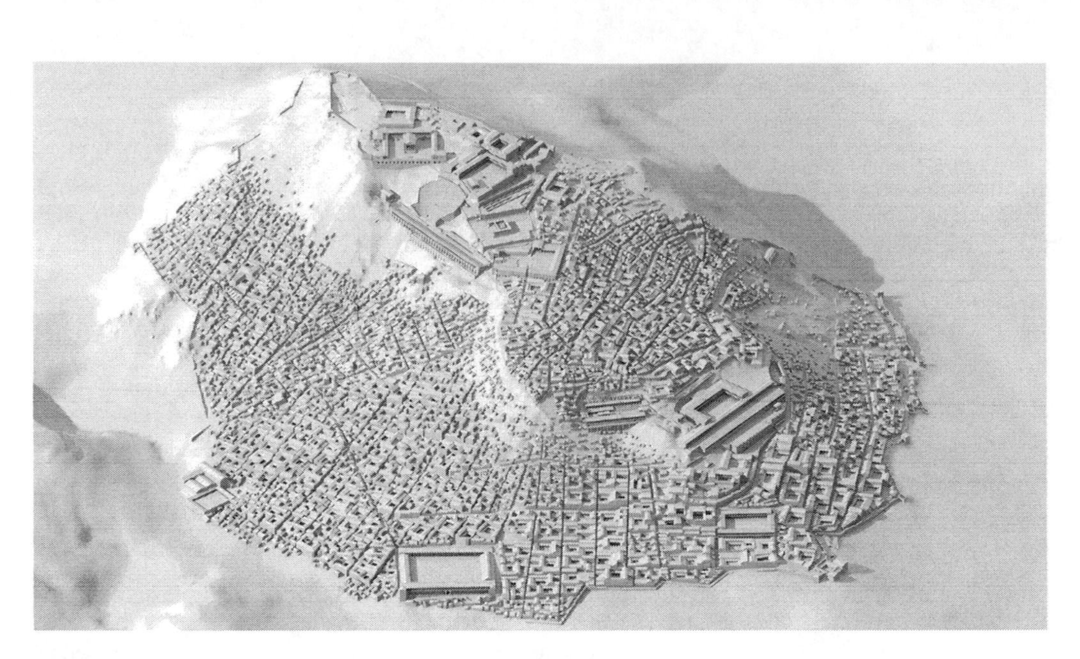

Figure 1. Bird's eye view of Pergamon around 200 AD.

Figure 2. Sanctuary of Athena in Pergamon around 200 AD with Dying Gaul and Ludovisi Group.

3.2 *Palatine Palaces in Rome*

The project of the Palatine Palaces in Rome (Figure 3 and 4, by Lengyel Toulouse Architects, Märtin & Wulf-Rheidt 2012) was funded by and elaborated in cooperation with the German Archaeological Institute in Berlin and was first exhibited in the Pergamon Museum Berlin in the major exhibition of the Excellence Cluster TOPOI – a cluster lead by the Freie University Berlin, Humboldt University Berlin, German Archaeological Institute, Zuse Institute Berlin and others, funded by the German Science Foundation DFG – and in many other exhibitions since.

Figure 3. Bird's eye view of the Palatine Palaces in Rom under Emperor Maxentius.

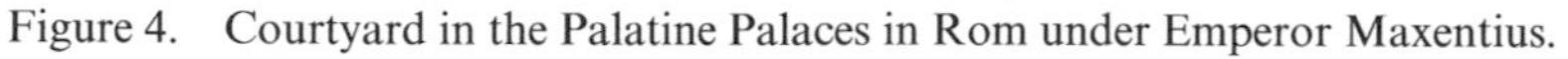

Figure 4. Courtyard in the Palatine Palaces in Rom under Emperor Maxentius.

3.3 *Cologne Cathedral and its Predecessors*

The project of Cologne Cathedral and its Predecessors (Figure 5 and 6, by Lengyel Toulouse Architects, Lengyel et al. 2011) was funded by and elaborated in cooperation with the cathedral hut. It was first exhibited in the major annual archaeological exhibition of the federal state Nordrhein-Westfalen and has been exhibited in the entrance area of the archaeological zone in the cathedral since 2011. In 2012 an illustrated book enriched the film's information by plans and sections of the building phases and an explanatory text by the Cathedral architect of

Figure 5. Cologne Cathedrals' predecessor around 1025 AD.

Figure 6. Cologne Cathedral around 1540 AD with its iconic construction crane.

the time, Prof. Dr. Barbara Schock-Werner. On the occasion of the 150th anniversary of the Cathedral's choir ensemble, we also visualized the interior of the building's choir at its state around 1856, that is shortly before the demolition of the wall that separated the choir from the future nave (Hardering 2016). The visualisations have been used in other publications of other authors for illustrative purposes (Dietmar et al. 2011) as well as for questions of authenticity (Rheidt 2017). The diagrammatic character of the changes of the urban context of the churches lead to a dedicated treatise about diagrammatic architecture (Lengyel et al. 2013).

REFERENCES

Dietmar, C. & Trier, M. 2011. COLONIA - Stadt der Franken. Köln vom 5. bis 10. Jahrhundert. 2. Aufl. Köln: DuMont.

Hardering, K. 2016. Die Binnenchorausstattung des Kölner Domes im 19. Jahrhundert. In Müller-Oberhäuser, C. & Klösges, S. (eds), *Die Musikaliensammlung Leibl. Neukatalogisierung der Musikalien der Kölner Domkapelle des 19. Jahrhunderts*: 66–77. Köln: Erzbischöfliche Diözesan- und Dombibliothek.

Hauser, C. & Loch, M. 2012. Museen. Forschung, die sich sehen lässt. Stand: Mai 2012. Bonn: BMBF, Referat Geistes-, Sozial- und Kulturwiss., Akad., Forschungsmuseen (Forschung).

Laufer, E. & Lengyel, D. & Pirson, F. & Stappmanns, V. & Toulouse, C. 2012. Die Wiederentstehung Pergamons als virtuelles Stadtmodell. In Grüßinger, R. & Kästner, V & Scholl, A. (eds), *Pergamon. Panorama der antiken Metropole*: 82–86. Petersberg: Michael Imhof Verlag.

Lengyel, D. & Schock-Werner, B. & Toulouse, C. 2011. *Die Bauphasen des Kölner Domes und seiner Vorgängerbauten. Cologne Cathedral and preceding buildings*. Köln: Verlag Kölner Dom.

Lengyel, D. & Toulouse, C. 2013. Die Bauphasen des Kölner Domes und seiner Vorgängerbauten. Gestaltung zwischen Architektur und Diagrammatik. In: Dietrich Boschung und Julian Jachmann (Hg.): Diagrammatik der Architektur. Paderborn: Fink (Morphomata, 6), S. 327–352.

Märtin, R.-P. & Wulf-Rheidt, U. 2012. Räume verwandeln. Der Palatin in Rom. In Exc. Cluster Topoi (eds), *Jenseits des Horizonts. Raum und Wissen in den Kulturen der Alten Welt*: 12–23. Stuttgart: Theiss.

Rheidt, K. 2017. Authentizität als Erklärungsmodell in der Baugeschichtsforschung. In: Christoph Bernhardt, Martin Sabrow und Achim Saupe (Hg.): Gebaute Geschichte. Historische Authentizität im Stadtraum. Göttingen: Wallstein Verlag, S. 78–93.

Science and Digital Technology for Cultural Heritage – Ortiz Calderón et al. (Eds)
© 2020 Taylor & Francis Group, London, ISBN 978-0-367-36368-0

Geographic information systems (GIS) and building information modeling (BIM): Tools for assessing the cultural values of mass housing neighbourhoods in medium-sized cities of Andalusia

D. Navas-Carrillo
Department of Urbanism and Regional Planning. University of Seville, Seville, Spain

J.C. Gómez de Cózar
Department of Architectural Constructions. University of Seville, Seville, Spain

M.T. Pérez Cano
Department of Urbanism and Regional Planning. University of Seville, Seville, Spain

ABSTRACT: This paper focuses on the urban growth experienced in the third quarter of the 20th century that was due to the generalized deficit of housing which characterized European cities mainly during the 20th century. This period was a time of greater growth of many cities and, therefore, one of higher architectural and urban production of its recent urban history. However, they are goods that, in most cases, lack generalized heritage recognition and, consequently, do not have comparable levels of protection to other residential types. These urban groups have been exhaustively studied in large cities, and research has found a gap in the analysis of intermediate ones. This work recognises social housing neighbourhoods in medium-sized cities as an asset to be protected and defends the generation of knowledge as a necessary resource for the cultural values assessment. Accordingly, GIS and BIM present as appropriate tools for clearly addressed a multi-scale analysis. They have allowed us to efficiently record and graphically represent the quantitative and qualitative heritage analysis carried out.

1 INTRODUCTION

The specific characteristics of the residential neighbourhoods of the second half of the 20th century make the application of the criteria most often used in heritage protection unviable. Its uniqueness within the architectural and urban production needs to define, from a contemporary perspective, the specific heritage values by which they must be recognised as a good to be protected. An important effort has been made to recognize mass housing neighbourhoods thanks to the elaboration of the records of the architecture of the Modern Movement (Pérez Escolano & Fernández-Baca Casares, 2012). However, it is based mainly on objectual considerations, based on its formal characteristics and, therefore, is closer to an outdated approach to the heritage event from the point of a monumentalist view of the 19th century (Mosquera Adell y Pérez Cano, 2011:404). Despite, this contrasts with the conceptual change of the notion of heritage the advance and breadth of the contemporary notion of heritage produced in the last decades, shifting the tutelary attention towards the subject that demands it (Castillo Ruiz, 2004). This fact consequently makes daily architectures like social housing excluded from the usual processes of legal protection. In this sense, this work aims to take a significant step in the process of heritagization of these urban pieces in cities of intermediate scale.

There are many authors who have study the construction of social housing in Spain from various disciplines and approaches (Moya González, 1983; Gaja Díaz, 1989; Ferrer i Aixalà,

1996; Sambricio, 2008; Gutiérrez Mozo & Caro Gallego, 2015; Monclús & Díez, 2015; Queiro Quijada, 2016). However, a range of cities that have not yet been studied. In this sense, the research focusses on the medium-sized coastal cities of Andalusia, a territorial domain with a high urbanisation index (Navas-Carrillo, et al., 2017). Although these cities have a lower number of promotions than regional centres, their strategic position within the territorial structure of the region also stands as recipients of these migration processes.

2 METHODOLOGY

The research seeks to make a sequential approach that addresses, from the general —the construction of the conceptual, theoretical and legislative frame that conditioned the construction of these urban complexes— to the particular, the analysis of the characteristics that define the social housing complexes of Sanlucar de Barrameda and that have resulted in its current urban configuration. For this purpose, GIS and BIM has been used as appropriate tools. The trajectory in the use of both tools and their usefulness in heritage studies has been analysed through authors such us Tobiáš (2016), Pinto Puerto, et al. (2017) or López et. al. (2018).

According to research goals and the possibilities of these tools, it has been started analysing the territorial structure to identify the network of medium-sized Andalusian coastal cities according to the regional land plan, namely POTA -Plan de Ordenación del Territorio de Andalucía- (Junta de Andalucía, 2006). Then, the population growth experienced by these cities during the third quarter of the 20th century has been studied using the historical series of the Municipal Register developed by the Institute of Statistics and Cartography of Andalusia (IECA). The third aspect analysed was the number of public and private housing built in this period of time. This information was obtained from the existing database at the Central Archive of the Ministry of Public Works. Finally, we have identified which of the founding cores of these cities have been declared as monuments. (Figure 1).

The following step has been the characterization of the set of residential neighbourhoods built between 1939 and 1985 in Sanlucar de Barrameda as the representative case of the whole sample concerning the construction of public housing between 1950 and 1980. The analysis of the whole of medium-sized cities in Andalusian Coast has made it possible to detect the importance of the total number of dwellings developed in Sanlucar (aprox. 2000 units), but also the number and their scale. A process of systematization and categorization of the data has been developed (Figure 2). Information that includes general information based in archive data as architects, location, developers, size and density, or the housing policies that regulated its construction. As well as specific information based in personal evaluation, such as the

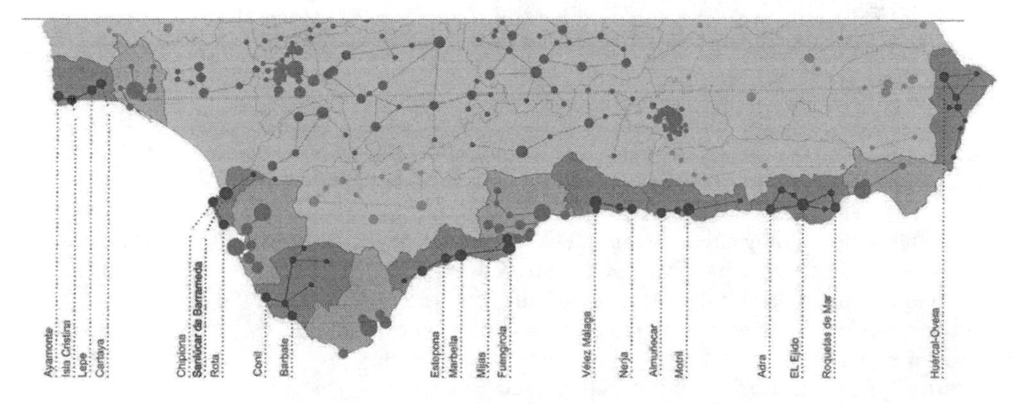

Figure 1. The medium-sized Andalusian coastal cities. Source: Compiled by the authors based on POTA, IECA and the central archive of the ministry of public works.

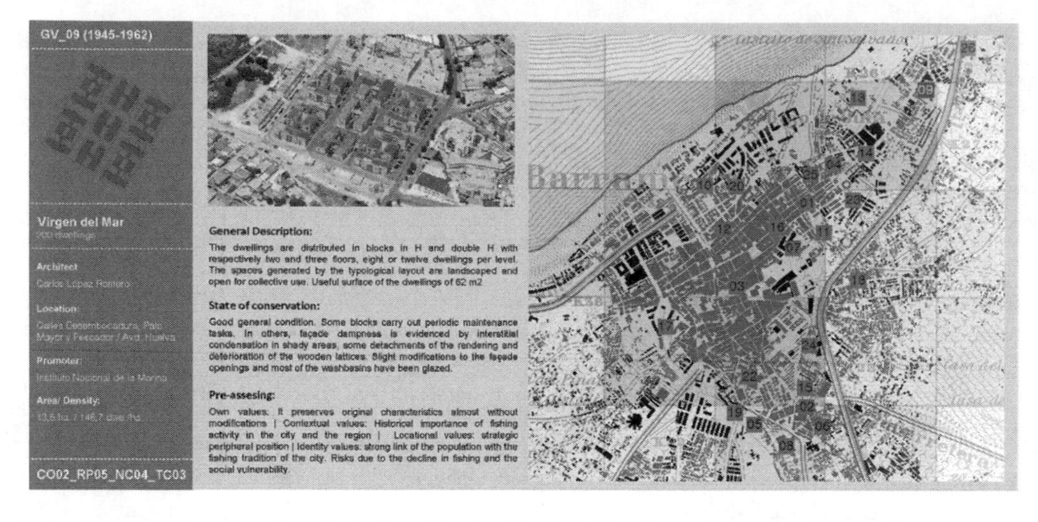

Figure 2. Neighbourhood analysis sheet. Source: Compiled by the authors using QGIS software.

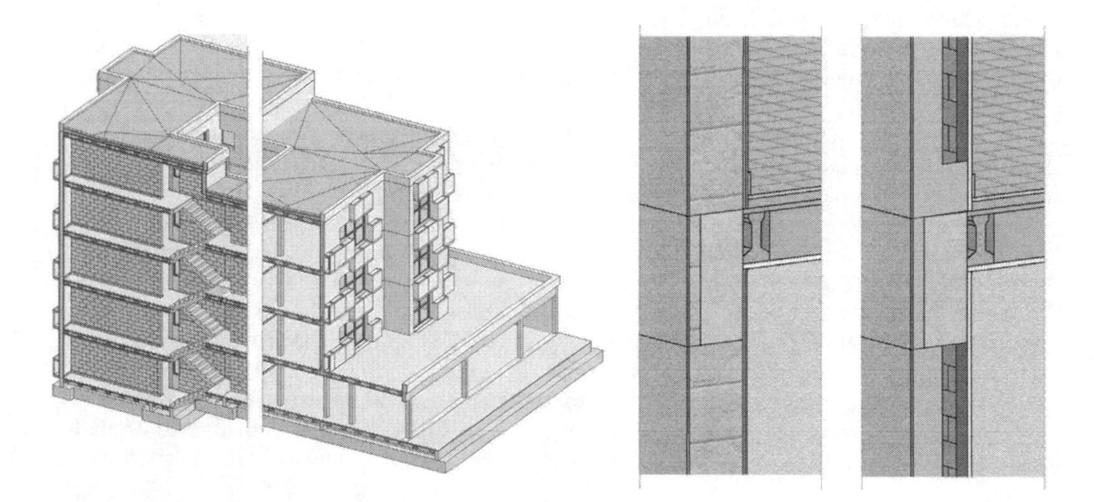

Figure 3. Modifications in the structural and construction systems. 3D Model and façade section detail. Source: Compiled by the authors using Autodesk Revit Software

architectural and urbanistic description of the neighbourhoods, the evaluation of the general state of conservation, or a cultural value pre-assessment based in previous works (Navas-Carrillo, et al. 2018, 2019). This multi-scale study has been clearly addressed using QGIS.

Ultimately, the typological and constructive solutions of the most representative neighbourhoods have been analysed by means of Autodesk Revit Software (figure. 3). The study started with the modelling of each type according to the requirements for LOD200. The ease of modifying the models has allowed us to register the evolution experienced by these types. This analysis recorded since the first designs to the last interventions.

Figures 3. Modifications in the structural and construction systems. 3D Model and façade section detail. Source: Compiled by the authors using Autodesk Revit Software

3 RESULTS AND CONCLUSIONS

These neighbourhoods are the result of the modes of urban production that have characterised the evolution of 20th-century European urbanism, and therefore they must be valued as part of the historical legacy of cities according to the Recommendation on the Historic Urban Landscape (UNESCO, 2011). They could also have an original or singular value due to their experimental character: functional, technical or social. This is related to the scientific or technical value in terms of the importance and relevance that their conception has had for society (UNESCO, 2004). They still retain the primitive residential function. This fact makes them have added values, on the one hand, the value of use, assumed as a criterion to evaluate the effort of conservation (Council of Europe,1975), and on the other hand a functional value. In the case under analysis, many of them have a strategic position in the current urban fabric: between historic centres and the latest peripheral developments, or in urban boundaries close to nature. Compared to other architectural typologies of the 20th century, social housing lacks acceptance on the part of the general public (Hernández, 2011:70). However, it is also recognised that there are social values that are intrinsic to some neighbourhoods. These values play an essential role in establishing the social and cultural identity of the population that live there. Values that will condition the interaction between its residents and will generate in them a relevant feeling of belonging to the place and the collective that inhabits it.

The work has demonstrated that the relevance of these urban groups lies not only in the fact that they are the most significant operations of the architecture of the Modern Movement in their respective cities, but also in the analysis of these achievements in their political, economic and social context, linked to the places where they are implemented and, especially, to the societies that inhabit these neighbourhoods. Intangible values that have a substantial impact on heritage conservation because they allow the construction of resilient societies that guarantee the survival of these groups.

In this research, GIS and BIM have been used as an independent recording and representation tools for the quantitative and qualitative heritage analysis carried out at different scales. The integrated use of these tools should, therefore, be considered a future line of research to work on.

ACKNOWLEDGMENTS

The authors are grateful for the financial support from the Research Plan of the University of Seville

REFERENCES

Castillo Ruiz, J. 2004. Los fundamentos de la protección: el efecto desintegrador producido por la consideración del Patrimonio Histórico como factor de desarrollo. *Patrimonio cultural y derecho*, 8: 11–36.

Council of Europe. 1975. *European Charter of the Architectural Heritage*. Amsterdam: Congress on the European Architectural Heritage.

Ferrer i Aixalà, A. 1996. *Els poligons de Barcelona*, Barcelona: Universitat Politècnica de Catalunya.

Gaja Díaz, F. 1989. *La promoción pública de la vivienda en Valencia (1939-1976*). Valencia: Consejería de Obras Públicas, Urbanismo y Transportes.

Gutiérrez Mozo, E.; Caro Gallego, C. (2015). La arquitectura de la obra sindical del hogar en la ciudad de Albacete: 1941-1981. *Al-Basit: Revista de estudios albacetenses*, 60: 123–170.

Hernández, F. 2011. Patrimonio arquitectónico y sociedad en América Latina. In: Hernández León, J.M; Espinosa de los Monteros, F. (coord.). *Criterios de Intervención en el Patrimonio Arquitectónico del Siglo XX*. Ministerio de Cultura. Madrid.

Junta de Andalucía, 2006. *Plan de Ordenación del Territorio de Andalucía*. Sevilla: Consejería de Obras Públicas y Transporte.

López, F.J.; Lerones, P.M.; Llamas, J.; Gómez-García-Bermejo, J.; Zalama, E (2018). A Review of Heritage Building Information Modeling (H-BIM). *Multimodal Technologies and Interact*, 2 (21): 1–29.
Monclús Fraga, J.; Díez Medina, C. (2015). El legado del Movimiento Moderno. Conjuntos de vivienda masiva en ciudades europeas del Oeste y del Este. *Rita*, 3: 88–97.
Mosquera Adell, E.; Pérez Cano, M. T. 2011. Refugios Conocidos. de Patrimonio de los Arquitectos a Patrimonio de Todos. In Hernández León, J.M; Espinosa de los Monteros, F. (Ed.). *Criterios de Intervención en el Patrimonio Arquitectónico del Siglo XX*. Madrid: Ministerio de Cultura.
Moya González, L. 1983. *Barrios de promoción oficial, Madrid 1939-1976: la política de promoción pública de vivienda*. Madrid: Colegio Oficial de Arquitectos de Madrid.
Navas-Carrillo, D.; Pérez Cano, T.; Rosa Jiménez, C. 2017. Mass Housing Neighbourhoods in Medium-Sized Andalusian Cities. Between Historic City Centres and New Peripheral Developments. *IOP CS: Materials Science and Engineering*, 245: 1–12.
Navas-Carrillo, D.; Gonçalves, M.; Pérez Cano, T. 2018. Los conjuntos de vivienda social como legado de la modernidad. In: Monclús, J.; Diez Medina, C. (Ed.) *Ciudad y forma: perspectivas transversales*. Vol. 1. Zaragoza: Prensas de la Universidad de Zaragoza.
Navas-Carrillo, D. 2019. Más allá de la connotación franquista. Una reflexión sobre la patrimonialización de las barriadas promovidas por el Instituto Nacional de la Vivienda. *Boletín PH*, 96: 246–248.
Pérez Escolano, V.; Fernández-Baca Casares, R. 2012. *Cien años de arquitectura en Andalucía: el Registro Andaluz de Arquitectura Contemporánea, 1900-2000*. Sevilla: Instituto Andaluz del Patrimonio Histórico. Junta de Andalucía.
Pinto Puerto F., Angulo Fornos R., Castellano Román M., Alba Dorado J.A., Ferreira Lopes P. 2018. Using BIM and GIS to Research and Teach Architecture. In: Castaño Perea, E.; Echeverria Valiente, E. (Ed.) *Architectural Draughtsmanship*. Springer.
Queiro Quijada, R. 2016. *Patronato Municipal y Real Patronato de Casas Baratas de Sevilla. Aportaciones a la conformación de la ciudad a través de la vivienda social: 1913–1986*. PhD. Universidad de Sevilla.
Sambricio, C. 2008. *100 años de historia de la intervención pública en la vivienda y la ciudad*. Madrid: Asociación Española de Gestores Públicos de Vivienda y Suelo.
Tobiáš, P. 2016. BIM, GIS and semantic models of cultural heritage buildings. *Geoinformatics FCE CTU* 15(2): 27–42.
UNESCO 2004. *Algunas reflexiones sobre autenticidad*. Paris: United Nations Educational, Scientific and Cultural Organization.
UNESCO. 2011. *Recomendación sobre el paisaje urbano histórico*. Paris: United Nations Educational, Scientific and Cultural Organization.

Science and Digital Technology for Cultural Heritage – Ortiz Calderón et al. (Eds)
 ISBN 978-0-367-36368-0

Dynamic identification for structural analysis of Santa Ana church in Seville

J. Aguilar Valseca, M.T. Rodriguez-Leon, P. Pachon Garcia, E. Vazquez Vicente & V. Compan Cardiel
Building Structures and Soil Engineering Department, Universidad de Sevilla, Seville, Spain

ABSTRACT: Santa Ana Church (1.266) is one of the main milestones in the heritage of Seville for being the first Christian church built outside the city walls and on the other side of the river, after the Reconquest of the city. The singularity of the geometry of its vaults together with its execution in brick and stone masonry make complex to determine the properties of materials. The impossibility of characterizing these materials by means of destructive tests leads to the application of other technologies that allow us to approach the behavior of the numerical models to the reality performed.

This paper tries to adjust the mechanical properties of the materials that constitute the structure of the church of Santa Ana, which are brick and stone masonry, through analyzing its dynamic behavior with environmental vibration (EVT) techniques and modal operations analysis (OMA). After having carried the campaign out using the methods of modal identification in the stochastic subspace and in the frequency domain, four vibration modes are obtained that adjust to the modal behavior of the current building, within a frequency range between 2 and 4 Hz.

This method adjusts a series of imperfections that the numerical model does not contemplate. Usually, the geometry has many ornaments that can modify the structural behavior. The boundary conditions are other properties very difficult to evaluate in a heritage building or another pathologies as cracking or humidity. OMA let us to approach the mechanicals parameters of a numerical model to adjust the behavior to the real structure

Initially, and thanks to new techniques such as photogrammetry, it has been possible to develop a finite element model (FEM) to analyze the structure with theoretical parameters of its materials.

With this model and through the environmental vibration technical it is possible to conclude with the suitability of the application of a non-destructive method in heritage buildings to extract the mechanical properties of the materials that constitute its structure through its dynamic behavior.

1 INTRODUCTION

Santa Ana church in Seville (*Figure 1*) is the first Catholic Church built in the city after the conquest (*Figure 2*). Its location outside the wall of the city makes the building work both as a Catholic temple and as a fortification because of having been built near the Boats Bridge and the Aljarafe zone.

The primitive building has influences from Cordoba and Burgos. It has 34 long rods, the central space is bigger than the laterals and the proportional relation between its spaces is around $\sqrt{2}$ (*Figure 4*). This proportional relations and the plane roof built due to possible battles make the vaults to be at the same height, and their geometry by imposition surfaces gets a better structural behavior. The primitive plant is shown without any indication of the existence of a crossing, and it has three naves separated by pillars that, in height, become pointed arches closed by a false triforium where the nerves of the vaults begin. Longitudinally, the vaults are crossed by a spine as it occurs in the Cistercian tradition. [6]

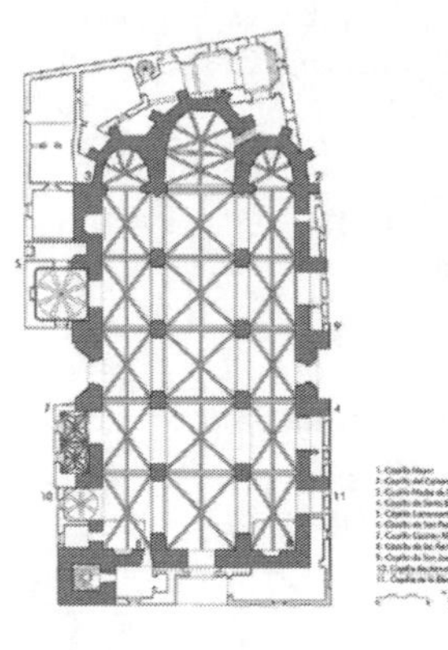

Figure 1. Santa Ana Church. Author's drawing.

Figure 2. Author's Sketch.

Figure 3. Brick and Stone Masonry.

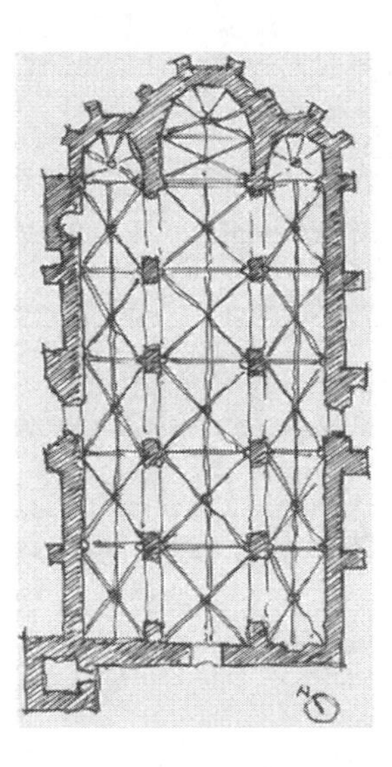

Figure 4. Primitive Building. Author's Sketch.

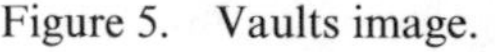

Figure 5. Vaults image.

The main building was built in brick and stone masonry, whose mechanicals properties are difficult to define for being non-homogenous and isotropic materials. Determining these mechanicals properties is going to be one of the main objectives in this work.

After the primitive construction, the building had numerous additions [6, 16], with the construction of small chapels that are annexed to the Epistle and the Gospel Naves. The tower suffered some interventions too.

From a structural point of view, we analyze those constructions that could condition the building behavior, depending of its dimensions and its disposition in the main building. These little Chapels could work as buttresses of the main building.

We analyze the most important little Chapels [15]:

- 1.507. Santa Bárbara Chapel. In the fourth section of the Epistle Nave between the buttresses of the primitive building. *(Figure 1.* Santa Ana Church. Author's drawing).
- Mid. s. XVI. In the second section of the Gospel Nave there is the Sacramental Chapel, built with square plant and cover with a hemispherical dome. (*Figure 1).*
- 1.587. Captain Monte Bernardo Chapel. In the last Gospel Nave section, built in a rectangular floor and covered with a Terceletes Vault. *(Figure 1.* Santa Ana Church. Author's drawing).
- Saint Joaquin. End of s. XVI. It is identical to the Santa Barbara Chapel and it is located in the second section of the Epistle Nave. (*Figure 1*)
- 1.614-17. The Baptismal Chapel is located in the last section of the Gospel Nave (*Figure 1*).
- In the last Section of the Epistle Nave, it is located the Divina Pastora Little Chapel. There is not information about its construction but it is possible to have been built around the year 1.865. (*Figure 1)*

2 OBJECTIVES

Traditionally, the classical theory of masonry buildings establishes premises or hypothesis to their study, such as the zero tensile strength of the masonry, low-tension levels and the impossibility of slippage failure between elements [4, 5]. This analysis is limited to force lines that work in a flat element. Another aspect considered is the safety of these buildings, which is established based on their geometry regardless of size [2]. For this work, numerical calculation models based on the finite element method (FEM) have been used. With this method, it is possible to analyze singular masonry structures such as Santa Ana Church, considering a three-dimensional behavior.

This work pretends to characterize the mechanical properties of the building's masonry by an experimental campaign to determine the dynamic properties of the building from the application of environmental vibration tests and modal identification of its structure (OMA). This type of technique is presented as a useful tool in the study of heritage building [12].

From a numerical model (*Figure 6)* get using the finite element method (MEF), we try to establish a relationship between the behaviors with real model for its validation. This work

Figure 6. Finite Elements Model.

Figure 7. Accelerometers.

will let us to do an advanced analysis to evaluate the structural damage and with so, calibrate the initial finite elements model.

As previously mentioned, the primitive building has suffered multiples additions and interventions through the construction of several Chapels next to the original building. The connection of these chapels to the main building is unknown and therefore the behavior of the actual building working as a whole could be questioned. Currently, there are no symptoms which show that the constructions could be disconnected from the main building (cracks or fissures), so the behavior could be considered as a whole. However, and considering the results that we have obtained, if these discontinuities could exist (micro breakage), they would not modify the dynamic results obtained based on the dynamic behavior analyzed.

The continuity and discontinuity could be analyzed as an Ultimate Limit State (ULS). The experimental campaign studies the dynamic behavior as a Service Limit State (SLS). After the first experimental campaign, the results show that it is possible to considerate the whole building as a unique structure due to the low level of excitation. In future campaigns, we will try to do an advance analysis with accelerometers located on the little Chapels' roofs.

3 METHODOLOGY

In order to get the experimental dynamic properties of the church, we need to previously make a finite element model (*Figure 6*). In order to do so, we make a traditional draw of the building and use photogrammetric software (*Figure 9*) to get the geometrical properties (vaults and nerves). We use Agisoft Photoscan [11] to elaborate the points cloud digital model. In order to do so we used a Laser Rugby 640G for levelling references targets and a Nikon D5000 for taking several photos in order to define the digital model. Once we have the digital model, we have to scale it to take measurements of the columns, nerves and vaults in order to define the numerical model geometry.

Before elaborating the numerical model, is necessary to do a geometrical one. In order to do this, it was necessary to work with Autocad and Rhinosceros softwares to generate complex geometries supported on photogrammetric works results.

Based on the geometric model and using Ansys ICEM to generate the finite element mesh, it is possible to analyze the structure with the ABAQUS CAE 6.13 [15]. In this software, the material properties are defined using solid elements for walls, columns, nerves and vaults, as

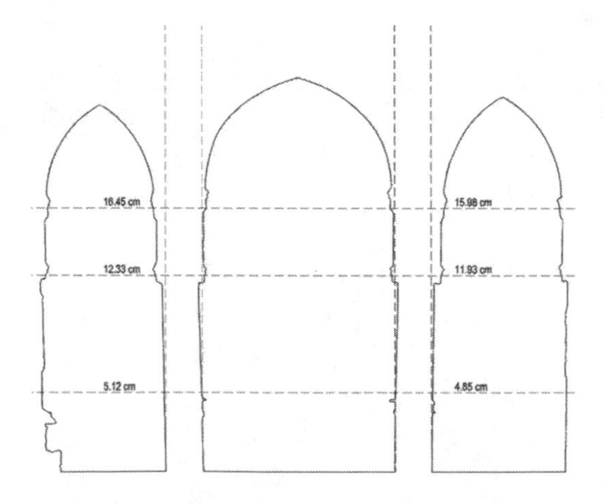

Figure 8. Displacements.

Figure 9. Columns Photogrammetric Job.

Table 1. Theoretical Mechanical Properties.

	ρ (N/m³)	E (N/m²)	ν (N/m²)	σ_c (N/m²)	σ_t (N/m²)
Stone Masonry	1.900	3e+09	0,2	4,0e+06	0,1e+06
Brick Masonry	1.800	1,5e+09	0,2	1,5-3,0e+06	0,1e+06
Filled Material	800	3e+08	0,3	-	

ρ: Density E: Young's Modulus. **ν**: coeff. Poisson. σ_c: compression stress. σ_t: tensile stress.

well as shell elements to describe the Little Chapels. The references of these materials were taken from specific bibliograph [1]

These are the parameters considered:

Once the numerical calculation model has been elaborated from the data obtained in the photogrammetry, we carry a campaign out by placing uniaxial and triaxial accelerometers on the roof, that provide us dynamic information of the building. These accelerometers were placed in the greater modal displaced points [12, 13], which were previously determined with the numerical model from the distribution of masses and stiffness of the building with an initial material properties in order to do the initial calculation (*Figure 7*).

$$f_n = \sqrt{\frac{K_n}{m}} \tag{1}$$

f_n= natural frequency
K_n= structural stiffness
m = mass

The position of the accelerometers [13] is determined thanks to an initial analysis of the maximum modal displacements using the numerical calculation model. With this, it is possible to start the experimental campaign. The vibration tests were carried out with environmental loads due to the complexity to generate forced loads in a singular structure like the one of our study

With the results, the data are processed with the Artemis software [14]. Through the methods of Modal Identification of Improved Decomposition in the Frequency Domain (EFFD) and Identification in the Stochastic Subspace (SSI), four modes of building vibration were identified *(Figure 11)*.

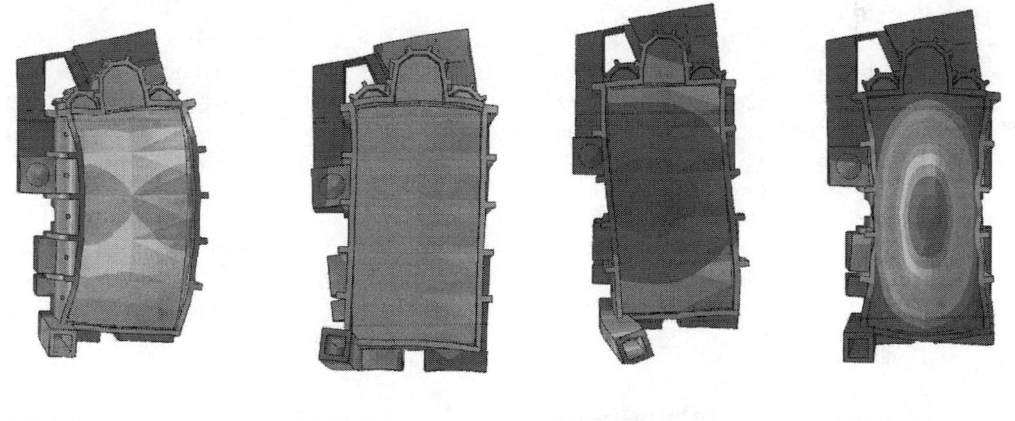

Figure 10. Main Vibration Modes.

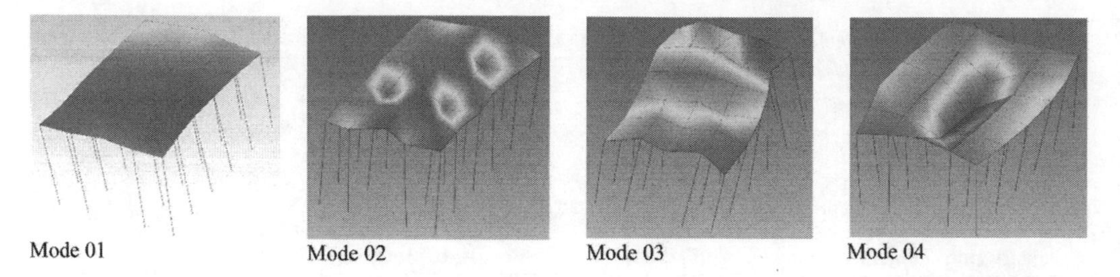

Figure 11. Vibration Modes Detected with Artemis Software Mode 02.

4 CONCLUSIONS

Thanks to this method, we have obtained four natural modes of vibration of the building similar to the modal behavior of the numerical model, getting frequencies with relative errors over the 5% [8].

MODE Nº	ARTEMIS (Hz)	ANALYSIS FEM FRECUENCY (Hz)	ANALYSIS FEM %
1	1,929	2,0251	4,75%
2	3,336	3,2034	4,14%
3	3,750	3,6543	2,62%
4	4,315	4,0872	5,57%

The presented method can be considered as a non-destructive highly technique for the dynamic evaluation of heritage buildings with masonry materials.

The knowledge of the dynamic behavior of the building through OMA allows us to evaluate the building against earthquakes with the elastic response spectrum in the numerical model.

After analyzing the frequencies results obtained in relation to those extracted from the numerical model, it is verified that the modal forms of both numerical and real models are similar. This fact will allow us to obtain a model calibrated in frequency and modal forms updating the mechanicals properties of their materials. With this, it is just possible to analyze the building structural behavior.

In futures works it will be possible to extend the experimental campaign with more strategical measuring points on the little Chapels roofs that will allow us to advance in the knowledge over the possible connections or disconnections between the main building and these Chapels, and their influence on the dynamic behavior of the whole.

We are currently using the updated numerical model to evaluate the structural behavior in a linear and non-linear domain.

Photogrammetry allows us to determine displacements values in the columns (Figure *11*) by using software to edit the points cloud and define complex geometries such as the one we are studying.

REFERENCES

[1] Augenti, N. P. (2012). MADA: online experimental database for mechanical modelling of existing masonry assemblages. In Proc. 15th World Conference on Earthquake Engineering, Lisbon (Portugal).

[2] Fernández, S. H. (2004). Arcos, Bóvedas y Cúpulas.

[3] Gentile, C. S. (2004). Dynamic-based F.E. Model Updating to Evaluate Damage in Masonry Towers, Proceedings of the 4th International Seminar on Structural analysis of Historical Constructions. Padova, Italy.
[5] Heyman, J. (1995). Teoría, Historia y Restauración de Obras de Fábrica.
[6] Heyman, J. (1999). El esqueleto de piedra.
Lampérez Romea, V. (1904). Historia de la arquitectura cristiana española en la Edad Media II. Ámbito Ediciones, Valladolid, 1904.
[7] M. Diaferio, D. F. (2017). A Nonlinear Numerical Model of a Historical Slender Structure.
[8] Manuel Romero, P. P. (2018). Operational Modal Analysis: A tools for assessing changes on structural health of historical constructions after consolidation and refoircfement work. Jura Chapel.
[9] Nieto Julián, J. E. (2013). La necesidad de un modelo de información aplicado al patrimonio arquitectónico.
[10] Pachón, P., Compán, V., & Rodríguez-Mayorga, E. (2016). Caracterización Diámica de Edificaciones Históricas Mediante Análisis Modal Operacional. Aplicación Sobre la Iglesia de Santiago de Jerez de la Frontera.
[11] PhotoScan, A. User Manual: Professional Edition. Version 1.2.6. (2016).
[12] Ramos, L. (2007). Damage Identification on Masonry Structures Based on Vibration Signatures (PhD Thesis). University of Minho.
[13] Ramos, L. F., Aguilar, R., Lourenço, P., & Moreira, S. (2012). Dynamic Structural Health Monitoring of Saint Torcato Church. Mechanical Systems and Signal Processing.
[14] *SVS. ARTeMIS Extractor 2010 release 5.0.* <http://www.svibs.com>. (2010).
[15] Users manuals. Dassault Systemes (2015), SIMULIA, ABAQUS v.6.13-3. (s.f.).
[16] Valseca, J. (2013). Traza, Geometría y Estructura de la Iglesia de Santa Ana de Sevilla. Diploma de Estudios Avanzados por la Universidad de Sevilla. (DEA).
[17] Valseca, J., & M.T. Rodríguez-León, C. C. (2017). Finite Elements Models for the Structural Analysis of Imposition Vaults: Santa Ana Church in Seville. International Conference on Mechanical Models in Structural Engineering. (CMMOST).
[18] Valseca, J., Rodríguez-León, M., Compán Cardiel, V. J., Vázquez, E., & Sánchez, J. (2018). Evaluación Estructural Mediante el Modelo de Elementos Finitos para el Estudio de Patologías de la Iglesia de Santa Ana en Sevilla. Euro-American Congress. Construction Pathology, Rehabilitation Technology and Heritage Managment. REHABEND.

Science and Digital Technology for Cultural Heritage – Ortiz Calderón et al. (Eds)
© 2020 Taylor & Francis Group, London, ISBN 978-0-367-36368-0

LOD BIM protocol for heritage interventions. Antonio Cavallini's wall paintings case study

F. Balbuena Marcilla
Architectonic Graphic Expression Department, University of Seville, Seville, Spain

ABSTRACT: The growing development of disrupting new technologies in Architectural fields, has allowed to set links between graphics and documental databases, as happens when using BIM software applications.

In this way, with the implementation of the Level of Development (LOD) specification guide, there seems to be accepted as the standard to be considered in design and works management of architectural design using BIM technology. Nevertheless, the potential application of these standards to a heritage digital model rises some questions.

This communication aims at identifying critical questions, issues and potential solutions to facilitate the establishment of new and specific LOD rules for Heritage Intervention in order to fit HBIM technology with HLOD specifications in a common platform besides documental and geometric definition

1 INTRODUCTION

This communication refers to the works covered by definition of monument present in article 1 of the Convention Concerning the Protection of the World Cultural and Natural Heritage (Paris from 17 October to 21 November 1972): "architectural works, works of monumental sculpture and painting, elements or structures of an archaeological nature, inscriptions, cave dwellings and combinations of features, which are of outstanding universal value from the point of view of history, art or science"

Since the UNESCO constitution was established in 1945 a variety of challenges and issues have emerged in relation to providing UNESCO world heritage nomination files (WHNF). The resource manual "Managing Cultural World Heritage" dated 2013, provides guidance for States Parties and all those involved in the care of World Heritage cultural properties on how to comply with the requirements of the World Heritage Convention. The main text of the Resource Manual explains what is involved in management for World Heritage, its context, its philosophies and its mechanisms. A set of appendices then offers guidance on how to put them into practice.

Oriented mainly to the management of museum heritage, but also applicable to any heritage element, there is a working group at CIDOC (ICOM international committee for documentation) aimed to describe the relationship between documentation and knowledge management. Leader: Maija Ekosaari

Because of the variety of agents involved in any action on historical heritage there are different initiatives aimed at developing document management models. In Spain, one of them quite significant is the initiative at Heritage Science Institute (Incipit) in CSIC, that develops an open language model named ConML (conceptual modeling language), in Object-oriented programming interface.

But there are not any rules about using BIM or GIS in Heritage Monuments and Sites.

The Level of Development (LOD) specification guide published by the American Institute of Architects in 2013, there seems to be accepted as the standard to be considered in design and works management of finished architectural products using BIM technology. This guide identifies the minimum content requirements for each element existing in the model, on a scale of progressive levels of development from 100 to 500

It is a reference that allows building agents to clearly articulate the content requirements of a BIM model for the different phases of the project process and its construction. The actual standard is based on controlling the level of development on four specific sections:

a) Level and precision in Geometric definition
b) Level of description and detailing of the elements of the Model
c) Specifications about non-graphic information and how is linked to the graphics database
d) Definition in budgets and cost estimation

It is thought and developed for its use concerning architecture projects, especially, works of new buildings and/or reform, but with difficult translate to HBIM, as discussed below. HBIM or Heritage Building Information Modelling, refers to the use of the BIM digital models to document and manage Heritage Monuments and Sites.

Due to the importance of BIM technology, several experiences similar to those explored in this work are being developed. I want to expressly cite the work done by the researchers Angulo Fornos, Roque (2015) and Antonopoulou, Sofia (2017).

2 METHODOLOGY TO THE INTERVENTION WITH HBIM MODELS

Following the recommendations and practice, we can identify the currently steps being followed in the interventions with HBIM models:

2.1 *First step for construction an HBIM Model: Documentation*

Starting from a good documentation is the crucial step to get an adequate intervention. It is necessary to follow the recommendations of the publication of "Operational Guidelines for the World Heritage Committee" and the resource manual "Managing Cultural World Heritage".

2.2 *Second step for construction an HBIM Model: Analysis and diagnostic*

With the documentation dossier, is necessary to elaborate an analytic dossier that allows to correctly identify the objectives of the intervention; at least has to compile: Report of the existing materiality, indicating the damages and pathologies present in the place, its possible causes and the proposals for intervention and consolidation. Ornament elements as mural painting, has to be considered part of this report. Historical report that considers the evolution of the place since its origin. Artistic report that identifies the movements and styles present in the place, including an assessment of the artistic quality of the same. Archaeological report that values the existing remains indicating the conditions of preservation and tasks to be considered

The Information regarding previous interventions, also must be linked to the graphic database, identifying each previous intervention separately and allowing its virtual reconstruction. It is important that the original, genuine and authentic elements can be identified in the mode. Finally, is necessary the existence of a study that verifies the authenticity of the sources used in the aforementioned reports is necessary

2.3 *Third step for construction an HBIM Model: Project of intervention*

The project intervention responds to the initial objectives set, giving solution to the problems raised in the analytical dossier.

2.4 *Fourth step for construction an HBIM Model: Execution of the intervention*

The execution of the intervention of a HBIM model presents critical levels of demand, due to the impossibility of correctly contemplating all possible situations in the same way that would be done with the work of a new building. This means the need to contemplate in the project the possibility of alteration of the same during the execution of work, through specific monitoring of several tasks.

2.5 *Fifth step for construction an HBIM Model: Final documentation of the intervention*

Once the work is finished, it is necessary to document what has been done with a complete structure that includes all the phases; it is not enough to edit a publication that collects what has been done.

2.6 *Sixth step for construction an HBIM Model: Monitoring and review*

The protocol of intervention in a monument should not end with the works carried out in it, so it is necessary to plan suitable future actions and scheduling of tasks that have been left pending

3 PROPOSAL FOR THE DEVELOPMENT OF A HBIM LOD PROTOCOL

The proposal presented here to formalize a protocol that allows to establish levels of definition (LOD) in interventions that are carried out on monuments, is advised by the need to reconcile new technologies and the use of BIM software instruments with the recommendations mentioned before of the publication of “Operational Guidelines for the World Heritage Committee” and the resource manual “Managing Cultural World Heritage.

The geometric model must be three-dimensional using BIM software and geo-referenced using GIS software, using the same structure of levels 100 to 500, of the usual specifications in the BIM projects, and controlling the level of development on same four specific sections:

a) Level and precision in Geometric definition
b) Level of description and detailing of the elements of the Model
c) Specifications about non-graphic information and how is linked to the graphics database
d) Definition in budgets and cost estimation

Nevertheless, also must be developed considering the six steps related before: 1, Documentation; 2 Analysis and diagnostic; 3, Project of intervention; 4, Execution of the intervention; 5, Final documentation of the intervention and 6, Monitoring and review

3.1 *HBIM LOD level 100*

Part A (documentation): The geometry is approximate and can be represented graphically by envelopes and generic volumes. The information is not necessary that it be linked to each element but must be compiled all known background: Historical antecedents, Previous interventions and Authors.

Part B (Analysis): Brief report justifying the intervention

Part C (Project Intervention): Only definition of basic objectives, because it is not possible to define any details of the intervention until the analysis of the documentation is completed

3.2 *HBIM LOD level 200*

Part A (documentation): The actual geometry is completely defined and has been got by photogrammetric procedures. The systematization of the elements and their classification is carried out. The no graphic information is not necessary that it be linked to the element but must be compiled all known background: Historical antecedents, previous interventions and authors and photo/video report.

Part B (Analysis): There are a complete report justifying the intervention based in analysis of documentation: archeology, material, artistic, and so on

Part C (project intervention): The geometry of intervention is defined. The construction elements are represented graphically in the model as generic systems, objects or assemblies with approximate indications of quantity, size, shape, location and orientation. The projected elements can be used to estimate costs by approximate techniques

3.3 *HBIM LOD level 300*

Part A (Documentation): The actual geometry is completely defined. Has been got by photogrammetric procedures. The systematization of the elements and their classification is carried out. All known background: Historical antecedents, previous interventions and authors are linked to each element. Photo/Video report: It is linked to the element

Part B (Analysis): There are a complete report justifying the intervention based in analysis of documentation: archeology, material, artistic, and so on

Part C (Project Intervention). The projected geometry is fully defined according to the final dimensions resulting from the calculation. The elements are represented graphically in the model as systems, objects or assemblies with specific indications of quantity, size, shape, location and orientation. All tasks that can be initially planned are temporarily scheduled. The projected elements have a sufficient definition to be used for estimating costs by requesting offers

3.4 *HBIM LOD level 400*

Part A (Updated documentation). The geometric information is updated as the intervention process progresses. Data linked to the elements can be modified by the intervention process. All known background and new dates are linked to the element. Photo/Video report: The intervention process is linked to the element

Part B (Updated Analysis): The new data obtained through the intervention process are analyzed and added to the existing documentation

Part C (Updated project of intervention): The geometry is adapted according to the new dates, getting final dimensions resulting from the new calculations. All kinds of modifications are collected as they occur. The elements reflect the current contract price. All kinds of modifications in estimations are collected as they occur

Part D (execution of intervention): The geometry executed is digitally verified. The elements can be used to verify the performance of the systems according to the requirements of the component. Project planning: Adaptation of the temporally established task planning.

3.5 *HBIM LOD level 500*

Part A (Final documentation): The elements are verified representations in the site and the works, in their final situation in terms of size, shape, location, quantity and orientation. All known background and new dates are linked to the element. Photo/Video report: The intervention process is linked to the element.

Part B (Updated Analysis): There are a complete report after the intervention based in analysis of all documentation used concerning the following items: Historical report, artistic report, archaeological report that values the existing remains indicating the conditions of preservation. The Information also is linked to the graphic database, identifying each previous intervention separately and allowing its virtual reconstruction. There is a study that verifies the authenticity of the sources used in the aforementioned reports, signed by author of document and project Manager. Revised by independent authority.

Figure 1. Antonio Cavallini's paints of Virgen del Pilar Chapel, in Santo Angel Church.

Part C (Scheduling pending tasks): Pending tasks must be classified with specific objectives. The geometry that may be affected must be defined. All tasks that can be initially planned are temporarily scheduled.

Part D (Monitoring and review): Appointment of review and supervision teams. Custody of the elaborated documents, maintenance and update of the digital files. Preparation of routine revision checklist of conventional elements (covers, walls, carpentry, rainwater collection system, etc). In case permanent measuring elements have been installed, monitoring, measurement and reporting. In case service infrastructures have been placed, elaborate the protocols of revision and the terms in which they must take place

4 ANTONIO CAVALLINI'S WALL PAINTINGS CASE STUDY

This case study refers to the restoration of the mural paintings made by Antonio Cavallini in the chapel of the Virgen del Pilar in Santo Angel Church of Seville.

The work pursues a documentary methodology prior to the intervention based on the application of BIM techniques for information management so that all the restoration processes are registered in the graphic database. Performing at HBIM LOD Level 200:

Part A (documentation): The actual geometry is completely defined and has been got by photogrammetric procedures.

The wall paintings are registered in high definition. The systematization of the elements and their classification is going carried out. The no graphic information is not linked yet to the element but are compiling all known background: Historical antecedents, previous interventions and authors and photo/video report.

Part B (Analysis): A complete report is being writing justifying the intervention based in analysis of existing documentation: archeology, material, artistic, and so on

Part C (project intervention): currently in execution phase

5 CONCLUSIONS

In conclusion, it can be agreed that the existence of a standard methodology for the application of BIM systems to interventions in historical heritage provides a common information management platform which is fully compatible with other developments (as ConML) and allows obtaining and consulting graphically all the information necessary in order to the get the best decisions possible.

REFERENCES

Angulo Fornos, Roque. 2015. Digital models applied to the analysis, intervention and management of architectural heritage. *Building Information Modelling (BIM) in Design*, Construction *and* Operations: 407-418. WIT Press.

Antonopoulou, Sofia. 2017. BIM for Heritage: Developing a Historic Building Information. *Historic England.*

Cavallini, Antonio. 1836-1905. Bissone, Tessin Canton, Swizzera.

CIDOC, ICOM international committee for documentation. http://network.icom.museum/cidoc/

ConML (conceptual modeling language) http://www.conml.org/

CSIC, Superior Council of Scientific Investigation. https://www.csic.es/

Level of Development (LOD) Specification is a reference that enables practitioners in the AEC Industry to specify and articulate with a high level of clarity the content and reliability of Building Information Models (BIMs) at various stages in the design and construction process. The LOD Specification utilizes the basic LOD definitions developed by the AIA for the AIA G202-2013 Building Information Modeling Protocol Form.

Maija Ekosaari. Secretary of ICOM CIDOC. She works as a Project Manager in Museum Centre Vapriikki in Tampere, Finland.

UNESCO. 1977. Operational Guidelines for the World Heritage Comitee.

UNESCO. 2011. Preparing world heritage nominations. *World Heritage Resource Manual.*

UNESCO. 2013. Managing cultural world heritage. *World Heritage Resource Manual.*

UNESCO. 2017. *Operational Guidelines for the Implementation of the World Heritage Convention.*

Science and Digital Technology for Cultural Heritage – Ortiz Calderón et al. (Eds)
© 2020 Taylor & Francis Group, London, ISBN 978-0-367-36368-0

Pilot experience of HBIM modelling on the Renaissance quadrant façade of the Cathedral of Seville to support its preventive conservation

R. Angulo Fornos, M. Castellano Román & F. Pinto Puerto
Department of Graphic and Architectural Expression, University of Seville, Seville, Spain

ABSTRACT: This conference report discusses the generation of a HBIM model for heritage information management, oriented to the preventive conservation of cultural interest assets, in this case, the renaissance quadrant façade of the Cathedral of Seville. The graphics elements of the model and its information parameters have been defined considering the methodology of work used by the company in charge of its preventive conservation, Artyco S.L., in order to enhance its regular controlling and monitoring activities on cultural assets. The research take part in the actions referred in the Project HAR2016-78113-R, supported by the Ministry of Science, Innovation and Universities I+D+i Plan. The company in charge of the building conservation is evaluating the results considering its suitability and viability for future applications. Furthermore, the results have been added to the experimental actions referred in the project mentioned, to be considered in its final general conclusions by the research team.

1 INTRODUCTION

HBIM research experiences differ profoundly in terms of objectives, methodologies, and results. From the modelling of complex parametric geometries oriented towards virtual reconstructions, to the manual or semi-automatic processing of point clouds to get the elements of the model, or the modelling as a part of the architectural survey process that may evolve according to the level of knowledge of the building. Those approaches are not necessarily mutually exclusive.

In fact, the experience presented has consisted of the generation of an HBIM model of a fragment on the *Renaissance quadrant* façade of the Cathedral of Seville from a point cloud processing. However, the model is also the result of the architectural analysis of the building, in order to define precisely the elements to model and the level of development of every single element.

This HBIM model is oriented towards the preventive conservation of the fragment of the building, which the company Artyco in charge of. The company has promoted this research in order to the future application of the results to its regular controlling and monitoring activities on cultural assets.

The so-called Renaissance quadrant of the Cathedral of Seville is a set of spaces that house the Sacristries and the Sacrum Senatum capitular rooms, resulting from a long process of projects, reforms and transformations that have been hidden behind an imposing wall that gives it the appearance of a unitary whole. Contained between this wall and the southeast façade of the cathedral, a group of heterogeneous buildings are amalgamated that follow one another and readjust with a certain continuity in time, until filling the available space. This wall was built in various irregular constructive impulses between 1528 and 1543 (Pinto 2013).

2 METHODOLOGY

This pilot experience has included the realisation of a digital model of the façade fragment, a database and a work procedure for the use of both linked elements, where the Artyco team could provide information, as well as obtain reports and infographic visualisations.

The essential aspects of the followed workflow are summarised below:

a) Photogrammetric survey of the façade fragment from the digital metric capture (DMC).
b) 3D modelling of the façade fragment.
c) Georeferencing of the fragment in relation to the cathedral.
d) Establishment of a spatial reference system and a common vocabulary for this element.
e) Structuring and relationship of the data required by Artyco so that they are associated with the model.
f) Development of the HBIM model.
g) Establishment of systematised and standardised input and output information protocols.
h) Implementation and evaluation of results with the Artyco team.

This methodology is based on previous experiences of the research group HUM 799 Heritage Knowledge Strategies (Angulo et al 2017). It is also one of the research lines developed in the Project HAR2016-78113-R, supported by the Ministry of Science, Innovation and Universities I+D+i Plan.

3 RENAISSANCE QUADRANT FAÇADE HBIM MODEL

3.1 *The reliability of the geometric basis for modelling*

Addressing the specific case of experience has allowed us to test the strategy outlined in the previous paragraphs. In this way, the workflow carried out has started from a massive data capture using photogrammetric techniques with a result or final dump that has consisted of a dense cloud of points with information about its position, chromaticism and orientation (X, Y and Z coordinates, colour RGB and normal Nx, Ny and Nz).

The information was processed for modelling by means of a set of vectorial graphical entities, flat CAD surfaces or NURBs, of the shell that surrounds the material volume of the object studied. Reverse engineering has been used to develop a semiautomatic work process within a specialised software consisting of three main phases:

- Conversion of the point cloud into an optimised mesh of triangles.
- Segmentation of the meshed surface.
- Manual conversion of regions into CAD vector surfaces and their editing by means of manual modification tools for the formalisation of the joints between the various surfaces and their corresponding dimensions.

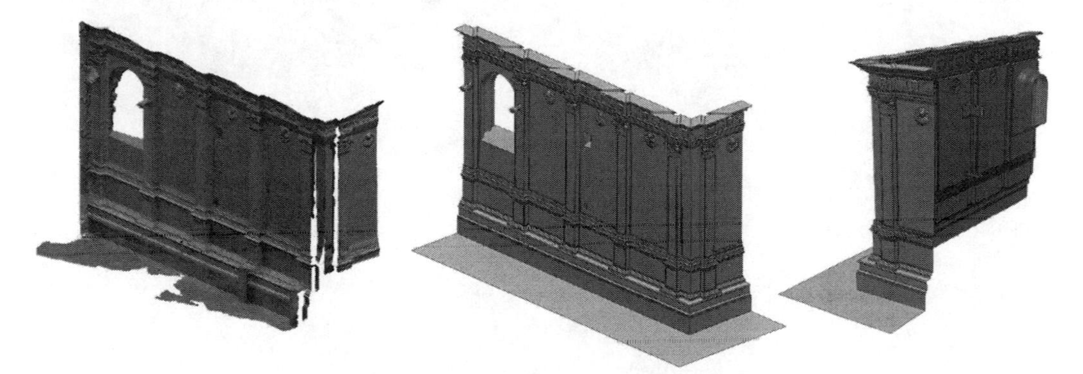

Figure 1. Axonometric view of the meshed surface obtained from the DCM (left). Exterior and interior views of the CAD vector model once edited (centre and right).

This last step is crucial when establishing the level of precision or accuracy of the final model with respect to the physical reality of the object dealt with, controlled by consulting the deviations produced between the mesh and the resulting CAD model.

3.2 *Modelling and its stratification. an open model*

At the software level, the modelling process has become independent of the process of creation and management of the HBIM model. The modelling software used facilitate the transformation of data from digital capture in three-dimensional entities that are complex enough to be a model of accurate and rigorous information at all levels once imported into the BIM modelling software.

In this sense, the HBIM model constitutes a living interpretation of the building, continuously fed back with information about the object, but it is in no way a substitution of reality; it will always be an abstraction based on the partial recording and capture of a reality of incomprehensible complexity.

The modelling process has followed the following sequence in the case study:

3.2.1 *Transformation of the DCM into a composite skin according to section 3.1*

3.2.2 *Generation of a solid type entity*

Using the previously modelled skin as a mould, the solid generated would inherit the geometry and exact shape of the wall that limits the scope of study. This process was carried out in conventional CAD software, specifically Autodesk® AutoCAD®. The choice of this software had to do with three fundamental aspects: the graphic power that it makes available to the user, the extent of its use in the discipline of architecture and the perfect connection at the export-import level with the selected BIM managing software: Autodesk® Revit®.

3.2.3 *Realisation of a duplicate model from the generated solid, eliminating figurative elements and formal configurations of a certain complexity in order to guarantee the necessary geometric base to have different levels of detail in the HBIM model.*

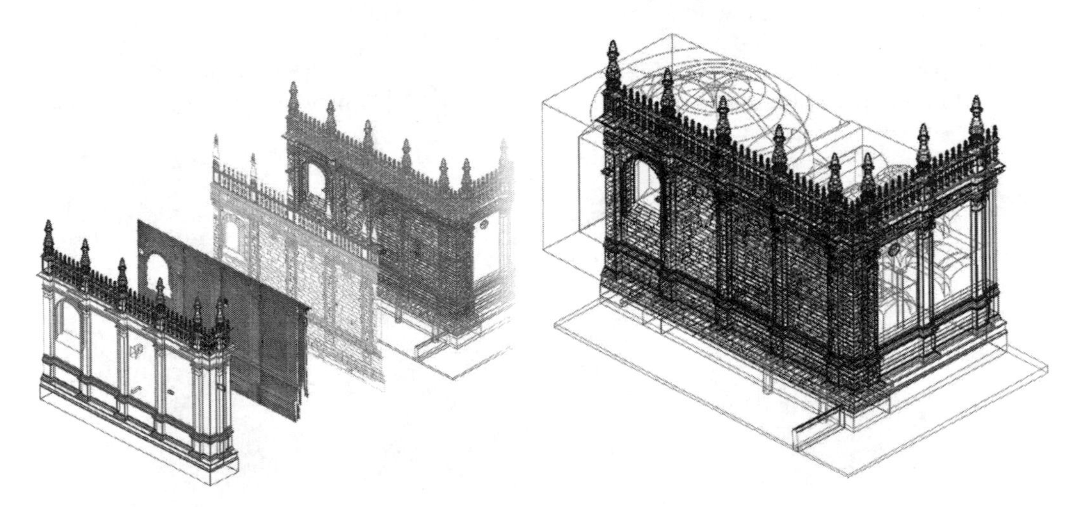

Figure 2. CAD editing process. Simplified solid, reference orthophoto, tracing of the exploded view in 2D and its application to the final exploded view in solid three-dimensional entities (left). Graphic model of complete solid CAD (right).

3.2.4 *Cutting of the detailed solid using as trace the different orthophotos obtained from the photogrammetric capture*

As previously discussed, the information model should be a container of varied information, from simple geometric measurements to the location of masonry marks or a detailed description of material and surface pathologies through complex chronological assignments.

This process involves implicit decisions of some importance in terms of knowledge of the object being treated and its implication on the structure of the model information. In this specific case, sufficient records or documents were not available to precisely deduce the thickness of the different layers that make up the structure of the façade and the existence of stones arranged like keys that, passing through the supposed internal layer of calicastrado (lime-crust walls), join the outer leaves of stone. The lack of certainty in this type of question, both geometric or formal and associated information, does not prevent the possibility of modelling the entities based on well-founded hypotheses and assigning them initial values that will be updated when the level of knowledge is increased.

3.3 *The structure of the information and its parameterization*

The structure of the information associated to the model has been oriented towards the preventive conservation of the element. The required parameters have been assigned according to the levels of development of the HBIM model, which includes two unique entity models, different in their level of geometric detail, and a cut model. These levels of development can be considered as Levels of Knowledge (LOK) for the purposes of the heritage use of the BIM (Castellano-Román, 2019). Parameters that refer to the element as a whole are associated to the unique entity models: identification, basic heritage information on the stylistic ascription and main authorship, figures of legal protection that affect it, registration of interventions and related dissemination actions.

In the cut model, each element is subject to registration and, therefore, parameters associated with a higher level of knowledge, that required to manage the preventive conservation of the element, can be associated. Therefore, each element is assigned a unique identifier and is characterised with identification parameters, damage, treatments and actions.

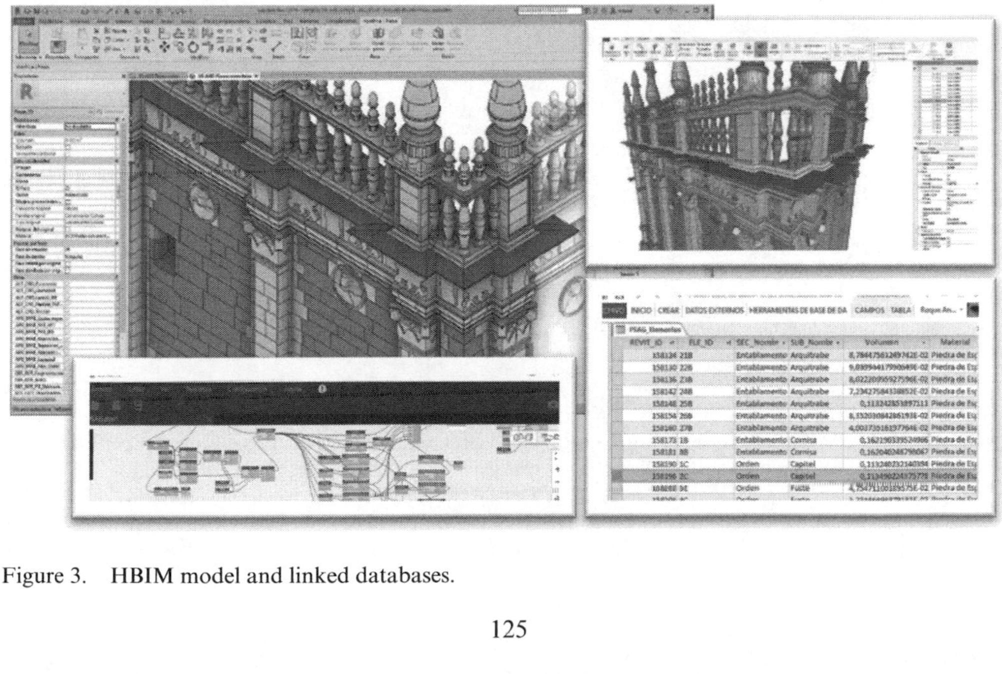

Figure 3. HBIM model and linked databases.

The HBIM model information is linked to an external database that allows editing outside the environment of the modelling programme. In this way, the technicians responsible for the preventive conservation of the element do not require specific knowledge about the BIM modelling software, but simply the basic training to interact with the tables and forms of the previously configured database.

3.4 *The HBIM model*

The HBIM model includes the graphic model produced in the BIM editing platform, the information associated with the parameters of the model and the linked databases for the management of information and associated documents. These associated documents are constituted as a documentary repository which is parallel to the HBIM model and which will be fed by the research and conservation actions that take place on the element studied.

4 CONCLUSIONS

A model of HBIM heritage information has been produced on the façade of the Renaissance quadrant of the Cathedral of Seville in order to serve as an assistant of the actions of preventive conservation to be undertaken by the company Artyco S.L.

The graphic model is based on a photogrammetric digital metric capture that is analysed and interpreted with the methodological keys of the architectural survey. Consequently, the model arises from an interpretation proposal of the metric data based on the archaeological and architectural knowledge of the element studied, differentiating those elements whose definition arises from a certain knowledge, from those modelled according to hypotheses pending confirmation.

Three levels of development of the model are established: two of them are unique entity models, including the whole façade as a single element. Their differences are in level of geometric details considered. In the third one, derived from the cutting of the unique-entity-models, every single constructive unit of the set has been delimited.

Unique entity models incorporate the information parameters related to the element as a whole. This information is completed with the parameters of each piece of the exploded model, where the parameters associated with preventive conservation actions are incorporated.

The protocols of systematised and standardised input and output of information are produced through the external databases linked to the graphic model, and the visualization of it through free viewers of great usability, allowing their use by technicians without specialised training in HBIM methodology.

ACKNOWLEDGEMENT

We would like to thank the company Artyco SL for the resources made available to us during the restoration work carried out in this building, and for the trust placed in our team.

REFERENCES

Angulo Fornos, Roque, Francisco Pinto Puerto, Jesús Rodríguez Medina, and Antonio Palomino. 2017. 'Digital Anastylosis of the Remains of a Portal by Master Builder Hernán Ruiz: Knowledge Strategies, Methods and Modelling Results'. *Digital Applications in Archaeology and Cultural Heritage* 7 (December): 32–41. https://doi.org/10.1016/j.daach.2017.09.003.

Brumana, Raffaella, S. Dellatorre, Daniela Oreni, M. Previtali, L. Cantini, Luigi Barazzetti, A. Franchi, et al. 2017. 'HBIM Challenge among the Paradigm of Complexity, Tools and

Preservation: The Basilica Di Collemaggio 8 Years after the Earthquake (L'aquila)'. *International Archives of the Photogrammetry, Remote Sensing and Spatial Information Sciences - ISPRS Archives* XLII (September): 97–104. https://doi.org/10.5194/isprs-archives-XLII-2-W5-97-2017.
Castellano Román, Manuel, Francisco Pinto Puerto, 2019. 'Dimensions and Levels of Knowledge in Heritage Building Information Modelling, HBIM: The model of the Charterhouse of Jerez (Cádiz, Spain)'*Digital Applications in Archaeology and Cultural Heritage* (August).https://doi.org/10.1016/j.daach.2019.e00110.
Macher, Hélène, Tania Landes, and Pierre Grussenmeyer. 2017. 'From Point Clouds to Building Information Models: 3D Semi-Automatic Reconstruction of Indoors of Existing Buildings'. *Applied Sciences*, no. 7. https://doi.org/10.3390/app7101030.
Palestini, C., A. Basso, and L. Graziani. 2018. 'Integrated Photogrammetric Survey and Bim Modelling for the Protection of School Heritage, Applications on a Case Study'. In *International Archives of the Photogrammetry, Remote Sensing and Spatial Information Sciences - ISPRS Archives*, 821–828. Riva del Garda. https://doi.org/10.5194/isprs-archives-XLII-2-821-2018.
Pinto Puerto, Francisco. 2013. La Sacristía de Los Cálices: Aportaciones Desde El Análisis de Sus Fábricas y Los Sistemas de Control Formal. Libro de Actas de La XX Edición Del Aula Hernán Ruiz. Sevilla: Cabildo Catedral de Sevilla, pp. 165–232.
Tommasi, C., C. Achille, and F. Fassi. 2016. 'From Point Cloud to BIM: A Modelling Challenge in the Cultural Heritage Field'. *International Archives of the Photogrammetry, Remote Sensing and Spatial Information Sciences - ISPRS Archives* 41 (B5): 429–436. https://doi.org/10.5194/isprsarchives-XLI-B5-429-2016.

Science and Digital Technology for Cultural Heritage – Ortiz Calderón et al. (Eds)
© 2020 Taylor & Francis Group, London, ISBN 978-0-367-36368-0

Application of BIM for the representation of the constructive evolution of the apse of the church of San Juan de los Caballeros of Jerez de la Frontera

J.M. Guerrero Vega, G. Mora Vicente & F. Pinto Puerto
Universidad de Sevilla, Seville, Spain

ABSTRACT: The recent restoration of two chapels attached to the chevet of the parish church of San Juan de los Caballeros has led to the construction of a digital model of the building. This has been aimed at providing support for the information generated both during the archaeological research phase and over the course of the intervention work. The use of this type of tool for the representation of the evolution of the building over time is explored and, based on the experience accumulated, it is proposed how the diachronic reading of architecture should be present in the generation of the digital model itself. In this way, it establishes the need for a rigorous process of previous analysis of the building, which takes into account the different historical transformations and that serves as a support when making decisions in the modelling process.

1 INTRODUCTION

The use of digital models of historical building has shown great potential in different aspects of their guardianship. In this sense, among the very different possibilities offered by BIM (Building Information Modeling) tools are the management of a large amount of information in heritage building (Simeone, Cursi, Toldo & Carrara 2009; Dore & Murphy 2017) and, in addition, the generation of images with great communicative potential that describe the state of the building in the successive historical stages, facing the transmission to the general public (Angulo Fornos 2012).

Based on these two benefits in the use of digital models, the chevet of the church of San Juan de los Caballeros has been suggested as an architectural object in which to explore these possibilities. This contribution tries to provide a series of reflections on the model construction process instead of presenting a culminated experience.

2 CASE STUDY

The foundation of the parish church of San Juan de los Caballeros, located inside the old walled enclosure of the city of Jerez de la Frontera (Cádiz, Spain), dates back to the conquest of the city by the Kingdom of Castile in the 13th century. It is a complex building due to the numerous construction phases that have occurred throughout its history, which is common for this type of building typology in which the successive transformations and juxtapositions have left us a fragmented and stratified building.

The chevet of the temple is a clear exponent of a decisive period in the history of the city, and its border status favoured the appearance and social ascension of local elites belonging to the aristocracy. Some lineages that promoted the construction of funerary spaces within the consolidated churches became important between the 14th and 16th centuries. These actions generated a series of singular buildings, such as the two square-floor chapels that were attached to the apse of San Juan. The Capilla de la Jura (Mora Vicente & Guerrero Vega 2014) was built at the beginning of the 15th century and that of the Carrizosa family was built at the end of the same century (Romero Bejarano 2014). Both oratorios present common characteristics and other particular ones that in both cases involve an advancement in the local architecture of their time and allow a very clear historical route on how the construction and the architectural tastes evolved.

The recent restoration intervention carried out on these chapels has been an opportunity to expand our knowledge about them. Work was carried out for the rehabilitation of the aforementioned chapels between 2014 and 2016.

3 DIFFERENT PERSPECTIVES IN THE PROCESS OF KNOWLEDGE

The previous and parallel studies to this research have been approached from different perspectives or disciplines. In this specific case, four fields of knowledge (architecture, history, archaeology and the restoration work itself) have been defined, without the intention of establishing a closed classification. But it is obvious that these visions can be different or be complemented with additional ones depending on the proposed objectives, the available resources and the characteristics of each building.

It is possible to make superimposed and interconnected readings as you progress in each of these fields, generating a conceptual model that allows you to incorporate knowledge about the building (González-Pérez 2018: 3-8). In this sense, the research carried out now is proposed as a first proposal that could, depending on the knowledge generated, advance from the general to the particular.

From the point of view of the architectural analysis, those characteristics that defined the different elements were recognised: materials, constructive solutions, decorative and spatial features. The geometric analysis of the covering systems was also addressed, which meant that the layout and construction methods of the vaults could be better known. An inventory of the most significant elements was completed in addition to a record of stonemason's marks. Within this field, the structural analysis also allowed the origin of the main damage that the state of conservation presented prior to the intervention to be identified.

On the other hand, in the consultation and critical review of historical sources, the reading of the original testaments of the patrons of the chapels should be noted, as well as the connection of their emblems with their decorative details and the use of iconographic languages in the 14th century, in which decoration of the intrados of the arch with rampant lions stands out. As has already been pointed out, all this was related with the development of local elites and their formulas of social affirmation. The role of the patrons, true promoters of the works, will be decisive when promoting novel solutions of one space with respect to the other, for example, in light processing. In addition, the solutions used in these chapels could be related to other references that were not always close and that illustrate the transfer of knowledge in medieval construction.

Archaeological studies at the level of architecture and subsoil have provided knowledge about both the use of the site prior to the chapels and the spatial relationships between these and the apse of the church. This has resulted in chronological and typological sequences with results

that have allowed a reading of the transformation of the building throughout its history, and have allowed us to get closer to the original image of its construction.

Finally, the restoration and consolidation process provided new information about the material configuration of the building, recovering ornamental elements that had disappeared in previous stages. On the other hand, constructive solutions and applied conservation treatments have consequently led to the modification of the previous state, generating valuable information for future interventions.

The data obtained from the different fields have been managed in the BIM model itself, but it is also possible to establish connections with other database management systems, so that communication between these tools can be done both ways (Angulo Fornos 2015). Based on the information incorporated into the model, consultations can be made, linking and crossing information of a different nature, carrying out more complex analyses. In this way, the digital model can be understood as a common tool that favors a dialogue between different fields of study.

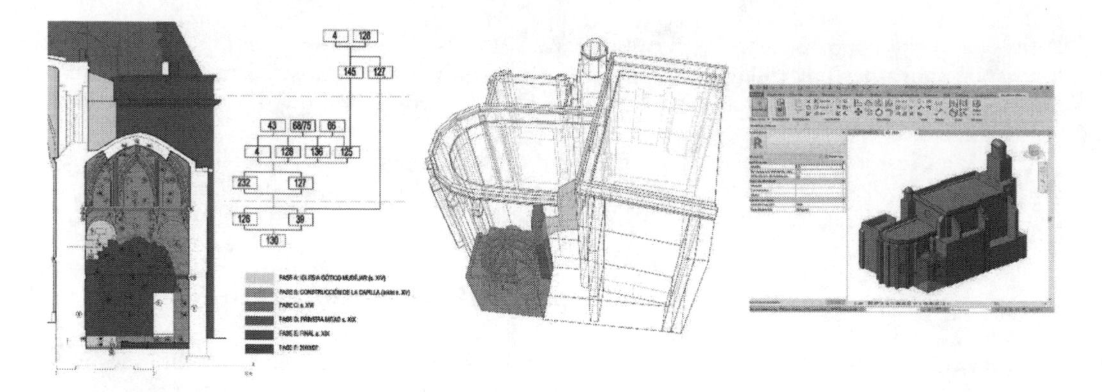

Figure 1. Different phases of the work process: Análisis estratigráfico, análisis topológico, construcción del modelo.

4 ANALYSIS AND INTERPRETATION

The need arises for a process of analysis and interpretation that addresses a general classification in constructive stages as a step prior to the construction of the digital model. Making an analogy with the traditional drawing would correspond to a set of drafts prior to the execution of the final result (Figure 1).

From the information obtained from different fields of knowledge, the main architectural units and their interrelation are identified, being able to recognize them as topological and chronological, based on their position in space or time. These relations between the different units detected serve as support when making decisions about the construction of the model.

Sometimes the division of the whole is immediately obtained from the construction process. This is not so evident at other times and more complex relationships are detected, as happens as the chapels are incorporated into the central body of the temple; for example, one of the abutments of the apse forms part of the enclosure of the Capilla de la Jura, and in turn it serves as support for one of the flying buttresses built in the late Gothic reform of the adjacent section of the church.

Among the possibilities offered by the BIM tools is the assignment of different attributes and properties to the elements that make up the model. From the point of view of time, the assignment of each constructive element to a specific phase allows continuous visualisation of the transformation sequence of the building.

In this way, in the case of the chevet of San Juan, the sequence has included the entire building (Figure 2). It would be a question of proposing a temporal sequence that can be modified in depending on the appearance of additional information on the chronology of specific elements.

5 CONCLUSIONS

The potential of this type of tool for the representation of the evolution of the property over time has been demonstrated, with display properties being selected for the elements belonging to a specific stage. From the expertise developed it is also proposed how the diachronic reading of architecture should be present in the generation of the digital model itself.

In this way, the need for a previous analysis of the building is established, which takes into account the different historical transformations and which serves as a support when making modelling decisions. This model may serve as a repository of the information generated in the research from the different fields considered. Like any model, the objective pursued in each case will determine the type of information to be entered and its

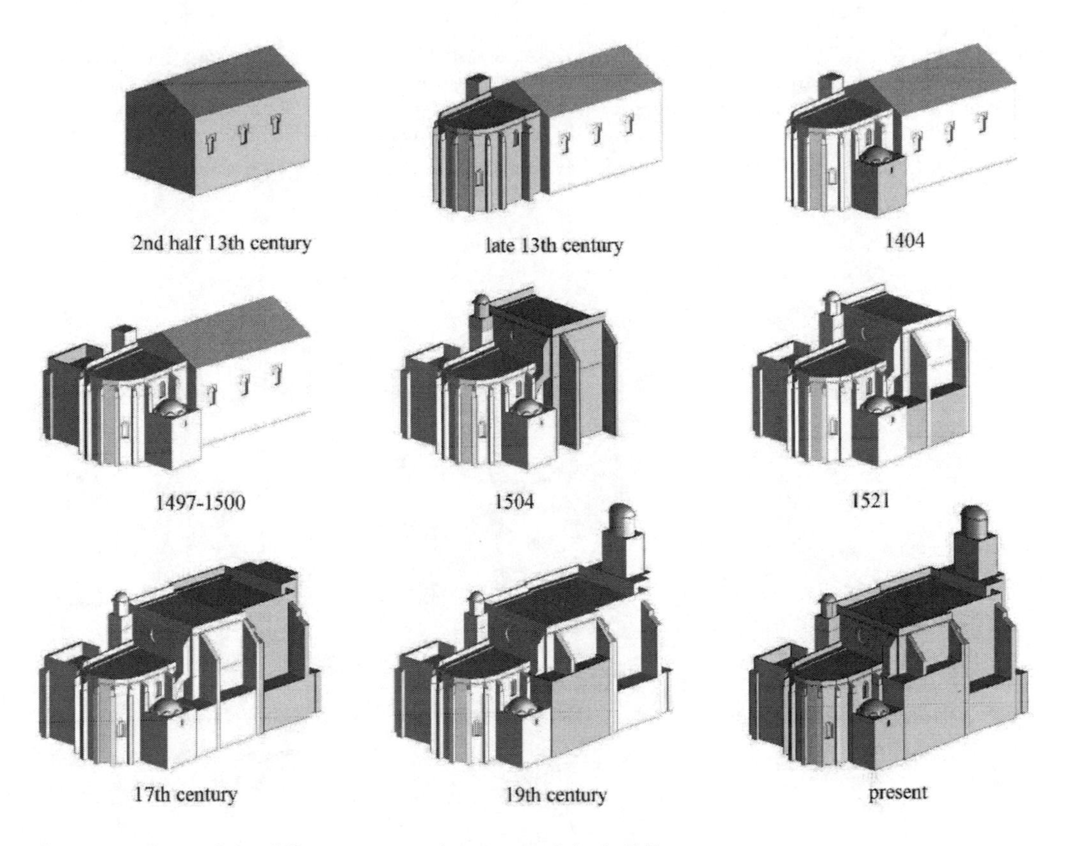

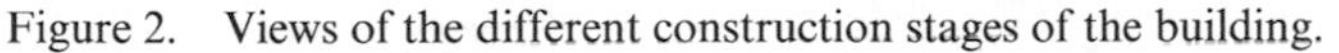

Figure 2. Views of the different construction stages of the building.

classification. Finally, it should be noted that an open model is pursued to include information from further research.

REFERENCES

Angulo Fornos, R. 2012. Construcción de la base gráfica para un sistema de información y gestión del patrimonio arquitectónico: Casa de Hylas. *Arqueología de la Arquitectura* 9: 11–25.

Angulo Fornos, R. 2015. Digital models applied to the analysis, intervention and management of architectural heritage. In *Building Information Modelling (BIM) in Design, Construction and Operations*: 407–418. Bristol: WIT Press, no. 149.

Dore, C. & Murphy, M. 2017. Current state of the art historic building information modelling. *The International Archives of the Photogrammetry, Remote Sensing and Spatial Information Sciences*, Volume XLII-2W5, 185–192.

González-Pérez, C. 2018. *Information Modelling for Archaeology and Anthropology: Software Engineering Principles for Cultural Heritage*. Springer.

Mora Vicente, G. & Guerrero Vega, J.M. 2015. Traza y proceso constructivo de la capilla de la Jura de Jerez de la Frontera. In Santiago Huerta y Paula Fuentes (ed.), *Actas del Noveno Congreso Nacional y Primer Congreso Internacional Hispanoamericano de Historia de la Construcción, Segovia, 13 a 17 de octubre de 2015*, 2: 1121-1113. Madrid: Instituto Juan de Herrera, Escuela Técnica Superior de Arquitectura de Madrid.

Romero Bejarano, M. 2014. Del mudéjar al gótico. Arquitectura religiosa a finales del XV en Jerez. In Manuel Antonio Barea Rodríguez y Manuel Romero Bejarano (coord.), *Actas del Congreso Científico Conmemorativo del 750 Aniversario de la Incorporación de Jerez a la Corona de Castilla: 1264–2014*: 437–458. Jerez: Ayuntamiento de Jerez.

Simeone, D., Cursi, S., Toldo, I., & Carrara, G. 2009. B(H)IM - Built Heritage Information Modelling. Extending the BIM approach to historical and archaeological heritage representation. In Thompson, Emine Mine (ed.), *Fusion - Proceedings of the 32nd eCAADe Conference - Volume 1, Department of Architecture and Built Environment, Faculty of Engineering and Environment, Newcastle upon Tyne, England, UK, 10–12 September 2014*, 1: 613–622.

Science and Digital Technology for Cultural Heritage – Ortiz Calderón et al. (Eds)
 ISBN 978-0-367-36368-0

A proposal for improving the conservation of the Asturian *Hórreos* by using the new technologies

I. Martínez & J. Rivas
Departamento Pintura y Conservación-Restauración. Facultad Bellas Artes UCM. Madrid, Spain

ABSTRACT: Most of the pieces that constitute an Asturian *hórreo* - only wood - have multiple features: they are resistant, the y must bear, transmit and distribute loads, resulting in a totally articulated and extraordinarily flexible assembly (Cobo, 2007).

In spite of being outdoors, the main problem of conservation of these specimens lies in the absence of a global register that includes all the *hórreos*, the lack of detailed information of each one, the bad regulations in terms of legislation, the rural exodus and the limited knowledge of what these constructions mean.

This article presents some of the solutions proposed to reduce these problems and protect this heritage as it deserves, by using new technologies which offer solutions within everyone's reach.

1 INTRODUCTION

The Asturian *hórreo* constitutes an outstanding movable heritage of medieval origins (Graña & López, 1986) that has been a privileged protagonist of the daily habits and cultural uses of its community throughout the centuries, to the point of becoming an irrefutable image and symbol of Asturias. But the rural exodus that took place during the 20th century led to the loss of its primary function as a barn and harvesting warehouse, giving rise to the gradual abandonment of the conservation habits.

According to current Asturian Cultural Heritage legislation, all *hórreos* erected prior to 1900 are considered *Movable Cultural Goods*, and must be covered by a system of integral protection that should be under the authority of the competent public institutions.

In spite of the existence of regulated protection by the Asturian Government, the fact is that such a large set of *hórreos* is unmanageable both from the point of view of its correct conservation and comprehension of its social value. Furthermore, there is an initial problem: the absence of a global register that includes all the exemplars, and the lack of detailed information on each one of them. In order to solve many of the conservation problems we are facing, it is interesting to use the new technologies, which offer us precise solutions within everyone's reach.

2 OBJECTIVES

- Organize in a hierarchy the whole collection of *hórreos*, considering that not all specimens have, necessarily and indiscriminately, the same degree of protection.
- Propose solutions for their conservation and value enhancement through the use of new technologies.
- Draw up a methodological guide that serves as a reference when registering these pieces from the point of view of their conservation.

- Awaken awareness about the importance of the *hórreos* within Asturian popular architecture, highlighting its documentary value, which to this day is still capable of explaining the history of this territory and the relationship that exists between man and environment.

3 METHODOLOGY

- Organize this Heritage in a hierarchy by establishing three different levels according to their importance as ethnographic testimonies.
- Elaborate a register of all existing *hórreos* through the use of new technologies, by geolocalizing *in situ* each one of them.
- Design a mutual database linked to the cartographic application.
- Using social networks to inspire a greater sensitivity and recognition for this type of popular artwork, in order to reach younger generations.

4 HIERARCHY OF THE COLLECTION

In our opinion, one of the main enemies for the *hórreos* belonging to the *Principality* of Asturias is, paradoxically, the legal overprotection to which absolutely *all* these specimens are exposed, without a clear differentiation of the intrinsic values of each. In fact, they all are protected at the same level of strictness. These circumstances contribute to turning them into an unembraceable Heritage that is difficult to protect. However, and again paradoxically, much worse than the above is the fact that in most of the cases there is not any kind of control over the interventions that are carried out on them.

In view of all the above anomalous circumstances, we propose that the collection of Asturian *hórreos* be hierarchically established at three different levels of importance, in order to be able to apply different protection measures in the future. In doing it so, we would be able to focus in the first instance on a smaller number of specimens, precisely in those that have a more relevant importance as ethnographic witnesses of the Asturian people. In order to determine the proposed levels (A, B and C), criteria have been established based on the ethnographic importance of each specimen, which will characterize each individually:

4.1 *Level A*

I) To be a determining example of the style to which it belongs because of its constructive characteristics or the motifs of its decorations.

II) To be prior to 18th century, currently maintaining much of its original structure.

III) To be an exemplary sample of its time by maintaining the original character of its polychromies or carvings in a visible way.

IV) To present singular forms or decorative motifs that are not common in the present styles.

V) To suppose an icon -individually or as part of an ensemble- for the village where it is erected.

4.2 *Level B*

VI) To present dated inscriptions or/and signatures of the craftsman that can be used to study and to date others from the same period or artist.

VII) To conserve remains of original painting, although the motifs are not entirely appreciable.

VIII) To maintain the original structure in its entirety, or with punctual documented restorations that have not distorted the whole.

IX) To present remarkable carvings, either in an isolated way or in the corridor.

4.3 *Level C*

The remaining specimens, which do not meet any of the nine commented criteria, will belong to Level C.

5 ELABORATION OF A METHODOLOGY FOR THE REGISTRATION OF THE ASTURIAS *HÓRREOS*

It is estimated that there are between 10,000 and 12,000 *hórreos* throughout the Principality of Asturias. Ideally, a new database should be set up and all the specimens should be recorded in a unified form. Finally, a new restoration methodology must be created (Valeriano, 2008).

The first step for the inscription consist of pure field work, since the whole Principality must be covered by filling in two basic files of the work (registration and evaluation of the state of conservation), apart from trying to contact the owners to know the history of the specimen and to know clearly to whom it belongs; besides, general and detailed photographs of each one of the constructions should be taken.

5.1 *Inventory No. Assignment*

It is essential to assign an inventory number to each specimen. A five-digit number has been used, preceded by two capital letters designating the council in which it is located, and followed by a lowercase letter indicating the Level to which it belongs within our hierarchy. We consider that the simplest form for the assignment of the number is by order of registration, reason why the digital code will be assigned of ascending form as they are documented.

5.2 *Geolocation*

A register of the location of the 142 copies catalogued has been accomplished so far by using Google Maps. This application allows us to create a personalized map in which it is possible to add markers (according to the longitude and latitude registered *in situ*) with a description and photographs. They have been marked with their inventory number, adding photographs

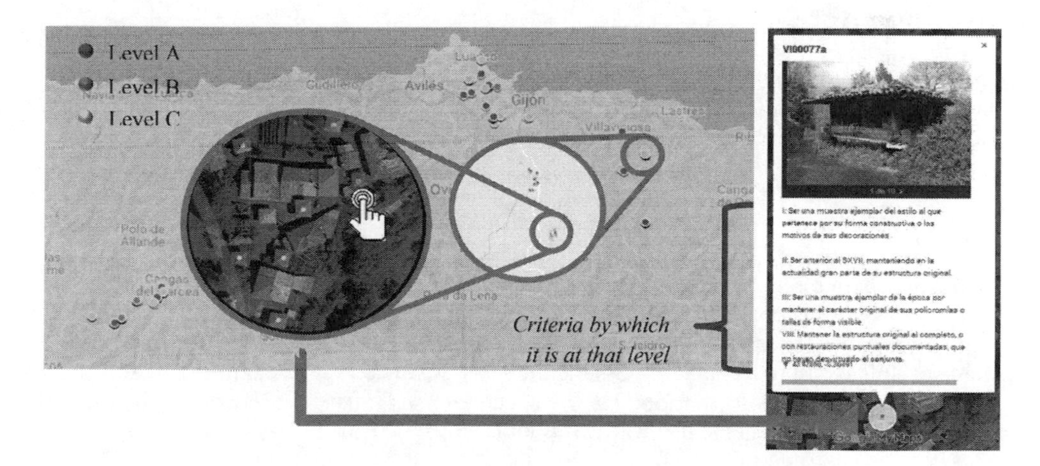

Figure 1. Sequence of the map 'Registro horreosypaneras.com' in Google Maps. Markers of each one of the specimens, with a color code by levels; in the right, the drop-down that appears when clicking one of the dots, with photographs and a list of the criteria to belong one level or another.

and writing the criteria they meet to be at one level or another. From now onwards, this task will be much simpler, as it will be possible to upload the information to the application *in situ.*

5.3 *Database creation*

It is necessary to create a database that contains all the information gathered in a clustered and structured way. Besides, this database must be accessible to add any kind of upgrade.

The database, which is still in the process of creation, should contain the data sheets and the state of conservation of the work, a direct link to the registration map and the restoration reports of the works that have been intervened. In addition, it must make it possible to search for the desired specimens using the inventory number.

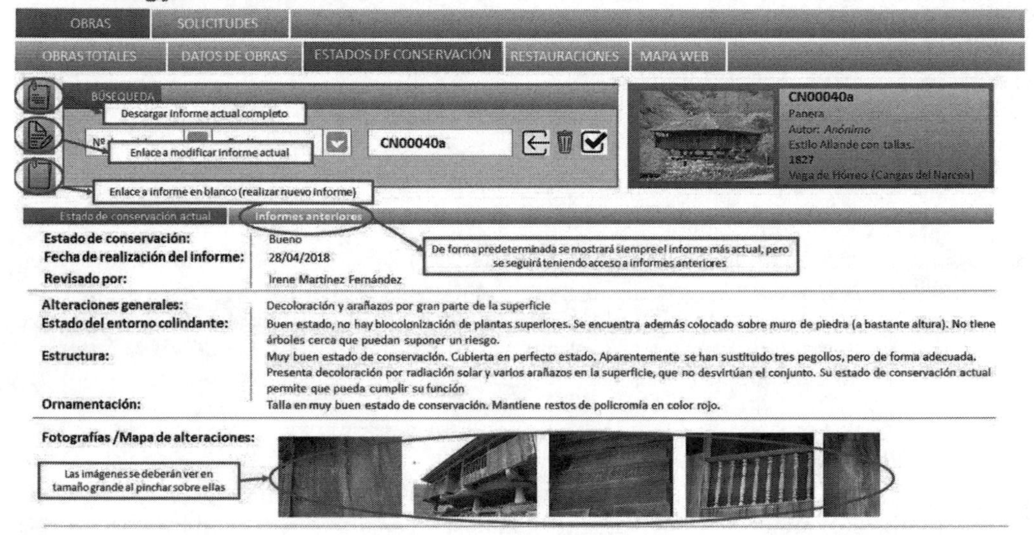

Figure 2. Database capture.

6 ADDING VALUE

In order to inspire a greater sensitivity and recognition towards this type of popular works of art and making them understandable for the people, social networks have been used for these purposes, achieving currently a good result with a high response rate.

7 CONCLUSIONS

The *hórreos* are an essential heritage to understand the customs and ways of life of the Asturian people. The anthropogenic agents have caused globally the biggest damages and problems for their proper conservation. It is necessary to carry out a hierarchical registration of all the existing specimens, using to make it possible the most advanced GPS technologies and gathering *in situ* cartographic information. The new technologies must be used to inspire both the sensitivity and acknowledgement towards this kind of popular art in the younger generations.

REFERENCES

Cobo, F. 2007. La restauración de los hórreos de Espinaredo. In *Intervenciones en el Patrimonio Cultural Asturiano*: 371–397. Asturias: Gobierno del Principado de Asturias.

Graña, A. & López, J. 1986. Dos nuevas vías para el studio del hórreo asturiano: una hipótesis sobre su prgen y una clarificación sobre sus decoraciones. In Frankowski, E. (ed.) *Hórreos y palafitos de la Península Ibérica*: 455–509. Oviedo: Itsmo.

Valeriano, J. 2008. Desarrollo de una guía metodológica sobre criterios generales para los proyectos de rehabilitación de hórreos. In *12th Internacional congress on Project Engineering*: *1736–1747*. Zaragoza: Asociación Española de Ingeniería de Proyectos.

Science and Digital Technology for Cultural Heritage – Ortiz Calderón et al. (Eds)
 ISBN 978-0-367-36368-0

Geographic information systems for the study of bunkers on the north shore of the Strait of Gibraltar

A. Atanasio-Guisado
Research Group. Strategies for Heritage Knowledge. University of Seville, Seville, Spain

ABSTRACT: During the Second World War, a fortified complex was built in the province of Cádiz whose mission was to protect the coastal strip linked to the Strait of Gibraltar. Composed of almost four hundred bunkers along more than one hundred and twenty kilometres, today they remain forgotten and unidentified, in spite of their value as an architectural reference of a turbulent epoch that marked the world event of the second half of the 20th century. Its dispersed configuration introduces a very clear territorial component, where Geographic Information Systems (GIS) is established as an essential work tool. This contribution synthesizes some of the procedures used with GIS tools to understand and record the fortified system on the north shore of the Strait of Gibraltar, and more specifically the complex of machine gun nests and anti-tank gun emplacements: what was constructed in the past; and what presently remains.

1 INTRODUCTION

In 1939 the Spanish Civil War ended and World War II began, with the Strait of Gibraltar as an area of great strategic value. It is then when Franco's government created the "South Coast Fortification Commission" (*Comisión de Fortificación de la Costa Sur*), with the mission of planning and constructing a fortified system on the north shore of the Strait. There were two objectives:

- To establish a defensive line to prevent enemy landings,
- To close navigation through the Strait at the will of the Spanish army.

To fulfil the first objective, about 400 nests were constructed for machine guns and anti-tank guns, most of them on the coastline. It is the type of construction usually associated with the term "bunker" or, in Spanish, "*fortín*". For the second objective, coastal batteries are constructed–with the mission of firing on ships that intend to cross the Strait–, and lighting posts–to light the passage at night while enabling coastal batteries to do their work–.All the nests, batteries, and lighting shelters are protected by anti-aircraft emplacements and linked by kilometres of roads and highways. It was, in short, a system that fortified the entire southern coast of the province of Cádiz.

However, after the Second World War, the interests of Spain had changed, and the military installation was practically abandoned, as it is up to present times, where there is still no complete inventory or any social recognition.

2 BACKGROUND

This contribution comes from a recently read thesis (Atanasio, 2017).The objective of the research was to understand the implications of the heritage of this defensive system, for which four objectives were established:

- To provide the defensive system with protective support as a heritage element, for which we rely on the figure of the Defensive Architecture Plan of Andalusia (PADA), which states that all Andalusian defensive architecture should be considered as an Asset of Cultural Interest, the highest level of protection in Spain.

- To provide the defensive system with theoretical support as a heritage element, either by analysing its modes of perception and representation or delving into the discomfort resulting from the mere presence of the elements that comprise it.

- To carry out systematic characterization of the bunkers of the Campo de Gibraltar as a heritage element, which is the briefly developed objective of this contribution. It is justified on the basis of a clear argument: groups of *fortines* are the substitutes for fortresses. Their multiplication is motivated by continuous advances in artillery, and where before there was only one defensive position, visible and prominent in the territory, in the 20th century dozens or hundreds had to be built, underground or camouflaged within the terrain, so that enemy fire could not locate them or destroy them when hit. Therefore, and as the fundamental idea, it is necessary to conceive the defensive complex of the Campo de Gibraltar as a system with a single purpose, where each of its components is part of the same network.

- Finally, to perform an architectural analysis of the bunkers that comprise the defensive system against coastal landings. It must be an individual and specific analysis of each of the *fortines*, without forgetting that they belong to a larger group that provides them with meaning. This purpose, which requires the previous one, is preliminary to the cataloguing of the bunkers of the north shore of the Strait.

3 SYSTEMATIC CHARACTERIZATION

It was proposed from the use of Geographic Information Systems (GIS), including all available information of the defensive complex of the north coast of the Strait on a properly georeferenced base. Our objective was to generate the tool from which to carry out the entire process of study, cataloguing, and future intervention of the bunkers that comprised the defensive system of the northern coast of the Strait; focusing on the defence system against coastal landings, the most numerous, consisting of nests for machine guns and/or anti-tank guns.

Now, we were faced with a dual objective. On the one hand, we had to understand what was constructed, on the other, to determine what presently remains.

3.1 *What was constructed, a snapshot of the complex in 1945*

For this phase, five joint projects were incorporated into the GIS database. For the network for lighting the Strait, we obtained the coordinates from a specialized bibliography (Sánchez de Alcázar, 2006).The other four (anti-aircrafts network, coastal batteries network, roads network and bunkers network), we obtained from different projects of the Southern Seville Intermediate Military Archive (AIMS).The plans of these projects are usually of considerable length, folded and bound to folio size. The possibility of working with photographs of these documents was infeasible because the scale was lost and the folds distorted the proportion, thus, the most interesting plans had to be scanned. Once scanned, they were georeferenced in GIS and then digitized, i.e., inserted as point-specific vector elements subject to parametrization.

In any case, these plans are nothing more than location drawings with territorial representation at a schematic level. It was, therefore, necessary to add a background planimetric base to our *database*, and that is where the "Cartographic Map of Andalusia from the German General Staff" with a scale of 1:50,000, drawn between 1940 and 1944 and entered into GIS by the Cartographic Institute of Andalusia, plays a leading role. (Figure 1).

We, therefore, have a planimetric base from the same time period for our joint projects, providing the result with a certain chronological coherence. It is that snapshot of the defensive complex we were looking for; the what was constructed. Obviously, the degree of reliability is moderate, since, at that time, it was drawn by hand and each of the bunkers was represented using a "large point" that far exceeded the real perimeter of the works. Even so, the degree of

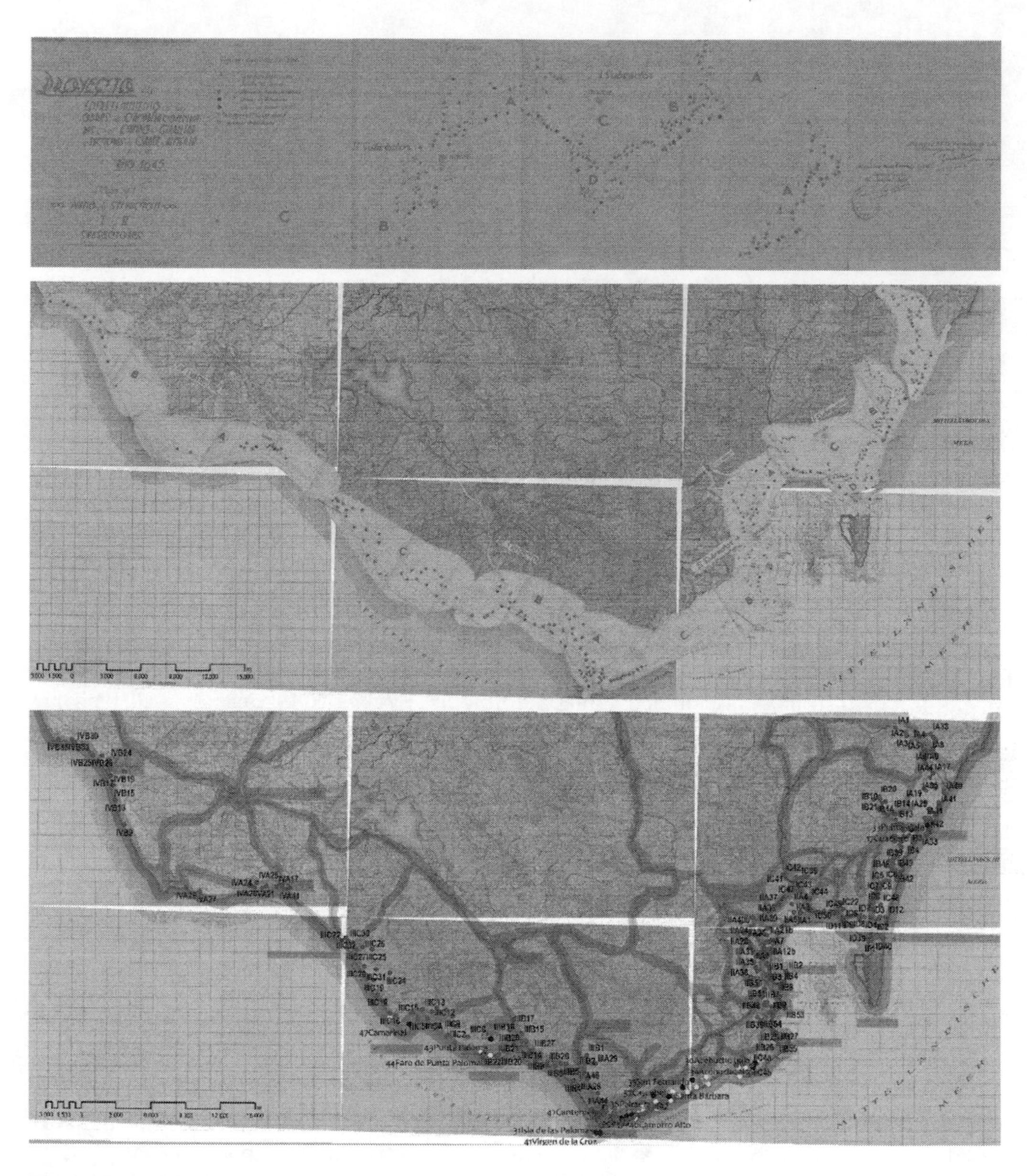

Figure 1. Digitalization sequence. Above, location drawing of the "*Proyecto de entretenimiento de Obras*" (AIMS, 1945), original scale 1:50,000.Original dimensions: 123x30 cm. In the central band, georeferencing on the German map. Below, digitization, including all subsystems studied (pillboxes against coastal landings in red).

precision is acceptable, and we have been able to verify that in the worst cases there are deviations of approximately five hundred meters.

3.2 *What presently remains*

The 1945 snapshot was the main tool with which the next phase began, a first step toward the cataloguing of bunkers that still exist on the north shore of the Strait. Given the difficulty of

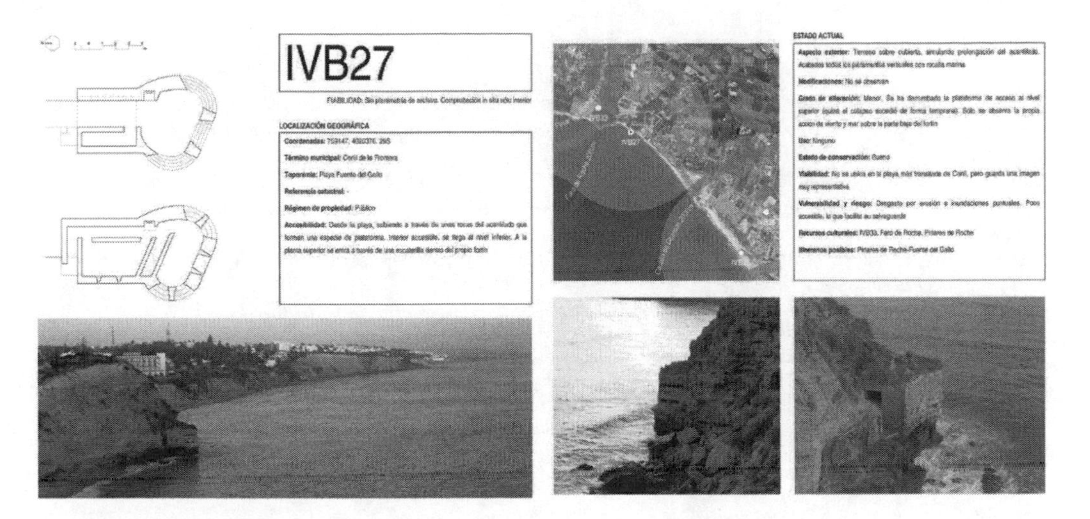

Figure 2. Main page and third page (current state) of the cataloguing of the bunker IV.B.27, for two machine guns and one anti-tank gun, located in Conil de la Frontera.

including the system as a whole, for the thesis, it was decided to select a specific area, centred on the municipalities of Conil de la Frontera, Vejer de la Frontera, and Barbate. Of the thirty-five bunkers constructed in that area as of 1945, fourteen have not been located. It has been verified that some have disappeared definitively. Others may still be there but have not yet been discovered. In any case, the missing *fortines* for which there exists some information–planimetry or location in previous orthophotos–have also been included in the catalogue, thus, only six out of thirty-five are not included.

The organization of the catalogue aims to be consistent with the rest of the research, meaning, a systemic understanding of the bunkers complex with the GIS as a support tool. It is divided into four blocks: location, on the main page; initial state, or what was constructed; current state, or what presently remains; and graphic annex (Figure 2).

4 CONCLUSION

As of today, the number of bunkers that remain standing is unknown, but the figure can be established at approximately three hundred. Of these, only one is specifically protected: a pillbox located in the municipality of Barbate, which has been included in the Catalogue of Protection of the General Urban Planning Program. Although we applaud this initial step, it is somewhat incoherent, given that just five hundred meters from this catalogued strongpoint there is another very similar one that has gone unnoticed by the planning drafters. In fact, in Barbate, at least twenty more bunkers were built, and of these, we have confirmed the existence of twelve.

This example summarizes what has happened with the *fortines* on the northern coast of the Strait at the heritage level, and we consider this contribution as appropriate: GIS is just a tool, a very powerful tool that enables us to advance in research for the protection of scattered architectures like the one under study, but does not propitiate that protection by itself. In this sense, we are of the same opinion as that of Professor Castillo Ruiz, who points out the need to understand heritage through current and subjective reading, ""*since the real reason that enables and bases heritage recognition of an asset is the importance or meaning that it acquires for society (or is provided by society)*"(Castillo, 2007).

REFERENCES

Archivo Intermedio Militar Sur de Sevilla.1945. Proyecto de entretenimiento de las obras de campaña construidas en el Campo de Gibraltar y sectores de Cádiz y Almería para el año 1945. Comandancia Ingenieros y Obras de Cádiz. 3982/1; 18/57–18/60.

Atanasio Guisado, A. 2017. *Arquitectura defensiva del siglo XX en el Campo de Gibraltar. Implantación territorial, análisis tipológico y valor patrimonial de los búnkeres*. Non published thesis. University of Seville.

Castillo Ruiz, J. 2007.El futuro del Patrimonio Histórico: la patrimonialización del hombre, *Revista electrónica de Patrimonio Histórico*, 1:3–35.

Sánchez de Alcázar García, C. 2006.*La artillería de costa en el Campo de Gibraltar.El RACTA nº5*.Valladolid: AF ediciones.

Science and Digital Technology for Cultural Heritage – Ortiz Calderón et al. (Eds)
© 2020 Taylor & Francis Group, London, ISBN 978-0-367-36368-0

Virtual reconstruction and constructive process: The case study of the dome of San Carlo alle Quattro Fontane

M. Canciani & M. Pastor Altaba
Architecture Department, Roma Tre University, Rome, Italy

ABSTRACT: This contribution focuses on the methodologies used for virtual reconstruction of cultural heritage with the aim of representing the different phases of the construction process.

These methodologies take part of the main research that relates the theory of the architectonical project with the execution of the real building, throughout the analysis of the construction of baroque oval domes and the temporary constructions and centrings used during the process. Considered the tool to ascertain the geometric correspondence between the theoretical form and the realized form, centrings are the first reality in which the project becomes before the real building is done.

Being adaptable to different cases of study, this methodology allows to extend the knowledge of ancient structures and offers the tools for a better readability of the monuments, through the integration of models and digital images with virtual and real augmented reality systems.

1 THE MAIN TOPIC

This contribution focuses on the methodologies used for virtual reconstruction of cultural heritage with the aim of representing the different phases of the construction process.

Particularly focused on temporary constructions and centrings, this research analyses the different techniques used during the construction process of baroque oval domes.

In this context, we can define centrings as the 1to1 scale reproduction of the geometry that should be [re]produced with the real building and, in the case of domes, centrings can be considered as the geometry that sustains the mix of materials that, once they are amalgamated, become construction.

2 THE METHODOLOGY APPLIED. THE DATA COLLECTION AND THE VIRTUAL RECONSTRUCTION

The methodology applied to this research uses a historic-comparative approach, by making a specific comparison of historical documents and surveys of the monument considered.

An effective comparison of information could only take place by the re-elaboration of two dimensional designs and three dimensional models, which are, in turn, the result of the study and analysis of the fonts, translated in a common language of representation.

According to this methodology, the virtual reconstruction hypotheses of a monument are the result of a process that starts with the collection and organization of all the relative information of each case of study.

This procedure to collect and organize the information has been divided in two phases. The first one consists on the acquisition of the onsite information (if possible) with methods of laser scanning and photogrammetry detection, in order to attach all the data to an existing geo-referenced reality; The second phase instead, includes the study of the ancient sources and the analysis of the most recent regarding research.

This methodology exploits the experience gathered before in the archaeological field, with the project of virtual [re]construction of Titu's Arch at Circus Maximus, in Rome.

In that occasion, the contribution of our group consisted on the development of the methodology that supported the virtual reconstruction of the arch, based on three-dimension virtual mesh modeling of the fragments found onsite and their connection with a multilevel database that combined all the different historical sources that described general and specific features of the monument with the main architectonical elements of the arch, described by a detailed geometrical model of each one. (Figures 1 and 2)

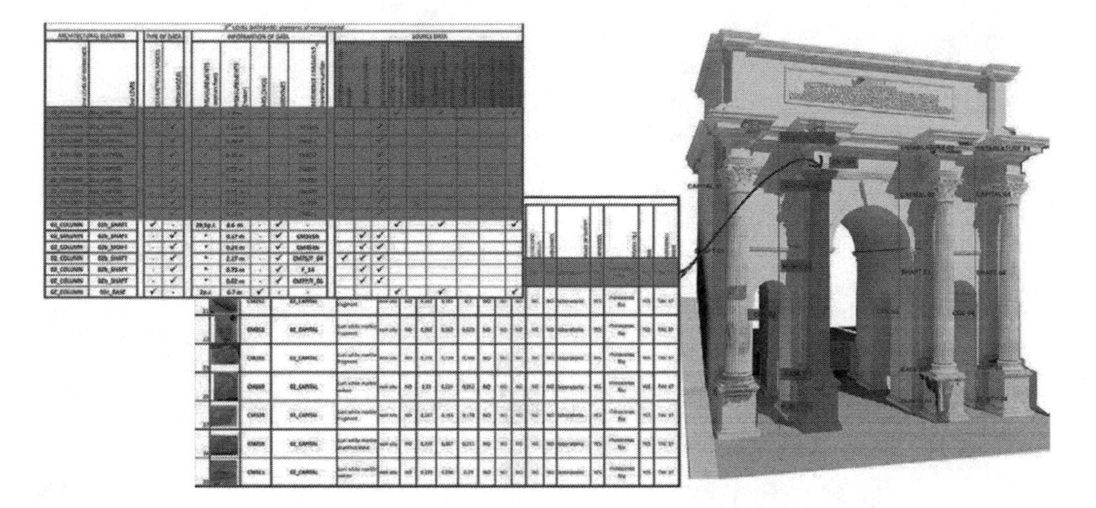

Figure 1. Reconstruction of Titu's Arch. Collaboration between Roma Tre University Architecture Department, and 'Sovrintendenza capitolina ai beni culturali' at the main project of Requalification f th environment and enhancement of archeological discoveries at Circus Maximus and surrounding public spaces of Via dei Cerchi and Porta Capena Square in Rome, that took place between the years 2012 and 2016.

Figure 2. Virtual Reconstruction of Titu's Arch, superimposed to the archeological site. Circus Maximus. Rome. The working group about Virtual Reconstruction of Roma Tre University, Department of Architecture, is formed by: M. Canciani (coordinator), C. Falcolini, M. Saccone, G. Romito, B. Mammì, M. Pastor Altaba. The working group of Sovraintendenza Capitolina ai Beni Culturali is formed by M. Buonfiglio, S. Pergola, A. Coletta.

3 CENTRINGS FOR THE DOME OF SAN CARLO ALLE QUATTRO FONTANE

In the specific case of this contribution, centrings for the construction of oval domes are the main topic, and the reproduction of the construction process of the dome of San Carlino alle Quattro Fontane is the main case study.

The beginning of this research took place in 2011, and follows all the previous studies of the geometry and static structure considerations of the dome. (Portoghesi P, 2001), (Alonso García, E, 2003), (Bellini, 2004).

Our studies started with the work of analysis of the drawings of F. Borromini, finding in them the geometric-constructive alignments that guided the realization and construction of the dome.

Being an ongoing research, at this moment we are focusing on some partial aspects, that are the collection, organization and analysis of the fonts, the survey of the cases study and the elaboration of the first digital restitutions.

To make evidence of the general panorama of the different nature of this kind of data, we have divided the information in two principal groups: the first one is composed by directly related information of the case study taken into consideration; and a second one, made up by not directly related information.

At the same time, directly related information has been divided in three main parts: the project's drawings of the architect, the surveys of the monument and their digital restitutions and the further analysis and studies made by other authors.

As part of the directly related information, in these synthetic drawings (Figure 3) we can see some of Borromini's drawings, compared with different horizontal sections of the surveyed model. These sections also show us how the oval shape varies slightly in height, and how the choice of the support points of the lantern convey the weight of the structure along the curvature of the bowl following precise directions.

At this moment we are realizing new 3D surveys, by laser scanning and photogrammetry methods, and their digital restitutions, to have a current copy of what it is actually built (Figure 4). The utility of these new surveys, made some years after the previous ones, is to obtain a more accurate representation of what is built, and to verify them, in order to take them into consideration as the basis for the drawing of the centrings.

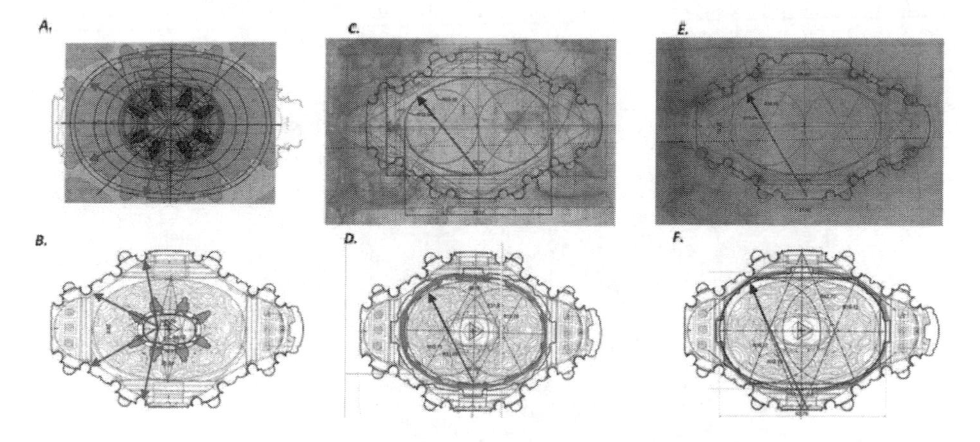

Figure 3. **A.** Plan of the dome, lantern height, church pavement and AzRom190 superimosed; **B.** Plan of the dome, lantern height (quote +21.92) and church pavement quote; **C.** AzRom175; **D.** Plan of the second level of coffering of the dome (quote +19.20); **E.** AzRom 170; **F.** Dome's plan from beneath, impost height (quota +16,38). Drawings and analysis published by the author Canciani M. in Drawing, geometry and Construction. The dome of San Carlino alle Quattro Fontane (1634-1675) by Francesco Borromini. In Amoruso, G., (ed). *Handbook of Research on Visual Computing and Emerging Geometrical Design Tools. 2016,* and in *Il disegno della cupola del San Carlino alle Quattro Fontane di Borromini: ovale canonico? In DISEGNARE CON. 2015, Vol. 8, 15:* 12.1-12.22. (http://disegnarecon.univaq.it/).

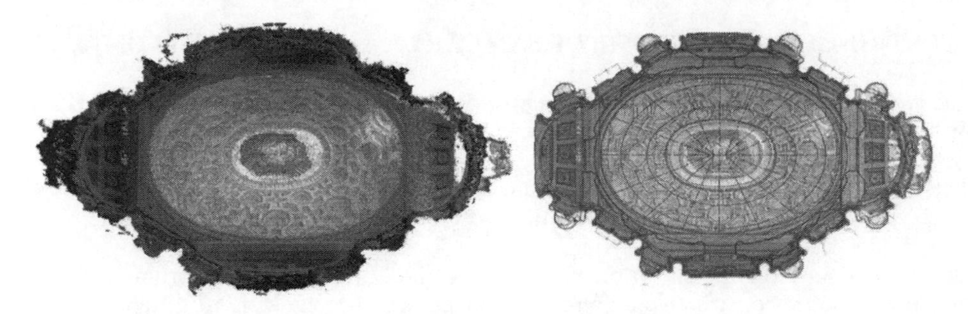

Figure 4. A. Phogrametric survey raw point cloud; B. Superposition of point cloud and bidimensional restitution of dome's plan.

As the second group of data, instead, we have mainly considered the information found in the architectural treaties written from sixteenth to eighteenth centuries, related to construction techniques of domes (that include proportion rules, geometric traces and constructive rules, depending on the materials used) and centrings and temporary constructions.

3.1 *The construction of domes from sixteenth to eighteenth century*

In this historical period, the criteria for the construction of domes followed the medieval and renaissance tradition, ruled by empiric notions and based on arithmetic proportions. Indeed, most of the times, the drawings found fixed the optimal proportions between diameter and thickness at the impost of the dome and at the oculus of the lantern. This proportion reduces time by time thanks to a better awareness and knowledge of structural behavior of domes.

In these two drawings of Carlo Fontana (Figure 5), he not only fixes the proportional relation, but combines it with a constructive notion of the material used. And he also defines and explains all the necessary elements for the construction, using a simple geometric rule that allows drawing the traces.

As an example of the different constructive techniques used to the construction of domes, we can also see Figures 6a and 6b, in which not only the technique, but also the materials used are detailed, and Figures 6c-6d, in which only the constructive technique, but also the structural elements of the carpentry for the centring are represented. (Villani, M. 2008).

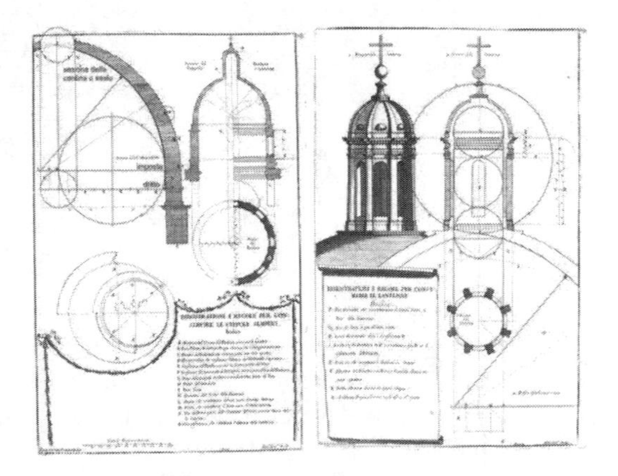

Figure 5. Carlo Fontana *'Il tempio Vaticano e sua origine…'* 1694; *'Demostratione e Regole per costruire le cupole'* (pp. 367-368). Pages of the treaty of Carlo Fontana, with the evidence of the construction lines of the arch for de dome.

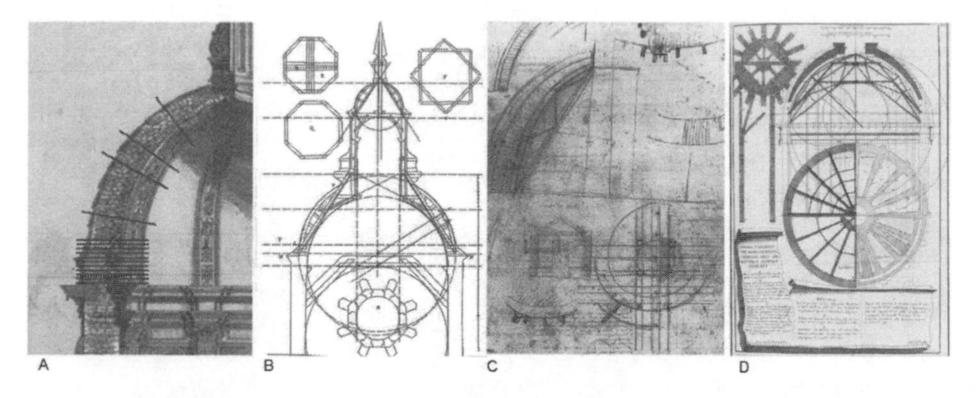

Figure 6a. Domenico Fontana, *Capella sistina di Santa Maria Maggiore*, Roma. 1590. Fabric Constructive technique; 6b. Fray Lorenzo de San Nicolás. *'Arte y uso de la Arquitectura'*, Madrid. First publication 1633. *Wooden Dome 'encamonada';* 6c. Gian Lorenzo Bernini. *Centrings for the dome of Santo Tommaso church at CastelGandolfo. Italy 1659.* Evidence of the geometric traces; 6d. Carlo Fontana. *'Pianta e profili che dimostrano il composto dell'armature e cuppola vaticana'* 1694. (p. 321). Plan and section to evidence the armors for Vatican dome.

3.2 *A hypothesis for a centring. The reproduction of the construction process*

From the combination of the two kind of data explained, taking into consideration the main specific documents (Fra Juan di Buenaventura, 1650-1655), (Num. Primo, 1634-1655), the studies of the construction (García Montijano, J.M., 1999) and baroque worksites (Marconi, N., 2004), and considering also the restoration works of the church, carried out from 1997 to 2006 and published in (Degni, P. 2007), we can synthesize the construction techniques and the process followed for the construction of the dome of San Carlino and have already obtained the first hypothesis of the centrings.

In this drawing (Figure 7) we see the longitudinal section of the dome, with the main centring that consists of a rectilinear element (the chord in the geometric construction) that connects the lower point of support at the impost plane and the one above at the junction with the lantern; by a curvilinear element, (the curvature of the centring and the dome); and finally by other secondary elements that reinforce and subdivide the structure.

In the plan it is possible to see that the structure of the columns of the lantern is geometrically aligned with the lower structure of the Church. This creates a simple geometric and static scheme that goes from the top to the impost level of the dome.

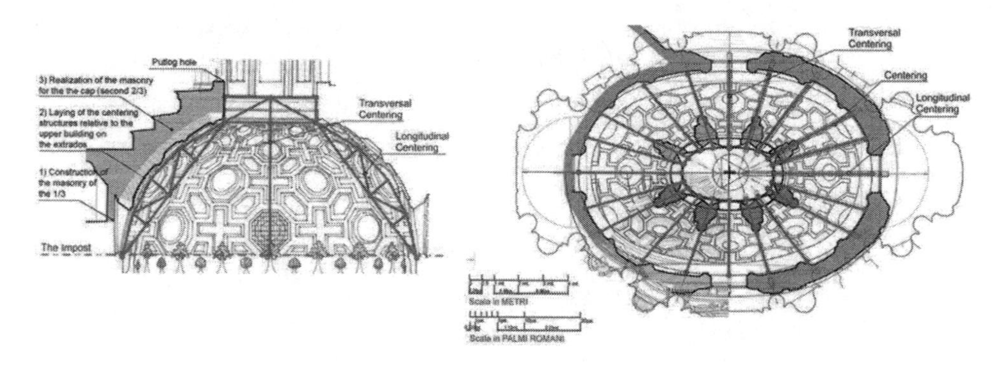

Figure 7. Longitudinal section and plan of San Carlino's dome. Survey hand drawings of M. Michelini. Evidence of the main structure for the centring hypothesis.

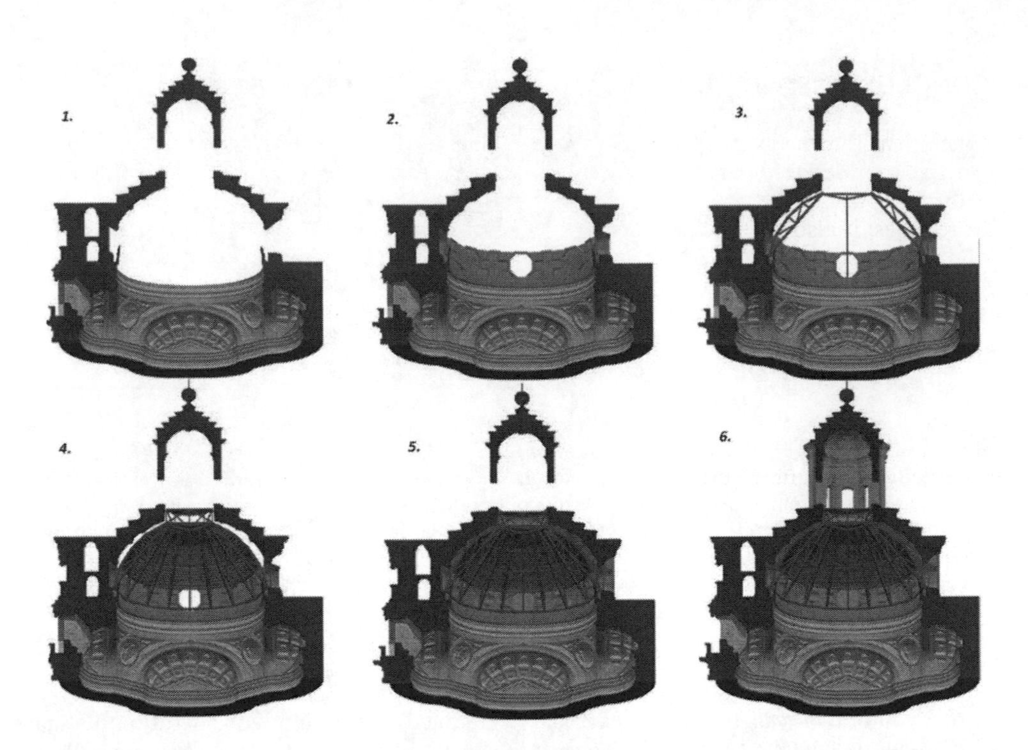

Figure 8. Phases of the construction process of the dome. Virtual digital model renders of M. Canciani, M. D'Angelico and M. Saccone. Models also published in: Canciani, M., Spadafora, G., Pastor Altaba, M., Formica, G., D'Angelico, M. & Lebbroni, C.2018. Geometric constructive traces in drawings by Francesco Borromini. In ICGG 2018. Proceedings of the 18th International conference on geometry and graphics 40th anniversary. Milan: 3-7 August 2018: 111-111.

With these models, we can synthesis (Figure 8) the construction of the dome of San Carlino:

The first phase consists on the construction of the first third of the dome. In this part the masonry is carried out following horizontal rows that lay on the levels beneath;

The second phase is related to the provisional centring, necessary only to the masonry construction of the remaining two thirds of the dome. On the pre-existing first row of coffers, taking advantage of the compartments of the four windows to anchor the wooden structure, the two main transverse and longitudinal centrings are built.

Then, the other twelve centrings are placed, following the geometric construction explained before and the planimetric system established in the project; Once the main structure is place, the planks are superimposed.

The following step, consists on the the construction of the tiburium and the risers.

As pointed out, the masonry gets stretcher along the curvilinear surface, with variation of the thickness, according to the horizontal reference planes we saw before and to the vertical section to which they belong. Above the risers, the 8 twin columns of the lantern are placed in the directions determined by the centrings.

To end with, the centrings are dismantled and a second structure is constructed to apply the finishing intrados layers. But, this structure does not necessarily follow the geometry of the dome.

4 CONCLUSIONS

To conclude with, we would like to list the threefold aim of the methodology used: it allows to extend the knowledge of ancient structures, strongly transformed, most of the times, over time, throughout the integration of different three-dimensional models that can combine different realities; it also shows how drawings and geometry can provide a pragmatic constructive and static solution; And in third place, being adaptable to different cases of study, it is useful to deepen the learning of monuments, and to offer the tools for a better readability of cultural heritage, through the integration of virtual and augmented reality systems, which can be viewed and consulted by everyone.

REFERENCES

Fra Juan di Buenaventura. 1650-1655. *Relazione e Fabrica del Convento di San Carlo alle Quattro Fontane (Ms.77a)*. Rome: Archivio di San Carlino alle Quattro Fontane. (A.S.C.Q.F.).

Num. Primo, 1634-1655. Rome: Archivio San Carlo alle Quattro Fontane.

Montijano García, J.M. 1999. *San Carlo alle quattro fontane di Francesco Borromini nella 'Relatione della fabrica' di Fra Juan de San Buenaventura*. Milan: Il Polifilo.

Portoghesi, P. 2001. *Storia di San Carlino alle Quattro Fontane*. Rome: Newton&Compton.

Alonso García, E. 2003. *San Carlino. La máquina geométrica de Borromini*. Valladolid: Universidad de Valladolid.

Bellini, F. 2004. *Le cupole di Borromini. La 'scienza' costruttiva in età barocca*. Milan: Electa.

Marconi, N. 2004. *Edificando Roma Barocca Macchine, Apparati, maestranze e cantieri tra XVI e XVIII secolo*. Rome: Edimond.

Degni, P. 2007. *La 'Fabrica' di San Carlino alle quattro fontane. Gli anni del restauro. Bollettino d'arte*. Rome: Istituto Poligrafico e Zecca dello Stato.

Villani, M. 2008. *La più nobil parte. L'architettura delle cupole a Roma 1580-1670*. Rome: Gangemi Editore.

Science and Digital Technology for Cultural Heritage – Ortiz Calderón et al. (Eds)
© 2020 Taylor & Francis Group, London, ISBN 978-0-367-36368-0

Fusion of 3D digitization technologies for the virtual exploration of re-covered archaeological remains

E. Pérez, P. Merchán, M.J. Merchán & S. Salamanca
Department of Electric Engineering, Electronics and Automatics, University of Extremadura, Badajoz, Spain

ABSTRACT: Occasionally, during the course of building new infrastructures works, some he-ritage pieces are unearthed. Given this situation, and when the finding cannot be properly preserved once excavated it is always needed a documentation process for the resilience of heritage since the discovered pieces are forced to be hidden again, covered by the planned infrastructure. Nowadays, the available technology allows to store a 3D digital copy of the heritage pieces and also gives the opportunity to people to visualize the site with accurate scale and color. In this article, our group (3D-CoViM) presents an example of such case that occurred when refurbishing an old road in Fuente del Maestre (Spain). We carried out the 3D digitalization of the found he-ritage site and also designed Augmented and Virtual Reality (AR and VR) applications.

1 INTRODUCTION

Technology has become a perfect ally to the Cultural Heritage (CH). Using scientific and technological techniques has allowed improving this field not only in research terms but also if we talk about conservation and preservation (Cozzani (2017), Zhou (2012)). This progress has come with a great saving in working hours -what implies researchers can advance more quickly- together with a more careful treatment of the objects and the creation of repositories to ensure the resilience of works of art. But, uttermost, it becomes an effective method of diffusion and transfer of results to society, which receive an understandable model far from the, in most cases, uninte-lligible fragments they can see. This possibility is even more important when the archaeological ruins must be covered and are not visible to visitors (Tait (2016)).

In this sense, and agreeing with the last idea, we present an AR and VR experience created to let the users visit and understand those archaeological sites that are hidden from the public view. To do this, we have designed a procedure to merge different acquisition techniques in order to create digital 3D models. This work allows us both to preserve and to disseminate CH through the creation of repositories that ensure its resilience and VR experiences (Pérez (2018)), which constitute a great vehicle for society to get involved with its own heritage.

2 ARCHAEOLOGICAL SITE "LA MATILLA"

During the course of the refurbishment works of the EX 360 road, near the town of Fuente del Maestre (Spain), the existence of various archaeological remains was discovered in the place known as *La Matilla*. After the corresponding study, it was determined that the main structures dated from the Roman era and that they were composed of two types of constructions: on the one hand, remains of walls of what could have been administrative buildings and, on the other hand, the most striking findings: remains of ovens whose basic structure (walls and arches) had been preserved in very good conditions. A sample of this finding can be observed in Figure 1.

Figure 1. (a) 3D acquisition of the interior of ovens using the Artec MHT 3D scanner; (b) Virtual scene of the outer are of the archaeological site where the rests of administrative buildings are observed.

3DCOVIM carried out the 3D digitalization of the entire site, with the objective of permanently document it and also designed and programmed some basic applications for the visualization of the data.

In the following sections the procedure that we have followed is explained.

3 ACQUISTION AND RECONTRUCTION PROCEDURES

3.1 *Acquisition procedure*

When carrying out a digitization work, one must take into consideration the basic differences between the different technologies available for this purpose. Among them, for this work we decided to apply three technologies: a Faro LS880 Laser-Scanner, an Artec MHT 3D structured-light scanner and a Nikon D60 digital camera.

Each of the technologies was applied to the different elements found in the site, considering the topology of these elements and the application to which they are aimed.

With respect to the typology, it was decided to digitize the entire outer area of the La Matilla using the laser scanner and the photogrammetry. On the other hand, the structured light scanner was applied to digitize the interior of the ovens (Figure 1b) that were the most characteristic elements, better preserved and that required a higher resolution.

According to the application, the laser scanner was applied only for documentation, while the other two technologies were used for the documentation as well as for visualization applications. This choice is made to optimize the processing time in one of the elements in particular: the outer area. By using two different technologies in that area, laser scanner and photogrammetry, two different resolutions will be obtained: the high-resolution data produced by the laser scanner, wanted when documenting CH; and the medium-low resolution data provided by photogrammetry, which are necessarily demanded when designing real-time visualization applications.

3.2 *Reconstruction procedure*

The data obtained with each of the acquisition technologies were treated practically independently, and using specific software in each case, to obtain different 3D models. In this section we explain the work scheme followed to generate those models, depicted in Figure 2.

The raw data offered by the Artec MHT 3D scanner consists of a wide set of textured meshes, which require: a registration in a common reference system; a filtering to eliminate the noise; a fusion operation to obtain a single mesh, and a filling process; occasionally, a surface retouching has also been required; and, to finish, the repair stage of the mesh is inescapable. The first 3 stages were carried out using the Artec Studio software. Next, the well-known open source software Meshlab was used to fill and repair the mesh. The surface

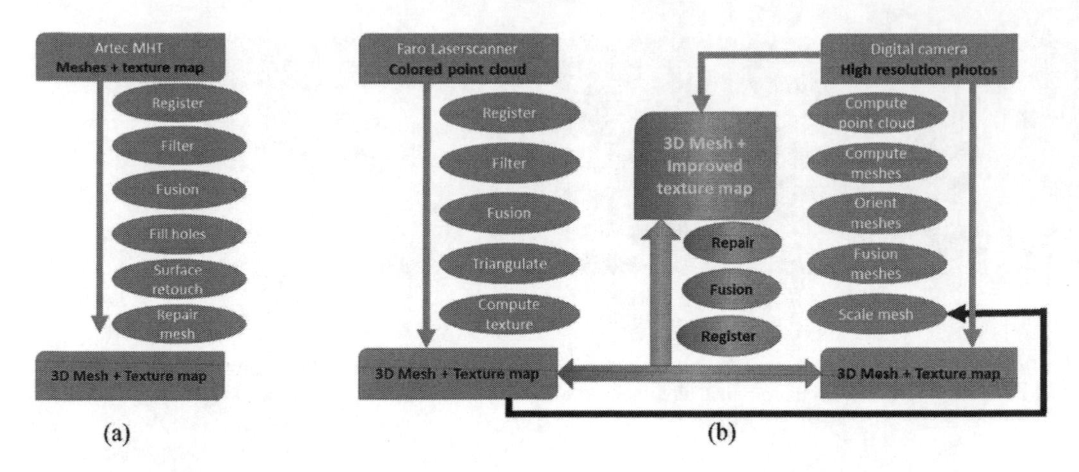

Figure 2. Work scheme followed to generate 3D textured models: (a) from Artec MHT 3D scanner, and (b) from Faro Laser scanner and Nikon D60 digital camera.

retouching was done with Blender software. Also, in specific cases, it was necessary to make some adjustments of texture equalization, which was achieved thanks to the use of GIMP software.

In relation to the data generated by the Faro laser scanner, the following steps are required: registration of point clouds; filtering the noise of the acquisition process; fusion in a single point cloud; triangulation of the cloud to obtain a meshed surface; and, finally, the calculation of the texture image that is applied on that surface. The first three stages were carried out using the Faro Scene software and the last two using Meshlab.

On the other hand, from the list of high-resolution photos taken by the Nikon camera, and thanks to the Agisoft Photoscan photogrammetry software, the point cloud was calculated and then the mesh for each subset of photos. Next, all the meshes were oriented and merged into a single mesh. Finally, the resulting model was scaled taking as reference some measurements taken in the 3D model obtained from the laser scanner data.

At this point, two models of different resolution were obtained from the entire external surface of the deposit, one in high resolution from the laser scanner data and another one in medium-low resolution from the photogrammetry. Although, with respect to the resolution, the first model resolution is considerately higher than the second, the same does not occur with respect to texture, since the images obtained by photogrammetry far exceed, in quality and number, those obtained with the laser scanner.

For this reason, and to incorporate some specific details not obtained through the Faro, it was decided to merge the two models. To do this, both are registered in a common reference system, the meshes are merged and the unavoidable repair stage is applied. Moreover, thanks to the Blender software, some of the photographs of the Nikon D60 camera are applied as a texture, resulting in a 3D mesh of the entire area of the deposit with an improved texture.

To produce a complete model of the deposit, which includes both the external area and the interior of the ovens, it is necessary to integrate all the data obtained through the three technologies to create a single global mesh.

After generating all the models in real scale, the following necessary process is the registration of the ovens with the outer surface, to position them in the location where they are located. This adjustment was made using the user-assisted registering tools, selecting points that were common in both models as a reference. In particular, part of the external entrance of the ovens were also digitized with Artec, which were also acquired through the other technologies, so as to have avai-lable overlapping zones that are essential to perform the registration.

4 VISUALIZATION APPLICATIONS

4.1 *Data adaptation*

Regarding VR and AR applications, it should be noted that the data will require an adaptation to be used in them, since, due to hardware limitations, both types of applications accept 3D with limited resolution and optimized texture.

To perform this adaptation, we divided the 3D into the following 4 categories: category A (laser scanner data); category B (Artec scanner data); category C (photogrammetry data); and, category D (3D modelled synthetic data)

Since we do not need high-resolution models, but the opposite, we discard the data obtained by the laser scanner.

The adaptation is performed, in general terms, considering three fundamental parameters: type of 3D models, resolution and color.

With respect to the type of model, that is, point cloud or 3D mesh, since we do not use the elements of category A and the rest of devices generates 3D meshes, an adaptation is not necessary.

With regards to resolution, as well as to texture and color, it is mandatory to reduce it in the elements of category B to be usable by VR application. The same occurs for both, the resolution and the color, with the elements of category C.

Finally, the elements of category D were already modeled considering VR requirements.

Since the AR application runs on a much less powerful device, it is necessary to further reduce both the resolution and the color.

4.2 *Augmented and virtual reality applications*

We have developed a proof of concept of AR without external markers using Unity and ARCore Android libraries. For the time being, it is only compatible with high performance mobiles, powerful enough to run 3D objects with a relatively high number of polygons.

Due to the big size of heritage 3D model, and to maintain a 3D resolution that let the user observe a certain level of detail, it was needed to split the 3D model into several parts.

In addition, a basic VR application has been developed also in Unity. This is aimed to explore the archaeological site both the external area and the interior of the ovens. It is compatible with a multitude of VR devices connected to a PC with affordable specifications. Apart from the digitalized heritage elements, a certain decoration was added to increase the realism of the environment represented in the 3D virtual scene. Figure 1b shows a screenshot of the VR application.

5 CONCLUSIONS

A case of documentation necessary for the resilience of cultural heritage has been explained.

The procedure followed to combine different acquisition technologies has been exposed, and two possible applications of visualization of the 3D data have been shown.

As future work, the improvement of applications through the insertion of labels and explanations is considered. In addition to that, we plan to use of real markers located at the site to position the 3D visualization over the approximate real location of the buried site. Moreover, a better optimization of 3D resources is also a pending task.

ACKNOWLEDGEMENTS

This work has been supported by the project IB16162 from Junta de Extremadura and European Regional Development Fund (ERDF) “A way to make Europe”.

REFERENCES

Cozzani, G. & Pozzi, F. & Dagnino, F.M. & Katos, A.V. & Katsouli, E.F. 2017. Innovative technologies for intangible cultural heritage education and preservation: the case of i-Treasures. In P. Thomas (ed.), *Personal and Ubiquitous Computing* 21(2): 253-265.

Pérez, E. & Merchán, M.J. & Moreno-Rabel, M.D. & Merchán, P. & Salamanca, S. 2018. Touring the Forum Adiectum of Augusta Emerita in a virtual reality experience. In M. Ioannides et al (eds.): *Digital Heritage. Progress in Cultural Heritage: Documentation, Preservation, and Protection. Lecture Notes in Computer Science* 11196: Springer.

Tait, E. & Laing, R. & Grinnall, A. & Burnett, S. & Isaacs, J. 2016. (Re)presenting heritage: Laser scanning and 3D visualisations for cultural resilience and community engagement. In A. Foster & P. Rafferty (eds), *Journal of Information Science* 42(3): 420–433.

Zhou, M. & Geng, G. & Wu, Z. 2012. *Digital Preservation Technology for Cultural Heritage*: Springer.

Science and Digital Technology for Cultural Heritage – Ortiz Calderón et al. (Eds)
© 2020 Taylor & Francis Group, London, ISBN 978-0-367-36368-0

GIS methodology and SDI for risk analysis in medieval defensive earth architecture: Territorial characterization through spatial analysis, Delphi method and analytic hierarchy process

E. Molero Melgarejo
Department of Urban Planning, University of Granada, Granada, Spain

D. Casado
University of Granada, Granada, Spain

M.L. Gutiérrez-Carrillo
Department of Architectural Constructions, University of Granada, Granada, Spain

ABSTRACT: Within the framework of the R+D+I project entitled "Sustainable methodology for the conservation and maintenance of the Medieval defensive earth architecture in the South-East Iberian Peninsula: diagnosis and prevention against natural and human risks (PREFORTI)", the task of creating a catalogue of defensive earth architecture located in the Spanish southeast, incorporating fact sheets on material, constructive and structural characteristics, was addressed. Subsequently, a selection is investigated more thoroughly in order to develop systematic databases that enable us to create charters of preventative measures, related to the risk charters recommended by the National Plan for Defensive Architecture and Preventative Conservation of Cultural Heritage.

1 INTRODUCTION

PREFORTI geodatabase compiles all the architectural elements covered by this project and establishes recognition and cataloguing criteria that are integrated into an open source GIS. Cases that merit more detailed study are selected using the intersection of territorial information (natural and anthropogenic agents) and the precise location of the items listed. This selection is investigated more thoroughly in order to develop systematic databases that enable us to create charters of preventative measures, related to the risk charters recommended by the National Plan for Defensive Architecture and Preventative Conservation of Cultural Heritage.

Recent studies have linked territorial variables with the location of the earthen architecture (Mileto et al, 2019), but only to relate the construction technique and the specific properties of the territory. Geographic information systems (GIS) are also being widely used in a wide variety of fields for the study of environmental risks (Bolteanu, 2010; Chiessi, 2010 & Giupponi, 2006). When the number of risks considered in the study is high, the comprehensive assessment is achieved through the application of probabilistic models, multicriteria analysis and thematic map overlay where the Delphi method occupies a prominent place (Asprone, 2010; Halpern, 2007 & Coppolillo, 2004). The application of GIS for the study of risks related to Cultural Heritage has been described in various articles (Lanza, 2003 & Canuti, 2000), being the most complete methodological approach, applied to the analysis of Cultural Heritage risks, the Charter of Risks of Italy (Baldi, 1995), model on which the design of this methodology is based.

Following the chronotypological classification established in the IPCE's National Plan for Defensive Architecture, the defensive architecture built in the southeastern Iberian Peninsula is mainly made up of medieval fortifications that have been continuously exposed to the effects of human and environmental agents since their construction. Vulnerability to these agents is one of the serious problems affecting this type of rammed earth construction, particularly in the southeast of the peninsula. Risk maps can be produced by cross-referencing the georeferenced items and the selected natural and human variables. The present work deals with the choice of open source database management systems (DBMS), as well as their attributes and normalisation, the data model used, and the gis methodology for the risk maps generation.

2 HYPOTHESIS AND OBJECTIVES OF THE STUDY

It is based on the hypothesis that, from the framework of the preventive conservation of this patrimonial elements, the study of the relationship between the danger that territorial variables infer to the patrimonial elements and their degree of intrinsic vulnerability would allow us to better understand the origin of the pathologies observed and would serve to predict the effect produced by each natural/anthropic agent (seismic, landslides, floods, snow/ice/rain, fires ..) and prioritize preventive, urgent and restoration actions.

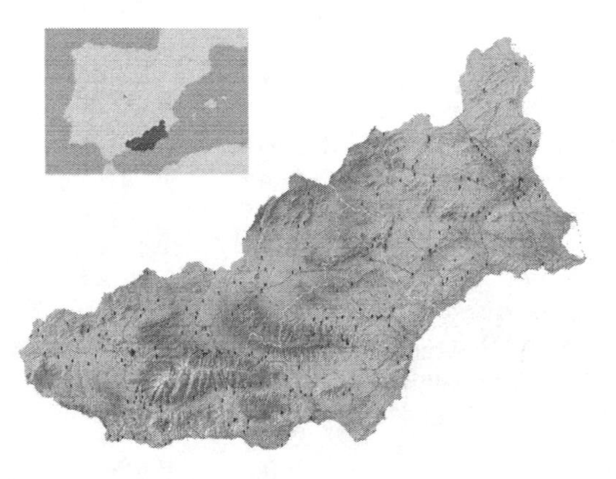

Figure 1. PREFORTI GIS Study area and items locations.

3 METODOLOGY

This inventory of medieval defensive earth architecture in the South-East Iberian Peninsula (252 items) and the catalogue selected (80 item) subsequently meet the INSPIRE conditions and each of the territorial information layers used are generated in the contextual framework of the Spanish Spatial Data Infrastructure (SDI)

The geographic information system chosen for the geodatabase was Qgis. This open source application, licensed as GPL (GNU Public License), can be used, distributed, studied and freely improved. For the geodatabase (database with a spatial component), PostGIS is used as a module for the PostgreSQL data engine, which adds support for geographic data types.

The use of remote web services (OGC services) and various public spatial data sources (IE-CA, REDIAM, IGN ...) allow us to map the variables that characterize the territory and show the different types of risk that threaten our heritage, both of natural origin (earthquake, landslide, fire, flood ...) and anthropic, and then assess their degree of vulnerability to them.

Each of the territorial variables contemplated is normalized in intervals to which they are assigned a value of 1 (minimal hazard) to 5 (maximum hazard) according to the hazard value that contributes to the elements to be protected. A team of experts using the Delphi method contribute their experience and decision-making capacity for this phase. Reclassification is performed in raster format of each of the layers of information that will then be superimposed using map algebra. Subsequently, with the intention of generating a global hazard indicator, the hierarchical analytical process (AHP) is used to calculate the weights of each of the layers.

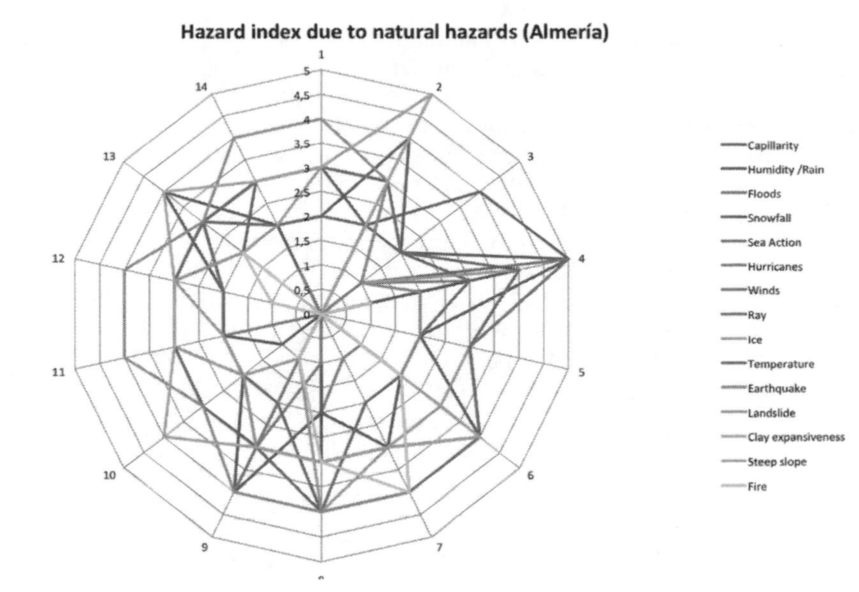

Figure 2. Hazard index due to natural hazard (Almería).

The comparison in pairs, in diagonal matrices, using the Thomas Saaty scale, allows a simple matrix calculation to estimate the multiplicative coefficients that accompany each normalized layer in the final algebraic expression. This methodology has been used in heritage for the study and valuation of industrial real estate (Claver et al. 2016), for the analysis of the useful life of a building (Macias, Calama & Chávez de Diego, 2014), for the analysis of risk, vulnerability and durability in the Archdiocese of Seville (Macias, 2012), in the evaluation of potential risks of destruction of archaeological sites (Márquez, 1999) or as strategic support for heritage conservation (Yaolín, 2006).The hazard index obtained and the hazard graphics generated for each variable (Figure 2) make up one of the innovations of this project.

At the same time, the study of the vulnerability of each patrimonial element is carried out, also by the Delphi method, which will allow to evaluate, together with the hazard indexes, the risk that will be reflected in the corresponding risk charts.

4 RESULTS

The results will be linked to the GIS in order to obtain Risk Charts from the creation of new cartographies of the territorial areas with the greatest impact on microzoning. The Risk Charts enables the analysis of the incidence of damage in relation to its risks. Based on the study based on macrozoned territorial cartographies, a territorial model for each risk - natural and/or anthropic - has been established.

The result is a battery of hazard cartographies that, crossed with the intrinsic vulnerability of each patrimonial element, will make it possible to demonstrate the cause-effect relationship that can give rise to the pathologies detected in this material.

5 CONCLUSIONS

The complexity in the conservation of certain cultural assets requires specific and complex tools, very different from those developed so far, for the application of preventive conservation strategies. It is in this aspect where the spatial data infrastructures can become a valuable tool for the modernization of cultural heritage management (Anguix, 2016).

The use of geographic information systems and spatial data infrastructures allow us to jointly manage the multiple variables that characterize the natural and anthropogenic risks that threaten our medieval defensive earth architecture and enable the analysis of the existing correlation between hazard-vulnerability and the pathologies detected through field work. Based on the data provided by the inventories and catalogues, and after creating the corresponding Risk Maps, we will have more precise knowledge of the state of defensive cultural heritage in Spain.

To achieve the appropriate hazard gradients it has been necessary to perform a combined risk analysis on a more detailed scale by generating interpolated surfaces and very precise digital terrain models. The methodology presented together with this double look at the environment of each heritage element, analyzed through techniques linked to the spatial analysis provided by the Geographic Information Systems, allows us to better understand the complex relationships of danger - vulnerability - pathology - conservation actions, and can become a Useful decision support system for the actors responsible for its conservation.

ACKNOWLEDGEMENTS

This work (BIA2015-69938-R) has been financed by the State Research Agency (SRA) and European Regional Development Fund (ERDF).

REFERENCES

Anguix, A. et al. 2016. IDE del Instituto de Patrimonio Cultural de España. JIIDE16.

Asprone, d.; Jalayer, F.; Prota, A. & Manfredi, G. 2010: Proposal of a probabilistic model for multi-hazard risk assessment of structures in seismic zones subjected to blast for the limit state of collapse. Structure Safety; 32: 25-34.

Baldi et al. 1995. Models and methods for the construction of risk maps for Cultural Heritage. *J. Ital. Statist. Soc.*: 1-16.

Bolteanu, D.; Chende, V.; Sima, M. & Enciu, P. (2010): A country-wide spatial assessment of landslide susceptibility in Romania. *Geomorphology* 124: 102–112.

Canuti et al. 2000. Hydrogeological hazard and risk in archaeological sites: some case studies in Italy". Journal of Cultural Heritage 1: 117-125.

Chiessi et al. 2010. Geological, geomechanical and geostatistical assessment of rockfall hazard in San Quirico Village (Abruzzo, Italy). *Geomorphology* 119: 147-161.

Claver, J. & Sebastián, M. 2016. *El Proceso Analítico Jerárquico. Aplicación al estudio del patrimonio industrial inmueble*. UNED. Isbn: 978-84-362-7171-3.

Coppolillo, P.; Gomez, H.; Maisels, F. & Wallace, R. 2004. Selection criteria for suites of landscape species as a bases for site-based conservation". Biological Conservation; 115: 419-30.

Giupponi, C. & Vladimirova, I. 2006. Ag-PIE: A GIS-based screening model for assessing agricultural pressures and impacts on water quality on a European scale". *Science of the Total Environment* 359: 57-75.

Halpern, B.S.; Selkoe, K.A.; Micheli, F. & Kappel, C.V. 2007. Evaluating and ranking the vulnerability of global marine ecosystems to anthropogenis threats". *Conservation Biology*, 21: 1301-15.

Lanza, S.G. 2003. Flood hazard threat on cultural heritage in the town of Genoa (Italy). *Journal of Cultural Heritage*. 4: 159–167.

Macías, J.; Calama, J. & Chávez, M. J. 2014. Modelo de predicción de la vida útil de la edificación patrimonial a partir de la lógica difusa. *Informes de la construcción*, 66 (533): 1-11. doi: http://dx.doi.org/10.3989/ic.12.107

Macías, J. 2012. *Modelo de predicción de la vida útil de un edificio: una aplicación de la lógica difusa*. Tesis doctoral. Universidad de Sevilla.

Márquez, H. 1999. Métodos matemáticos de evaluación de factores de riesgo para el Patrimonio Arqueológico. Una aplicación Gis del método de jerarquías analíticas de T. L. Saaty. *SPAL: Revista de Prehistoria y arqueología de la Universidad de Sevilla* (8): 21-38.

Mileto, C.; Vegas, F.; Villacampa, L. & García-Soriano, L. 2019. The influence of geographical factors in traditional earthen architecture: the case of the Iberian Peninsula. *Sustainability* 2019, 11, 2369; doi: 10.3390/su11082369.

Yaolín, Z. 2006. An application of the AHP in Cultural Heritage Conservation Strategy for China. *Canadian Social Science*, 2(3): 16-20.

Science and Digital Technology for Cultural Heritage – Ortiz Calderón et al. (Eds)
© 2020 Taylor & Francis Group, London, ISBN 978-0-367-36368-0

A pilot study of space syntax applied to historical cities. Studying Seville and Lisbon in the 16th-17th centuries

P. Ferreira-Lopes
DEGA, ETSA Universidad de Sevilla, Seville, Spain and The Digital Humanities Lab, IHC FCSH NOVA Lisboa, Lisbon, Portugal

ABSTRACT: During the transition to the Modern Age, the "Reconquista" influenced the territorial consolidation. In this framework, two cities were considered the main "capitals" of the Iberian Peninsula and one of the principal axes for the transfer of materials, people and ideas: Seville and Lisbon. Over the years, many methods and tools of spatial analysis have been developed for modelling real-world phenomena. Space Syntax models the morphology of urban spaces by using a graph representation. Although its application is extended in urban analysis, there is still a need for exploration of these tools for analysing historical urban spaces. The main methodological phases in the present research were: 1) historical source acquisition and digitization; geo-processing of data sources; 2) network model preparation and analysis; 3) statistical study and analysis; 4) maps visualizations. In this paper, we present the results of the parameters of connectivity, global integration, local integration [R3] and intelligibility.

1 INTRODUCTION

It is essential to carry out a study in which the historic aspects could be combined with digital technology and urban methods for historical cities research development (Griffiths, 2012) not only for the analysis of the parameters itself and its measurements but as well as to improve a cyclical synergy and an intersection during the work-flow between Humanities and Computing. Thus, the purpose of this work is also to show, by researching a concrete case of study, this interdisciplinarity in the study of historical cities and its importance for boosting our knowledge.

In this paper, we present an exploratory study in which we apply the basic concepts, methods and tools of a methodology already widespread: the space syntax (Hilliar & Hanson, 1983). Space syntax is applied in order to study urban function, using topological distances and calculating how each space is accessible from the other spaces in the city. For this, the city is studied as a system, with a high focus in its connections.

The goal of our research is to enhance the knowledge of the functioning of historical cities as spatial conformations by testing some of the parameters defined by the space syntax theory against two important cities in the Iberian Peninsula during the 16th and 17th centuries: Seville and Lisbon. This paper presents very briefly some findings of a research started at the University of Seville (Ferreira Lopes & Pinto Puerto, 2018; Ferreira Lopes, 2018) and linked with the R&D projects HAR2016-78113-R and HAR2016-76371-P.

The paper is organised as follows. In the next section, we will describe first the historical context of these two cities to briefly introduce a historical background. In section 3, we present the methodology that was applied. In Section 4, we show some of the analysis made. Finally, in section 5, we draw some conclusions and outline future work that should be done in order to unfold new insights.

2 SEVILLE AND LISBON DURING THE 16TH-17TH CENTURIES

During the 16th century, both the Kingdom of Castile and the Kingdom of Portugal were the scene of major political, social, cultural and economic transformations due to the process of Christian "Reconquista" and the discovery of America, among other events. Therefore, the great economic expansion that took place in the 16th century led to the emergence of new urban spaces, as well as the growth and consolidation of other existing ones. Among the main urban spaces of that historical moment, Seville and Lisbon had, in particular, some similarities due to their economic activities at the time and their common transfer networks that made their operation quite similar in a certain way (Chaunu, 1983). Therefore, with this pilot study we want to observe if their urban spaces had also some similarities.

3 METHODOLOGY

The key point of this research was to generate digital models of both cities, that represent their urban configuration in a specific historical period, in order to compare the morphology and functionality of these cities using the spatial syntax method. For this, a cyclical methodological process was designed and executed. The main workflow stages were: 1) historical data acquisition and digitization; geo-processing of data sources; 2) axial map model preparation; 3) statistical study and analysis; 4) maps visualizations and interpretations (Figure 1).

Firstly, the information of the urban layout contained in historical cartographic sources was digitised, transcribed and rectified to generate a GIS model of the historic map of the cities.

For the city of Lisbon, the "City of Lisbon" map produced by João Nunes Tinoco in 1650 (Tinoco, 1650) was used as the main document. This, together with the engraving by George Braun and Frank Hogenberg (Civitates Orbis, vol. V, 1598), the work of María Calado (Calado, 1993) and the current map of the city in *shapefile* formed our basis for the elaboration of the historical layout of the city of Lisbon. For the model of the city of Seville, the model already transcribed and rectified by Algarín Veléz (Algarín Vélez, 2000) was used,

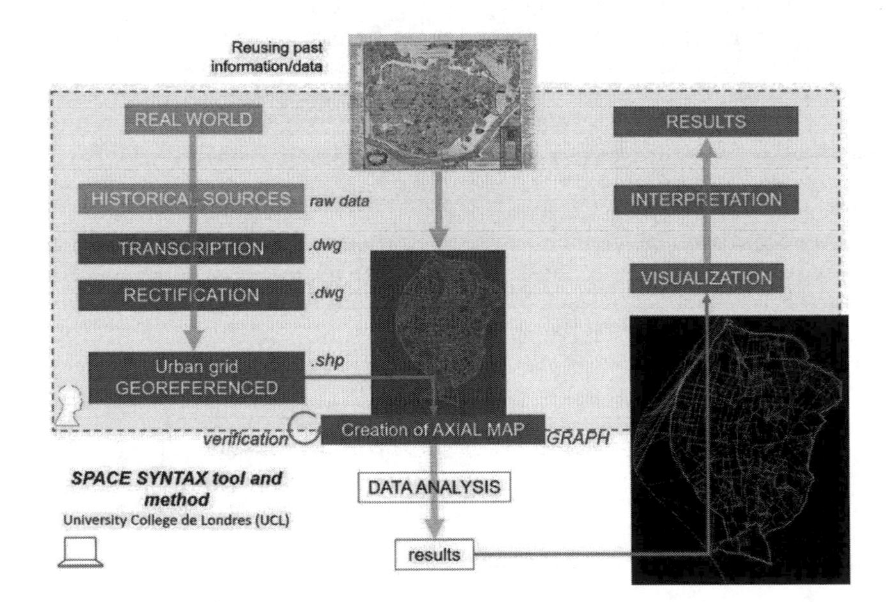

Figure 1. Diagram showing the phases of the workflow.

which had the "Plano topográfico de la M(uy) N(oble) L(eal) ciudad de Sevilla" as its main document, better known as the "Plano de Olavide" (Olavide, 1771).

Once the models were generated in *.shp* format (old cities' planimetry, within blocks, buildings and streets), we proceeded to verify the possible errors regarding the vectors of the models using QGIS Space Syntax Toolkit plugin tool (Gil, J. et al., 2015). After verification and correction of the errors found, the axial map was created through the Depthmap 0.35 tool. The Depthmap software generates the axial maps with three possible representations: (1) Total axial map; (2) Fewest line map (subset); and (3) Fewest line map (minimal). We have based the analysis of this paper on the Fewest line map (minimal) model generated. Once this map was automatically generated by the Depthmap tool, we manually verified the axial lines. In this study, about 10% of the lines had to be edited (deleted, added or modified) manually in order to generate an axial map model more in line with the definition and concept itself (Hillier & Hanson, 1993).

The next step of the process was the realisation of the analyses, the comparison and interpretation of the data. In this paper we present the results of the parameters of connectivity, global integration, local integration [R3] and intelligibility.

4 INITIAL ANALYSIS

Regarding the global scale, a basic static parameter used by Spatial Syntax is connectivity. Connectivity measure the number of other lines with which an axial line intersects (Hillier & Hanson, 1993). Comparing the results, we could observe that both cities show similar values of connectivity. Also, we can observe that each city has a nucleus area with a higher rate, in Lisbon the area of Rossio square, and in Seville the area of Alameda square. The Global Integration analysis, "calculates how close the origin space is to all other spaces" (Hillier & Hanson, 1993), showed that Lisbon presents a slightly higher values of integration than Seville. If we compare both global integration map, we can observe some "corridors" in both cities that were more integrated. Figure 2 shows that these corridors were linked with the nucleus observed in the Connectivity analysis and that they also had an expansion towards the river line (Figure 2).

Regarding the local scale, the Local Integration parameter, "values of axial lines at the radius 3, root plus two topological steps from the root" (Hillier & Hanson, 1993), showed similar values in both cities. We can observe in each city a nucleus spot concerning to the Rossio square in Lisbon and the Alameda square in Seville. Intelligibility was measured here by scattergrams as a correlation between global integration and local connectivity and informs that Lisbon has R2 = 0,1842 and Seville, R2 = 0, 2097 (Figure 3).

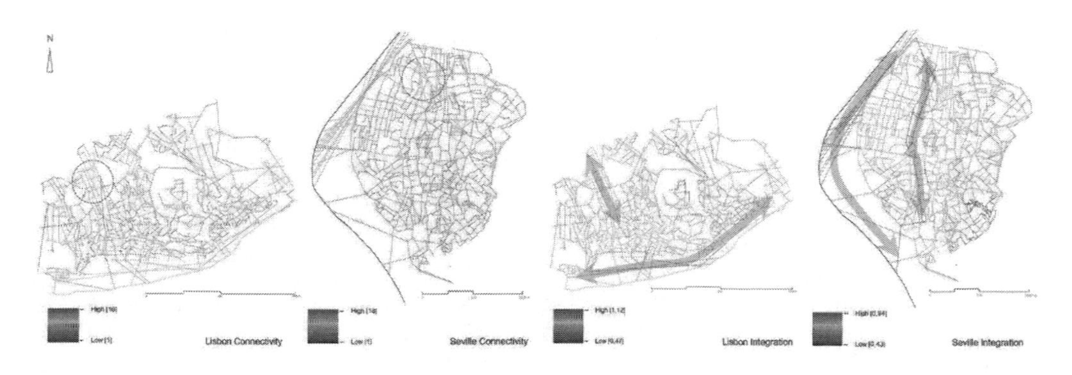

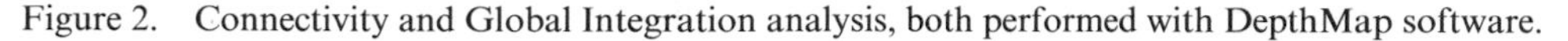

Figure 2. Connectivity and Global Integration analysis, both performed with DepthMap software.

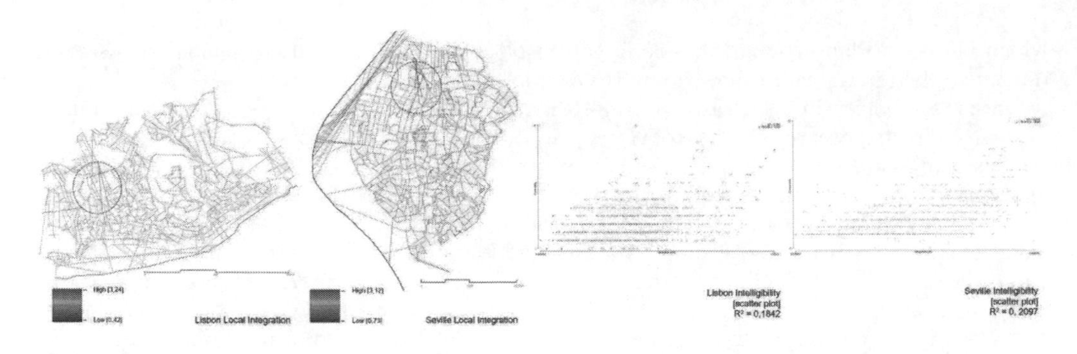

Figure 3. Local Integration analysis and Intelligibility scattergrams, both performed with DepthMap software.

5 CONCLUSIONS AND FUTURE WORKS

Based on this initial phase of the research project it was possible to see that the ability to do analysis using Spatial Syntax method is entirely relevant to study historical cities since it could combine the historical data treatment, spatial analysis and visualization. While the theory and methods of Spatial Syntax provide a positive approach to historical research these should not be restricted to a single spatial analysis "method" but rather be integrated within a multifaceted framework in which their potential could be combined and developed. Such an integrated approach can support and extend historians and urban historians knowledge of the built environment. In this sense, this pilot study provided a starting point to stablish a dialog between space syntax theory and urban historian to advance understanding historical modes of spatial organization. Of course, the comparison realised in this research between Lisbon and Seville is an approximate one.

Regarding future work to be carried out, the research aims to progress in two directions: i) Deepening in the historical study of both cities in order to improve the interpretations of the models and create new ones with different chronologies to understand and analyses the evolution of these cities in a wider time interval. ii) Implement the analysis of other parameters of the Space Syntax methodology also applying segmented axial maps.

ACKNOWLEDGEMENTS

The author gratefully acknowledges two parallel collaborative R&D projects HAR2016-78113-R and HAR2016-76371-P both funded by the Ministry of Science, Innovation and Universities of Spain. Furthermore, the author would like to thank the help of the professor Ignacio Algarín Vélez especially during the initial phase of the present research.

REFERENCES

Algarín Vélez, I. 2000. *Método de transcripción y restitución planimétrica*. Sevilla: Focus Abengoa.

Braun, G. & Hogenberg, F. 1598. *Civitates orbis Terrarum*. Cologne: Peter Brachel & B. Buchholtz.

Calado, M. 1993. *Atlas de Lisboa: a cidade no espaço e no tempo*. Lisboa: Contexto.

Chaunu, P. & Sánchez Mantero, R. 1983. *Sevilla y América: siglos XVI y XVII*. Sevilla: Universidad de Sevilla, Secretariado de Publicaciones.

Ferreira-Lopes, P. & Pinto Puerto, F. 2018. GIS and Graph Models for Social, Temporal and Spatial Digital Analysis in Heritage: The case-study of Ancient Kingdom of Seville Late Gothic Production. *Digital Application in Archaeology and Cultural Heritage* 9:1–14. DOI: 10.1016/j.daach.2018.e00074.

Ferreira-Lopes, P. 2018. La transformación del proceso de investigación en Historia de la Arquitectura con el uso de las tecnologías digitales. *Artnodes* 22: 61–72. http://dx.doi.org/10.7238/a.v0i22.3218

Gil, J., Varoudis, T., Karimi, K. & Penn, A. 2015. The space syntax toolkit: Integrating depthmapX and exploratory spatial analysis workflows in QGIS. In *SSS 2015-10th International Space Syntax Symposium* (Vol. 10). London: Space Syntax Laboratory, The Bartlett School of Architecture, UCL (University College London).

Griffiths, S. 2012. The use of space syntax in historical research: current practice and future possibilities. In *Proceedings of the Eighth International Space Syntax Symposium* pp. 1–26.

Hillier, B. & Hanson, J. 1993. *The social logic of space*. Cambridge: New York.

Olavide, P. 1771. *Plano de Sevilla*. Biblioteca Virtual del Patrimonio Bibliográfico. http://bvpb.mcu.es/es/consulta/registro.cmd?id=423028

Tinoco, J. N. 1650. *Planta da cidade de Lisboa*. Museo de Lisboa, MC.DES.1084.

Science and Digital Technology for Cultural Heritage – Ortiz Calderón et al. (Eds)
© 2020 Taylor & Francis Group, London, ISBN 978-0-367-36368-0

Strategic control of castramental space and geospatial analysis of visual prevalence. Implementation of GIS methodologies

J.J. Fondevilla Aparicio
Architect. Chief of the Department of Conservation of Historical Heritage. Territorial Delegation of the Counseling of Culture and Historical Heritage in Huelva. Researcher assigned to the research group HUM799 of the University of Seville, Seville, Spain

ABSTRACT: It is recognized that the strategic control of the historical territory constitutes a functionality inherent to the very origin of the conception of defensive architecture, which is materialized through the selection of high hypsographic prevalence locations endowed with wide viewsheds.

A way to study this topic is the multivariate (MCE) geospatial analysis implemented through Geographical Information Systems (GIS) in the context of the materialized research on the late medieval fortifications of the Galician Band, located in the north-western limit of the Sevillian Alfoz, allowed to endorse certain historiographical hypotheses regarding the strategy of territorial implantation of these passive defenses and establishing new theories about the hierarchy of those castral spaces.

In conclusion we can say that the intervisibility analysis between these castles located on the border between the kingdom of Seville and the domains of military orders, confirms that they formed a network of interrelated fortifications structured around the nuclear castles.

1 STRATEGIC CONTROL OF CASTRAMENTAL SPACE

One of the main functionalities of military and defensive architecture corresponds to the strategic control of the territory assigned under its guardianship and defense. The studies of viewshed, cumulative viewshed and visual prevalence (Wheatley, D. and Gillings, M. 2000) through GIS methodologies constitute a powerful geospatial analysis tool as we can see in the example of Santa Olalla del Cala Caltle (Figures 1-2).

Not only is it necessary to know the geographical space perceived from a castle, but to analyze the level of sharpness with which it is perceived, a concept that is linked to the fuzzy viewshed (Higuchi, T. 1983, Fisher, P. 1994) that is related to weather conditions and distance to the point of Observation derived from the attenuation of visibility by interaction with atmospheric particles that produce energy losses by dissipation as we can see in Figure 1.The geostatistical analysis of the directional distribution of the viewsheds allows defining where the castles direct their gaze, calculating through GIS systems parameters such as the angle of rotation or the standard deviation ellipse that allow a graphic visualization of this concept. The compactness of the viewshed is measured in relation to the middle center of the viewshed of each fortification, illustrating through the geometric construction of the sphere of standard deviation the referential space in which the visual beams are concentrated. Data that complements the value of the extension of the visual basin with that of its spatial compactness index

The results shown in the figures 3-4 let us consider that the Galician Band in the Late Middle Ages formed a coherent defense of the northwestern limit of the Sevillian alfoz, structured around an intervisibility network that hierarchized the castral space around its castles, defining satellite watch towers that allow the optical link with the prominent castles, providing a strategic control of the historical territory around its main roads of space penetration. The spatial density analysis of the viewshed is approachable through geostatistical studies such as

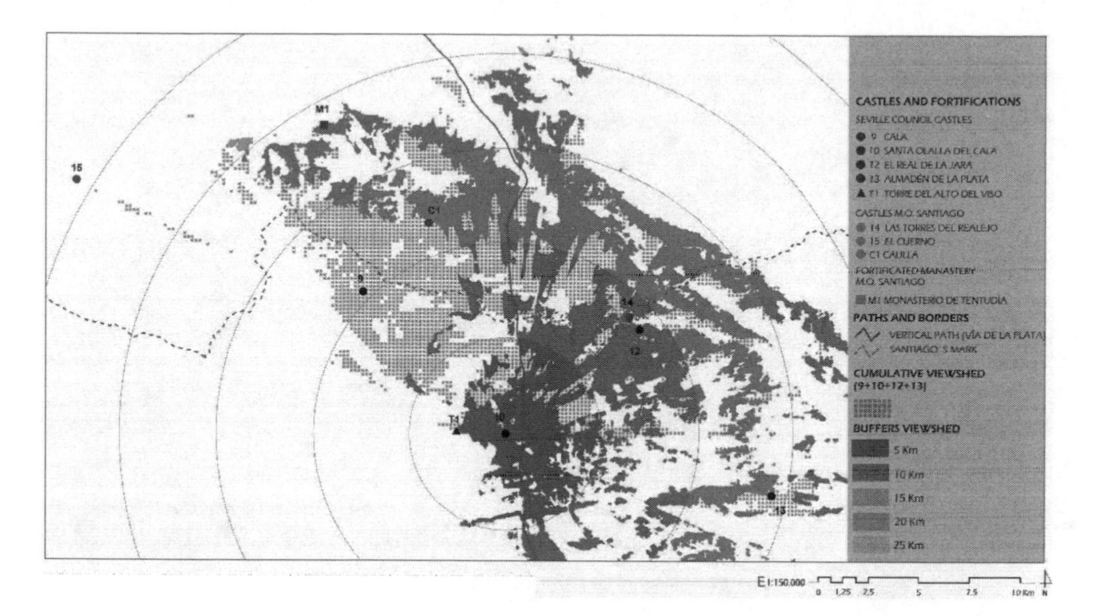

Figure 1. Santa Olalla Del Cala Castle buffers viewshed analysis.

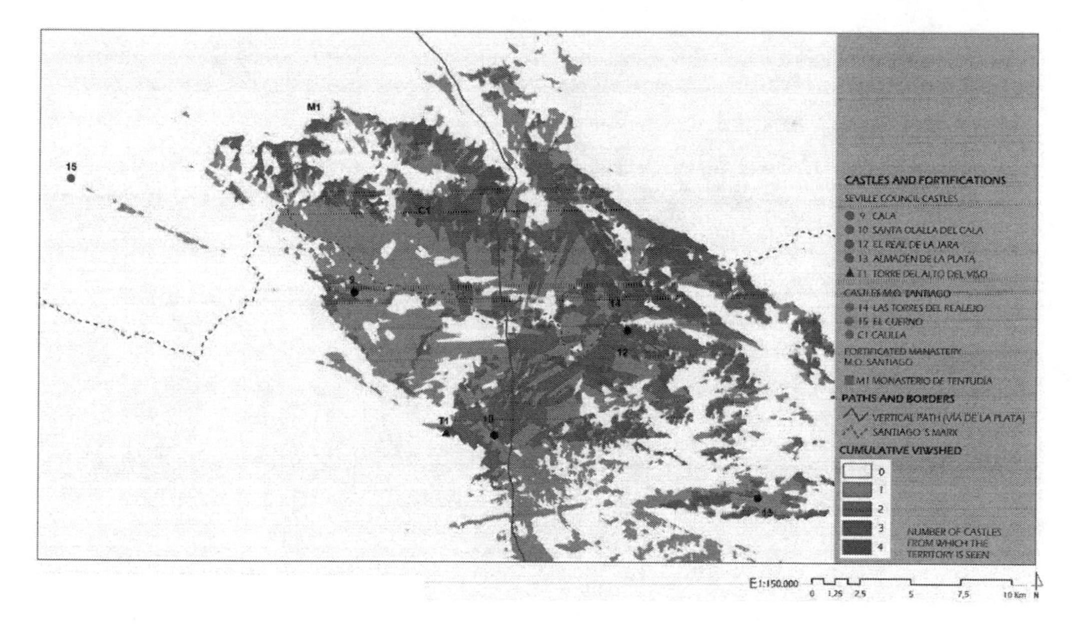

Figure 2. Cumulative viewshed castles (9+10+12+13).

Kernel density (Fiz, I. 2013) that show how the territory visually controlled by the defensive network gravitates around castles and historic roads (Figure 4).

The production of analytical cartography related to these concepts starts from the definition of a specific methodology that allows us to infer the existence of a solidarity behaviour of the castles for the joint defense of the referential geographic space, linked to each other around an intervisibility network that conditions territorial implantation of these late medieval castles as we can see in Figure 3.

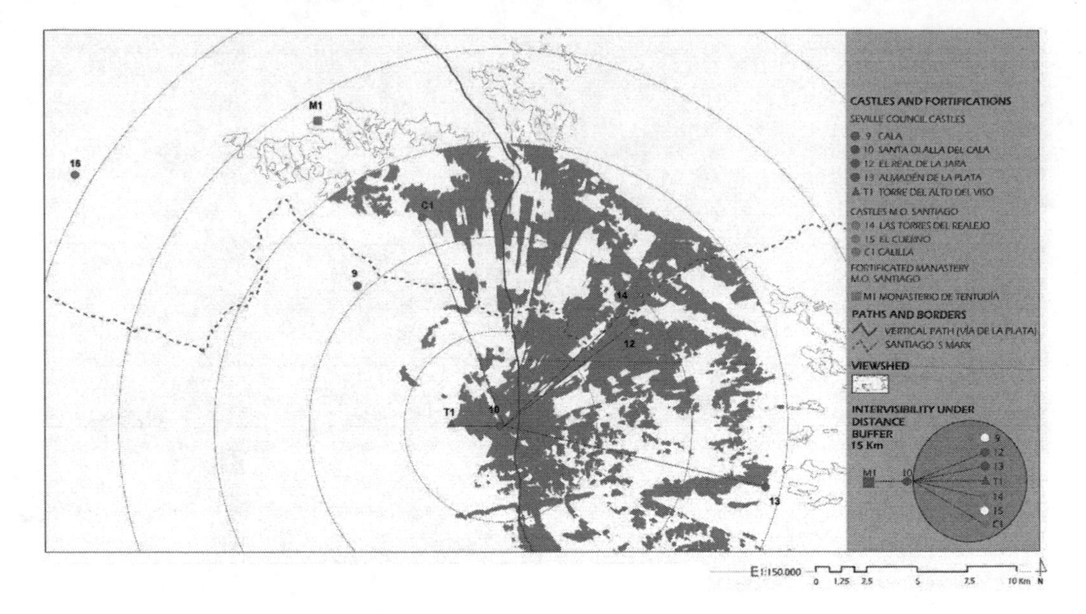

Figure 3. Intervisibility bewtween fortifications from Santa Olalla Del Cala Castle. 15Km.

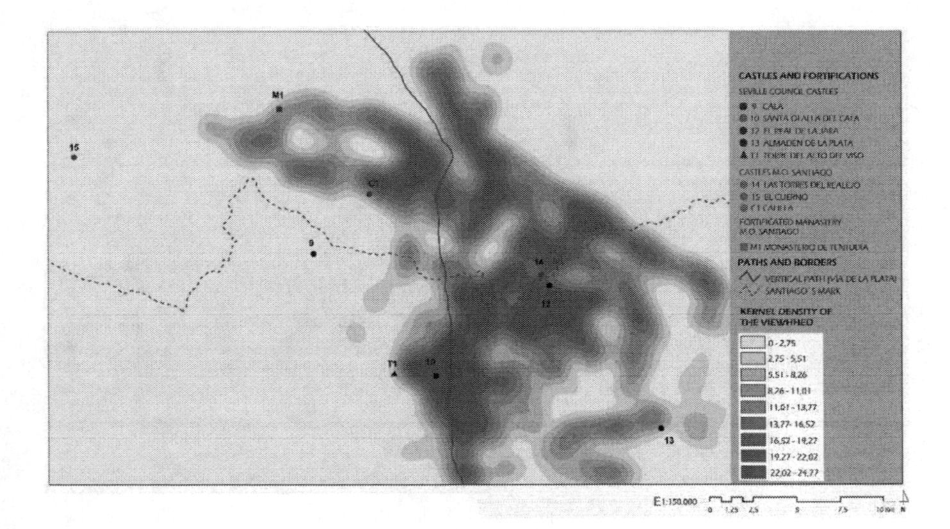

Figure 4. Kernel density anlaysis of Santa Olalla Castle viewshed.

2 GEO-SPATIAL PATTERNS ANALYSIS OF FORTIFICATION OF THE TERRITORY: GIS IMPLEMENTATION OF MCE METHODOLOGIES

Innovation in knowledge management of Defensive Architecture through the commitment to the georeferencing of its spatial delimitations, the modeling of its referential geographical space and the use of geographic information technologies (GIT) for the analysis of their interactions at territorial scale, allows to delve into the detailed study of the fortification patterns of the territory. Those studies emphasize the intimate relationship between the functional dimension of these military architectures and the physical environment on which they are based. Methodologically, the incorporation of the multicriterial decision paradigm into spatial analysis allows the definition of specific analytical instruments based on the implementation through Geographic Information Systems (GIS) of Multicriteria Evaluation (MCE) methodologies (Chen, S. *et al* 2012).

Those instruments makes possible to perform locational analysis, through map algebra, based on the overlay of thematic geoinformation linked to determining factors in the spatial characterization of the guard and defense patterns of the territory under study. Those overlays requires the previous normalization, and weighting of such factors, based on the assignment thereof of logical consistency indices, for the definition of paired comparison matrices of such analytical parameters.

The application of these quantitative GIS techniques, for territorial analysis, requires the previous modeling of the referential geographic space, assigning entities according to their spatial dimension to a raster or vectorial format in consideration of their discrete or continuous nature and in relation to the type of analysis spatial or geostatistical to which we must submit such geoinformation (Malczewski, J. 2006).

The weighted linear combination (WCL) allows the spatial combination of discrete boolean factors, as well as the consideration of continuous variables. The geostatistical treatment of this information allows us to undertake an exploratory analysis of these spatial data for the structural analysis of the territory in which such previously modeled spatial entities interact. The GIS implementation of MCE methodologies allows the elaboration of analytical cartography in order to evaluate the aptitude of a territory for its passive defense based on intrinsic and extrinsic factors to the historical territory, graphing the spatial relationships between continuous spatial entities and discrete.

3 CONCLUSIONS

The analytical potentials of GIS are determinant in studies of spatial interaction of the fortifications in the territory to which they provide guard and defense, evaluating their hypsographic prevalence, their spatial accessibility or the visual control of the territory, that allow a complex multivariate geospatial analysis, leading to discernment of the spatial patterns of fortification of the territory.

In conclusion we can say that the geostatistical analysis instruments allowed to stablished patterns of territorial implantation of the castles that comprised the Galician Band in the Late Middle Ages, seeking strategic control and the passive defense of this frontier territory. The use of MCE methodologies (Eastman *et al* 1993) to elaborate predictive cartography that served as the basis for the location of new fortifications whose existence was not known or whose location was uncertain, allowing their precise location and spatial georeferencing.

ACKNOWLEDGMENTS

The present article subscribed is part of the R+D+i Project (HAR2016-78113-R) funded by the Ministry of Economy and Competitiveness for the period 2017-2019 whose principal investigator is Mr. Francisco Pinto Puerto in the context of the Doctoral Thesis directed by Mr. Antonio Luis Ampliato Briones, Professor of the Department of Architectural Graphic Expression of the School of Architecture of Seville.

REFERENCES

Chen, S., Jjang, Y.& Diao, C. 2012. Cost constrained mediation model for analytic hierarchy process negotiated decision making. Journal of Multicriteria Decision Analysis, 19, pp.3–13.

Eastman, J.R. *et al.* 1993. Gis and Decision Making. United Nations Institute for Training and Research (UITAR), Ginebra.

Fisher, P. 1994. Probable and fuzzy models of the viewshed operation, in: Worboys, M. (Ed.) Innovations in GIS, pp. 161–175 (London: Taylor & Francis).

Fiz, I., 2013. Statistical methods and GIS functions: a model proposal of the Tarraconensis ager settlement. Archeological Spanish Archive, 86, pp.91–112.

Higuchi, T. 1983. The Visual and Spatial Structure of Landscapes. Ed. MIT Press, Cambridge, MA, 1a ed., 232 p.

Malczewski, J. 2006. GIS-based multicriteria decision analysis: a survey of the literature.

Wheatley, D. and Gillings, M. 2000. Vision, Perception and GIS: developing enriched approaches to the study of archeological visibility. In: G. Lock (eds), Beyond the Map, Amsterdam, 1-28.

Management and sustainability of the Cultural Heritage information. Social value, policies and applications about standardization and protocols

Science and Digital Technology for Cultural Heritage – Ortiz Calderón et al. (Eds)
© 2020 Taylor & Francis Group, London, ISBN 978-0-367-36368-0

Preventive conservation of built cultural heritage in Southwestern Europe: HeritageCare

M. García-Casasola Gómez, B. Castellano Bravo, B. Del Espino Hidalgo, M. López-Marcos & R. Fernández-Baca Casares
Instituto Andaluz del Patrimonio Histórico, Seville, Spain

ABSTRACT: Conservation and maintenance of buildings with cultural and historic value in Southwestern Europe lack of a systematic and regular approach. Facing this scenario, the HeritageCare project (Interreg-SUDOE) was launched to implement a methodology for the preventive conservation and maintenance of built cultural heritage in Portugal, Spain and France. Through a consortium joining public and private partners and associated entities, the main objectives and expected results of the project are: the definition of an integrated methodology for inspection, diagnosis, monitoring and preventive conservation of heritage constructions within the region; the assemblage of advanced technologies for the implementation of this methodology; the standardization of methods and tools through the development of guidelines for preventive conservation, and the creation of an online 4D platform for information exchange and assets management. Major findings include the need to create technical recommendations in user-friendly language or to implement periodic revisions, among others.

1 INTRODUCTION AND MISSION OF THE PROJECT

The preservation of buildings of historical and cultural value in south-west Europe is currently not a regularly organised activity (Oliveira & Masciotta, 2019:16). Even if numerous guidelines on this have been raised at the international and national levels (ICCROM, 2000; IPCE, 2015), the actions generally start after the first ailments are found in the building (Rogerson & Garside, 2017). In order to face this situation, HeritageCare project aims to implement the first joint, systematic and uniform preventive heritage conservation strategy in south-west Europe by means of: designing an advanced methodology that integrates different technologies for the inspection, monitoring, follow-up and management of historical constructions and their integrated movable assets; implementing a management system for the preventive conservation of historical and cultural heritage, based on a set of services provided by a non-profit entity to be created in Spain, France and Portugal; and, finally, advising owners or managers on good conservation practices.

2 SCOPE AND PARTNERSHIP

HeritageCare has been granted by the Interreg-SUDOE -Southwestern Europe- program, co-funded by the European Regional Development Framework (ERDF), and has lasted 3 years. The SUDOE space is an operational division of the EU made up of all the Spanish autonomous communities and cities (except the Canary Islands), the 6 regions of south-west France (Aquitaine, Auvergne, Languedoc-Roussillon, Limousin, Midi-Pyrenees and Poitou-Charentes), the continental regions of Portugal, Gibraltar and the Principality of Andorra. The SUDOE space is particularly rich in terms of listed heritage buildings. With a total population of 77 million inhabitants and a territory area of 82467 km2, it includes 31,163 listed buildings protected by cultural or heritage administrations (Figure 1), which is a proportion of 429 protected buildings per million inhabitants. Therefore, the impact of the project is

Figure 1. Listed buildings by Region on each of the three countries of the SUDOE space. Source: The authours, based on HeritageCare project partnership's data.

broad not only due to the possibilities of being exported to other countries but also because of the wealth and diversity of cultural heritage of the regions integrated into this international space.

The consortium is composed of 8 partners (Figure 2): Universidade do Minho (leader team), Direção Regional de Cultura do Norte, and Centro de Computação Gráfica, from Portugal; Fundación Santa María la Real del Patrimonio Histórico, Universidad de Salamanca and Instituto Andaluz del Patrimonio Histórico, from Spain; and Universities of Clermont-Auvergne and Limoges, in France. It incorporates 27 researchers and collaborators, as well as 11 associated partners (which do not receive a direct budget, but they benefit from the results). Including the partner institutions and the associated beneficiaries, HeritageCare incorporates 6 universities and research centres (5 partners, 1 associated), 4 public institutions of heritage management (2 partners, 2 associated), 5 representatives of the private sector (1 partner, 5 associated), and 6 agents from society (2 partners, 4 associated). Hence, it is a multi-sectoral, transversal project. Altogether have been working jointly not only on the implementation of the project but, furthermore, on the diffusion, awareness and outreach to all the potential social actors and stakeholders implied on cultural heritage-related works, so dialogue and joint efforts are encouraged.

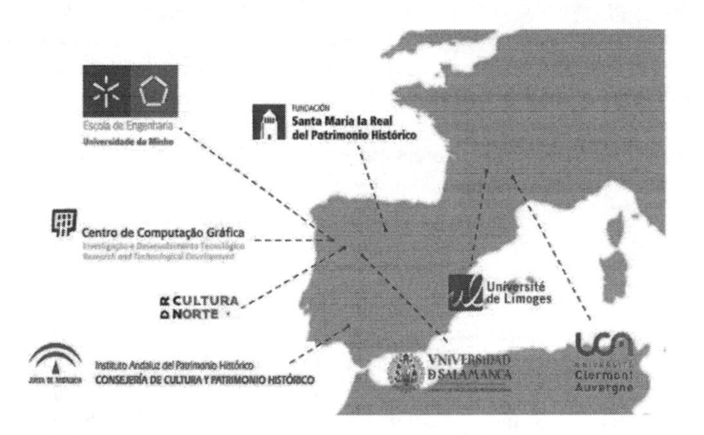

Figure 2. Institutional affiliation and location of the 8 partners which constitute the project consortium. Source: The authours, based on HeritageCare project partnership's data.

3 OBJECTIVES

The final goal of the project was the creation of a joint non-profit, self-sustaining entity that would use the methodology and tools in both in public and privately-owned buildings (either listed or not) throughout a coordinated strategy in the three countries. In order to reach this aim, specific short and mid-term objectives were planned: definition of an integrated methodology for inspection, diagnosis, monitoring and preventive conservation of heritage constructions within the SUDOE space; an assemblage of advanced technologies for the implementation of the methodology; standardization of methods and tools through the development of guidelines and rules of "good practice" for preventive conservation; and creation of an online 4D database for information exchange and assets management.

Also, further long-term aims have been formulated to make a contribution to the economic, political and scientific development according to a sustainable strategy by means of raising public awareness of the social and economic benefits associated with the preventive conservation of historical heritage, involving them in its conservation; and involve society, in particular, the technical-scientific community, public authorities and institutions, as well as the conservation and restoration sector, to drive a more efficient and sustainable preventive conservation framework for the protection of historical and cultural heritage.

4 PROJECT METHODOLOGY

Through all the planned works, HeritageCare seeks the principle "prevention is better than cure". This involves regular inspections and providing users and managers of cultural property with tools and recommendations that they can apply on a daily basis and at a low cost, keeping their assets in a constant good condition, minimising damage and possible deterioration. The methodology has been designed according to three service levels of growing complexity:

In Service Level 1 (StandardCare), qualified teams have performed detailed on-site inspections (Figure 3) at regular intervals and elaborated specific reports for the owners with the aid of an application for mobile devices that includes specific fillable forms. Previously, an atlas of damages to both immovable and immovable cultural heritage assets has been drawn up in order to systematize and facilitate the task of the experts who have carried out the inspections.

In Service Level 2 (PlusCare), more complex information about the building has been incorporated in the 4D database, such as full 3D photogrammetric surveys, data from monitoring systems and an inventory of assets. This has made it possible to expand information on a selection of assets included as case studies by incorporating in situ measurements aggregated to a point cloud-based survey.

Figure 3. Level 1 inspection work on a listed building developed by a qualified team. Source: The authors.

In Service Level 3 (TotalCare), management models have been developed with the help of hBIM technologies (historical Building Information Modelling) to assist decision making regarding future interventions that should be carried out, after the inspectors' advice, by specialized professionals that would transcend owners and manager's usual maintenance activities.

Each implementation phase was followed by the proposal of a feedback questionnaire to be answered by the owner or manager of each of the buildings inspected. The answers obtained have made it possible to know the usefulness and scope of the proposed methodology with a view to its reformulation for future development through the aimed non-profit entity.

5 RESULTS AND CONCLUSIONS

Project's results have included the execution of the abovementioned methodology over 60 case studies at no cost to owners during the project implementation phase. Previous phases implied the development of an Atlas of Damages in four languages (EN, PT, ES, FR) as well as the creation of questionnaires for owners and managers of buildings with historical and cultural value. The damage atlas has been developed to include in the database the most typical damages that inspectors may find both in buildings and assets. After having integrated this database within the virtual tool, 60 case studies have been inspected to test the methodology and its tools for Service Level 1. Some aspects and problems detected during the process have been improved in order to carry out the inspections for Service Level 2, which involved 15 case studies, and Service Level 3 on a sample composed by 3 case studies, each one them located in a different country.

As an outcome of the project, an owner's friendly affordable digital system has been created to allow the integration of regular inspections and reports, including forms, e-forms, and an app to help the visual inspection labours. The use of drones and photogrammetry to aid 3D representation, as well as the integration of hBIM have also been combined. All the results have been merged onto the common 4D database, incorporated into a platform that provides interoperability. Based on this, work is being done on the future development of the revised methodology through the creation of non-profit organisations that can offer their services to owners and managers, both public and private, of any built heritage asset.

Finally, the validation of the methodology through the incorporation of feedback questionnaires and the participants' own experience has shown findings that lead to some conclusions:

First inspections' experiences led to the need for a simplification of the technic vocabulary and an express indication of the periodicity and routines to be developed by owners and managers of the historic buildings. This has been incorporated as an outcome by means of the edition and publication of a document including guidelines for best practices in preventive conservation, which has been delivered to the general public and potential stakeholders, both public and private.

The interest shown by the owners and managers of constructed assets has been generally greater in the case of small buildings. Large complexes, particularly those belonging to public bodies and administrations, generally already have maintenance systems in place. In addition, they require long and complicated processes to allow inspection by personnel outside the institution.

As regards the continuation and application outside the project phase, the main barrier has been the difficulty of creating a joint international non-profit organisation, so three associations, each based in one country, are currently being created and will work in a coordinated manner.

REFERENCES

ICCROM-International Centre for the Study of the Conservation of Cultural Property. 2000. Towards a European Preventive Conservation Strategy. Helsinki: Institute of Art and Design.

IPCE – Instituto del Patrimonio Cultural de España. 2015. Plan Nacional de Conservación Preventiva. Madrid: Ministerio de Educación, Cultura y Deporte.
Oliveira, D.V. & Masciotta, M.G. 2019. HeritageCare: "Prevenir mejor que curar". Revista PH 96: 16–18.
Rogerson, C. & Garside, P. 2017. Increasing the profile and influence of conservation—an unexpected benefit of risk assessments. Journal of the Institute of Conservation 40(1): 34–48.

Science and Digital Technology for Cultural Heritage – Ortiz Calderón et al. (Eds)
 ISBN 978-0-367-36368-0

Documenting knowledge: Document management of intervention projects on immovable heritage

P. Acosta Ibáñez
Instituto Andaluz del Patrimonio Histórico, Seville, Spain

ABSTRACT: The need to document the actions of intervention in conservation and restoration of cultural assets stems, first, from the demand for exhaustive knowledge of the object of the intervention and, secondly, from the responsibility of recording the actions carried out for its maintenance and future conservation. This need, which appears in all areas of intervention in Cultural Heritage, generates a precious amount of documentary heritage that is not always well managed in the case of actions in immovable heritage. The documents generated throughout these projects guarantee the preservation of the intervened assets, and provide a basis and foundations for further interventions as well as sources of information for research in related topics. What should be done with these documents? How to manage and archive them? Are they available? How to integrate them into knowledge management? These and many other questions depend on the document management strategies applied to work processes and information systems. Knowledge management is only possible after a proper information management, and this is only possible if data and documents are available. A transversal, underlying issue is usually forgotten, that is, the preservation and conservation of these last two elements, which should be contemplated in any system of information and knowledge management before designing it. The objective of this paper is to expose the process of document management and archiving of documents generated throughout intervention projects on conservation and restoration of immovable heritage; and to demonstrate the need to integrate them in the management of these projects from the beginning, as part of the necessary tasks to guarantee the management, disposition, access and preservation of the information and knowledge generated through the process

1 INTRODUCTION

> *Moreover, every intervention on the artwork must be preceded by a written report that documents the artwork and explains the motivations for the work to be done (last section Art. 5). During the course of the work a journal must be kept and will be followed by a final report. This will contain photographic documentation showing the conditions before, during and after the work was completed. They will include documentation of all scientific tests and research done with the aid of such disciplines as chemistry, physics, microbiology and other sciences. A copy of these reports will be kept in the archives..."* *(RESTORATION CHART, 1972).*

The need to document interventions in the conservation and restoration of cultural assets firstly stems from the demand for exhaustive knowledge about the object of the intervention and, secondly, from the responsibility of recording the interventions carried out for its maintenance and future conservation. The documents generated throughout these projects guarantee the preservation of the assets subject to intervention, and provide a basis and foundations for further interventions as well as sources of information for research in related topics. The presence of the document is essential to know the cultural assets and to justify any action on them.

In the case of immovable heritage, the process of documenting actions of intervention in conservation and restoration, which is applicable to any area of Cultural Heritage, in turn generates a documentary heritage of great value for the asset itself, which, together with the complexity of the projects, makes the design of guidelines for its management, organisation and archiving recommendable. The documentation of these projects will be the scientific basis that sustains the actions of intervention on the asset. (Quiñazo 2006).

Aware of the responsibility as a producer of part of said documentary heritage, the Andalusian Historical Heritage Institute, through collaboration between the Projects Department and the Archive, is designing the document management process and archive of the documents generated by the projects of intervention in conservation and restoration of immovable heritage with the ultimate goal of improving the management of information and the transfer of knowledge of the work carried out.

Next, the key points for the development of this process of Document Management are gathered and must be specified in a Guide for the Document Management of Projects of Intervention in Immovable Heritage. The ultimate aim is that these guidelines can be implemented in any Information System and serve as a basis for the management of the knowledge generated.

2 THE PROCESS OF DOCUMENT MANAGEMENT OF PROJECTS ON CONSERVATION OF IMMOVABLE HERITAGE

The Process[1] of Document Management of Projects on Conservation of Immovable Heritage is a process that is based on and basic criteria in two other processes put into effect previously in the institution: on the one hand, the Process of Document Management of the institution[2] and, on the other hand, the Process of Intervention in Immovable Heritage.

The following objectives were set as a starting point for the design of said process: first, to guarantee the adequate exploitation and use of information, ensuring its current and future availability; second, to optimise the resources and the completion dates for the execution of the works and the provision of the documents; third, locate and access information in a much faster, effective and accurate way; fourth, to control the use and flow of information, ensuring compliance with rights and duties, if applicable; fifth, decrease the risks and threats of loss of data; and sixth, to value the results of the works and the documentary heritage generated.

Below are some of the key aspects of the sequence of activities designed for the process and the complete sequence of tasks is shown in the diagram in Figure 1, among them: the identification of project documents, times and actors in the process, the main activities of the documentary management, and the access and dissemination of the project as a final result.

The identification of the documents that are going to be generated is one of the basic tasks of document management of a project. A lot of documentation is managed in projects on conservation of immovable heritage, but not all of it will be documentary production nor will it require the same level of processing. Therefore, we must distinguish between two large blocks of documents: one, the sources of information that we consult to support our actions and decisions; and two, the documents received and produced in the development and execution of the projects. The sources of information are not generated by the project: they are copies and they have a support and information function, they are contingent and do not require responsibility for their future conservation. For its part, the documents generated by the project are original, comply with a function of processing, communication or resolution, are necessary and require responsibility for their future conservation. Documentary management actions fall on

[1]Process is understood as *the organised set of activities that are carried out to produce a product or service: it has a delimited beginning and end, involves resources and leads to a result*. (Cruz Mundet, 2011 p. 293).

[2]This process has been published and is accessible through the website with the title The Process of Document Management in a Public Body: design and implementation in the Andalusian Historical Heritage Institute. https://repositorio.iaph.es/handle/11532/320348 [Accessed: 29/04/2019)

Figure 1. Alcázar de Sevilla. Fondo Becerra. IAPH archive.

the latter. Despite this, it is also important to give some criteria for the organisation of information sources in order to ensure access to them during the execution of the work when necessary.

In order for the Process of Document Management of Immovable Heritage Intervention Projects to function and achieve its objectives, the following questions should be taken into account in relation to the times and the actors involved: first, the preparation of document management guidelines and tasks should be considered at the start of the project; second, it is interesting to have a person in the team, a specialist in document management and archives, who can prepare these guidelines and do the follow-up during the execution of the project; third, before starting them, they must be approved by the project managers; fourth, once approved by the project managers, these should be communicated to all the team, especially those actions that are related to the identification of documents, use of information, access and dissemination; fifth, after the completion of the project, the "project file" will be transferred to the Archive of the institution that will make it available to the general public through the consultation service and the institutional repository.

For its part, the main activities of document management that should be considered in the sequence of tasks will be: first, the registry of documents and sources of information to know what documents we have, what their date is and the person/s responsible for them; secondly, the classification of the documents according to the previously defined tasks and areas of action; third, the conservation and digital preservation that should take into account both the different media and formats, as well as the specificity of the graphic materials; fourth, the distribution and access, that is, what, how and who can access the project information; and finally the dissemination that guarantees the transfer of the knowledge generated.

Access and dissemination are an especially important area of activity considering that the documents generated have a marked technical and scientific component and should be considered to ensure knowledge transfer. Information will be accessed at least taking into account the current regulations on files, data protection and intellectual property. It will be necessary to consider some specific rules for the management and preservation of the graphic material, especially the photographs since they represent an important volume of information which is overlooked most of the time. On the other hand, it is proposed that documents and information of a scientific and technical nature should be disseminated on an open access basis through the institutional repository once the project is completed, in compliance with the policies of transparency and access to public information.

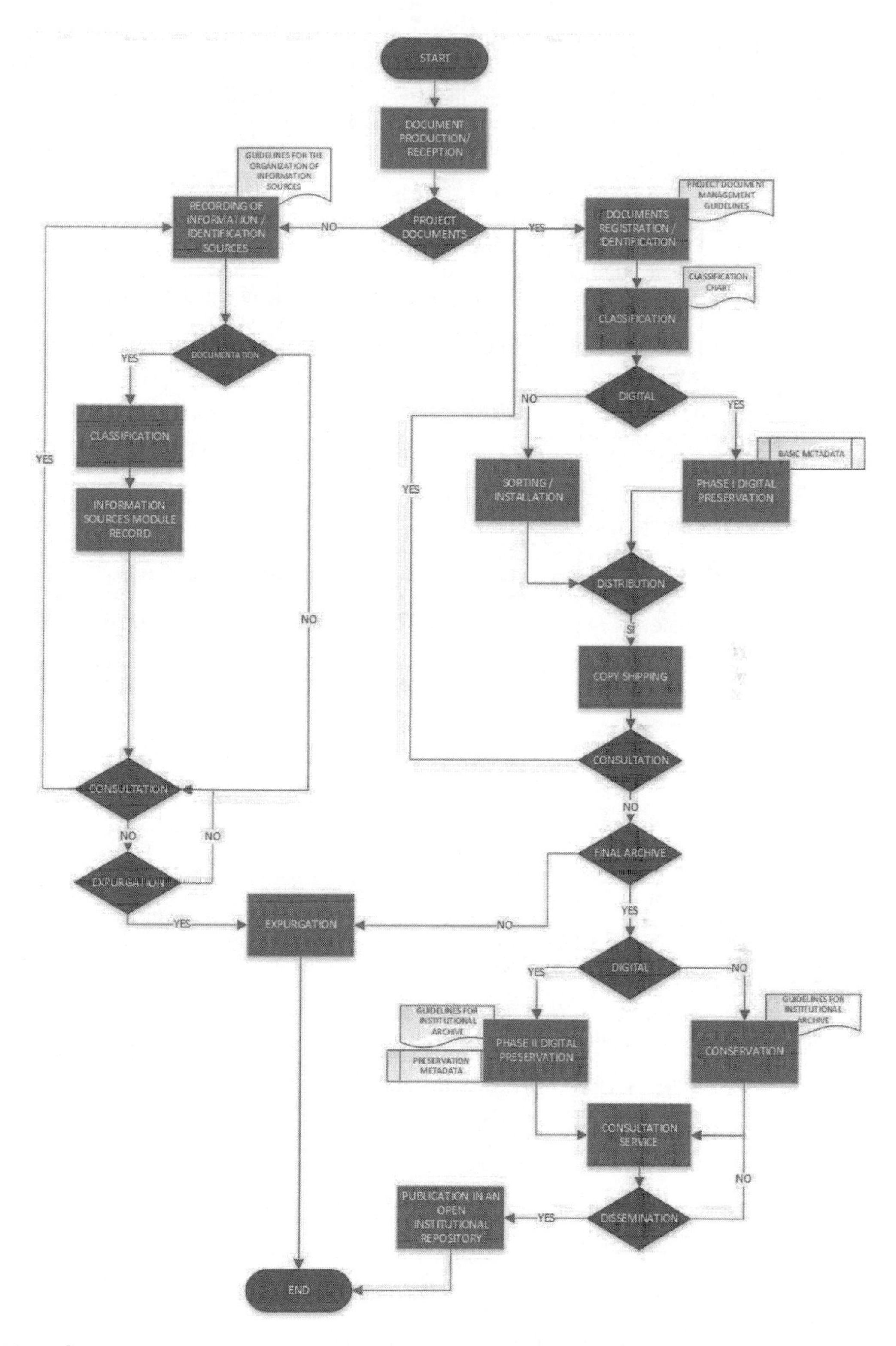

Figure 2. Flow chart of the process of document management of projects on intervention in immovable heritage.

3 DISCUSSION

A warning call to the need of considering documental management as a cross-cutting issue has been attempted. Most of the new management and information systems do not consider documental management as a cross-cutting layer assuring the comprehensive knowledge produced. Furthermore, most of the times, digital preservation needs are not taken into account, resulting the risk of information loss in the short term. The question is: will we be able to access to the scientific and technic information produced at present by this type of projects in the future? Has this scenario been taken into account? Has the digital information and documentation fragility been taking into account? Those questions should be answered if the access to the information and knowledge produced from the documents is required and to assure future conservation actions in cultural assets. It is necessary to design management and information systems, also integrating the management and future preservation needs of the documents produced as real project assets.

4 CONCLUSIONS

- Documental management must be integrated in any information system of cultural assets conservation if the preservation and conservation of the information and documentation produced is required.
- Information systems and platforms currently under development do not often consider this layer and it should be implemented from the outset.
- The information and data-packages transfer to institutional Archive and Preservation Information Systems should be envisaged.
- A metadata and preservation policy according to international standards should be considered[3].

BIBLIOGRAPHY

Acosta Ibáñez, Pilar. 2016. El proceso de Gestión Documental en un Organismo Público: diseño y puesta en marcha en el Instituto Andaluz del Patrimonio Histórico. *Revista TRIA*. Nº 20. 355–3378.

Cruz Mundet, José Ramón. 2011. *Diccionario de Archivística*. Madrid: Alianza Editorial.

Cruz Mundet, José Ramón; Díez Carrera, Carmen. 2016. Sistema de Información de Archivo Abierto OAIS: luces y sombras de un modelo de referencia. *Investigación Bibliotecología*. Nº 70. 221–257.

Ministerio de Instrucción Pública, Gobierno de Italia. 1972. *Carta del Restauro*.

Quiñazo Rivas, Pilar. 2006. La documentación como fuente de información para la restauración arquitectónica. *RECOPAR*. Nº1. 1–17. 1.

[3]There is a preservation standard accepted by the majority of the international community, based in OAIS System ISO-14721. It consists of a conceptual model which propose a scheme of the different elements and procedures that should be involved in the preservation process and offers a structure and common language in order to ensure the conservation and the access to digital documents.

Science and Digital Technology for Cultural Heritage – Ortiz Calderón et al. (Eds)
 ISBN 978-0-367-36368-0

Historical small smart city protocol for sustainable regeneration of the small historic centers: A technological planning tool for multiple scenarios

V. Pica, M. Cerasoli & M. Canciani
Department of Architecture, University of Studies of Roma Tre, Rome, Italy

ABSTRACT: Conserving the beauty and the identity value of small historical centers of the zones in condition of marginality, together with guaranteeing their good-quality livelihood is one of the main challenges of urban planning nowadays. New instruments that incorporate computing and communication technology can contribute to address the inefficiencies of predictive top-down traditional urban planning tools. The Protocol "Historical Small Smart City" is an innovative tool based on a bonus certification system to stimulate the small municipalities to submit regional funding integrated projects. It prescribes the "modulation of the protection" to achieve the goal of conserving the built minor heritage and also allowing certain degrees of energy retrofit.

This tool can calculate different scenarios simulating possible developments linked to different decisions. GIS Data Analysis is the core basis of this system, set through a Geo Big Data and Local Data portal that allows calculating different indicators to orient political decision-making.

1 INTRODUCTION

The paper presents the design and applicability processes of an interactive Geographic Information System (GIS)-based tool referred to as the "Historical Small Smart City" Protocol (HISMACITY). It is a certification system for the "smartness" of small historic towns in rural areas. This quality standard is defined as the result of summing a series of levels of performance measured through quantitative and qualitative indicators, which are connected to several evaluation criteria. Through a performance-based reward system, the tool encourages sustainable strategic planning and smart development of small municipalities of the internal areas, far from the major centers. It has been designed for Italy, where these areas are managed through the National Strategy of Internal Areas (ATC, 2017), and Spain, where the Law for the Sustainable Development of the Rural Environment 45/2007 and the Strategic National Plan of Rural Development have enhanced their strategies of management (Esparcia y Noguera, 2001; Gozálvez Pérez, 2001). The case study of application is being conducted in the small historical center of Sutri in the province of Viterbo, Italy. Its European scalability is going to be analyzed by applying it to a second case study in the town of Berga in Catalonia, Spain.

The main goal of the Protocol is to counteract depopulation of rural areas (Pica, Cerasoli, 2018). To address this objective, it leverages e-participation processes, supported by new technological opportunities, in collaboration with citizens who are aware of the concept of landscape as "common good" (Minervino, 2016). In this way, the Protocol also deals with compensating for the fact that national and regional legislative apparatuses have not fully considered today's challenges that rural towns face (Aristone & Palazzo, 2000; Pica, 2018a).

The protocol is constituted by three kinds of contents[1]. The first group consist in six tables containing the lists of the evaluation criteria, encompassed in six priority action areas (Mobility, Economy, Environment, Heritage, Living, Governance).

Secondly, there is the webpage of the project, that allows to visualize an online Observatory[2]. The portal also gives access to reading the Geographic Information System (GIS) dataset to define and measure the indicators (Pica *et. al.*, 2019). Finally, there is the Guidelines document, i.e. a technical manual that explains the quality standards of the key performance indicators that convey to the credits assignment.

2 THE HISMACITY PROTOCOL'S STRATEGY: CONTEXT AND OBJECTIVES

The tool's design process takes on the position that a dynamic and that supports e-participation is needed to protect the rural landscape, while managing the built heritage reuse and valorization (Vázquez & Vicente, 2019). This topic has been at the center of technical-scientific debates in Italy for years (ANCSA, 1971; Gabrielli, 1982). Both Italian and Spanish local policies currently demonstrate their growing disconnect from the civil society (Egan, Nugent, Paterson, 2010). Moreover, the public sector is often unprepared to satisfy local needs and an effective development strategy, which Information and Communications Technology (ICT) could supply (Ferguson, 2019). Meanwhile, minor municipalities are becoming more independent from larger urban centers, which is also due to ICT. This contributes greatly to making local communities more aware of their condition and the possibilities offered by the new globalized era. Of note, a "multi-level governance" approach has been considered by national politics and is taken into account in our research. It has been addressed by the legislative *path* of the Bassanini Law No. 59/1997 to the recent Law No. 1/2003 in Italy of the administrative simplification through the assignment of management tasks to the regions and the local authorities. In Spain, independent management of municipalities has been set by the Constitution in 1978, article 140, and through the successive numerous laws that laws that determine the competences of autonomous communities (Fernández, 2013).

In Italy, the governmental National Strategy for the Internal Areas and the recent Regional Laws for Urban Regeneration demand to be attentive not to make only the interests of the private investors while respecting the constraints and guaranteeing the civil right to affordable housing and efficient services (Barbanente, 2017). In Spain, a similar scenario is prescribed by the Royal Legislative Decree 07/2015 (BOE, 2015). Additionally, there is a general lack of vision for the future and the absence of a shared and integrated strategy of development from the public sector (Mangialardo & Micelli, 2017).

Within this context, the HISMACITY Protocol aims at generating a common vision while interconnecting local stakeholders and administrators through its web portal and certification system, useful to guide decision-making and financing projects. Also, many of its Guidelines are beneficial if focused on shareable urban quality improvement procedures between various stakeholders and participatory actors.

3 METHODOLOGY

The strategic framework of the Protocol is technically grounded on a long-term and elastic management plan organized through the structure of the six priority areas of action. Its strategy uses an integrated methodologic approach that connects the areas of action regarding the Cultural Historic Environment protection (MIBACT, 2018) with the socio-economic

1. As it has been requested by the financing program of the research, the Marie Sklodowska-Curie Actions IF Fellowship, released by the European Commission.
2. Retrievable from https://www.arcgis.com/apps/MapSeries/index.html?appid=8870d0be46bc46acad6de2066e1e1a4b.

development (SEPI, 2011). This framework fosters the tertiary sector and utilizes a human-centered focus for sustainable development and capacity building for the Local Community Development (LCD) (UNDG, 2014; Weitz, 1981; Barberis, 2011; Wei, 2014; Streitz, 2019). The still in progress construction process of the Protocol Strategic Framework has been developed in three phases: priority areas definition and input data definition; multiple criteria, alternative interventions and Tables definition; key factor indicators calculation and Guidelines drafting.

The first step has been carried out on the basis of a SWOT matrix analysis matrix (Strengths, Weaknesses, Opportunities, Threats), integrating different aspects of the actual conditions of Sutri and other Italian and European minor centers. The strategy is oriented at the enhancement of opportunities and lines of force, while decreasing the critical aspects and threats. The set of the priority areas have also been defined by a comprehensive analysis of the literature and the main projects concerning Smart Cities[3] (Nocca, 2017). The structure of the framework has been designed according to the recent Italian concept of the Sustainable Historical Smart City (CDP, 2013; Forum PA, 2014). The construction of the contents of the Tables and the Guidelines of the tool are being carried out after completing the dataset framework realized through local surveys and Geo Big Data collection campaigns. The multiple criteria definition phase provides the construction of the alternative solutions or integrated interventions and their standards of quality, which leads decision makers through the MCDA (Multiple Criteria Decision Analysis) to set the weights that can be assigned in the calculation formula of the composed indicators, supported by the GIS "weighted overlay" technique, i.e., an overlapping procedure of different feature classes with linear combinations to obtain weighted indices (Cherubini, *et. al.*, 2006). At this point local decision making is crucial, and can be combined with coordinated bottom-up processes. Decisors (city councils and local stakeholders) don't have only the opportunity to choose the weights of the indicators but they can also manage the planning program by choosing the priority alternative actions.

Therefore, the system has been designed to leverage on a set of variable alternatives by the use of the AHP (Analytic Hierarchy Process) of the MCDA (Multiple Criteria Decision Analysis), through which the Protocol's Data Framework becomes a Decision Support System (Saaty, 1980). The alternative solutions are the integrated interventions that prefigure different scenarios and performance standards. They are evaluated by the certification system through the AHP hierarchical structure. This is built by a Work Breakdown Structure (WBS), created to unpack large action fields, i.e. the objectives and the evaluation criteria, or milestones. At least two alternatives unfold from each criterion. Both alternatives are evaluated by using the AHP method of "comparison in pairs". The decision maker will choose the option with the highest level of satisfaction. This process contributes objectivity and transparency to the decision maker's judgment (AVCP, 2011).

The method outlines that action effectiveness is not only measured between the intervention's beginning and end but also is progressively evaluated through the GIS software for every single move put into place in order to allow constant feedback about the success of the intervention model (Figure 1).

Decision makers can modify their strategies by choosing the different alternatives according to the local preferences. In order to obtain a basic certification standard, the decisor must undertake at least on alternative intervention for each criterion.

3. See the project "Bee Smart City" available online: https://hub.beesmart.city/smart-city-indicators/(accessed on 13 February 2019).

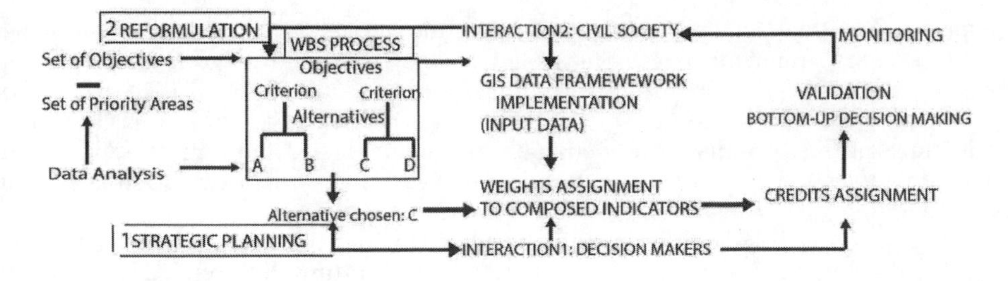

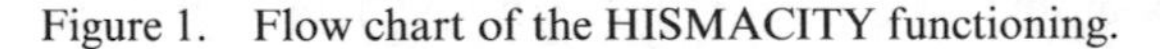
Figure 1. Flow chart of the HISMACITY functioning.

4 EXPECTED RESULTS

The operational procedures of the applicability of the protocol could follow various lines. The strategy of the framework foresees that the Italian Recovery Plans, instituted by the Law 457/1978, and the Spanish PERI (Internal Special Recovery Plans), could be flanked by this tool in order to optimize interventions, at both the executive and economic level.

The project management approach through the MCDA (Multi Criteria Decision Analysis) prioritizes the interventions and set their levels of feasibility. The results of the decision-making processes are integrated with a final process analysis that goes from the most immediate and easy actions to the most difficult, following the evaluation of the operative, economic, and highly critical issues.

Regarding Cultural Heritage, the tool supports the application of the "modulation of the protection", a procedure already included in the technical standards for implementation (NTA) of the Recovery Plan of the historical center of Formello (Cerasoli, 2010). It prescribes interventions for cultural heritage conservation in the dataset by joining the buildings' feature class with tables of attributes to indicate the levels of transformability in the historical fabric (Pica, 2018b).

With regard to economic feasibility of the integrated interventions, the framework includes several compound reward measures, such as the reduction of building permit fees; the increase in volumetry outside the historical center's perimeter; simplified paperwork for change of intended use; lowering of taxes, and better loan conditions. It is important that the long-term aid is granted in a context of sustainable development so that partners and institutions do not invest money only for it to be used up. Each institution should participate in the wider system of control and scrutiny with their own budgets.

This strategy encourages the Union of Municipalities (Article 30 of the Italian Legislative Decree 267/2000), which is the Italian institutional form that allows municipalities to enter into preferential rankings for access to EU and national funding. Thus, the union has a leverage effect. This is why actions undertaken within a union are awarded more bonuses. Among them: Sustainable Mobility Plans (PUMS), Programmes of Urban Regeneration (PRU), and Action Plans for Sustainable Energy (PAES). A municipality alone is less effective, and its actions can perform only basic scores. The best rankings can be achieved only within a union.

5 CONCLUSIONS

The Protocol's integrated approach can provide some solutions to reconnect the civil society with the public administration, by using an integrated strategy that leverages on ICT application. Its MCDA valuation methodology provides a certain level of flexibility of the system, that can allow local decision-makers to design strategies adapted to their preferences. The database framework is grounded on recent findings on traditional community-shared building rules, codes, and the overall building process that is inclusive of decision making. The

Protocol has been conceived in order to design a new model that can be scalable and useful in rethinking the foundations of actual planning legislation. It could be used in conjunction with traditional planning tools, moreover Recovery Plans, yet also for solving problems that such tools cannot address.

It is also intended for developing planning tools and regulations that allow for maintaining the cultural continuity of the built environment. Finally, the Protocol could be adopted throughout the Italian country within the National Strategy for Internal Areas as a valid method for allocating EU funds.

In line with this strategy, the Protocol also envisions and supports a networking project management for the territorial revitalization of which the municipalities can become protagonists through the Union of Municipalities. The most rewarded interventions and planning actions can be sustainable and feasible through this convention.

REFERENCES

Aristone, O. & Palazzo, A. L. 2000. *Città storiche, interventi per il riuso*. Milano: Il Sole 24 Ore.

Authority for the Supervision of Public Works Contracts (AVCP) 2011. *Linee guida per l'applicazione dell'offerta economicamente più vantaggiosa negli appalti di servizi e furniture*.

Agency for Territorial Cohesion (ATC). 2017. *Strategia nazionale per le Aree interne: definizione, obiettivi, strumenti e governance. Accordo di Partenariato 2014-2020*. Settembre 2017. Available online: http://www.agenziacoesione.gov.it/it/arint/index.html#accept.

Barbanente, 2017. Emilia Romagna: rigenerazione urbana per chi? Relazione al convegno Privatizzare l'urbanistica? Bologna, 15 novembre 2017. *Il Manifesto*, december 2017. Available online: http://www.ilmanifestobologna.it/wp/2017/12/emilia-romagna-rigenerazione-urbana-per-chi/.

Barberis, W. 2011. Ciudad urbótica contemporánea: urbanística y nuevas tecnologías en el espacio y los servicios urbanos. In: *ACE: Architecture, City and Environment*, 6 (17): 95-108.

Official State Gazette Agency (BOE). 2015. Real Decreto Legislativo 7/2015, de 30 de octubre, por el que se aprueba el texto refundido de la Ley de Suelo y Rehabilitación Urbana, *BOE*, 261, 31/ 10/2015.

Esparcia, J. y Noguera, J. 2001. Los espacios rurales en transición. In Romero, J.; Morales, A.; Salom, J. y Vera, F. (coords), *La periferia emergente*. Barcelona: Ed. Ariel, pp. 343-372.

CDP (Cassa Depositi e Prestiti). 2013. *Smart City – Progetti di sviluppo e strumenti di finanziamento*, Bologna. Available online: http://osservatoriosmartcity.it/smart-city-progetti-di-sviluppo-strumenti-di-finanziamento-lindagine-di-cassa-depositi-prestiti/.

Cerasoli, M. 2010. Il recupero dei centri storici: la modulazione della tutela. In: Atti della XIII Conferenza della Società Italiana degli Urbanisti "Città e crisi globale: clima, sviluppo e convivenza". Rome, February 25 th-27th, 2010. *PLANUM - The journal of Urbanism*.

Cherubini C., Orlando G., Reina A., Torre C.M. 2006. La sinergia tra analisi multicriteriale (AMC) e GIS nella valutazione del rischio di esondazione: il bacino della lama Baronale-Picone. (Provincia di Bari). *Giornale di Geologia Applicata* 3:109-113.

Egan, M. P., Nugent, N., Paterson, W. E. (eds) 2010. *Research agendas in EU studies: Stalking the elephant*. Palgrave Macmillan, Basingstoke.

Ferguson, M. 2019. How public sector IT can exploit the last 'big things'. *Public Sector Executive* (PSE), April/May 2019. Available online: http://www.publicsectorexecutive.com/Robot-News/how-public-sector-it-can-exploit-the-last-big-things.

Fernández, E. A. 2013. Los principales periodos de desarrollo del Estado autonómico. *Anuario Jurídico de La Rioja*, (2), 121-144.

Forum PA. 2014. Soluzioni per la Smart City a confronto: pratiche di trasferimento e riuso. Sharing Lab dell'Osservatorio Nazionale Smart City di ANCI. Available online: http://www.forumpa.it/expo-2014/soluzioni-per-la-smart-city-a-confronto-pratiche-di-trasferimento-e-riuso.

Gabrielli, B. 1982. Punto e a capo per il riuso. *Urbanistica*, 74: 2-8.

Gozálvez Pérez, V. 2001. Breve historia del desarrollo rural. In Martínez Puche, A. (coord.), *El desarrollo rurallocal integrado y el papel de los poderes locales*. Alicante: Ed. Servicio de Publicaciones de la Universidad de Alicante, pp. 31-39.

Mangialardo, A., & Micelli, E. 2018. From sources of financial value to commons: Emerging policies for enhancing public real-estate assets in Italy. *Papers in Regional Science*, 97(4): 1397-1408.

MIBACT 2018. *Carta Nazionale del Paesaggio. Elementi per una Strategia per il paesaggio italiano*. Roma, Italy: Gangemi Editore.

Minervino, G. 2016. “Il Progetto Artena: rivitalizzazione di un borgo attraverso una metodologia sistemica.” In: Anele R. (ed) *Rigenerazione urbana e progetto sociale. Quaderni del laboratorio di urbanistica e architettura*, n. 1. Map Design Project, Cosenza, 83-90.
National Association of Historic-Artistic Centers (ANCSA) & Sassaro, L. 1971. Per una revisione critica del problema dei centri storici. *Proceedings of the Study Seminar of the National Association of Historic-Artistic Centers*. Gubbio, 5-6 settembre 1970. Genova: Grafica LP.
Nocca, F. 2017. The role of cultural heritage in sustainable development: Multidimensional indicators as decision-making tool. *Sustainability*, 9(10): 1882.
Pica, V.; Cerasoli, M. 2018. Protocol of integrated sustainable interventions for historic small smart cities: the mitigation of disaster risk. In *Proceedings of the VI International Conference Heritage 2018, Heritage and Sustainable Development*. Barcelos, Portogallo: Green Lines Institute for Sustainable Development. In press.
Pica, V. 2018a. Beyond the Sendai Framework for Disaster Risk Reduction: Vulnerability Reduction as a Challenge Involving Historical and Traditional Buildings. *Buildings* 2018, Volume 8, 50. https://doi.org/10.3390/buildings8040050.
Pica, V. 2018b. Certification protocol for integrated interventions “Historical Small Smart Cities”: an instrument for the sustainable regeneration of minor historical centers. *Journal of Biourbanism* (JBU), (2)7: 97-112.
Pica, V., Cecili, A., Annicchiarico, S., & Volkova, E. 2019. The Historical Small Smart City Protocol (HISMACITY): Toward an Intelligent Tool Using Geo Big Data for the Sustainable Management of Minor Historical Assets. *Data*, 4(1): 30.
Saaty, T. L. 1980. *The analytic hierarchy process*. McGraw-Hill.
SEPI, M. 2011. Opinion of the European Economic and Social Committee on ‘The need to apply an integrated approach to urban regeneration’ (exploratory opinion). *Official Journal of the European Union. Resolutions, recommendations and opinions*, 8.
Streitz, N. 2019. Beyond ‘smart-only’cities: redefining the ‘smart-everything’paradigm. *Journal of Ambient Intelligence and Humanized Computing*, 10(2),791-812.
United Nations Development Group (UNDG) 2014. United Nations Development System- A Collective Approach to Supporting Capacity Development (PDF). Archived from the original (PDF) on 9 February 2014. Retrieved 7 July 2014.
Vázquez, A. N. & Vicente, M. R. 2019. Exploring the Determinants of e-Participation in Smart Cities. In *E-Participation in Smart Cities: Technologies and Models of Governance for Citizen Engagement* (pp. 157-178). Springer, Cham.
Wei, G. Responsive Urban Simulation. 2014. An approach towards real-time evaluation of urban design projects. Master’s Thesis, Polytechnic of Milan, Milan, 2014. Available online: https://www.politesi.polimi.it/handle/10589/98116 (accessed on 6 June 2018).
Weitz, R. 1981. *Desarrollo rural integrado: el enfoque de Rejonet*. Ed. Consejo Nacional de Ciencia y Tecnología (2ª edición), México, 118 pp.

Science and Digital Technology for Cultural Heritage – Ortiz Calderón et al. (Eds)
 ISBN 978-0-367-36368-0

SILKNOW. Designing a thesaurus about historical silk for small and medium-sized textile museums

A. Léon
Garin S.A., Moncada, Spain

M. Gaitán, J. Sebastián, E. Alba Pagán & I. Insa
Art History Department, Universitat de València, Valencia, Spain

ABSTRACT: The cultural heritage domain in general and silk textiles in particular, are characterized by large, rich and heterogenous data sets. Silk heritage vocabulary comes from multiple sources that have been mixed up across time and space. This has led to the use of different terminology in specialized organizations in order to describe their artifacts. This makes data interoperability between independent catalogues very difficult. Moreover, the interaction level of existing resources is low, most complex queries are not possible and results are poorly shown. In this regard, a recent EU-funded research project titled SILKNOW is building a multilingual thesaurus related to silk textiles. It is being carried out by experts in textile terminology and art historians, and computationally implemented by experts in text mining, multi-/cross-linguality and semantic extraction from text. This paper presents the rationale behind the realization of this thesaurus

1 BACKGROUND

The cultural heritage domain is characterized by large, rich and heterogeneous data sets where different organizations use different terminology to describe their objects (Amin et al., 2010; Owens and Atherton Cochrane, 2004). GLAMs and Cultural Heritage institutions aim to conserve and disseminate their collections; in order to do so, prior knowledge is extremely important. The registration of a cultural asset in an inventory or its inclusion in a catalog assumes its recognition as an element that requires conservation and protection. Controlled vocabularies stand out as essentials to provide access to museum collections not only to inside users (curatorial staff, conservators, education department), but also to external users who wish to know more about a subject without knowing the specific term of its search (Baca, 2004). Nowadays, Cultural Heritage institutions strive to obtain controlled vocabularies based on their own collections (Schreiber et al., 2008), such as the The Textile Museum Thesaurus from the Textile Museum in Washington, D.C. or the Museon Arlaten (Arles, département des Bouches-du-Rhône). While specific thesauri are useful for each institution, the result is a multitude of vocabularies in different languages that are difficult to standardize, especially when there is a strong need not only to make heritage accessible to the general public but also to facilitate exchanges across collections and institutions (Isaac et al., 2007). In this regard, some efforts have been made such as the Getty AAT (https://www.getty.edu/research/tools/vocabularies/aat/) or the UNESCO thesaurus. Concerning silk heritage terminology, it comes from various sources that have changed across time and space. Moreover, it changes according specialties (weavers or historians), nationalities (Europe or North America), or disciplines (ethnographic specialists vs art historians), etc. (Anderson, 2006). For example, local variations of a term are rarely taken into account (e.g. *espolín* has different meanings in some regions of Spain). To meet these challenges, SILKNOW is building a multilingual thesaurus dedicated to the specific vocabulary of historic silk textiles, which will also include local term variants. The thesaurus will help heritage institutions to provide access to and preserve silk heritage in the digital environment. Participating and collaborating institutions will radically improve their cataloging practices and digital data retention. In addition, the thesaurus will serve as an example of the benefits of shared cataloging frameworks and data interoperability.

2 THE SILKNOW THESAURUS

2.1 *Objectives*

The SILKNOW thesaurus aims to improve silk knowledge by building a web portal that will allow advanced search and representation of information related to this heritage. It intends to use as many scientific references as possible, in order to generate quality research that eventually will allow investigators to expand their studies. This thesaurus is targeted to researchers, students and cultural heritage professionals. For example, a researcher may use the thesaurus to connect terms that she or he could have found in historical documentation, providing updated, standard naming for these terms. In addition, when planning exhibitions, a curator often needs to either write the exhibition catalogue or ask for a loan to another museum. In these cases, artworks coming from other countries can be identified in their vernacular languages. Having a thesaurus will help curators to standardize terminology. Art history and conservation students will be familiarized with a variety of textile terms. It will also help to disseminate scientifically accurate terms that come from a standardized vocabulary whose final purpose is to improve the understanding, conservation and dissemination of silk heritage.

2.2 *Methodology*

To compile the thesaurus, inductive and deductive methods were undertaken (Nielsen 2004). Around 80% of inductive methodology was used. This means that terms were included in the thesaurus as soon as they were found in the literature. The other 20% was deductive, based on museum records and previous knowledge from the researchers. In order to be as accurate as possible, the compilers consulted a number of sources, mainly written sources that can be divided into: specialized textile dictionaries that cover the domain of silk heritage; general dictionaries that were useful when defining Preferred Terms (PT) and synonyms; theses related to historical fabrics; books about silk and textiles that had glossaries with specific terminology; historical sources that helped the compilers to define historical terms and their evolution over time and their current use; and, finally, other thesauri that, albeit quite generic, in some cases they were useful to specify certain terms. These terms were added manually by the researchers and adapted into SKOS format by automatic, computer-based extraction. A RDF model was considered, however it was decided to use the Simple Knowledge Organization System as it was more appropriated.

Next, terms and concepts were controlled and described by adding scope notes, qualifiers and synonyms. Regarding concepts, a PT was chosen to refer a unique concept, these terms are based in specialized literature. Whenever a term had more than one meaning (polysemy), all its meanings were included. In some cases, polysemy was not easy to undertake, since some concepts may refer to a weaving technique and to a fabric at the same time, such as *satin* or *twill*. Adding qualifiers to those terms was the best solution for these problems. Since there are no specialized silk thesauri, when entering terms the degree specificity was as the highest possible. Terms were included based on an estimation about those that will most likely appear in museums' catalogues, taking into account that not all museums catalog the same way, and that the thesaurus may not be used only by curators, but also by conservators. In order to make clearer what those concepts meant, scope notes were added following specialized literature. Special care was taken to follow the same syntactic style. As far as possible, compilers tried to add the etymology of the words and describe their uses in time and space. Finally, these definitions were reviewed by international experts.

The next logical step was to categorize those terms. The SILKNOW thesaurus is based on the Getty AAT structure. Following it, we have established three different relationships.

1) Hierarchical: when the relationship between terms is broader and narrower, SILKNOW used the 7 facets and subfacets of the AAT. When possible, parents were also placed according to the AAT structure. As silk heritage terminology is extensive and not easy to classify, compilers had to add new guide terms and subfacets in order to place some terms with maximum accuracy.
2) Equivalence: this relationship was appropriate when different names refer to the same concept as they are synonyms or quasi-synonyms.
3) Associative relationships: used when some terms are conceptually very related, but not hierarchically.

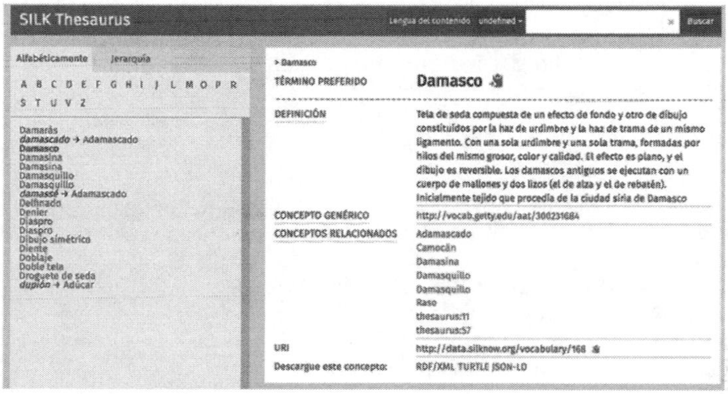

Figure 1. The SILKNOW thesaurus.

2.3 *A multilingual thesaurus*

The SILKNOW thesaurus is symmetrical, which means that all terms needed to be translated. First steps included the direct translation of terms, synonyms and associated concepts. As it was done with the master spreadsheet, translators (who also are textile specialists) used specialized sources, which in some cases provided translations in other languages (such as the Castany Saladrigas dictionary, 1949). Some issues arose when translating some terms; for example, some terms only existed in one language, or needed more than one translation.

2.4 *Extension*

The current version of the SILKNOW thesaurus has 545 terms as per May 2019. The implementation of the multilingual web-based thesaurus is due in 2020. When the thesaurus is completed it will be shared with the AAT in order to include new terms to their vocabulary. A first coverage calculation was made with 2 different Spanish databases in order to find out if the terms included in the thesaurus could be found in museum records. The result was a 60.8% of matches between the thesaurus terms and those two databases.

3 CONCLUSIONS

Conservation of cultural heritage includes a set of direct and indirect actions in order to ensure the physical persistence of objects; therefore, prior knowledge is extremely important. The registration of a cultural asset in an inventory or its inclusion in a catalog assumes its recognition as an element that requires conservation and protection. Nowadays, museums around the world are generating tools that allow the development of a systematic and coherent cataloging of museum collections and the improvement of a standardized model, in order to avoid the lack of common criteria when dealing with these kinds of records.

The SILKNOW thesaurus is designed to generate synergies across European textile museums. It will allow not only to connect various textiles located in diverse collections but also to trace similar artistic trends. Moreover, it will help to standardize a wide and fragile heritage, hence enhancing its conservation. Understanding that conservation requires proper management is essential. SILKNOW, as an open access tool, seeks to have a strong impact not only for museum collections and researchers, but also among the general public. If society becomes more engaged with its heritage, this will lead to the development of a true awareness and defense of heritage, turning it into an agent of innovation (Bakhski et al., 2008). Society must become guardian of its own heritage, facilitating the transmission between the past and the future (Andreu Pintado, 2014).

This is the first step to generate a common framework of knowledge and dissemination of historical textiles in general, and silk heritage in particular.

REFERENCES

Amin, A., Hildebrand, M., van Ossenbruggen, J., Hardman, L., 2010. Designing a thesaurus-based comparison search interface for linked cultural heritage sources. Proc. 15th Int. Conf. Intell. User Interfaces IUI'10, 249–258.

Anderson, C.G., 2006. Talking about Textiles: The Making of the Textile Museum Thesaurus. Textile Society of America Symposium Proceedings 302.

Andreu Pintado, J., 2014. Arqueología en directo: canales de comunicación y transferencia de resultados en la investigación sobre patrimonio arqueológico: la ciudad romana de los Bañales un castillo en Zaragoza. Visibilidad Divulg. Investig. Desde Las Humanidades Digit. Exp. Proy., 17-41.

Baca, M., 2004. Fear of Authority? Authority Control and Thesaurus Building for Art and Material Culture Information. Cat. Classif. Q. 38, 143–151.

Bakhski, H., McVittie, E., Simmie, J., 2008. Arts and Humanities Research and Innovation. Arts and Humanities Research Council.

Isaac, A., Zinn, C., Matthezing, H., Van de Meij, H., Schlobach, S., Wang, S., 2007. The value of usage scenarios for thesaurus alignment in cultural heritage context, in: Cultural Heritage on the Semantic Web Workshop at the 6th International Semantic Web Conference (ISWC 2007).

Nielsen, Marianne Lykke. 2004. Thesaurus Construction: Key Issues and Selected Readings. Cataloging & Classification Quarterly 37, no. 3–4 (1 January), 57–74.

Owens, L.A., Atherton Cochrane, P., 2004. Thesaurus Evaluation. Cat. Classif. Q. 37, 87–102.

Schreiber, G., Amin, A., Aroyo, L., van Assem, M., de Boer, V., Hardman, L., Hildebrand, M., Omelayenko, B., van Osenbruggen, J., Tordai, A., Wielemaker, J., Wielinga, B., 2008. Semantic annotation and search of cultural-heritage collections: The MultimediaN E-Culture demonstrator. Semantic Web Chall. 20062007 6, 243–249.

Science and Digital Technology for Cultural Heritage – Ortiz Calderón et al. (Eds)
© 2020 Taylor & Francis Group, London, ISBN 978-0-367-36368-0

Software developments in hyperspectral image capturing for the automatic generation of false-colour composites

J.P. López, C. Vega & J.M. Menéndez
Universidad Politécnica de Madrid, Madrid, Spain

ABSTRACT: Hyperspectral imaging has been a particularly useful technique for pigment characterizing on paintings and other types of art manifestations in previous years, thanks to its non-invasive character and because it is complementary to chemical analysis, providing high efficiency in results. This technique consists of capturing a number of images in the visible spectrum and the near infrared band using frequency intervals between 400 and 1700nm. False-colour composites facilitate the contextualization of images at different wavelengths as opposed to the image obtained by direct light, which is useful to highlight the information not easily seen by the human eye. One of the main problems for the generation of this type of composites lies in the capturing conditions of the images, obtained through cameras with different resolutions and scales, and influenced by the position, angle and distance to the portrayed object. This makes it difficult for the overlapping of RGB (red, green, and blue) channels on the direct light image with new channels from images captured with different techniques and camera conditions. Thus, the purpose of this research is the design and implementation of software tools that allow the automatic generation of false-colour composites that take into account the different conditions of the images being captured. The application substitutes each of the RGB channels of direct light photograph through the information obtained after applying techniques such as X-ray and hyperspectral imaging. The result is an artificial composite that facilitates researchers' work in characterizing the pigments needed to create a more complex and improved interpretation of the artwork and its evolution in the restoration process. These interpretations will be of great support in the analysis, restoration and conservation of cultural heritage. Our approach provides a simple intuitive tool for quick preliminary analysis using non-invasive techniques, enabling the statistical analysis and subsequent segmentation. Preliminary tests on 14th Century altarpieces and Flemish paintings demonstrate improvement in the rendering of the paintings; as a result of the visible spectrum provided by these particular composites, the identification of pigments brought to light hidden information in the paintings.

1 INTRODUCTION

Different imaging techniques are applied to conservation and preservation of cultural heritage (CH) assets due to their non-invasive character and because they are easy and fast to implement with the use of modified cameras or filters (Cosentino, 2016). These methods are useful in art examination to help researchers in different tasks, such as pigment identification, image mapping or inpaint detection. Among the most popular imaging techniques for art examination, CHSOS (2016) highlights the following ones:

- Technical Photography (TP): this set of methods offer high-quality photographic documentation composed of a collection of images in the spectral range of 360 to 1000nm, taken with a digital camera with modified sensors, and provides information for different purposes. Firstly, this technique includes the visible (VIS) or direct light sources, which presents the image in the visible range of wavelengths. This set of techniques also includes

more sophisticated methods of image capturing, such as Polarized Light photography (PL) or Raking Light photography (RAK), also corresponding to the visible spectrum but taken with specific characteristics. Additionally, other techniques allow the capturing of information about the CH assets from the non-visible spectrum, such as Ultraviolet Fluorescence Photography (UVF), when the image is captured after being exposed to ultraviolet radiation; Infrared Photography (IR), when the image is captured by the emission of IR radiation by filtering the visible wavelength in order to reveal underdrawings, amongst others.
- Multispectral Imaging (MSI) and Hyperspectral Imaging (HI), defined as effective non-destructive methods for pigment identification, while considering a high variety of frequencies of the spectrum in a set of aligned images (Pamart, 2017).
- Infrared Reflectography (IRR). IRR usually uses InGaAs cameras, which allow images to be more transparent than IR capturing, therefore improving the detection of pigments.
- Reflectance Transformation Imaging (RTI) and Reflectance Spectroscopy (RS).

Hyperspectral Imaging (HI) stands out because it offers a high variety of data for pigment characterization (Hayem-Ghez, 2015). HI works in the spectrum at wavelengths between 400 and 1700nm in the visible spectrum and the near infrared band using frequency intervals, creating a mosaic of images at different bandwidths allowing information to show that would normally be hidden to the human eye.

False-colour composites is a technique for preliminary data processing, which consists of the creation of hybrid images with data belonging to captured images obtained through different methods, including technical photography or multispectral imaging. For that reason, captured images are merged by substitution of each RGB channel in order to help in the contextualization of the information obtained. False-colour composites enable the interpretation of images taken by different techniques so as to highlight information that is difficult to observe at a glance. As false-colour images are experimental hybrid methods whose results confront images captured with different types of techniques, they provide much more information in helping researchers in the field of art examination. An example of false-colour composites is included in Figure 1.

However, the multiplicity of techniques for obtaining data from CH assets generates an inconvenience that is the keystone of this research. As the variety of capturing techniques is high, the capturing conditions of the images is highly diverse. Capturing techniques employ cameras with different capturing conditions of image resolution, scale, as well as position, angle and distance to the object. Thus, the combination of RGB channels for obtaining false-colour composites corresponding to images captured in diverse conditions is not trivial, and the research requires the preprocessing of data for a correct image alignment of sources.

Software tools for raster graphics edition, such as Adobe Photoshop (Cosentino, 2014) are commonly used for the creation of false-colour composites but these methods lack real automatic methods for image alignment and force the researchers to do this manually. Computer vision techniques combined with deep learning techniques help in this task, improving the result and offering high accuracy in the resulting images.

Figure 1. Capturing techniques: VIS photography (left), UVF photography (center left), and an example of false-colour composite (center right) and K-means clustering (right).

2 FALSE-COLOUR COMPOSITES GENERATION

The objective of this research is the design and implementation of an intuitive tool for quick preliminary analysis using non-invasive techniques for conservation and restoration of cultural heritage, which is able to facilitate the examination of art objects, based on objective measurement, statistical analysis of the pixels, and segmentation. This tool allows creating false-colour images by combining the RGB channels of data belonging to images of a CH asset captured with different techniques. The requirements of the false-colour composites include the adaptation to different conditions of images obtained from cameras with separate resolutions and scales. Additionally, the tool will be able to develop further analysis of combined pixels to generate object segmentation, histograms of distribution of information. Finally, it will also be able to test CH assets with images captured from different sources for drawing conclusions

The alignment (Zampieri, 2018) is based on the detection of singular points and tracking of the objects and, consequently, vectors alignment of the images (or channels of the image) involved. The process is similar to image stitching, with the use of sliding windows to find the minimization of error through a set of metrics: SAD (Sum of Absolute Differences), MSE (Mean Square Error) and Correlation.

For singular point detection, different techniques are applied depending on the type of content included in the image. First, considering that most of the paintings include human faces, the use of a neural network based on Umdfaces dataset (Bansal, 2017) detects the facial keypoints and the system proceeds to perform the alignment based on minimization of error metrics, creating vectors of singular point connections. For images that do not include natural human faces, the system uses traditional computer vision techniques, such as the Harris corner detector, which is useful for detecting the corners in the painting framework, or SIFT (Scale-Invariant Feature Transform) for finding other singular points beyond human faces or framework corners. These vectors allow the alignment of positions from different sources, solving problems of scaling, shifting or rotating in the set of images. Once the images are aligned, the user proceeds to change the combination of RGB channels among the different capturing images, finding the best solutions that help in the interpretation of the artpiece.

However, image alignment is not the only problem that needs solving. The human eye can be deceiving for several reasons, not only due to the capacity of the human eye itself, but also due to other factors, such as the calibration of the monitor where the image is screened. For this reason, it is advised to include objective tools that respond reliably to the needs of researchers, such as colour histograms, vectorscopes or statistics of pixels. Pixel segmentation techniques, such as K-means clustering (Dhanachandra, 2015), and statistical distribution of pixels, offer extra information by clustering and automatically classifying pixels, as shown in Figure 1 (right).

3 TESTS AND RESULTS

To develop the methodology for image alignment presented in this paper and to demonstrate the validity of false-colour composites, the first step consists of selecting a set of CH assets for assessing the combination of images captured with different techniques in order to draw conclusions about the pigments. The CH assets chosen for art examination are the following ones:

- *Virgin with child*, 15-16th Century Flemish painting (details shown in Figure 1 and 2). Lázaro Galdiano Museum, Madrid (Spain).
- *Saint Martin Altarpiece,* 13th Century (detail shown in Figure 3). Pilgrimage and Santiago Museum, Santiago de Compostela (Spain).

Tests developed over these two CH items demonstrated the improvement in the interpretation of the paintings and the identification of pigments thanks to the highlighting of hidden information that comes up in the context of the visible spectrum as an effect of these

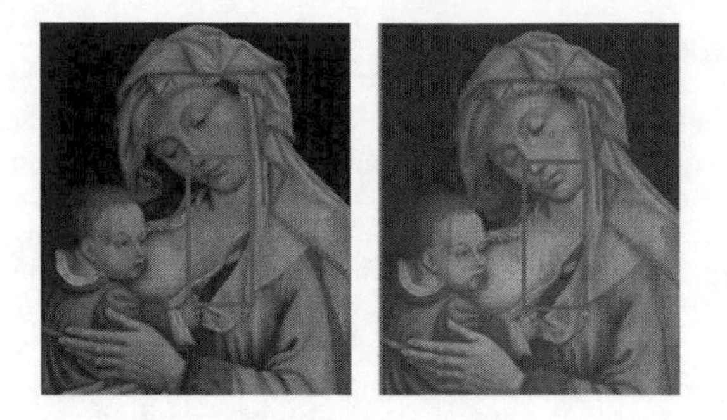

Figure 2. *Virgin with child* VIS image (left) and false-colour composite with UV photography data (right).

Figure 3. *Saint Martin Altarpiece* VIS image (left) and false-colour composite (right).

particular composites. Throughout this improvement, false-color composites allow the detection of inpaints on the painting and, on the other hand, the mapping of pigments. For instance, Figure 2 highlights the inpaints found in different restorations undertaken on the work of art, otherwise impalpable in the VIS image; and Figure 3 highlights different pigments in dress and in framework, which demonstrate the authenticity of the pigments used in framework that differs from the ones used in the clothing, which correspond to subsequent restorations.

4 CONCLUSIONS

The creation of false-colour composites that combine images captured through different techniques are powerful tools for art examination in research environments. The use of this tool produces automatic results comparable with those of commercial software, saving time and improving accuracy. The interpretations based on false-colour composites aid in analysis, restoration and conservation of cultural heritage, as demonstrated with tests and results.

The software approach provides an intuitive and automatic tool for quick preliminary analysis using non-invasive techniques, enabling statistical analysis and subsequent segmentation. The tools for image alignment adjust the scale and position of frames from different sources, allowing the addition of information. Furthermore, automatic generation is a time-saver for researchers and improves the results of false-colour composites. Overall, it permits conclusions to be drawn regarding the origin, deterioration and detection of processes of restoration undertaken on paintings and other CH assets, using non-invasive techniques.

ACKNOWLEDGEMENT

This study has been supported by the Spanish Cultural Heritage Institute (IPCE), with the right to use of images of art captured with the most innovative techniques.

REFERENCES

Bansal, A., Nanduri, A., Castillo, C. D., Ranjan, R., & Chellappa, R. 2017. Umdfaces: An annotated face dataset for training deep networks. In 2017 IEEE International Joint Conference on Biometrics (IJCB) (pp. 464-473). IEEE.

Cosentino, A. and Stout, S. 2014. Photoshop and Multispectral Imaging for Art Documentation. e-Preservation Science, 11, 91–98, 2014.

Cosentino., A. 2016. Infrared Technical Photography for Art Examination. *e-Preservation Science*, 13, 1-6, 2016.

Dhanachandra, N., Manglem, K., & Chanu, Y. J. 2015. Image segmentation using K-means clustering algorithm and subtractive clustering algorithm. Procedia Computer Science, 54, 764-771.

Hayem-Ghez, A., Ravaud, E., Boust, C., Bastian, G., Menu, M., & Brodie-Linder, N. 2015. Characterizing pigments with hyperspectral imaging variable false-color composites. *Applied Physics A*, 121 (3),939-947.

CHSOS (Cultural Heritage Science Open Source). Applications in Art examination. https://chsopensource.org/infrared-false-color-photography-irfc/.

Pamart, A., Guillon, O., Faraci, S., Gattet, E., Genevois, M., Vallet, J., De Luca, L. 2017. Multispectral photogrammetric data acquisition and processing for wall paintings studies. ISPRS-International Archives of Photogrammetry, Remote Sensing and Spatial Information Sciences (Vol.42, pp. 559-566).

Zampieri, A., Charpiat, G., Girard, N., & Tarabalka, Y. (2018). Multimodal image alignment through a multiscale chain of neural networks with application to remote sensing. In Proceedings of the European Conference on Computer Vision (ECCV) (pp. 657-673).

Science and Digital Technology for Cultural Heritage – Ortiz Calderón et al. (Eds)
 ISBN 978-0-367-36368-0

Protocol for cataloguing earthen defensive architecture from the perspective of risk

M.L. Gutiérrez-Carrillo
Department of Architectural Constructions, University of Granada, Granada, Spain

I. Bestué Cardiel
Department of Graphic Expression in Architecture and Engineering, University of Granada, Granada, Spain

E. Molero Melgarejo
Department of Urban Planning, University of Granada, Granada, Spain

J.J. Fondevilla Aparicio
Researcher Group HUM799, University of Seville, Seville, Spain

ABSTRACT: This paper presents a model for the analysis of the earthen defense heritage of southeast Spain which has been developed by the Spanish Ministry of Economy and Competitiveness PREFORTI R+D+i Project. The protocol configures a geo-referenced inventory which corresponds to the National Heritage Protection Plans. The catalogue is structured into two sections on the basis of the risk diagnosis.

Phase I: Information regarding the asset´s heritage on the basis of standard parameters.

Phase II: Information regarding its material, construction and structural characteristics, its state of conservation, the risk factors to which it is exposed, and its degree of vulnerability is incorporated into a GIS.

This model explores, firstly, how heritage assessment and risk diagnosis can be improved with the inclusion of information from multidisciplinary sources and secondly, the effect that the implementation of ICT in heritage management has on sustainable preventive conservation strategies.

1 INTRODUCTION

The value ascribed to defensive architecture in social, political and administrative contexts has varied in the past, from its being regarded as an important symbol of history to its relegation to that of an obscure relic. However, this kind of heritage holds significant intrinsic value (PNAD, 2006; Gómez Robles, 2010) and so it has been granted the maximum level of protection. Nevertheless, it remains seriously at risk despite such attempts to safeguard it (Decree for the Protection of Spanish Castles 1949; Spanish Historical Heritage Law of the 25th of June 1985; Andalusian Law 2007). This is in part due to its unique history and historical trajectory, when it was patently ignored at times, consequently abandoned and then became so deteriorated that most of its essential features were lost. These unique circumstances therefore demand that information from broader, multidisciplinary sources is applied to its appraisal and conservation.

2 THE MATTER IN QUESTION

The structures from which the sample has been drawn can be found throughout the Granada, Almeria and Murcia provinces and of such number that their study and classification has been consequently heterogeneous in nature; however, these are all medieval defensive structures built rammed earth, a fact which implies the need to regard this subgroup as having its own, unique identity. Despite being officially recognised by the administration, there have been obstacles to their becoming more widely known (Cobos & Castro, 2007), such as their remote nature, the inherent difficulty in accessing them, the fact that few types and examples remain, and because their complex archival documentation needs to be revised. The systematic cataloguing of fortified heritage constructed for the purposes of protection began in Spain in the middle of the 20th century. The 1949 Decree for the Protection of Spanish Castles specifically included a requirement to catalogue Spanish fortified structures. However, at that time, this requirement was separate from heritage protection; this was later rectified in 1968, with the publication of the Inventory Plan of the Protection of European Cultural Heritage, Spain. Although it could be considered basic, limited and lacking in homogeneity, it was the first that met the objectives outlined in the 1949 decree. This was followed by the Ministry of Culture Inventory 1980 and one produced several years later by the same institution, the Ministry of Culture Inventory 2008, which was comprised of 787 catalogued records, basically put together by the Association of Friends of the Castles (*Asociación Amigos de los Castillos*). Also worth mentioning is the inventory model for the autonomous area of Andalusia which was drawn up as part of the Andalusian Plan of Defensive Architecture (PADA, 2004) and was intended to be the defining instrument for the protection, conservation and awareness-generation of fortifications and other types of military and defensive heritage (Fondevilla, 2007).

3 OBJECTIVES OF THE PROPOSED CATALOGUING MODEL

The PREFORTI R+D+i Project incorporates the proposed model for cataloguing and not only furthers knowledge about this specific heritage but also more closely examines aspects regarding its preventive conservation. This allows for some of the national and international heritage protection models´ protocols to be implemented and tested. The following are included amongst the wider objectives and more specifically associated with methodological aspects:

- Systematise the analytical models in order to facilitate assessment of the degree of vulnerability and risks from natural and anthropic agents (which have the most profound implications) for a specific heritage.
- Develop preventive action maps which with the risk maps and damage maps, can be used to deliver an effective and coherent prevention programme for these structures.
- Create a sustainable heritage maintenance model for assets where physical constraints make intervention difficult, which will in turn help reduce conservation management costs.

4 DEVELOPMENT OF THE PROPOSED MODEL FOR CATALOGUING

After each specific case was studied using an inductive method of observation, a system of records was created which made the fast retrieval of information and objective case comparisons possible. This led to the creation of a database structured on two levels or phases, Inventory and Catalogue.

- Phase I - Inventory. Contributes towards what is known about the heritage asset using standard parameters: Identification, Location, Characterization, Formal Description, State of Conservation and Images.

This first stage examines the asset from a general perspective which must be both versatile and able to be integrated into any management system, either public or private (Figure 1). This examination is broadened by examining the findings from previous research, which include identification and location information, heritage protection status, geo-referencing, its

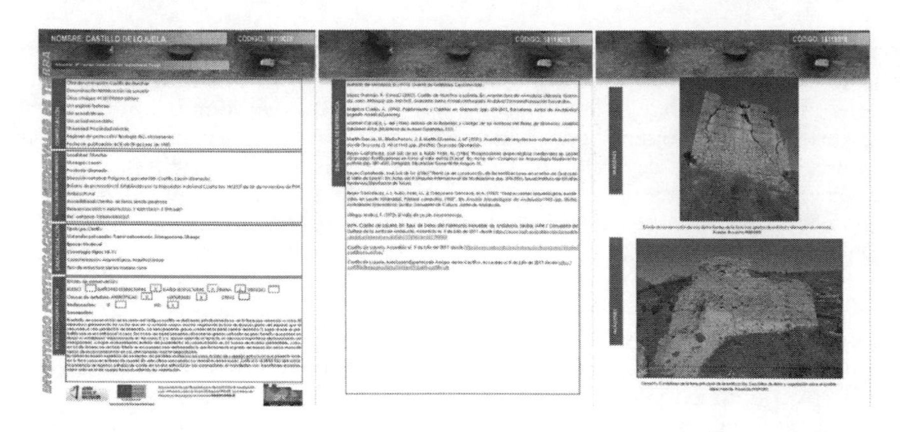

Figure 1. Inventory record model. PREFORTI project.

historical, artistic and sociocultural significance, a summary of its state of conservation, and images for identification and graphic purposes. This first cataloguing phase was carried out for 229 cases distributed between the Granada (104), Almeria (54) and Murcia (71) provinces. Generic information was gathered for each asset which allowed for the initial results for the overall state of conservation, associated physical location and anthropic factors to be compared.

– Phase II - Catalogue: information about material, construction and structural characteristics, state of conservation, risk factors and the degree of vulnerability. This is integrated into an open source GIS so that the correlate risk maps can be extracted. The territory is further explored using GIS tools for geo-referencing and geospatial analysis of the records, and a Geo-referenced Catalogue is configured in which alphanumeric identifiers can be linked to geospatial coverage for study at a micro-territorial scale (Figure 2).

During this second cataloguing phase, the focus is on evaluating the asset´s state of conservation and determining which factors have a more direct effect on its ongoing deterioration. Consequently, it becomes possible to determine which actions should be prioritised as well as identify conservation and sustainable preservation strategies. This process was carried out with a total sample of 39 cases located throughout the same geographical areas as in Phase I. The subsequent findings for comparison were, the most commonly found pathologies and the systematic and reoccurring risk factors found for the majority of cases.

This unique heritage is an inherent and defining feature of the territory and surrounding landscape, therefore using geographic tools to catalogue it is justified. Additionally, the effects of the territory and environmental conditions on these assets need to be better understood, so the use of ICT in the analysis of the territory becomes even more important. This is done by incorporating the existing risk planimetry of the area held on various official platforms into the database and adding microzoning information. This allows for detailed study of the environmental and anthropic agents affecting them (Figure 3). This system allows for database records to be standardised in line with European directives (EU 2007; EU 2010) and to be linked to the territory, which can be treated as an additional variable in the catalogue of risks; significant examples can be included according to their heritage characteristics, territory and the incidence of risk -natural or anthropic-. Consequently, it provides objective systematised and indexed data for each element in the database, it expedites analytical and information management tasks, recent information about its features can be updated and integrated, conclusions and strategies can be proposed and finally, graphic and cartographic findings can be reproduced.

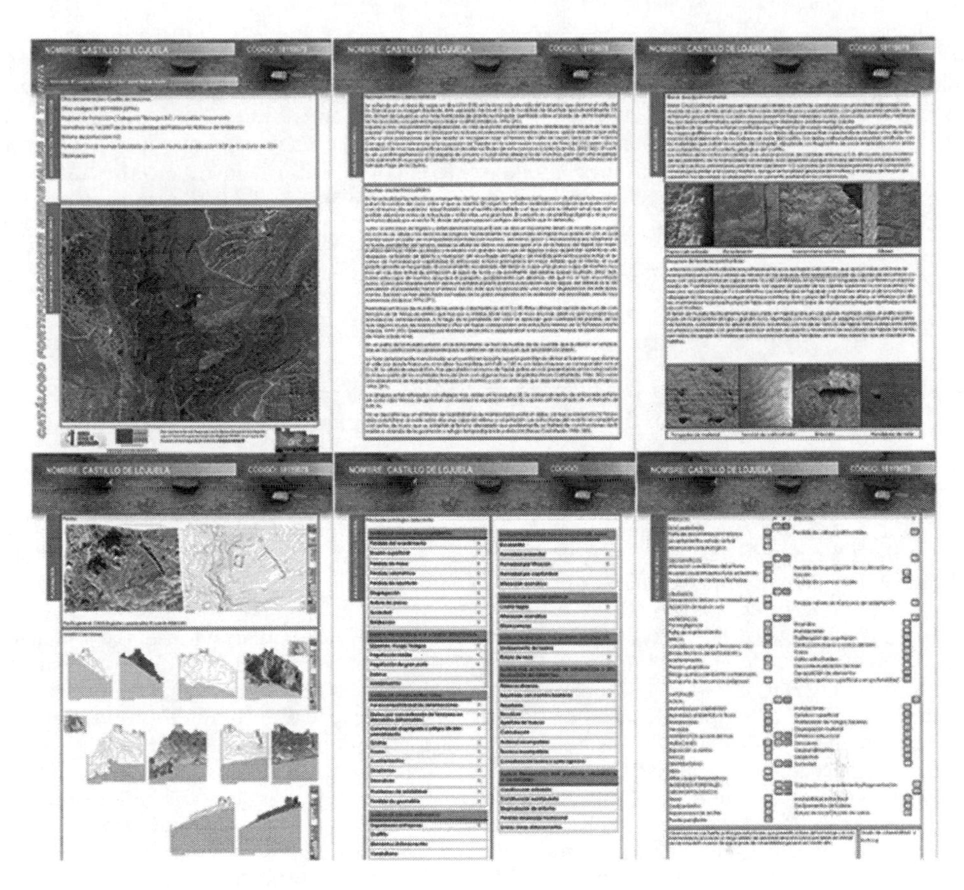

Figure 2. Catalogue record model. PREFORTI project.

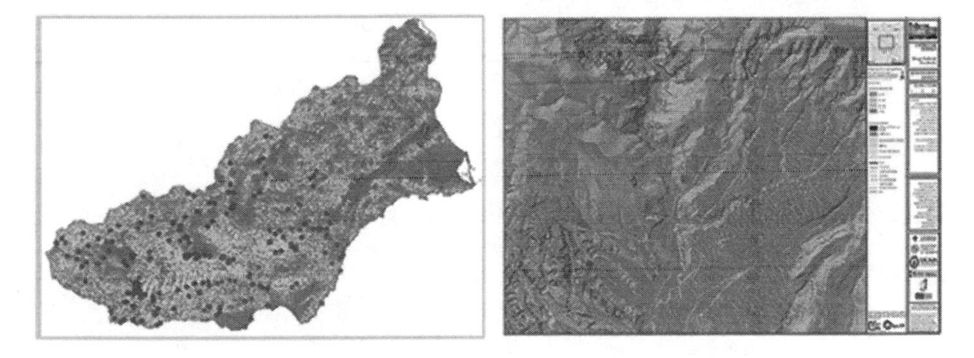

Figure 3. Macro and microzoned risk maps. Lojuela Castle. Sloped terrain models. PREFORTI project.

5 CONCLUSIONS

The application of this model for cataloguing fortification heritage paves the way for new approaches to its preventive and sustainable conservation. This will have considerable benefits for those administrations and agencies involved in their stewardship, given that every asset can have its own preventive action map associated with a specific risk map. At the same time, any priority actions can be identified and an efficient maintenance programme can be established. By integrating information which closely explores the risk factors into the heritage catalogues, information which is very important to planning their preventive and sustainable conservation is improved. Past experience has shown how not taking this information into account leads to substantial expenditure going towards emergency repairs which could have been minimised had the information been taken into account. Improvements to the future economic sustainability of these assets -of importance to both public and private management- will be expedited as the overall and multidisciplinary knowledge about each cultural asset improves.

ACKNOWLEDGEMENTS

This study is part of the PREFORTI project (BIA2015-69938-R) has been financed by the State Research Agency (SRA) and European Regional Development Fund (ERDF).

REFERENCES

Cobos, F. & Castro, J.J. 2007. Medieval Walls, Castles and Towers. In *Arquitectura Militar*. Vol. 2, Colección Patrimonio Arquitectónico de Castilla y León.

Decree for the Protection of Spanish Castles 1949.

EU, European Union. 2007. Directive 2007/EC/2 of the European Parliament and the Council of 14 March 2007 establishing an Infrastructure for Spatial Information in the European Community (IN-SPIRE), *Official Journal of the European Union*, L 108, vol. 50.http://inspire.jrc.ec.europe.eu.

EU, European Union. 2010. "Commission Regulation (Eu) No1089/2010 of 23 November 2010 imple-menting Directive2007/2/EC of the European Parliament and of the Council as regards inter-operability of spatial data sets and services".

Fondevilla Aparicio, J.J. (coord) 2007. *The defensive architecture of the Huelva province. Territorial keys for establishing a Border Territory*. Sevilla, Consejería de Cultura de la Junta de Andalucía.

Gómez Robles, L. 2010. The value of a restored monument. An approach to scientific restoration. *Revista Boletín PH*, nº75, Sevilla, Instituto Andaluz del Patrimonio Histórico: 80–93.

PADA 2004. *Andalusian Plan of Defensive Architecture*. Consejería de Cultura. Junta de Andalucía.

PNAD 2011. *National Plan for Defensive Architecture*. MECD.

Spanish Historical Heritage Law of the 25th of June 1985.

Science and Digital Technology for Cultural Heritage – Ortiz Calderón et al. (Eds)
© 2020 Taylor & Francis Group, London, ISBN 978-0-367-36368-0

Traceability of contemporary building culture as a key to the future heritage consideration

M. Pelegrin-Rodriguez
Department Architectural Projects, E:T:S: Arquitectura, University of Seville, Seville, Spain

ABSTRACT: We call "Traceability of contemporary building culture", to a methodology proposal for documenting the process of design and construction of a group of recent architectural works, that identifies a series of cultural identity factors, such as who builds, where do material come from, how long does it take incorporate them on site, that define a culture heritage. while the initial tool is used to aid our decision making as architects by facilitating the better comprehension of these factors, later on by encouraging greater consideration of the material, economical and human resources, within the vicinity of the project site, renders it implications within the building culture embodied in the process. Conclusions show that the idea of traceability is a tool to document during the design and construction stages implications like actors involved, the materials used and their origin, that is contemporarily part of the building culture as a key to the future heritage material and immaterial consideration.

1 INTRODUCTION

1.1 *Interest of the study field*

After the fall of the Berlin Wall in 1989, Europe's political and economic map changed, and architecture as a discipline, it underwent a process of redefinition of new guidelines for architecture design tools and logics, which despite sharing certain genealogical and epistemological traits with some preceding experiences, involved a change of course regarding the relevance of the discipline. The research focusses on a series of case studies with a pragmatic, non-discursive and committed attitude toward the discipline, a design, theoretical and practical approach which are all especially relevant to the design and building practice with in the urban and cultural context. This research aims to explore the building culture embodied in the construction process, through the collection, management and representation of information about who contributed to each project, what they do, where they came from, when they were on site, and most importantly how far they had to travel. As well as incorporating project stakeholders (through participation) throughout the construction process, this methodology incorporates local knowledge and global techniques that affect the use, ownership and maintenance of each building.

In the last two decades it is possible to recognize and identify an emergent sensibility that proposes a way of making pragmatic architecture that is technically simple and ′aware′ of the means of production, where dialogues with the social and cultural context in which buildings are inserted may lead to encounters with dominant forces. We speak of an architectural practice supported by some built examples, that share a persistent concern not to cleave the profession of intellectual reflection, to emphasize the material practice and empirical experience, to internalize and understand the technical and creative process of project development, as well as the construction and incorporation of architecture into the natural and social environment in which it is situated.

Knowing that architecture participates in a contemporary 'cultural status', these practices highlight the value of the processes of material production of 'space'. From this new sensibility these spaces are considered to be places of intermediation with the territory and landscape. They are points of compatibility between the public, private, collective. They are places of production and everyday life. In this sense, to explore and explain these processes allows the agents involved to know and understand architecture as an agent in the production of this intermediation. As such, it becomes of interest to study and understand the materials, technology and context of the architecture, which- even in the European context- has global technologies (standardized serial, industrialized) and local techniques (trades, operators, local professionals).

Thus, within this interest for material practice as a cultural fact, that is, not merely technical or object; from a concept of architectural production as open rather than closed we propose a more comprehensive and complex understanding of sustainability. There is not only concern for the energy balance of the construction and operation of a particular architectural space, but also the social aspects and cultural context in which this space is inserted. The motivation for this research is based on the necessity to rebuild relations between "conception and practice, technique and expression, craftsman and artist, between practice and theory" (Sennett, 2009).

1.2 *Research Context*

The term "traceability" was used initially in the agri-food sector, although its scope has expanded to other areas. This term is defined by the Committee of Food Security AECOC as those pre-established and self-sufficient procedures to provide information about the history, location and trajectory of a product or batch of products along the supply chain at any given moment through certain tools. Its application in production processes are defined by ISO as the "ability to trace the history, application or location of that which is under consideration" (UNE-ISO 9000:2005), so that its adaptation to field construction focuses on the ability to identify the origin and the different stages of a process of production and distribution of products and agents. So the notion of traceability is of the outmost importance for the development of criteria and methods of built heritage conservation in the future, but not even discussed as such at the UNESCO WH 1972, neither among the several fields treated in reference manuals (as reference, Stanley-Price, N., Price, N., Talley, M. K.; & Vaccaro, A. M., 1996; Richmond, A. & Braker, A..2010).

In this context we understand that the procedure applied to the processes of production and materialization of architecture allows for spatial and temporal linkages, while enabling us to see the 'material culture' and 'tacit knowledge' of construction agents involved in the building as a constituent of a building´s "social capital": Knowledge and skills that are accumulated and transmitted through social interaction, from their places of origin to the place of production of the building. The procedure not only allows us to see the agents or quantitatively assess the environmental impact of the resulting product; but also to visualize and rebuild relationships that determine their incorporation into building in terms of a socio-cultural perspective, as an aid in decision making.

1.3 *From theoretical framework to case studies for testing the method*

This work explores new visualization tools and guidance for making decisions regarding construction and the environment in which it occurs, as a contingent, creative and open contribution towards incorporating sustainability parameters in the material production of architecture.

To try the method we have applied the analysis method to buildings developed within the professional practice Mediomundo arquitectos: Library in Cañada Rosal (Sevilla), Theatre in Vejer (Cádiz), Dwellings in Conil (Cádiz), Cyber centre in Macarena Tres Huertas (Sevilla), and the Faculty of Health Sciences in Granada (Granada), as they were laboratory test sample.

After the analysis and discussion of the case studies, which have been carried out through interviews, site visits, the documentation of bibliography, photographs as well as drawings, and completing or contextualizing this by studying preceding and even coetaneous production both in the realm of architecture and culture. We have been able to confirm the existence of certain shared design traits and tools to be traced back in the spatial and technical resolution of the architectural works studied. We can synthesize them as: the design of enclosures, the design of guidelines, the organization and articulation of spaces, and the implementation of an industrial, standardized and assembled materialization in their construction.

From the common ground of academic and professional interests, we set out to explore and get to know how to frame out the risks and contradictions which were taken, with a certain attitude, by a sector of architecture practices during the two decades of the turn of the century, all with the aim of searching for wider cultural value of architectural works. Our interest lies in understanding and organizing the guidelines, techniques and tools that redefined this approach and development of the architectural project. Hypothesized as future heritage contributions to take into account, the characteristics that are inherent to the architectural project which, without losing their sense of historical criticism, but allow extracting instrumental knowledge from them. They can be applied, also critically, as a tool of analysis and design both in the classrooms and in the professional practice itself.

By looking at artistic practices we can observe how the work-subject-space relationship that they experimented with over the previous decades. Artistic practices had been detecting and questioning their incorporation into the cultural industry with works that had approached the issue.

2 HYPOTHESIS

Within this context, the possibility of underlining at some architectural works the tangible and intangible inputs by means of the traceability of some of the factors and agents involved, makes it possible to consider to a way of practicing and understanding architecture within the production and thought during the period between the 20th and 21st centuries disposed as Contemporary Cultural Heritage.

3 OBJECTIVES

The objective of the research is to frame the theoretical background and the concepts that would define a tool that allows: To visualize and analyse the interrelation of actors, materials and systems involved in material production of an architectural work, emphasizing the impact of distance and time of origin of involvement of the different actors, materials and technologies and their impact on development of the work; To generate a tool for analysis and support for the decision-making in processes of project and construction management; To make a comparative assessment of the buildings studied and the interrelationships between material agents and location, and to define the variables that determine these results.

3.1 *Summery of methodology*

For the development of the tool, the following methodology is proposed, based on the identification of opportunities and concerns involved in their design. In the first instance we proceed to the documentation of data, where information is collected relating to staff (technicians and operators) involved in the construction of a particular work, stating their origin (distance travelled to work), the duration of their involvement (days, weeks, months), the material, system or task that they undertake or develop, and the price involved in material implementation. By adding the times for each agent we obtain a total time employed or spent in work, of which each participant takes a percentage.

In the second instance, with the data obtained we proceed to the preparation of corresponding graphs. The data obtained is made into a circular graph where the portion of time spent by each agent or system is represented by an arc.

On the same graph the distance from the origin of each actor responsible for the implementation of a task or system is described. These distances are represented by the length of each

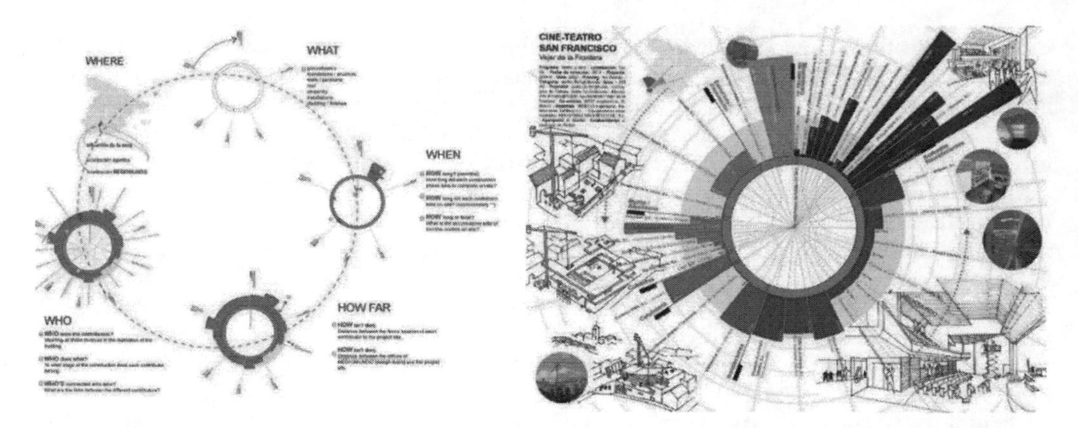

Figures 1 and 2. The application of the tool to diagrams about the buildings developed by the professional practice of the studio MEDIOMUNDO architects.

arc's radius, which is proportional to the distance from origin of each agent. Thus each agent is represented by a portion of the circle determined by the time spent and distance from origin. This is a measured visualization of the intensity of time employed as well as the movement of the agents involved in each job or unit of work (subcontracting and operators).

WHERE. Geographic location of the project and the urban-territorial context of the intervention; WHAT. Tasks, systems or materials involved in the work; WHO are the actors involved in the development of various tasks. What contribution did each actor make? What relationship is maintained during construction?; TIME. Working hours on site for each agent in respect to the work.; DISTANCE. Origin and movement of each agent.

Samples:

4 DISCUSSION OF RESULTS

This research has identified space-time interrelations between material and technical agents that occur in the case studies. As a result of this work we can see that there are several factors affecting the results such as the functional program (use) of the building, its size, complexity and relative location with respect to networks, production, and concentration of agents, materials and technologies. We not only consider the protocols that incorporate the distance factor as a topic of interest, but also identify the influence of the displacement from the sites of origin to its use in construction, in the insertion and interaction within the construction process and the environment in which a building is situated.

The results show the relationship between the type of work, its scale, complexity and programmatic uniqueness, characteristics of the environment, the scale of the place and its territorial status. We understand that this determines the possibilities of management and incorporation of media and techniques with varying degrees of ease and availability both in terms of supply and capacity and to the disposal of the management and the main contractor of the work. In this sense it can be shown that larger scale projects of greater complexity, greater specificity and greater restriction have implications on the type and size of the contractor team.

So if the population is small or at certain peripheral location, and the project is of a certain scale and budget, processes are likely to be executed by exogenous contractors and managed by agencies or companies remote to the town. We are referring, for example, to cases where there may be requirements for specialist equipment such as a theatre stage or a programme with complex sanitary needs; where functional and technical requirements impede the availability of technical, material and systems within close proximity.

5 CONCLUSIONS

We understand that the main use of these tools lies in the ability to analyse and study the spatial-temporal relationships involved in the processes of building construction materials and technological agents. In this sense, the "Projective Traceability" study is an important tool and an important indicator in the evaluation of architecture, that documented as such, means a heritage embodied within the architectural designs and production. To apply the documentation method to the architecture building process (design and construction) would add scientific consistency to the research in the field of future cultural heritage consideration of works of architecture it self. So that we understand this type of practice, it is necessary to define the performance protocol based on criteria for recovery and rebuilding of relations between the practice of architecture and local social-cultural values.

REFERENCES

Richmond, A. & Braker, A. Eds. 2010. Conservation, Principles, Dilemas and Umconfortable truths, Routledge, London.

Sennet, R. 2010. *El Artesano*, Anagrama Madrid.

Stanley-Price, N., Price, N., Talley, M. K.; & Vaccaro, A. M., Eds 1996. Historical and Philosophical Issues in the Conservation of Cultural Heritage,Getty Conservation Institute, Los Ángeles.

Unesco. 1972. Convention on the protection of the world, cultural and natural heritage. General Conference of the United Nations Organization for Education, Science and Culture, París, 17-21 nov..

Science and Digital Technology for Cultural Heritage – Ortiz Calderón et al. (Eds)
© 2020 Taylor & Francis Group, London, ISBN 978-0-367-36368-0

Testing the HeritageCare methodology for preventive conservation of built cultural heritage in Southwestern Europe: Case studies in Andalusia for service level 1

B. Castellano Bravo, A. Mayoral Moratilla, A. Bouzas Abad, J.L. Gómez Villa, B. Del Espino Hidalgo, M. López-Marcos & M. García-Casasola Gómez
Instituto Andaluz del Patrimonio Histórico, Seville, Spain

ABSTRACT: The HeritageCare project (Interreg-SUDOE) has been launched to implement a methodology for the preventive conservation and maintenance of built cultural heritage in Portugal, Spain and France. Through a consortium joining public and private partners and associated entities, an integrated methodology for inspection, diagnosis, monitoring and preventive conservation of heritage constructions is being developed and tested. The methodology is designed according to three service levels of growing complexity: in Service Level I, qualified teams perform detailed on-site inspections. Specific reports with information about the building, an assessment of its condition and a preventive conservation record and schedule are elaborated and delivered to the owners. This paper addresses the selection of ten case studies in Andalusia conducted by the research team of the Andalusian Institute for Historical Heritage (IAPH), the implementation of the methodology, the elaboration of reports for the owners and the aspects and problems found during the process.

1 HERITAGECARE PROJECT: A THREE-LEVEL METHODOLOGY

Currently, no systematic policy for the preventive conservation of built cultural heritage exists in the South-West Europe. The existing approaches for inspection, diagnosis, monitoring and curative conservation are often intermittent, unplanned, overpriced and lack a methodical strategy (Oliveira & Masciotta, 2019:16).

The HeritageCARE project – monitoring and preventive conservation of historic and cultural heritage – arises in response to this need. Its ultimate goal is the creation of a non-profit self-sustaining entity which will keep supervising the accomplishment of the methodology and the sustainability of the results once the project is concluded. Thus, HeritageCare entity shall operate according to a common methodology based on a system of services organised in three levels (Table 1). All information from SL1, SL2 and SL3 are integrated and stored in a 4D database. Although all levels provide high-quality services suited to serve the purpose they are conceived for, specific fieldwork, tools and final outputs vary for each of them. Hence, different names are associated with the three service levels, depending on their functionality.

2 METHODOLOGY: SERVICE LEVEL 1. STANDARDCARE

According to the European Standard UNI EN 16096, the condition survey is the first step to undertake in order to plan appropriate and effective preventive conservation measures on built cultural heritage, thereby minimizing future damage and deterioration processes. The condition survey is carried out by HeritageCare professionals through on-site inspection(s), appropriately preceded and followed by pre- and post-inspection phases. Three phases form the core of Service Level 1, which can be schematized as follows (Table 2):

Table 1. Synthesis of the methodology designed for the HeritageCare project. Source: HeritageCare, 2017.

Service Level	Designation	Functionality
SL1	StandardCare	Provision of what is essential for the primary health and ordinary maintenance of the heritage building
SL2	PlusCare	Provision of what is necessary for the primary health, ordinary maintenance and thorough screening of the heritage building along with its integrated and movable assets
SL3	TotalCare	Provision of what is necessary for the primary health, ordinary maintenance, thorough screening and enhanced management of the heritage building along with its integrated and movable assets

Table 2. Works to be developed on each phase of Service Level 1. Source: HeritageCare, 2017.

Pre-inspection	During-inspection	Post-inspection
Owner interview Building ID Historical Information List of assets and integrated objects Inspection kits	Inspection protocol Condition and risk classification On-site work	Database update Web plataform/databe synchronisation Report and recommendations Owners Feedback

As a final result of these three phases, the end user is given a document (inspection report) that synthesizes the information of the building, its state of conservation and recommends a series of preventive conservation actions whose development is planned over time. In addition, throughout the period of implementation of the methodology in the Interreg SUDOE territory, a series of recommendations have been established for its application (HeritageCare, 2017), among which the following stand out:

- SL1 must be simple, feasible, low-cost and should allow a rapid condition screening of the heritage construction
- Each heritage building/asset must have an ID card
- A historic analysis of the building should be performed prior to on-site inspections in order to get acquainted about construction phases, previous structural interventions, maintenance works, actual use, etc.
- Inspections must be carried out by at least two inspectors able to give a general overview of and assess both structure and integrated assets. Based on this first inspection, the real need of having a second inspection beyond the scope of preventive conservation by experienced professionals, will be evaluated.
- The inspections will follow a protocol with a precise checklist of items to inspect which will be also used for the report with recommendations to owners.
- The report establishes four levels for the general condition of the building (good, fair, poor or bad). For each of these, priority actions are defined as urgent, short term, medium term and long term. For each of the building systems, the condition is evaluated, damage is defined and concrete conservation recommendations are indicated.
- The periodicity of inspections may vary from case to case, depending on the health condition and size of the building, but a time span of 3 years is fixed for 'regular' buildings, although annually is recommended.
- Re-inspections must repeat the same steps of the first inspection, following the same checklist so as to easily compare current and previous building conditions. The re-inspections should also highlight whether or not owners have followed the recommendations provided during the first inspection.

3 IMPLEMENTATION AND CASE STUDIES SELECTION

The effectiveness and validity of the HeritageCare methodology for preventive conservation of built cultural heritage is being verified through its application to real case studies to choose among the great variety of heritage buildings spread over the SUDOE territory. The first service level of the methodology is currently being applied and validated through sixty case-study buildings (twenty per country). Fifteen out of sixty case studies (five per country) will be then selected for the implementation of the second service level of the methodology. Finally, one case study per country will be further exploited to put into practice the third and last level of the HeritageCare methodology. During the project time frame, the implementation of the methodology has relied on the voluntary collaboration of end users. The Andalusian cases have been selected on the basis of the typological variety of the heritage built in this region (Table 3):

Among the 10 case studies to which the Service Level 1 has been already implemented in Andalusia, 4 examples have been chosen with diverse typologies in order to summarize the results and conclusions of the first pre-inspection, inspection and post-inspection processes.

- Centre for the Documentation of Scenic Arts in Andalusia, formerly Santa Lucía church (Gómez de Cózar et al., 2000). A Gothic Mudejar church representative of a type of religious architecture very relevant in the local context, from the first half of the 14th century, deconsecrated in the 19th century and inaugurated as a public documentation centre in 2012, so the state is good.
- Alberto Lista library. A former residential building from the mid-twentieth century, it was rehabilitated in to be used as the public municipal library of a neighbourhood in the historic centre of Seville. Historical and graphic information was hard to be found due to the low level of protection, even if it is a listed building by the urban plan (Ayuntamiento de Sevilla, 2006). Thus, managers collaboration from the Municipality was fundamental.
- La Moharra farmhouse. This example of agricultural heritage was chosen as particularly interest due to the fact is privately owned, what perfectly fitted the HeritageCare project conditions. Built in the 20th century in an isolated rural area, the damages and conditions are slightly different to those of the urban buildings. Historic or graphic information were missing, except for a catalogue of agricultural heritage (Junta de Andalucía, 2009).
- Gran Poder Basilica. It is a private-owned church, currently in use. Built in the 60s of the 20th century, specially interesting due to the use of reinforced concrete in its structure, covering a large octagonal dome. Also, it houses, among a large number of movable assets, the figure of a Christ of great devotion and historical and artistic value, dating from the seventeenth century, as well as a figure of a virgin of the eighteenth century. Furthermore, a large a mount of well structured information is available (Gran Poder, 2019).

Table 3. Distribution of typologies within the case studies selected for HeritageCare project implementation in Andalusia. Source: Own elaboration.

Typology	Percentage	Nr. of cases	Service Level 1	Service Level 2	Service Level 3
Religious Architecture	20%	2	2	1	-
Military Architecture	<1%	-	-	-	-
Civil Architecture	20%	2	2	1	-
Industrial Architecture	10%	1	1	-	-
Garden Architecture	<1%	-	-	-	-
Agricultural Architecture	10%	1	1	-	-
20th Century Architecture	10%	1	1	-	-
Others	30%	3	3	-	-
Total	100%	10	10	2	0

4 RESULTS AND CONCLUSIONS

The results of the four selected case studies are summarized below:

The former church of Santa Lucía has a generally good state of conservation. It seems to have had a correct preventive conservation dynamic. The only damages found are the presence of humidity on the roof and graffiti on the external envelope. The Alberto Lista library, on the other hand, was in a fair state of conservation, according to the established classification. The most relevant problem of capillary humidity has not been solved by recent interventions. As a consequence, the walls have detached finishes. It has been found abundant biological colonization in roofs. This could easily be solved following the recommendations given by the inspectors to the managers. The farmhouse La Moharra was generally well preserved, although its condition as an isolated rural building has caused minor damages to the building envelope, mainly biological colonization on roofs and detachment of coating materials. A large number of movable assets was found, although none of them was listed. The Basilica of the Great Power presented the best state of conservation of all the selected cases. Only some minor damages were found on the roofs, all in acceptable condition and without risk. The inspection focused on making the owner aware of the importance of preventive conservation in order to avoid worsening existing damage or the appearance of new ones.

In almost all cases, the conservation of the buildings is adequate, although problems are encountered due to the lack of permanent maintenance. In the pre-inspection, the collaboration of owners and managers has been fundamental given the general scarcity of documentary and graphic information about the goods. During the inspection, most of the localized damage is due to the presence of humidity or biological colonization, mainly on the roof, as well as slight detachments of coating material on walls. Post-inspection has been fundamental in detecting knowledge gaps on the part of those in charge of building maintenance.

In general terms, the experience of these four case studies of the application of Service Level 1 has been positive, as it has allowed the verification of the methodology, its usefulness and the drawing of conclusions for the future. Particularly, IAPH has benefited of the technical training in conservation, preventive conservation and maintenance, in order to advance in the sustainable management of Andalusian built heritage. The progress made in risk prevention through the systematization of information after carrying out inspections, by means of the tools provided by the project, has been relevant in terms of saving costs and working hours. Further research lines will be developed in order to implement the automatic permanent monitoring of the conservation conditions through the integration of measurement points in a digital 3d model. This will involve a further degree of rigour in the process, with a positive impact on the efficiency.

REFERENCES

Ayuntamiento de Sevilla, 2006. Plan General de Ordenación Urbanística de Sevilla.

Gómez de Cózar, J. C., Rodríguez-Liñán, C., & Rubio de Hita, P. (2000). Geometrías concertadas. Las cabeceras de las iglesias gótico-mudéjares de la ciudad de Sevilla. In Tercer Congreso Nacional de Historia de la Construcción (2000), p 397–403. Instituto Juan de Herrera, CEHOPU, US.

Gran Poder, 2019. Basílica de Jesús del Gran Poder. Web page available at: http://www.gran-poder.es/patrimonio/patrimonio-inmueble/basilica-de-jesus-del-gran-poder/ (Last accessed 06/05/2019).

HeritageCare, 2017. General methodology for the preventive conservation of cultural heritage buildings. HeritageCare project. Intern report.

Junta de Andalucía, 2009. Cortijos, haciendas y lagares. Provincia de Sevilla. Sevilla: Consejería de Fomento, Infraestructuras y Ordenación del Territorio.

Oliveira, D.V. & Masciotta, 2019. 1965. HeritageCare: “Prevenir mejor que curar”. Revista PH 96: 16–18.

Science and Digital Technology for Cultural Heritage – Ortiz Calderón et al. (Eds)
 ISBN 978-0-367-36368-0

Between two worlds: Colors and techniques of the mural paintings preserved in a colonial house in Chajul (El Quiche department, Guatemala)

M.L. Vázquez de Ágredos-Pascual & C. Vidal Lorenzo
Art History Department, Universitat de València, Valencia, Spain

K. Radnicka & J. Źrałka
Institute of Archaeology, Jagiellonian University, Cracow, Poland

J.L. Velásquez
Instituto de Antropología e Historia, Ciudad de Guatemala, Guatemala

ABSTRACT: In the Guatemalan village of Chajul there are colonial houses that in some of its rooms conserved mural paintings of great historical and artistic interest. The study of the materials and artistic techniques of these works through a project of scientific collaboration between the *Jagiellonian University* of Poland and the *Universitat de València* (Spain) has allowed identifying a crossroads of traditions and knowledge technicians in them. Most of the coloring materials characterized in the pictorial film are of local origin, and coincide with those that were used by the painters of Maya murals since Late Preclassic times (ca. 300 BC-AD 300), especially in the architectureal paintings from the Maya Lowlands. However, in the preparation and technical treatment we observed differences with respect to the pre-Hispanic past, which are related to the new artistic procedures that came to the New World from the Old World after the sixteenth century. This synthesis of knowledge and traditions is also notable in the limestone preparation bases of the Chajul murals. Optimized techniques in the characterization of organic and mineral components in works of art were used to study the materiality of these paintings, including optical microscopy (LM), Scanning Electron Microscopy-X-Ray microanalysis (SEM-EDX), X-Ray Diffraction (XRD) and Gas Chromatography-Mass Spectrometry (GC-MS). The study provides scientific data of interest to interpret in its correct historical-cultural dimension this mural painting. Finally, this study has been completely necessary to elaborate the preventive conservation plan that, after the intervention of these murals, should be put in practice to guarantee the best conservation of these works in its space of origin, a house of colonial era that continues to be inhabited by a Maya family today.

1 THE HOUSE OF ASICONA FAMILY: BACKGROUND TO THIS ARCHAEOMETRIC AND CULTURAL STUDY

The *House of Asicona family* (also named *House 3*) was built from adobe bricks bound with mortar containing an admixture of pine needles. This surface was covered by white stucco and painted with scenes whose iconography responds to the new aesthetics of the times of the Spanish Colony, very far from the pre-Hispanic antiquity (Figure. 1-2). In 2003, an *Institute of Anthropology and History of Guatemala* (IDAEH) representative commission was sent to Chajul and drew up a report about the colonial mural paintings conserved in several houses, not only the *House of Asicona family*. Lars Frühsorge visited Chajul in 2007 and published preliminary information concerning these paintings (Frühsorge 2008). Three years later, in 2011, houses with murals, including *House 3*, were visited by members of the *Nakum Archaeological Project* who carried out a preliminary photographic documentation of almost all of the

Figure 1-2. (Figure 1: to the left) Maya mural paintings conserved at the *house of Asicona family* (Chajul-Guatemala), 2015. (Figure 2: to the right) The women who inhabit this house of Chajul. Mural paintings are part of their everyday environment, 2011.

murals (Źrałka & Radnicka 2014). In 2014, thanks to permission given by Lucas Asicona, *House 3* and its murals were scanned by Bogumił Pilarski with a *Faro Focus 3D scanner*. Next, in 2015, Polish scholars Katarzyna Radnicka and Tomasz Skrzypiec executed the conservation of murals from *House 3*, which was made possible thanks to the permission of IDAEH and private funds (Radnicka, Skrzypiec & Velásquez 2016). The restoration of these paintings was the starting point of its physical-chemical study, which was carried out with the following objectives: (a) to identify the materials and techniques of the mural painting conserved at *House 3*, (b) to contrast the results with the technical tradition of pre-Hispanic Maya mural painting, (c) to link the results with the incorporation of pictorial materials and techniques from the Old World to the New World, and (d) to establish the basis for a future preventive conservation project.

2 MATERIALS AND METHODS

Sixteen samples of pigments were taken from the *House of Asicona family* (Table 1). The physicochemical analysis techniques that were used in the study were as follows:

(a) *Optical microscopy (LM): we use* a Leica DMR optical microscope with an incident/transmitted light system and polarization system in both cases.
(b) *Scanning Electronic Microscopy combined with energy dispersive X-ray (SEM-EDX)*: The sample has been analyzed using a Hitachi Variable Pressure Scanning Electron Microscope (VP-SEM) model S-3400N, equipped with a Bruker dispersed X-ray energy spectrometer model XFlash® with a silicon drift droplet detector (SDD), with Dura-Beryllium window (8 µm) and energy resolution of 125 eV @5.9 keV. The analysis was performed using 20 kV, a working distance of 10 mm, and pressure inside the chamber of the chamber of 60 Pa.
(c) *X-ray powder diffraction* (XRPD) was carried out on randomly oriented samples after grinding the powder samples in an agate mortar. A Bruker D8 Advance system, operating in q:q mode was used; generator setting 40 kV, 40 mA, Cu anode ($Cu\text{-}K\alpha$ = 1.5418 Å), Ni filter, 2q range 5-80°, step size 0.01°, scan speed 0.5° min^{-1}. Qualitative phase determination was carried out using QualX2.0 software and the correlated COD database. Quantitative Phase Analysis (QPA) was carried out using Quanto software.
(d) *Fourier Transform Infrared spectroscopy (FTIR):* Fourier Transform Infrared Spectroscopy analyses of potential organic binders were obtained with a Thermo Nicolet Nexus spectrophotometer. Spectra were acquired between 4000 and 400cm^{-1}, 64 scans with 4cm^{-1} of resolution. FTIR coupled with ATR (Attenuated Total Reflectance) was also performed in the same analytical conditions.

Table 1. Pigments identified by physicochemical analysis in the House of Asicona family (Chajul).

SAMPLE	COMPOUNDS IDENTIFIED	PIGMENT
CH-1	Iron (Fe), aluminum (Al), silica (Si), potassium (K), and titanium (Ti)	Red earth/Fe-based pigment
CH-2	Iron (Fe), aluminum (Al), silica (Si), potassium (K), and magnesium (Mg)	Red earth/Fe-based pigment
CH-3	Iron (Fe), aluminum (Al), silica (Si), and calcite (Ca)	Red earth/Fe-based pigment
CH-4	Calcium (Ca), phosphorus (P) Iron (Fe), aluminum (Al), silica (Si), potassium (K), and other minor compounds	Calcium apatite + Red earth/Fe-based pigment
CH-5	Iron (Fe), aluminum (Al), silica (Si), potassium (K), and phosphorous (P)	Red earth/Fe-based pigment
CH-6	Iron (Fe)	Hematite Fe_2O_3
CH-7	Iron (Fe), aluminum (Al), silica (Si), and calcite (Ca)	Red earth/Fe-based pigment
CH-8	Calcium (Ca), phosphorus (P) Iron (Fe), aluminum (Al), silica (Si) and magnesium (Mg)	Calcium apatite + Red earth/Fe-based pigment
PM-1	Iron (Fe), aluminum (Al), silica (Si), potassium (K), and magnesium (Mg), titanium (Ti), chlorine (Cl). As minor compound we found chromate (Cr)	Red earth/Fe-based pigment
PM-2	Iron (Fe), aluminum (Al), silica (Si), potassium (K), and magnesium (Mg), and titanium (Ti). As minor compound we found chromate (Cr)	Red earth/Fe-based pigment
WM-1	Iron (Fe), aluminum (Al), silica (Si), and titanium (Ti) –the Ti in very low quantity-	Red earth/Fe-based pigment
WM-2	Iron (Fe), aluminum (Al), silica (Si), and titanium (Ti)	Ilmenite ($FeTiO_3$).
WM-3	Calcium carbonate (CaCO3) Iron (Fe), aluminum (Al), silica (Si)	Calcite + Red earth/Fe-based pigment
WM-4	Iron (Fe), aluminum (Al), silica (Si), potassium (K), and magnesium (Mg)	Red earth/Fe-based pigment
WM-5	Iron (Fe), aluminum (Al), silica (Si), potassium (K), and titanium (Ti)	Red earth/Fe-based pigment
N	Indigo + palygorskite	Maya blue pigment

(e) *Gas chromatography-mass spectrometry* (GC-MS): a technique specializing in the characterization of organic substances. The gas chromatograph used in the characterization of the samples was an Agilent 6890N (Agilent Technologies, Palo Alto, CA, USA). The chromatograph is docked with an HP 5973 mass detector. The column used was an HP-5MS-5% phenyl and 95% polydimethylsiloxane). The oven temperature program was 60-220 ºC, with an increase of 1 °C/min; the temperature was maintained at 220 °C for 3 min. The injector temperature was at 250 °C. The injection volume was 1 μ L (95:5), with an inlet pressure of 7.96 psi. The carrier gas was He. The interface temperature was 280 °C. For the mass detector, the ionization temperature was set at 230 °C. The GC-MS database (NIST Library version 2002) was used for the possible identification of the organic components.

3 RESULTS AND DISCUSSION

The studies carried out in the laboratories of the *Scientific Park* of the Universitat de València indicate that the materials and techniques applied in the preparation of the murals in the *House of Asicona family* continued almost unchanged since the pre-Columbian times. Physicochemical analysis of the samples studied led to the identification of a colour palette typical for

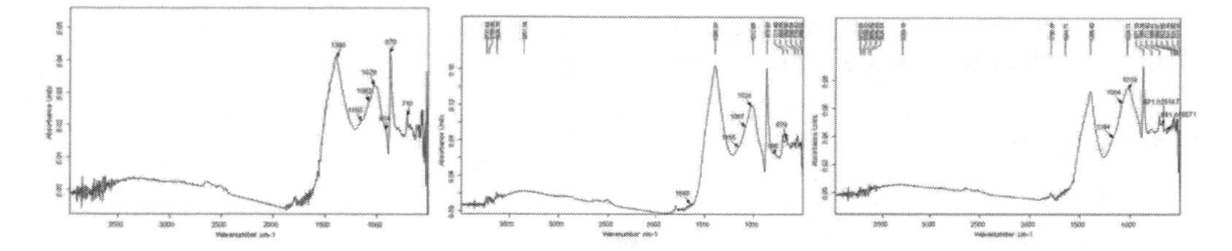

Figure 3 a-c. FTIR spectrums of different earth pigments of the mural paintings conserved at the house of Asicona family (samples CH-1, CH-2 & CH-3.

pre-Columbian Maya mural paintings, at least in the Maya Lowlands: white lime ($CaCO_3$), hematite (Fe_2O_3), ocher and red earth (of possible local origin) (Figure 3a-c), and the so-called 'Maya blue' (sample N) present. The greyish hue of this pigment may have two causes: (a) conservation problems or (b) a decline in traditional knowledge concerning the preparation of this pigment. It is important to indicate that we consider that the remains of this blue pigment would initially correspond to a Maya blue because no blue or green mineral has been identified that could explain this colour (such as malaquite or azurite). As a consequence, it is highly probable that the blue pigment that was originally used here was 'Maya blue'. In addition, a magnesium silicate linked to this blue sample was identified, and this seems to correspond to the matrix of palygorskite that was used to prepare the colour since Preclassic times. In this sense, it should be remembered that Maya blue can be considered as a hybrid organic-inorganic material resulting from the attachment of indigo, a blue dye extracted from *Indigofera suffruticosa* and other species, as well as local clay, palygorskite. Natural indigo is formed by indigotin (3H-indol-3-one, 2-(1,3-dihydro-3-oxo-2H-indol-2-ylidene)-1,2-dihydro), a quasi-planar molecule of approximate dimensions 4.8 × 12 Å.6. The inorganic component of Maya blue, palygorskite, is a fibrous phyllosilicate with an ideal composition of $(Mg,-Al)_4Si_8$ $(O,OH,H_2O)_{24},nH_2O$ (Doménech *et al.* 2009). The ancient Maya prepared this pigment by crushing dry indigo with palygorskite, with subsequent (or parallel) heating at temperatures below 250°C. At more than this temperature the rings of the indigo break and the preparation of the pigment fails. The result is a more greyish colour and very similar to the one we have in the *House of Asicona family*. That is why it is so important to document this blue-grey pigment in the aforementioned colonial house. We believe that its chromatic hue stems from the fact that (1) the pre-Hispanic Maya blue was still in use but (2) the technology that guaranteed optimal preparation was already in decline, as a result of its progressive substitution with other blue colours of mineral origin. On the other hand, the chemical-analytical results indicate that there is another technological aspect of the paintings that is associated with the Pre-Hispanic mural paintings of the Mayan Lowlands: the composition and stratigraphy of the underlying plaster. The colour of the mural was spread in two calcareous layers. The lower mortar is much thicker and more irregular than the upper plaster. The main component of both levels/layers is calcium carbonate ($CaCO_3$) of local origin mixed with aragonite ($CaCO_3$) obtained by grinding shells. This could indicate the survival in the Maya area of a pictorial tradition that was very widespread during the Postclassic, especially in coastal regions and cities (Vázquez de Ágredos 2010).

4 CONCLUSIONS

From a technical point of view, the mural painting preserved in the *House of Asicona family* represents the survival of the pictorial tradition of the Maya Lowlands. The underlying stucco and the colours used in this work of art have the same origin and composition as those of pre-Hispanic times. The materials used are probably of local origin and could have been prepared by the Maya who lived in Colonial times. While the iconography shows clear differences with the pre-Columbian Maya and reflects the new power group and its ideals, the technique still manifests a socio-cultural Maya pre-Hispanic identity.

BIBLIOGRAPHY

Doménech A. *et al.* 2009. "Correlation between spectral SEM/EDX and electrochemical properties of Maya blue: A Chemometric study". In: *Archaeometry* 51 (6): 1015–1034.

Frühsorge, L. 2008. Murals from the Ixil area. In: *Mexicon* XXX (2): 31–32.

Radnicka, K., Skrzypiec, T., & Velásquez, J. L. 2016. *Proyecto "Conservación de los Murales de Chajul"—COMUCH.* Unpublished report submitted to the Institute of Anthropology and History of Guatemala, Guatemala City.

Vázquez de Ágredos-Pascual M.L. 2010. *La pintura mural maya. Materiales y técnicas artísticas.* México: CEPHCIS-UNAM.

Źrałka J. & K. Radnicka 2015. "Between the past and the present: the Ixil Maya and the discovery of rare mural paintings in Guatemala". In: *Estudios Latinoamericanos* 33/34: 169–185.

Science and Digital Technology for Cultural Heritage – Ortiz Calderón et al. (Eds)
© 2020 Taylor & Francis Group, London, ISBN 978-0-367-36368-0

Coloring materials and technical recipes in the mural paintings of the Hermitage of "Mas de Tous" (La Pobla de Vallbona, Valencia). Physicochemical and cultural study

E. Alba Pagán, M.L. Vázquez de Ágredos-Pascual & L. Rojo Iranzo
Art History Department, Universitat de València, Valencia, Spain

L. Arbolí Valls & M. Navarro Font
Centre d'Art d'Època Moderna, Universitat de Lleida, Lleida, Spain

ABSTRACT: The painters who worked in the Europe of the late nineteenth century and the beginning of the twentieth century had to synthesize in their works very different materials and artistic techniques. Pigments that had a long tradition, such as hematite ($Fe_2 O_3$) or carbon black, were used together with the new coloring materials that appeared between the 18th and 19th centuries, such as cobalt blue, also named *Blue Thénard* ($CoO.Al_2O_3$), or ultramarine artificial blue $Na_{6-10}Al_6Si_6O_{24}S_{2-4}$ (approximately) (Plesters 2002:55).The colors of organic origin, and especially those from the plant world, were also used in this period for its physical and luminous properties, in accordance with previous studies on Contemporary Art. This was the case, for example, of many Valencian painters, such as José Benlliure, Ignacio Pinazo, Salvador Abril, Antonio Muñoz Degrain or Joaquín Sorolla, among others. The physicochemical study of four pictorial panels preserved at the Hermitage of "Mas de Tous" (La Pobla de Vallbona), and dated in this period, has revealed important data to contribute to the understanding of Contemporary Valencian painting. Moreover, the technical study has allowed identifying some similarities and differences with the work of other Contemporary Valencian painters, as relevant as Joaquín Sorolla. The physicochemical techniques that have been used for this purpose in the *Laboratory of Analysis and Diagnosis of Work of Art* of the Universitat de València, and the *Scientific Park* of the same institution, have been the Optical Microscopy (LM), the Scanning Electron Microscopy- X-ray microanalysis (SEM-EDX) or Fourier Transform Infrared Spectroscopy (FTIR), among others, such as visible spectrophotometry (UV-vis) or Raman Spectroscopy (RMS).

1 THE PAINTER AUGUSTO DANVILA JALDERO AND ITS ARTISTIC CIRCLE: THE CLOSENESS WITH JOAQUÍN SOROLLA

Between 1905-1910 Augusto Danvila Jaldero (1853-1935) developed an interesting cycle of mural paintings for the small Hermitage of the *Virgen de la Abundance* (Pobla de Vallbona, Valencia). In these works, Danvila was close to the thought associated with the naturalist and symbolist new school of Spanish Art, away from the History Painting. This painter achieved recognition as a scholar of merit of the *Academy of Fine Arts in San Fernando*, and in Valencia was formed at the *School of Fine Arts in San Carlos* with Ignacio Pinazo, one of the most significant artists of Valencia in the late nineteenth century. On the other hand, our painter was a friend of Joaquín Sorolla, who in his youth would dedicate several of his paintings "to my friend Danvila". This is the case of the canvas *Dos de Mayo* (1884, *Museo del Prado*) and also of the painting *Father Jofré defending a Madman* (1887, *Diputació de València*), so it seems probable that for the preparation of his canvas of historical affairs, Sorolla had the help and advice of his scholar friend. Both painters also shared long vacation stays in the farmhouse of the family Danvila, where Sorolla well could have helped Danvila in his compositions for the Hermitage of the *Virgen de la Abondancia.* In fact, Danvila possessed one of the best collections of color notes of Sorolla's landscapes. This close relationship between Danvila and Joaquín Sorolla represents the starting point of this research and its objectives: (1) to carry out the physicochemical characterization of the mural paintings of the Hermitage of the *Virgen de la Abundancia*, (2) to identify the pictorial technique employed by the painter, (3) to compare the color palette used by Danvila with that used by Joaquín Sorolla in his paintings, (4) set the foundations for a future intervention and preventive conservation project.

2 MATERIALS AND METHODS

A total of 47 samples from the four pictorial panels that decorate the walls of the Hermitage were taken (Table 1). The physicochemical analysis techniques that were used in the study were as follows:

(a) *Optical microscopy (LM): we use* a Leica DMR optical microscope with an incident/transmitted light system and polarization system in both cases.
(b) *Scanning Electronic Microscopy combined with energy dispersive X-ray (SEM-EDX)*: The sample has been analyzed using a Hitachi Variable Pressure Scanning Electron Microscope (VP-SEM) model S-3400N, equipped with a Bruker dispersed X-ray energy spectrometer model XFlash® with a silicon drift droplet detector (SDD), with Dura-Beryllium window (8 μm) and energy resolution of 125 eV @5.9 keV. The analysis was performed using 20 kV, a working distance of 10 mm, and pressure inside the chamber of the chamber of 60 Pa.
(c) *RAMAN Spectroscopy.* Raman analyses were performed using a HORIBA Jobin Yvon iHR320 spectrometer with a Peltier-cooled CCD for detection and 785 and 532nm doubled YAG laser as excitation. A 50x magnification LWD objective was used to focus the laser on the sample and collect the scattered light. The measurements were performed using a laser power between 10 and 30mW, an integration time of 120s and up to 5 spectral accumulations.
(d) *Fourier Transform Infrared spectroscopy (FTIR).* Fourier Transform Infrared Spectroscopy analyses of potential organic binders were obtained with a Thermo Nicolet Nexus spectrophotometer. Spectra were acquired between 4000 and 400cm^{-1}, 64 scans with 4cm^{-1} of resolution. FTIR coupled with ATR (Attenuated Total Reflectance) was also performed in the same analytical conditions.
(e) *Gas chromatography-mass spectrometry* (GC-MS): a technique specializing in the characterization of organic substances. The gas chromatograph used in the characterization of the samples was an Agilent 6890N (Agilent Technologies, Palo Alto, CA, USA). The chromatograph is docked with an HP 5973 mass detector. The column used was an HP-5MS-5% phenyl and 95% polydimethylsiloxane). The oven temperature program was 60-220 ºC, with an increase of 1 °C/min; the temperature was maintained at 220 °C for 3 min. The injector temperature was at 250 °C. The injection volume was 1 μ L (95:5), with an inlet pressure of 7.96 psi. The carrier gas was He. The interface temperature was 280 °C. For the mass detector, the ionization temperature was set at 230 °C. The GC-MS database (NIST Library version 2002) was used for the possible identification of the organic components.

Table 1. Historical Pigments identified on the mural paintings developed by Augusto Danvila Jaldero at the Hermitage of "Mas de Tous".

Sample	Colour	Coloring material identified	Chemical formula	Chronology
		PANEL 1: *The Adoration of the Magi*		
MT-1	$White_1$	Lime White	$CaCO_3$	Since antiquity
MT-2	$White_2$	Zinc White	ZnO	1834 by *Windsor and Newton*
MT-3	Black	Carbón black	Carbon Black	Since antiquity
MT-4	Grey	Zinc White + Carbón Black	ZnO + Carbón Black	ZnO: 1834 Carbon black: since antiquity
MT-5	Red_1	Hematite	Fe_2O_3	Since antiquity
MT-6	Red_2	Vermilion	HgS	Since antiquity
MT-7	Red_3	Hematite + Vermilion + Lime White	$Fe_2O_3 + HgS + CaCO_3$	All them: since antiquity
MT-8	$Brown_1$	Naples yellow + Hematite	$Pb_3(SbO_4)_2 + Fe_2O_3$	$Pb_3(SbO_4)_2$: The first recipes on their artificial elaboration are dated on 13th century Fe_2O_3: since antiquity
MT-9	$Brown_2$	Naples yellow + Hematite + Vermilion	$Pb_3(SbO_4)_2 + Fe_2O_3 + HgS$	$Pb_3(SbO_4)_2$: 13th century Fe_2O_3: since antiquity HgS: since antiquity
MT-10	Yellow	Naples yellow	$Pb_3(SbO_4)_2$	$Pb_3(SbO_4)_2$: 13th century
MT-11	Blue	Ultramarine artificial blue	$Na_{6-10}Al_6Si_6O_{24}S_{2-4}$	Since 1828 by *Guimet*
MT-12	$Green_1$	Malachite	$Cu_2CO_3(OH)_2$	Since antiquity
MT-13	$Green_2$	Malachite + Carbon Black	$Cu_2CO_3(OH)_2$ + Carbon Black	Both since antiquity
		PANEL 2: *Ecce Homo*		
MT-14	White	Lime White	$CaCO_3$	Since antiquity
MT-15	Black	Carbon Black	Carbon Black	Since antiquity
MT-16	Red_1	Vermilion	HgS	Since antiquity
MT-17	Red_2	Hematite + Vermilion	$Fe_2O_3 + HgS$	Both since antiquity
MT-18	$Brown_1$	Naples yellow + Hematite + Vermilion	$Pb_3(SbO_4)_2 + Fe_2O_3 + HgS$	$Pb_3(SbO_4)_2$: 13th century Fe_2O_3: since antiquity HgS: since antiquity
MT-19	$Brown_2$	Naples yellow + Hematite	$Pb_3(SbO_4)_2 + Fe_2O_3$	$Pb_3(SbO_4)_2$: 13th century Fe_2O_3: since antiquity
MT-20	$Brown_3$	Hematite + Carbon Black	Fe_2O_3 + Carbon Black	Both since antiquity
MT-21	Blue	Ultramarine artificial blue	$Na_{6-10}Al_6Si_6O_{24}S_{2-4}$	1828
MT-22	Green	Malachite	$Cu_2CO_3(OH)_2$	Since antiquity
		PANEL 3: *The Deposition of Christ*		
MT-23	White	Zinc White	ZnO	1834
MT-24	Black	Carbon Black	Carbon Black	Since antiquity
MT-25	Red_1	Hematite + Vermilion	$Fe_2O_3 + HgS$	Both since antiquity
MT-26	Red_2	Hematite	Fe_2O_3	Since antiquity
MT-27	$Brown_1$	Vermilion + Carbon Black	HgS + Carbon Black	Both since antiquity
MT-28	$Brown_2$	Hematite + Carbon Black	Fe_2O_3 + Carbon Black	Both since antiquity
M7-29	$Yellow_1$	Naples yellow + Hematite	$Pb_3(SbO_4)_2 + Fe_2O_3$	$Pb_3(SbO_4)_2$: 13th century Fe_2O_3: since antiquity
MT-30	$Green_1$	Malachite	$Cu_2CO_3(OH)_2$	Since antiquity
MT-31	$Blue_1$			1828

(*Continued*)

Table 1. (*Continued*)

Sample	Colour	Coloring material identified	Chemical formula	Chronology
		Ultramarine artificial blue	$Na_{6-10}Al_6Si_6O_{24}S_{2-4}$	
MT-32	$Blue_2$	Ultramarine artificial blue + Carbon Black	$Na_{6-10}Al_6Si_6O_{24}S_{2-4}$ + Carbon Black	$(Na,Ca)_8(AlSiO_4)_6(SO_4,S,Cl)_2$: 1828 Carbon Black: since antiquity
		PANEL 4: *Resurrection of Jairo's Daughter*		
M-33	$White_1$	Zinc White + Lead White	$ZnO + (PbCO_3)_2 \cdot Pb(OH)_2$	ZnO $(PbCO_3)_2 \cdot Pb(OH)_2$: Since antiquity
M-34	$White_2$	Zinc White	ZnO	1834
M-35	$White_3$	Lead White	$(PbCO_3)_2 \cdot Pb(OH)_2$	Since antiquity
M-36	Black	Carbon Black	Carbon Black	Since antiquity
M-37	Red_1	Hematite + Vermilion	$Fe_2O_3 + HgS$	Both since antiquity
M-38	Red_2	Cadmium red	CdSe + CdS	Its commercial production starts in 1910
M-39	Red_3	Cadmium red + Vermilion	CdSe + CdS & HgS	CdSe + CdS: 1910 HgS: since antiquity
M-40	Red_4	Vermilion + Cadmium yellow	HgS + CdS	HgS: Since antiquity CdS: 1829 by *Meleandri*
M-41	$Yellow_1$	Chrome yellow	$PbCrO_4$	1818 by *Vauquelin*
M-42	$Yellow_2$	Cadmium yellow	CdS	1829
M-43	$Blue_1$	Cobalt blue	$CoO.Al_2O_3$	1800 by *Thénard*
M-44	$Blue_2$	Prussian blue	$Fe_4(Fe(CN)_6)_3$	1710
M-45	$Blue_3$	Ultramarine artificial blue	$Na_{6-10}Al_6Si_6O_{24}S_{2-4}$	1828
M-46	$Green_1$	Chrome green	Cr_2O_3	1862
M-47	$Green_2$	Chrome green + Chrome yellow	$PbCrO_4 + PbCrO_4$	$PbCrO_4$: 1862 $PbCrO_4$ 1818

3 RESULTS AND DISCUSSION

The main component of the render of these mural paintings is calcium carbonate ($CaCO_3$) (Figure 5). This favored the execution of the painting cycle with the technique of the *frescco*. However, it was identified dry touch-ups in some colors incompatible with the process of carbonation of lime, including malachite $Cu_2(CO_3)(OH)_2$ or vermilion (HgS). The techniques FTIR and GC-MS diagnosed in these touch-ups the presence of oil linseed. The colors used in these paintings were white lime ($CaCO_3$), lead white $(PbCO_3)_2 \cdot Pb(OH)_2$, zinc white ZnO, carbon black, hematite Fe_2O_3, vermilion (HgS), cadmium red CdSe + CdS, Naples yellow $Pb_3(SbO_4)_2$, chrome yellow $PbCrO_4$, cadmium yellow CdS, malachite, ultramarine artificial blue, cobalt blue, chrome green Cr_2O_3, and Prussian blue $Fe_4(Fe(CN)_6)_3$ (Table 1). However, not all panels included these coloring matters equally. In three of them, the presence of the new contemporary pigments was scarce. We refer to the paintings of (a) *The Adoration of the Magi* (Figure 1), (b) *Ecce Homo* (Figure 2) and (c) *The Deposition of Christ* (Figure 3). Only the zinc white (1834) and the artificial ultramarine blue (1828) are part of these three panels, in which pigments of very old origin and artistic use are predominant. This differentiates them from the fourth and last panel: *The Resurrection of Daughter of Jairus* (Figure 4), which containing a greater number of contemporary coloring matters and novel technical recipes. The presence of cadmium red in this late panel dates its execution in 1910 (or shortly after), because the commercial production and use of this pigment began in the first decade of the twentieth century, and also suggests that this was the last panel

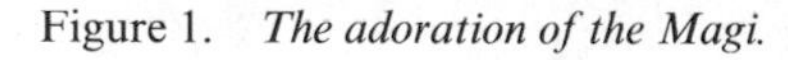

Figure 1. *The adoration of the Magi.*

Figure 2. *Ecce Homo* (following Antonio Ciseri).

Figure 3. *The deposition of Christ.*

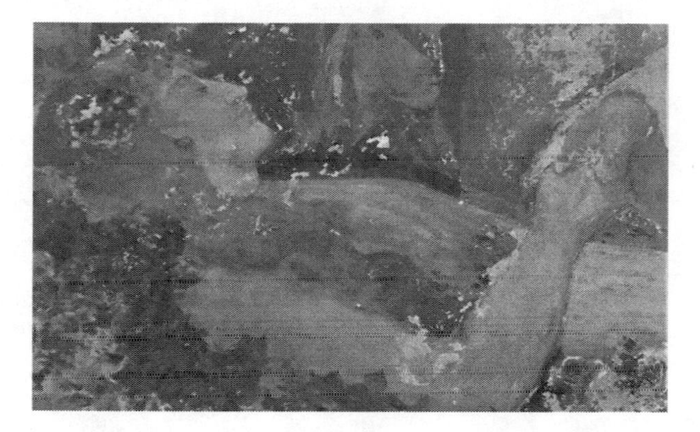

Figure 4. *The resurrection of Jairo's Daughter(detail)* (following Ilia Repin).

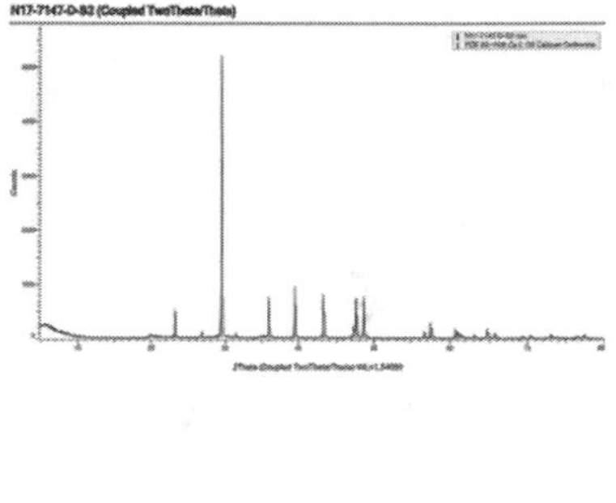

Figure 5. Render of calcium carbonate of the mural paintings.

painted by the artist, possibly in 1910. This same panel is the one that offers two aspects that directly connect the work of Danvila with Joaquín Sorolla (Ferrero *et al.* 2001): (a) whites prepared from a mixture of lead white and zinc white -pigment only present in this panel- and (b) the mixture of oils in the cooked state -especially linseed and walnut- to agglutinate the multiple touch-ups to dry vermillion. Both aspects are recurrent in the work of Joaquín Sorolla and are identified in his canvases.

4 CONCLUSIONS

The analysis of the mural paintings conserved at the Hermitage of the *Virgen de la Abundance* has confirmed that the paintings were executed with the technique of the *frescco* with dry touch-ups, which Danvila resolved with linseed oil, except when he applied vermilion on the panel of the *Resurrection of Jairo's Daughter*. In this case he used a mixture of walnut oil and linseed oil that refers to Joaquín Sorolla technique. In this same sense, the mixture of white lead and white zinc in the same panel is also recurrent in the work of Joaquín Sorolla. Finally, only cadmium red has been identified in this panel. The distribution and sale of this pigment for artistic purposes did not occur until 1910, and this suggests that this was the last panel painted by Augusto Danvila at the Hermitage of “Mas de Tous”.

BIBLIOGRAPHY

Ferrero, J.L. *et al.* 2001. “Análisis de pigmentos de la obra de Joaquín Sorolla mediante EDXRF”. In: *Actas del III Congreso Nacional de Arqueometría*, pp.127-134 (M.L Pardo *et al*, Ed.). Sevilla: Universidad de Sevilla.

Plesters, J. 2002. “Ultramarine Blue, Natural and Artificial”. In: *Artist's Pigments: A Handbook of their History and Characteristics*, vol.2, pp.37-65 (A. Roy, Ed.). London: Archetype Publications.

Science and Digital Technology for Cultural Heritage – Ortiz Calderón et al. (Eds)
© 2020 Taylor & Francis Group, London, ISBN 978-0-367-36368-0

Social opinion of monuments vulnerability in historical cities

J. Benítez, A. Tirado-Hernández, R. Ortiz, M. Moreno & P. Ortiz
Department of Physical, Chemical and Natural Systems, Pablo de Olavide University, Seville, Spain

I. Turbay
Faculty of Architecture, Fundación Universitaria de Popayán, Sede San José, Popayán, Colombia

M.A. Garrido-Vizuete
Department Applied Mathematics I, University of Seville, Seville, Spain

ABSTRACT: The present work try to analyze the concern of society in different historical cities about their cultural heritage in order to assess the relationship between the opinion of citizens and the vulnerability of monuments in historical cities. Anonymous surveys were carried out in four different cities: Seville, Osuna and Marchena (Spain) and Popayan (Colombia). According to the preliminary results, most of population asserts that it is important to preserve and to protect the cultural heritage of their cities; however, some of them believe that there is a low political involvement. The comparison of results between different cities allows beginning the assessment of the relationship between social diversity and the vulnerability of monuments. Nevertheless, further studies are necessary to establish the influence of reflective societies in the vulnerability of monuments in order to develop strategies for conservation of cultural heritage.

1 INTRODUCTION

The engagement of citizens to support cultural heritage is a key point for sustainable Historical Cities. Reflective societies, that encourage the conservation of their monuments, artworks and traditions, may reduce the vulnerability of their tangible and intangible heritage.

The citizens play a fundamental role in the conservation of the historical heritage, so it can determine the success or failure of the actions that are carried out in this direction (Morate Martín, 2007). Therefore, the opinion of the citizens must be studied in order to increase awareness for cultural heritage preservation.

2 METHODOLOGY

The surveys are a useful method to know preferences, tastes, perceptions and assessments of citi-zens on such diverse topics as politics (Chambers et al., 2019; Mendoza Tolosa et al., 2012), cultural heritage (Valdelamar Sarabia, 2010; Palacio & van der Hammen, 2007; Caraballo Perichi, 2008; Morate Martín, 2007) or science. The present work try to analyze the concern of society in different historical cities about the cultural heritage conservation in order to assess the influence of the citizens opinion in the vulnerability of monuments in historical cities, with this purpose, a specific survey was designed by the research team of ART-RISK Project (Artificial intelligence applied to preventive conservation of heritage buildings (https://www.upo.es/investiga/art-risk/)).

Anonymous surveys were used to ask the opinion to representative population of four different cities. The chosen cities were Seville, Osuna and Marchena, in Spain, and Popayán in Colombia Seville is the capital of Andalusia and has a population close to 700.000 inhabitants.

Table 1. Number of surveys by gender and age range.

	Seville		Osuna		Marchena		Popayán	
Age	Men	Women	Men	Women	Men	Women	Men	Women
18-29	30	29	38	33	35	32	53	52
30-44	52	53	49	49	54	53	60	61
45-59	50	55	55	54	50	50	42	50
+60	47	68	44	56	48	56	28	37

Osuna and Marchena are part of the countryside of Seville and their populations are close to 20.000 inhabitants. Popayán is the capital of the region of Cauca, Colombia, and its population is close to 290.000 inhabitants.

It has been observed that Spanish cities have aged populations with a large number of people over 60 years old, while the population of Popayán is younger, with a high percentage of people between 18 and 29 years.

The Table 1 shows the number of surveys by gender and age carried out in each city, according to their population structure.

The survey had closed multiple choice questions and a general test to know the citizens opinion about the main weathering agents, the level of monuments conservation in the city, as well as the role of citizens and their influence on conservation. The results allowed us an approach to assess the concern of the population about the heritage, the degree of knowledge of the heritage of the neighborhoods, and the concern of the local administration for the cultural heritage.

3 RESULTS

Figure 1 shows the result of survey. More than 75% of citizens from Seville, Osuna and Popayán are concerned about the conservation of monuments; while this percentage is slightly lower in Marchena (62.1%).

More than 50% of citizens considered that the monuments are not very well known by its inhabitants in Seville (61.5%) and Popayán (73.4%), while the trend is a bit more positive in Osuna (49.7%) or Marchena (43.0%).

Larger cities considered that the level of politicians concern for heritage conservation is very low; as less than 10% of surveys showed agreement (values 4 and 5). Medium size cities, as Osuna and Marchena, with less than 20.000 inhabitants, had an answer more positive, with

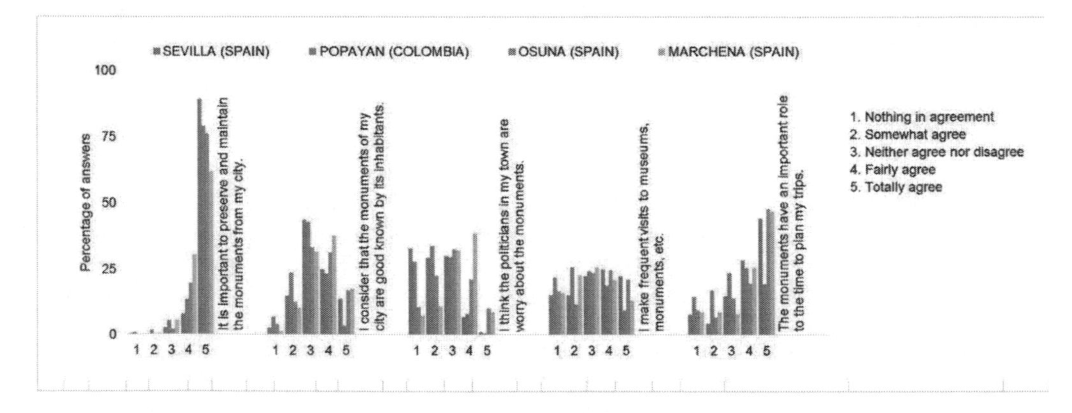

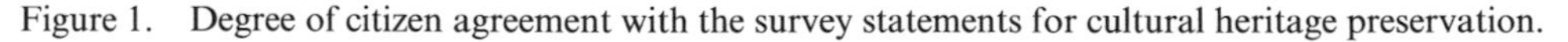

Figure 1. Degree of citizen agreement with the survey statements for cultural heritage preservation.

more than 25% and 40% respectively of surveys agree or strongly agree with the involvement of politicians.

Monuments play an important role in planning trips for citizens of Seville, Osuna and Marchena while the inhabitants of Popayán are not so interested. Also sevillians mention that they visit monuments more frequently than others.

Citizens from Spain seems to be more worried about their cultural heritage with an average of 3.5/5 (Seville: 3.5 ± 1.0, Osuna: 3.7 ± 0.7, Marchena: 3.7 ± 0.7) while, the average is slightly lower, 3.1 ± 0.9 in Popayán (Colombia).

Most of the Spanish citizens considered that the monuments are in a normal or good state of conservation (Figure 2), with an average over 80.0%, while the opinions about the state of conservation decrease in Colombia (66.4%).

Figure 3 shows the range of opinions about the main problem that affect the monuments in each city. The main concern of the citizens are human actions related to vandalism and theft in Seville 31.3% and Popayan 31.8%, while in Osuna and Marchena the environmental agents are the main problem according to the citizen opinions (Osuna 34%, Marchena 49.6%).

Popayan is located in Cauca region, and has high hazard of earthquakes and frequently suffers Niña and Niño phenomenon, so citizens are also worried about environmental agents (25.5%), structural problems (19.5%) and natural disasters (1.8%). Sevillians consider atmospheric pollution (14.8%) and lack of interest of population (19%) as important factors to take into account. Finally, citizens of Osuna and Marchena are also worried about lack of interest of politicians (14.9%, 9.3%) and structural problems (12.5%, 16.4%).

The surveys of the larger cities (Seville and Popayán) showed that social opinion had a low influence on the conservation of the monuments. In the other hand, citizens from Osuna and Marchena think that social opinion has a high influence on the conservation of the monuments in their cities.

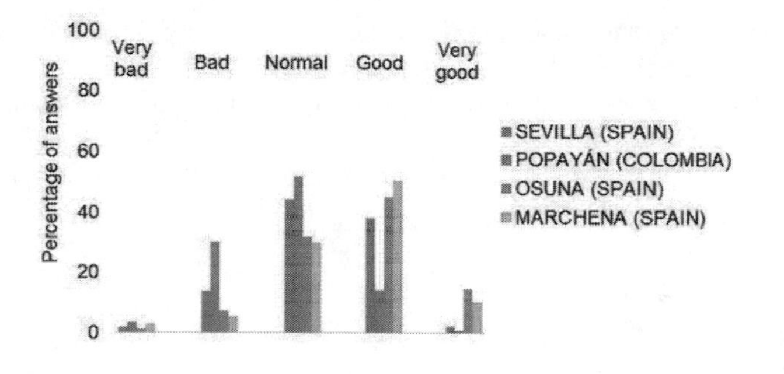

Figure 2. Opinion about the state of conservation of the monuments.

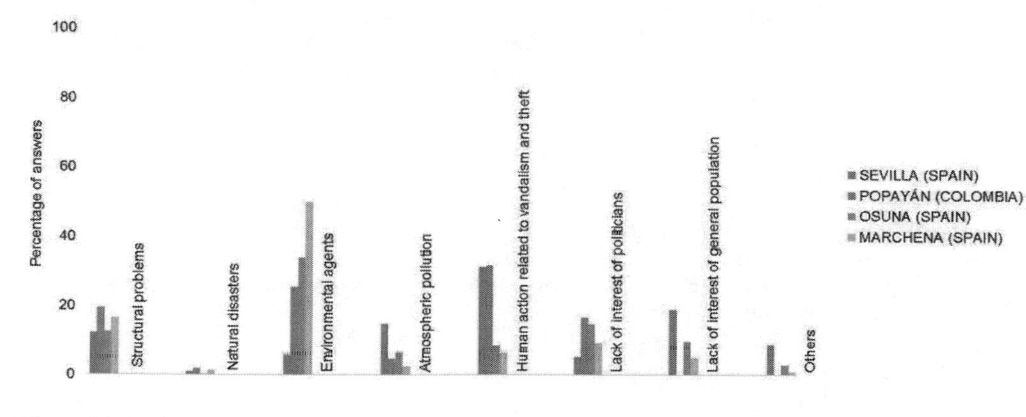

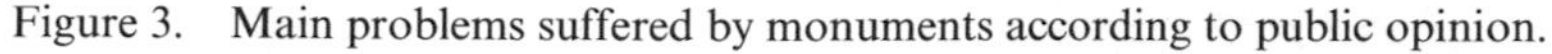

Figure 3. Main problems suffered by monuments according to public opinion.

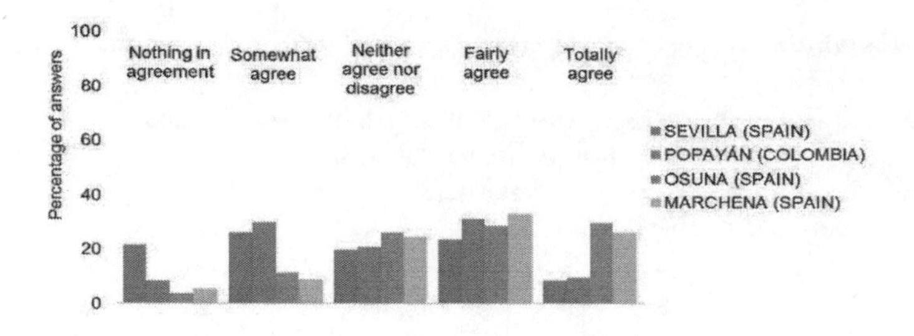

Figure 4. Opinion about the influence that public opinion has on the conservation of monuments.

4 CONCLUSIONS

The present work try to establish a initial relationship between social opinion and the vulnerability of monuments. Cities with more than 200.000 inhabitants (Seville and Popayán) showed that the level of politicians concern about heritage conservation is very low and their citizens considered that they cannot influence on the conservation of monuments, while smaller cities as Osuna and Marchena consider that citizens can influence on the state of monuments conservation and politicians are more worried about Cultural Heritage preservation. Further studies are necessary to assess to the opinion of the reflective societies in the preservation of Cultural Heritage in order to stablish the relationship between the conservation of monuments and the society in which they are located.

ACKNOWLEDGMENTS

This paper has been supported and based on the Methodology developed by two Projects: RIVUPH, an Excellence Project of Junta de Andalucía (code HUM-6775), and Art-Risk, a RETOS Project of Ministerio de Economía y Competitividad and Fondo Europeo de Desarrollo Regional (FEDER), (code: BIA2015-64878-R (MINECO/FEDER, UE)). Pilar Ortiz is grateful to Salvador Madariaga and the University of Oxford for her visit as researcher. Furthers thank Professor Xavier Coller (Universidad Pablo de Olavide) for his opinions to improve the design of surveys.

REFERENCES

Caraballo Perichi, C. 2008. *El patrimonio cultural y los nuevos criterios de intervención. La participación de los actores sociales*. Palapa. Revista de Investigación Científica en Arquitectura. Vol. III, número 1, pp. 41–49. Universidad de Colima, Mexico.

Chambers, I., Costanza, R., Zingus, L., Cork, S., Hernández, M., Sofiullaf, A., Htwe T. Z,. Kenny D., Atkins, P., Kasser, T., Kubiszewski, I., Liao Y., Chan Maung, A., Yuan, K., Finnigan, D., Harte, S. 2019. *A public opinion survey of four scenarios for Australia in 2050*. Futures. Vol. 107, pp. 119–132..

Mendoza Tolosa, H. A., Prieto Bustos, W. O., Barrero Nieto, C. A. 2012. *Encuesta de opinión para la evaluación de la gestión pública en Colombia: Una propuesta de medición*. Semestre Económico, vol. 15, número 32, pp. 77–102. ISSN 0120–6346.

Morate Martín, G. 2007. *Conocimiento y percepción del patrimonio histórico en la sociedad española*. Erph. Revista electrónica de Patrimonio Histórico. Número 1.

Palacio D. C., van der Hammen, M. C. 2007. *Redes heterogéneas del patrimonio. Los casos del Centro histórico y el humedal Córdoba, Bogotá (Colombia)*. REDES. Revista hispana para el análisis de redes sociales. Vol.13, 1, Diciembre 2007.

Valdelamar Sarabia, L. 2010. *Monumentos y conflictos en la construcción de identidades e imaginarios en Cartagena de indias: hacia un inventario simbólico*. Cuadernos de Literatura del Caribe e Hispanoamérica. Número 11, pp. 253–274.

Risk Assessment and monitoring of Cultural Heritage (Pollution, Climate Change, Natural Events, Microclimate)

Science and Digital Technology for Cultural Heritage – Ortiz Calderón et al. (Eds)
© 2020 Taylor & Francis Group, London, ISBN 978-0-367-36368-0

"Experiences" in the subterranean environment as a risk factor for cultural heritage. The case of Nerja Cave (Malaga, Spain)

C. Liñán, Y. del Rosal & L.E. Fernández
Research Institute, Nerja Cave Foundation, Malaga, Spain.
Andalusia's Research Groups RNM-308 and RNM-295

ABSTRACT: Additionally to its conventional touristic use, subterranean environments, as the caves, have been used as places to celebrate diverse events, recently called "experiences". Caves are extremely fragile sites where minimal environmental changes can produce irreversible damages in their ecosystem and in the patrimonial values they hold. This work exposes the case of Nerja Cave (Malaga, Spain), where an International Music and Dance Festival is annually celebrated. The environmental monitoring of the event along years has allowed to know its negative impact in the cave: anomalous increases in the air and rock temperature and carbon dioxide concentrations and high levels of fungi concentration in the air, considered one of the main rock art biodeterioration factor, amongst others. The Nerja Cave Research Institute has designed and implement a specific conservation protocol aimed to prevent or minimize the impact produced by the celebration of this "experience" in the subterranean environment. This protocol could serve as a model for the adequate conservation of other cavities or heritage sites with similar characteristics where this type of activities is celebrated.

1 INTRODUCTION

In the last years, some authors are critical about the celebration of recreational activities in protected natural areas (Ballantyne et al., 2014; Ballantyne & Pickering, 2015; Newsome & Lacroix, 2011) by the significant impacts arising from them. Newsome & Hughes (2018) refer numerous examples of these activities (live concerts, theatres, movies, competitive sporting events, reality TV productions...) that require a management control and appropriate policy development to guarantee the conservation of the area where they are celebrate. For these researchers "the increasing visitor numbers coupled with a trending 'fast food' entertainment style of experience compromises the conservation mandate". In the tourist caves -and other subterranean environment- has been also created new modalities of interaction with the visitors, basically lights show, concerts and adventure activities. But the impact that "new activities" or "experiences" produce on the protected area is not sufficiently investigated or evaluated (Newsome & Hughes, 2018).

The Nerja Cave is one of the most important tourist caves in Spain, with about 450.000 visitors annually and it may answer many questions related to the biology, geology and archaeology of the subterranean environment. The cave preserves a chrono-cultural and paleo-environmental sequence which makes it possible to carry out many interesting interpretations of prehistoric settings (Jordá Pardo, 1986) and more than three hundred rock art paintings be-longing to the Upper Paleolithic and Late Prehistory (Sanchidrián, 1994) (Figure 1). Therefore, the cave is categorized as Asset of Cultural Interest, in the Archaeological Sites category and is also an internationally recognized Heritage Sight of Geological Relevance (García Cortés, 2008). Nevertheless, these fragile ecosystems can be disturbed by many anthropogenic and/or natural factors. The anthropic impact in Nerja Cave have been widely measured and studied. More information about this topic and the environmental

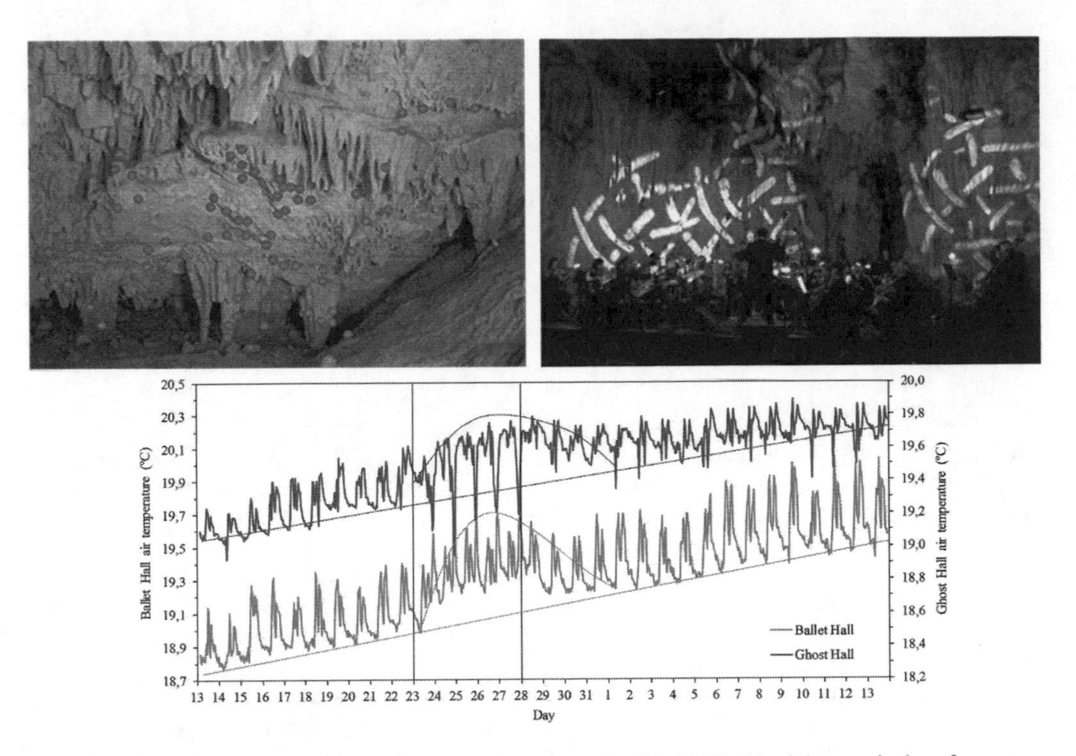

Figure 1. Up left: paleolithic pigment in the Ballet hall (PGI-CN, 2018). Up right: musical performance, during the cavity recovery time. Down: increases in air temperature during the festival (from 23 to 27 July) and several days after the end. Modified from Carrasco et al., (2002).

parameters of the cave can be consulted in Carrasco et al., (2002a), Del Rosal et al., (2007), Liñán & del Rosal (2015, 2016) and Liñán et al., (2018).

Since it was discovered in 1959, many human activities or "experiences" have been developed in the Nerja Cave in addition to the touristic visit. The aim of this work is to show that these "experiences" can represent an additional risk factor for the conservation of the caves and for Cultural and Natural Heritage they hold and the necessary application of specific protocols to avoid the anthropic damages. We presenter, as case of study, the annual International Music and Dance Festival of Nerja Cave.

2 "EXPERIENCES" IN NERJA CAVE. THE MUSIC AND DANCE FESTIVAL

The Music and Dance Festival entails the development of musical performances within the cave during several days of summer, normally in July. During the Festival period, the Nerja Cave Research Institute intensifies the environmental monitoring and vigilance of the cavity (I.I.C.N. 2002-2018), because:

(1) the Festival requires the installation in the subterranean environment of infrastructures such as the stage, sound systems, artistic lighting and seating for the public (Figure 2). These elements can damage the walls, speleothems and potentially, the rock art (accidental incisions, scratches, paintings, etc) when used as a direct support.
(2) the presence of these allochthonous objects in the cave induces the development of microorganisms, especially fungi, considered one of the main factors causing the biodeterioration of the rock art (Figure 2) (Sáiz, 2012; Martín-Sánchez et al., 2014).
(3) during the shows, the established habitual capacity and the average time of people in the cave is exceeded and the audience remains in the same hall (the Ballet Hall) for about two hours.

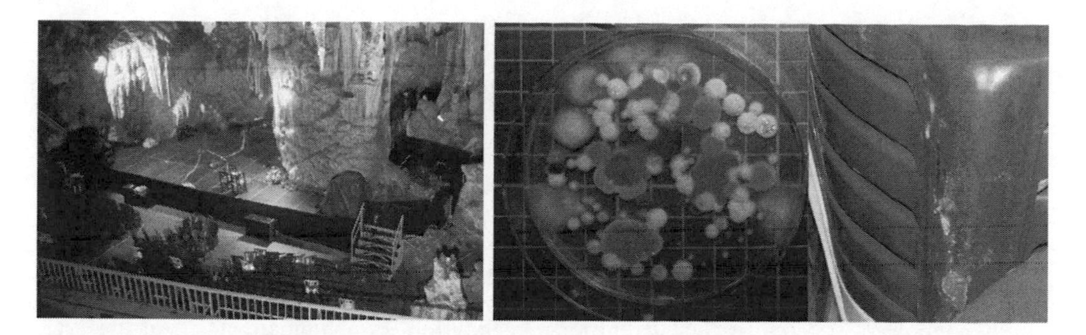

Figure 2. The festival requires the installation of infrastructures which can damage the walls, speleothems and the rock art (left). Petri dish with fungi colonies from a cave air sample (middle). Chairs with fungi colonies development (right).

Both facts generate an additional anthropic impact in the cave that adds to the daily impact produced by the tourists' visits, which reflects in anomalous increases registered in the air and rock temperature and carbon dioxide concentrations, amongst others, which persist in the cave during several days after the end of the event (Carrasco et al., 2002b) (Figure 1). All this indicate that the recovery capacity of the cave is not enough to eliminate the impact of the Festival during the night (closed period of the cave), in contrast to another days of normal visitability, in which the anthropic impact disappear during the night.

(4) the presence of the infrastructures installed for the Festival in the subterranean landscape damages to the institutional image of the Nerja Cave, alters visitor touristic experience and limit understanding of the natural and cultural environment (Zhang et al., 2009).

3 CONSERVATION PROTOCOL

To limit the risk involved in holding these festivals within the cavity, the Nerja Cave Research Institute has put in place a series of measures which form the Conservation Protocol:

(1) Preventative measures: the main aim of these measures is to prevent damage to the cave itself caused by the installation of the necessary infrastructure (such as the stage, seating, lighting, etc.), the presence of people and the development of potentially contaminating agents. The contemplated measures include: a reduction in the number of shows held in the cave; a physical protection for the most fragile elements, with the fragility being due to either the nature of the object (the rock art) and/or the actual location of the elements which may be close to the infrastructure installed; a reduction in the number of spectators admitted, thus diminishing the space needed for the audience; to maximize the cleanliness of the objects introduced into the cave as well as the concrete floor of the hall where the shows are held; to prohibit any material of a porous nature or made up of organic material and limit the time autochthonous elements can remain in the cave.
(2) Environmental control measures: these measures should be set up both in the days prior to the event and in the days after the event to specifically control the environmental and biological parameters in the air, water and substrate in order to identify possible alterations related to the event itself and its ongoing impact.
(3) Training measures: these measures are aimed at the workers in the cave; especially those involved in setting up the installation of the infrastructure for the events and the dismantling it afterwards. The objective is to involve workers in the conservation of the natural and cultural heritage of the Nerja Cave.

4 CONCLUSIONS

The uncontrolled development of events or "experiences" can represent a risk factor for the conservation of the caves and for the rock art, especially if appropriate preventive and/or corrective measures are not carried out. It has been registered diverse impacts on the subterranean environment linked to the holding of the Music and Dance Festival in the Nerja Cave. To minimize this impact and reduce risk to the conservation of the cave, the Research Institute has designed a protocol of measures specifically related to this type of "experiences". This protocol, in use since 2013, has allowed to diminish the damage caused by the Festival to the cultural heritage and the impact to the subterranean environment, although some damage continues to occur. Therefore, the current recommendation is to eliminate events of this type within caves, especially those which are part of our cultural heritage.

REFERENCES

Ballantyne, M., Gudes, O. & Pickering, C.M. 2014. Recreational trails are an important cause of fragmentation in endangered urban forests: a case-study from Australia. *Landscape and Urban Planning*, 130(1):112-124.

Ballantyne M. & Pickering, C.M. 2015. Recreational trails as a source of negative impacts on the persistence of keystone species and facilitation. *Journal of Environmental Management*, 159:48-57.

Carrasco, F., Vadillo, I., Liñán, C., Andreo, B. & Durán, J.J. 2002a. Control of environmental parameters for management and conservation of Nerja Cave (Malaga, Spain). *Acta Carsologica* 31/1, 9: 105-122.

Carrasco, F., Liñán, C., Durán, J., Andreo, B. & Vadillo, I. 2002b. Modificaciones de los parámetros ambientales de la Cueva de Nerja provocadas por la entrada de visitantes. *Geogaceta*, 31: 15-18.

Del Rosal, Y., Martínez-Manzanares, E., Marín, F., Liñán, C. & Simón, M.D. 2007. Análisis de la calidad microbiológica aérea en el interior de la Cueva de Nerja. In J.J. Durán, P.A. Robledo & J. Vázquez (eds.), *Cuevas turísticas: aportación al desarrollo sostenible*, Asociación de Cuevas Turísticas Españolas (ACTE), Madrid.

García-Cortés, A. (ed.). 2008. *Contextos geológicos españoles: una aproximación al patrimonio geológico español de relevancia internacional*. Instituto Geológico y Minero de España.

otherJordá Pardo, J.F. 1986. *La Prehistoria de la Cueva de Nerja. Málaga*. Patronato de la Cueva de Nerja.

I.I.C.N. 2002-2018. Memoria Científica Anual del Instituto de Investigación Cueva de Nerja (*unpublished*).

Liñán, C. & del Rosal, Y. 2015. Natural Ventilation of Karstic Caves: New Data on the Nerja Cave (Malaga, S of Spain). In B. Andreo, F. Carrasco, J.J. Durán, P. Jiménez & J. LaMoreaux (eds), *Hydrogeological and Environmental Investigations in Karst Systems*. Environmental Earth Sciences, 1. Springer, Berlin, Heidelberg.

Liñán, C. & del Rosal, Y. 2016. Procesos de ventilación natural Cueva de Nerja - Cueva Pintada (Nerja, Málaga). In B. Andreo & J.J. Durán (eds), *El karst y el hombre: las cuevas como Patrimonio Mundial*. Asociación de Cuevas Turísticas Españolas (ACTE), Nerja (Málaga).

Liñán, C., del Rosal, Y., Carrasco, F., Vadillo, I., Benavente, J. & Ojeda, L. 2018. Highlighting the importance of transitional ventilation regimes in the management of Mediterranean show caves (Nerja-Pintada system, southern Spain). *Science of The Total Environment*, 631-632: 1268-1278.

Martín-Sánchez, P., Jurado, V., Porca, E., Bastian, F., Lacanette, D., Alabouvett, C.& Sáiz, C. 2014. Airborne microorganisms in Lascoux Cave (France). *International Journal of Spelelolgy*, 43: 295-303.

Newsome, D. & Lacroix, C. 2011. Changing recreational emphasis and the loss of 'natural experiences' in protected areas: an issue that deserves consideration, dialogue, and investigation. *Journal of Tourism and Leisure Studies*, 17 (2): 315-334.

Newsome, D. & Hughes, M. 2018. The contemporary conservation reserve visitor phenomenon! *Biodiversity and Conservation*, 27 (2): 521-529.

Sáiz-Jiménez, C. 2012. Microbiological and environmental issues in show caves. *World Journal of microbiology and biotechnology*, 28(7): 2453-64.

Sanchidrián, J.L. 1994. *Arte Rupestre de la Cueva de Nerja. Málaga*. Patronato de la Cueva de Nerja.

V.V.A.A. 2018. *Memoria del Proyecto General de Investigación Cueva de Nerja*. Fundación Pública de Servicios Cueva de Nerja. Inédito.

Zhang, C. Z., Xu, H.G., Su, B.T. & Ryan, C. 2009. Visitors' perceptions of the use of cable cars and lifts in Wulingyuan World Heritage Site, China. *Journal of Sustainable Tourism*,17(5): 551-566.

Science and Digital Technology for Cultural Heritage – Ortiz Calderón et al. (Eds)
 ISBN 978-0-367-36368-0

Two shelters under review: Preliminary results from the Roman archaeological site of Complutum

C. Cabello Briones & J. Barrio Martín
Service for the Conservation, Restoration and Scientific Studies of the Archaeological Heritage, Universidad Autónoma de Madrid, Madrid, Spain

ABSTRACT: Shelters are nowadays one of the most commonly used solutions to preserve archaeological sites from further decay. They are able to reduce the impact of some of the decay factors that usually affect exposed sites such as direct sunlight and rainfall. However, it has been documented that the type of cover selected can have a considerably impact on not only the interpretation and appreciation of the site but also the conservation state of the remains. The Service for the Conservation, Restoration and Scientific Studies of Archaeological Heritage (SECYR) at Universidad Autónoma de Madrid is currently undertaking an extensive monitoring programme on the two shelters at the Roman archaeological site of Complutum (Alcala de Henares, Spain). The House of Hippolytus, originally located outside the Roman city and famous for its mosaics, was covered with an enclosed shelter in 1999. On the other hand, the House of the Griffins, decorated with mural paintings, has been protected with an open shelter since 2011. The aim is to evaluate the influence of different solutions on the preservation of the remains. This paper shows the methodology employed, based on the microclimatic monitoring inside and outside the shelters using small and easy to hide data loggers, together with some preliminary results.

1 INTRODUCTION

Shelters have been extensively considered preventive conservation measures for archaeological sites. These barriers can reduce the impact of the adverse environmental conditions, such as the effect of light damage or direct rainfall, while still providing with the possibility of visiting the site, as opposed to backfilling (Roby 2006). However, the construction of shelters may itself be damaging to archaeological sites. If built incorrectly, a shelter may not only provide with a reduced protection, but also exacerbate the deterioration mechanisms as a result of the creation of undesirable microclimates (Curteis 2018).

An environmental monitoring programme has being undertaken in Complutum (40° 28′ 26.146” N, 3° 23′ 16.49” W) since May 2018. Complutum is a Roman archaeological site and, as part of the historic precinct of Alcala de Henares, was included in the UNESCO World Heritage list in 1999. It was also declared of cultural interest under the Spanish legislation in 1992, which conferred it with special protection. The main construction materials are rammed earth, masonry and brick, covered with decorative elements such as mural paintings and mosaics.

Currently, there are two areas of the site covered with shelters: the House of Hippolytus and the House of the Griffins. The shelter of the first one was constructed in 1999 as part of a cultural regeneration programme of the area. This shelter covers an area of 1318 m^2 and can be described as a partially enclosed design (Figure 1). It is defined by strip footings, load bearing walls made of bricks in the lower part and galvanised metallic meshes in the upper part, and a cover of galvanized steel sheets and hydrophobic agglomerate wooden boards (Rodríguez Frade 2001). In the case of The House of the Griffins (Figure 1), the shelter of approximately 1300 m^2 was set in place in 2011. It is made of a dome-shape steel structure with lateral claddings of metallic meshes hanging from the structure without standing on the floor (Sánchez Montes & Rascón Marqués 2012). The final result could be classified as a partially-open shelter, a lighter structure that avoids deep foundations which could affect the archaeological remains, what is a reflection of the worldwide current trend (Cabello Briones 2016).

The service in charge of the management of the site carries out regular maintenance of the remains. This action is relevant for their conservation state as there are no relevant signs of long-term decay. However, it is possible to find some evidences of physical and biological deterioration outside the shelters, two types of weathering that could affect the site if it was not covered. In addition, the enclosed shelter at the House of Hippolytus shows some moisture stains on the walls that could be related to an inefficient draining system. This could increase the humidity levels inside the shelter and affect the ruins by enhancing biocolonization in the long run. The aim of the study is to evaluate the performance of the two shelters in Complutum regarding the preservation of the remains and to determine the difference between those types in their protective role. Although a visual survey can provide us with an initial approximation, any evaluation of existing shelters should involve a monitoring programme over a specific period of time (Pesaresi & Stewart 2018), so the previous hypotheses can be corroborated.

2 RESEARCH METHODS

This study presents the preliminary results corresponding to 4 months of monitoring (from 17 de May to 17 to September) so as the performance of the shelters with high mean temperatures can be compared. The data concerning the House of the Griffins is also presented as a complement for that obtained by Martínez Garrido et al. (2016) between December 2014 and July 2015.

The environmental conditions were monitored using small and easy-to-hide data loggers (Lascar Electronics, accuracy= 0.55°C, 2.25%RH), which provided hourly temperature and relative humidity readings. The loggers, synchronised at the same time, were placed inside the two shelters and an additional sensor was located outside the House of the Griffins to be used as a reference for the outer environment. The three of them were located on the top of site columns, therefore, they were not easily visible by visitors or thefts. The parameters thought critical for the preservation of the archaeological remains are temperature and relative humidity (RH) extreme values and fluctuations. These indicators are considered highly responsible for salt crystallisation, freezing-thaw and wetting-drying events, in addition to biological development (Torraca 2009).

Non-parametric Mann-Whitney-Wilcoxon tests were used to compare daily means and standards deviations of temperature and RH. This allowed determining if there were statically significant differences among the studied locations.

Figure 1. a) The house of hippolytus (left) and b) the house of the griffins (right).

3 RESULTS

The temperature and RH values recorded in Complutum in spring and summer are normal for this location and time of the year, with monthly mean temperatures increasing from around 18°C to 28°C, and RH values ranging between 65% and 35%. In addition, the variability of the data, represented by the standard deviations, is also within usual limits (Table 1).

Statistical tests on daily mean temperatures showed that temperature outside was significantly warmer than inside both shelters, the House of the Griffins (U= 764, p-value=1.843e^{-10}) and the House of Hippolytus (U= 32, p-value <2.2e^{-16}). On the other hand, the House of the Griffins had higher temperatures inside the shelter than the other cover (U= 5671, p-value <2.2e^{-16}). Similarly, outside presented more variability on the temperature data than the House of the Griffins (U= 21.5, p-value <2.2e^{-16}) and the House of Hippolytus (U = 0, p-value <2.2e^{-16}). If both shelters are compared, the temperature values fluctuated more in the first case (U= 0, p-value <2.2e^{-16}).

Regarding daily mean RH values, statistical tests showed that RH outside was not significantly different than inside the House of the Griffins (U=2503, p-value=0.2954), and RH was lower outside than inside the House of Hippolytus (U=1110.5, p-value=5.464e^{-08}). On the other hand, the House of Hippolytus had higher RH values inside the shelter than the other shelter (U= 5627, p-value <2.2e^{-16}). Additionally, outside presented more variability on the RH data than the House of the Griffins (U= 62.5, p-value <2.2e^{-16}) and the House of Hippolytus (U=110.5, p-value <2.2e^{-16}), presenting the first one higher RH fluctuations (U= 974, p-value= 4.472e^{-09}).

Another point to be considered is that outside had the most extreme values during days and nights, represented by maximum and minimum temperatures respectively, and both shelters were found to have lower maximum temperatures and higher minimum ones in comparison to outside. Although this may indicate that the shelters are working efficiently, mainly the House of Hippolytus, there were greater temperature values during the summer days than expected. In August, the number of times hourly temperature went up to 35°C, was 64 outside, 42 inside the House of the Griffins and 21 inside the House of Hippolytus.

Thermal fatigue associated with solar heating and night cooling events can be better described in terms of diurnal differences between maximum and minimum temperatures (Table 2).

Diurnal temperature variations were greater during summer because of the higher maximum temperatures, which increased more than 10°C. On the other hand, the differences between daily maximum and minimum temperatures inside were lower than outside, although relatively high in the House of the Griffins. However, a diurnal temperature range of around 12°C is probably negligible for thermal stress if compared with other studies on stone heritage (Al-Omari et al. 2014).

Table 1. Monthly means and standard deviations of the daily temperature and RH outside, and inside the shelters at the house of the griffins and hippolytus from May to September 2018.

	Temperature (°C)			RH (%)		
	House of the Griffins	Outside	House of Hippolytus	House of the Griffins	Outside	House of Hippolytus
May	18.39 °C (±3.24)	18.73 °C (±4.16)	17.96 °C (±2.79)	63.34 % (±14.85)	66.40 % (±21.47)	64.56 % (±14.26)
June	22.49 °C (±5.76)	22.83 °C (±6.31)	22.01 °C (5.41)	51.76 % (±16.63)	52.14 % (±18.80)	52.61 % (±16.20)
July	26.53 °C (±4.21)	26.65 °C (±4.47)	25.98 °C (±3.71)	36.17 % (±11.10)	36.31 % (±11.80)	36.94 % (±10.93)
August	27.81 °C (±4.54)	28.70 °C (±4.93)	27.18 °C (±3.98)	36.42 % (±11.62)	35.63 % (±12.84)	37.66 % (±11.84)
September	24.25 °C (±4.15)	27.04 °C (±7.05)	23.42 °C (±3.20)	48.67 % (1 ±2.42)	44.30 % (±16.40)	50.54 % (±11.67)

Table 2. Monthly means and standard deviations of the daily temperature differences (Tmax-Tmin) outside, and inside the shelters at the house of the griffins and hippolytus from May to September 2018.

	House of the Griffins	Outside	House of Hippolytus
May	8.43 °C (±2.33)	10.87 °C (±2.41)	6.77 °C (±1.87)
June	10.22 °C (±2.99)	11.55 °C (±2.35)	8.37 °C (±2.51)
July	12.08 °C (±1.94)	12.35 °C (±1.76)	9.92 °C (±1.54)
August	12.98 °C (±1.80)	13.47 °C (±2.00)	10.00 °C (±1.47)
September	12.92 °C (±2.88)	14.20 °C (±8.78)	8.02 °C (±1.55)

4 DISCUSSIONS AND CONCLUSION

The House of the Griffins, with a partially open shelter, and the House of Hippolytus, with a partially closed one, represent two different alternatives to protect the archaeological remains in the Roman archaeological site of Complutum. The data from May to September 2018 shows that the shelter at the House of the Griffins followed more closely the outer temperature probably due to its more open structure. Nevertheless, the high mean temperatures and diurnal temperature variations are not enough to consider an imminent damage for the remains. On the other hand, the House of Hippolytus was able to keep the environment more stable than the other shelter. However, it showed a slightly more humid environment, which could be related to an inadequate draining or water disposal system. The House of the Griffins showed more variability in the RH data but also lower RH means than the other shelter. This may indicate a drier environment and, therefore, lower risk to reach critical salt hydration and crystallisation thresholds. However, a further study considering the specific salt mixtures present in the remains is recommended.

The shelters avoid excessive solar heating on the remains, which is a benefit also seen in sites in Mediterranean countries (Cabello Briones 2016). The results also indicate that higher maximum temperatures, more frequent temperature and RH variations, and greater diurnal temperature differences outside the shelters could have led to the physical weathering observed in some of the exposed ruins, corroborating the initial hypotheses.

In conclusion, this study with regards to spring and summer conditions has demonstrated that an exhibition of the archaeological remains in Complutum without sheltering is probable to lead to further decay. Both types of shelters are mitigating decay factors keeping a more stable environment inside them in relation to outside, although the shelter at the House of Hippolytus would improve if the water disposal system was revised.

REFERENCES

Al-Omari, A., Brunetaud, X., Beck, K. & Al-Mukhtar, M. 2014. Effect of thermal stress, condensation and freezing-thawing action on the degradation of stones on the Castle of Chambord, France. *Environmental Earth Sciences*, *71*: 3977–3989.

Cabello-Briones, C. 2016. *The Effects of Open Shelters on the Preservation of Limestone Remains at Archaeological Sites*, Doctoral Thesis, University of Oxford..

Curteis, T. 2018. "The use of environmental survey and monitoring in the design and evaluation of archaeological shelters". In Aslan, Z. et al. (eds.), *Protective shelters for archaeological sites; Proceedings of SYMPOSIUM, Herculaneum, 23-27 September 2013*. Rome: The British School.

Martínez-Garrido, M.I., Ergenç, D. & Fort, R. 2016: "Wireless monitoring to evaluate the effectiveness of roofing systems over archaeological sites", *Sensors and Actuators A*, *252*: 120–133.

Pesaresi, P. & Stewart, J. 2018. "Shelters evaluation, monitoring and maintenance in the context of archaeological site management". In Aslan, Z. et al. (eds.), *Protective shelters for archaeological sites; Proceedings of SYMPOSIUM, Herculaneum, 23-27 September 2013*. Rome: The British School.

Roby, T. 2006. "The Conservation of Mosaics in situ. Preserving Context and Integrity". In Abed. A.B. (ed.): *Stories in Stone. Conserving mosaics of Roman Africa. Masterpieces from the National Museums of Tunisia.* Los Angeles: The Getty Conservation Institute.
Rodríguez Frade, J. P., Madariaga Méndez, A., & Rascón Marqués, S. 2001. "Cubierta para las ruinas arqueológicas de la Casa de Hippolytus Complutum" [Online], In *Memoria para el Premio Calidad a la Estética, Dirección General de Arquitectura y Vivienda de la Comunidad de Madrid.* Available: http://www.madrid.org/bdccm/premios/PDF/LIBRO2001-2000/T3.pdf (Accessed November 2018).
Sánchez Montes, A.L. & Rascón Marqués, S. 2012. "Musealización del proceso de excavación de la Casa de los Grifos (Alcalá de Henares, Madrid). La Casa de los Grifos en el contexto de los yacimientos musealizados en España". In *VIII Jornadas de Patrimonio Arqueológico en la Comunidad de Madrid, 15–16 November 2011.* Madrid: Comunidad de Madrid.
Torraca, G. 2009. *Lectures on Materials Science for Architectural Conservation.* Los Angeles: The Getty Conservation Institute.

Science and Digital Technology for Cultural Heritage – Ortiz Calderón et al. (Eds)
© 2020 Taylor & Francis Group, London, ISBN 978-0-367-36368-0

Integral heritage rehabilitation challenges in Southern Chile: The case of wooden cities, Valdivia

G. Saelzer, P. Gómez & A. Ruiz-Tagle
Universidad Austral De Chile, Valdivia, Chile

ABSTRACT: In the context of Southern Chile, historical buildings are under pressure of seeking sustainable integral rehabilitation. Cultural arguments and technological decisions are issues that these building could not solve together. A gap is observed in the national regulation and regional guidelines on heritage rehabilitation. The article references an empirical case study focused on restoration practices in Valdivia, bringing up tentative discussions on the national programs. Our objective is to link the cultural arguments and the legal framework of heritage restoration, considering building materials as part of history and architectural development of the area. The authors were able to identify seven key points, connecting history with the architectural material conditions that could be missing in the Chilean Public service. Professional experience for restoration practice allows the revision of Secretary programs that are used for heritage rehabilitation at medium and small size historical buildings.

1 INTRODUCTION

Chile is one of the world's most renowned earthquake zones. According to the earthquaketrack.com, Chile experiences averagely 684 earthquakes in every 365 days. The 9.1°R earthquake of May 22, 1960, is the largest ever instrumentally recorded, occurred off the coast of southern Chile. The rivers changed their course, new water bodies were born, historical buildings were destroyed and the geography was markedly modified. The evidence of slowly recovering from the giant earthquake after 59 years without a reconstruction plan particularly in the city of Valdivia, offers imperatives to analyze and check the Heritage Rehabilitation of architecture for the group of buildings built before 1960 in city. Programs for the heritage restoration and rehabilitation of public and historical buildings in Chile started in 2007 at the national level by the Ministry of Public Works (MOP)[1], followed by Ministry of Housing and Urban Planning (MINVU), and Ministry of Culture, Arts and Heritage (MICAP).

The MICAP program is oriented towards social sciences -historiography, social anthropology, and territorial identity- according to heritage regulatory framework and its legal focus (MINEDUC-CMN, 2011, p.6-8)[2]. The arguments based on these contents play a crucial role in the application of projects and the criteria for management plan evaluation (MICAP, 2017, p.5,13,18). Invariably, the orientation on social sciences downplays the architectural, technical and buildings materials which have been a source of identity. This does not ensure the integral rehabilitation of heritage buildings due to the fact that the impacts of technical capabilities and material compositions to history and culture have been overlooked in the heritage regulatory frameworks in Chile. Furthermore, the extent of these impacts and divergence with local trends, especially in southern Chile where cities built with wood -Wooden Cities- are attempting to rescue local values is not well considered. The research seeks to link the socio-cultural arguments with the legal framework of heritage restoration, considering building materials as

[1]"Puesta en Valor del Patrimonio" program.
[2]Ley de Monumentos Nacionales (LMN), n°17.288 1970 and 2011.

part of history and architectural development of the area. The aim is to improve restoration practices in regions of registered risks, directing the discussion to territorial specific planning and future service life of buildings. It also aims to contribute to the discussions and characteristics applied to Southern Chile regions with respect to public investments on heritage restorations.

2 INTEGRAL HERITAGE REHABILITATION: A REVIEW

The concept of heritage which initially embodied landmark historic buildings, has progressively expanded to include other types of building which conceivably are less aesthetically-appealing but of similar or greater value in terms of what they tell us about the society that created them (Bianca, 2010, SEMO-WHC, 2018, Nyseth & Sognnaes, 2012). Undoubtingly, the process for the restoration and rehabilitation of heritage buildings must give greater attention to context- the building as an integral part of its urban setting, ruling out any approach that takes the building as something separate from its context, historic and surroundings, and limited to façades. (SEMO-WHC, 2018; Sirisrisak, 2009).

In the case of southern Chile, settlements originated from the productions of a common eco-region with slight industrial inputs shaped its contemporary towns. Formulated as a nineteenth-century post-colonial characteristic, these historical settlements built of wood characterized the emergence of modernization in southern Chile; a significant historical architectural and urban experience was installed related to natural events and disasters (Guarda 1995; Saelzer, 2019). Heritage restoration processes started in the southern Chile barely 10 years ago despite the necessity some 60 years ago. However, the processes are far from being integral, since building materials stay far from preliminary evaluation and are left for individual interpretations during the restoration process and the guidelines that can be derived from the National Monuments Law (LMN) and the General Law for Urban Planning and Construction (LGUC) are historiographical and anthropological based. In fact, the lack of integral heritage rehabilitation (IHR) is due to social science arguments in the heart of heritage regulatory framework without technical guidelines to the fragile variable materials. In this sense, Torres (2014) presented that: *"in terms of policies, except for volume conservation guidelines, in the Chilean context there are no specific regulations for interventions in architectural heritage"*. Ideally, it is difficult to understand the scope and specifications of conservation and restoration of architectural projects, including interventions on the lifespan of a building (Prieto et. al. 2019).

3 RESEARCH DESIGN AND METHODOLOGY

The paper adopted a mix design approach (*both qualitative and quantitative*) focusing primarily on a desk review of national policy documents, legislation, regulations and institutional manuals and guidelines to understand the restoration requirements, environmental problems, anthropic risks and materials of heritage buildings in the Los Rios and Los Lagos Regions. Specifically, to understand the most fragile variables in Valdivia (foundations, timber and masonry facades and other perimetral walls) and two most critical factors (*Natural long-term degradation of the materials- due to exposure to the environment* and *Short-time dynamic actions- such as vibrations or impact loads)* (Kliukas, Kačianauskas & Jaras, 2008) which reduce the ability of building structures to retain their original properties and serviceability. Primary data were obtained through direct interviews with the heads of the institutions which have been identified as stakeholders for the implementation of heritage restoration programs. The institutions were the: Ministry of Public Works (MOP), Ministry of Housing and Urban Planning (MINVU) and Ministry of Cultures, Arts and Heritage (MICAP) to triangulate and facilitate the validity, consistency and comprehensiveness of the data while ensuring a high level of error minimization. The data inputting, processing and analysis involve the use of Computer-Aided Spatial-Analysis design tools and software. The results and findings were thoroughly discussed using maps, parametric and non-parametric techniques where applicable.

4 CASE STUDY

4.1 *Experimenting architectural restoration strategies in Valdivia*

The historical city center of Valdivia has always been subjected to natural hazards. First, it was damaged by fire in 1859 and 1909, followed by the 1960 (9.1°R) and 2010 (6.5°R) earthquakes, as well as floods in 1922 and 1960. With the enormous impact on GGP -Gross Geographic Product, the 9.1°R earthquake obliged the redefinition of materials and its structural strategies for new buildings, but the integration of historical monuments was not included in the planning. The 1850-1960 architecture was left to its own fate; without programmed financing and restoration guidance. Anthropic risks (e.g. Casa Werkmeister burnt in 1985, the plot still vacant) and seismic activity (2010, 6.5°R) continue to become evident on the extreme vulnerability of architectonic cultural heritage in Valdivia. For instance, the exposal of the basement and masonry structured to flooding leading to the continuous weakening of foundations, led to the restoration of Casa Ehrenfeld –Music Conservatory-1700 m^2, made up of three floors timber building with a metal coating. The restoration, which was done on a co-financing agreement between MICAP program and the owners was unable to afford a complete restoration project. The contributions of the owners were towards solving priorities like non-combustible heating, electricity reinforcement and fire-retardant networks, whilst the finance from MICAP was also not able to provide a complete structural restoration. The experience of stakeholders that historical architecture needs different approaches to finance infrastructure strategies informed the conception of the program for cultural infrastructure[3]. The repairing of Casa Hettich *–currently being used as Municipal Library*, and conditioning of basement of Casa Cau-Cau *–Wetlands Protection Center (originated in 1960)* provided the basis.

4.2 *The management of heritage rehabilitation processes in Valdivia*

4.2.1 *Towards integration of historiography background and structural typology of buildings*

Stratifying the historical town center of Valdivia into periods of experimental building strategies from 1850-1960, brings up the morphological timelines of building practices. As reported by PEC[4]: timber buildings (strongly ca.1850-1909)[5], characterized by metallic cladding (1860-1909), later coated façade cladding (1910-1960) as well timber buildings protected by masonry perimeter (1910-1960)[6], have been identified but are less analyzed. The latter (1910-1960) can also be divided into two main periods. The 1910-ca.1930 period presents timber platform mixed with diverse masonry elements for perimeters of the buildings[7] and ca.1930-1960, masonry perimeter and slabs completed less on timber parts, were also identified. The path of recent time contribution to the application of the program for the two first period, and valorise and discuss the third period, trace the need of complete integral rehabilitation (IHR) that considers service life criteria wide range of vulnerability issues (Prieto et. al. 2019).

4.2.2 *Interdependency factors regarding the behaviour of buildings*

IHR aspects were highlighted after the lack of results in protection and practical instruments announced by inhabitants at Tipical Zones (TZ), ruled by national heritage regulations, as

[3]"Fondo de Infraestructura Cultural" program.
[4]Working Program for Built Heritage and Context (Spanish: PEC)
[5]Casa Ehrenfeld, 1919, restoration 2013-2014. Casa Luis Oyarzún, 1886, restoration 2016-2017.
[6]Casas Reccius and Anwandter, 1910, staying Casa Central UACh, restoration 2017-2018.
[7]Convento de San Francisco, 1929, habilitation -enabling some use- 2019, previous to restoration.

PEC diagnosed in four of five TZs (Saelzer et. al., 2018)[8]. According to local analysis by Jeri (2015) and historic city centres degradation as presented by Zumelzu, et.al. (2016), there are defused identity of relevant goals to be achieved. If restoration is a relevant preservation event, the culture, legal framework, and technological practices approaches are interdependent factors determining physical recovery of buildings, their survival on time and improvement of life conditions. Technical answers to the Valdivian buildings of 1960s or older, and some under long lasted temporary facades[9], are not entirely comparable on national bases to provide capable operations to restore them and also their urban context[10] (Maldonado, 2019).

4.2.3 *Interdependency of factors regarding public finance and operations of restorations*

MICAP's program is focused on ownership of cultural and public services and provides partial donor funding for restoration of projects. Counterpart funding is compulsory in the MICAP programs, and definition of possible items to be covered by the maximum financing quota are also given. MINVU programs, on the other hand, cover the total cost that reaches five to fifteen times the amount that MICAP covers. Nevertheless, MINVU restoration projects face averagely a five-year late start. MICAP has a reliability in relation to the start of projects which are normally a year bound, and are compatible with a variety of economic backgrounds. In addition is the yearly eligibility to compete for additional justifiable funding for the extension of projects. However, private houses and retail buildings, which are an inseparable part of historical areas and contribute to cultural identity and economy, are left behind due to the Socio-Developing Secretary investment classifications and guidelines.

4.3 *Technical guidelines for ensuring integral heritage rehabilitation in Valdivia*

With reference to geotechnical land analysis for suitable restoration strategies in the historic center (Alvarado et.al, 2019), the authors identified seven key variables connecting historical background with material conditions of architecture that could be missing in the legal frameworks, as follows:

1) Historical Placement: validated by the recent geotechnical soil analysis and re-evaluation of seismic dangers.
2) Evaluation of Material selection and adequacy of practices when reacting to the several historical risks time lines of the same building.
3) Adressing earthquake conditions to correct the transformation of the original architecture.
4) Integration of current rehabilitation standards in accordance with public agencies during restorations.
5) Resolving the territorial identity through architecture and according to National and regional laws among which valorisation and restoration of buildings play a public role.
6) Management of plans in terms of making investment in restorations profitable, incorporating materiality, service life and occupation load of restored building.
7) A balance in the National Heritage Law on conditions to operate on lands regarding interests of social science disciplines (e.g archaeology) and engineering.

The gaps between the frameworks for integral rehabilitation need a clear tryst in order to standardize building qualities. From a technical viewpoint, the guidelines for HR in the region of Los Ríos –Valdivia, offers an incomplete comprehension of the architecture challenges regarding natural actions, environmental and anthropic aggressiveness. In addition, local strategies are essential to preserve content and enhance materials for cultural heritage for the

[8]Valdivia (at Cochrane bridge project including Casas Lopetegui, Da Bove and Von Stillfried) 2017-2018, La Unión (by Mayor's consultants at Municipality) 2018, Puerto Octay (through PEC consultancy) 2017-2018, Frutillar (a diagnosis by the director of Municipal Infraestructure) 2018.

[9]Casas Reccius and Holzapfel (current Casa Central UACh), 1910, damaged in 1960, temporary facades 1960-2016, restored in 2017-2018.

[10]Ex San Francisco Convent, 1929, habilitation 2019, further restoration programed from 2020.

society and heritage management. Therefor, culture and architectural technicalities are inseparable at material conservation level (Prieto et.al. 2019). In that regard, these seven points are essential to enhancing the compatibility of all building material components of restorations and rehabilitations. These are also essential to standardize diagnoses and procedures to cover the integral rehabilitations of historical heritage projects.

5 CONCLUSION

The study aimed at assessing the adequacy of the Chilean heritage regulatory framework and its legal focus on ensuring integral rehabilitation of heritage buildings. The objectives of the study have evolved from the focal complexities of the heritage regulatory framework which downplays the importance of architectural, technical capabilities and building material compositions which contribute immensely to history, culture and identity of building, as an integral part of the urban setting within which the restoration takes place. Within the limits of the study, it was established that the focus of the national regulations on historic architecture and the regional guiding instrument for heritage restorations impedes IHR due to the particular focus on historiographical and anthropological streak, as in its proved in the Los Ríos region. Eventually, there are gaps between the social science procedures, engineering and architectural suppositions with respect to restoration procedures of the fragile building materials. For restoration processes to be integral, preliminary evaluations must include building materials- prioritising the future service life of the building, in response to the integration with the immediate surroundings, original architectonic identity and the ease with which it can be culturally associated. This important step should not be left for individual interpretations during the restoration process. In light of this, seven key points connecting history with the architectural material conditions that could be missing in the Chilean Public service were identified. The harmonization of these points with heritage regulatory framework will improve the integral rehabilitation of heritage buildings in southern Chilean cities.

BIBLIGRAFICAL REFERENCES

Alvarado, D., G. Valdebenito, M. Burgos. (2019). *Evaluación de características dinámicas de los suelos de Valdivia empleando métodos sísmicos de prospección geofísica.* XII Congreso ACHISINA, Valdivia.

Bianca, S. (2010), Historic cities in the 21st century: core values for a globalizing world. In Managing Historic Cities, ed. Ron van Oers and Sachiko Haraguchi (Paris: UNESCO World Heritage Centre, 2010), 28.

CFC-MOP-SERVIU. (2017). *Estudios complementarios de Ingeniería de detalle para Puente Cochrane – Dirección de Vialidad.* Capítulo 10, Fase 2. Caracterización y Anteproyecto Conceptual.

Guarda, G., (1995). La tradición de la madera. Ed. PUC.

Jeri, T. (2015). *Impacto de Declaratorias de Zonas Típicas y Pintorescas en el Desarrollo del Centro Histórico: estudio de la gestión del patrimonio, organización ciudadana y urbanismo en tres barrios de la comuna de Santiago de Chile.* Santiago: Universidad de Chile.

Kliukas, R. R. Kačianauskas, A. Jaras (2008): A monument of historical heritage – Vilnius Archcathedral belfry: the dynamic investigationJournal of Civil Engineering and Management, 14 (2) pp. 139–146.

Maldonado N., P. Martin, et.al. (2019). Historic Mansonry. Online First, IntechOpen DOI: 10.5772.

MINEDUC-CMN (2011). *Ley 12.288 de Monumentos Nacionales y Normas Relacionadas. Consejo de Monumentos Nacionales.* Santiago, Chile.

MICAP, (2017). Bases del Fondo del Patrimonio Cultural 2017, Fondos del Cultura CNCA, at: http://www.fondosdecultura.cl/wp-content/uploads/2017/04/bases-fondo-patrimonio-cultural-2017.pdf

MOP, (2010). *Diagnóstico del Patrimonio Cultural - Programa puesta en valor del patrimonio, Región de los Ríos.* Editores P. Barra, L. Calles, P. Sepúlveda, V. Fierro. Valdivia.

Nyseth, T., J. Sognnaes (2012): Preservation of old towns in Norway: heritage discourses, community processes and the new cultural economy, Cities, 31 pp. 69–75.

Prieto AJ, Ortiz R, Macías-Bernal JM, Chávez M-J and Ortiz P (2019) *Artificial intelligence applied to the preventive conservation of heritage buildings.* 4th TechnoHeritage 2019. Seville, Spain, 26–30 March 2019.

Regional Secretariat for Southern Europe and the Mediterranean of the Organization of World Heritage Cities (SEMO-WHC), (2018): The Guide on Heritage Rehabilitation, Cordoba. Available at www.ovpm.org
Saelzer, G. (2019). *Casa Central: Proyecto de Intervención Patrimonial.* Valdivia, Chile: Universidad Austral de Chile y Ministerio de las Culturas, las Artes y el Patrimonio.
Sirisrisak, T. (2009): Conservation of Bangkok old town, Habitat International, 33, pp. 405–411.
Zumelzu, A., R. Burgos, S. Navarro (2016). Expansión periférica y procesos de centralidad en Valdivia entre 1900 – 2015: Un análisis desde la perspectiva de la sintaxis del espacio. Revista AUS, nº19, p.24–30.

New technologies, products and materials for conservation and maintenance of Cultural Heritage

Science and Digital Technology for Cultural Heritage – Ortiz Calderón et al. (Eds)
© 2020 Taylor & Francis Group, London, ISBN 978-0-367-36368-0

Artificial intelligence applied to the preventive conservation of heritage buildings

A.J. Prieto
Instituto de Arquitectura y Urbanismo – Universidad Austral de Chile, Ernst Kasper, Campus Isla Teja, Valdivia, Chile

R. Ortiz
Departamento Sistemas Físicos, Químicos y Naturales – Universidad Pablo de Olavide, Seville, Spain

J.M. Macías-Bernal
Departamento Construcciones Arquitectónicas II, ETSIE, Universidad de Sevilla, Seville, Spain

M.J. Chávez
Departamento Matemática Aplicada I, ETSIE, Universidad de Sevilla, Seville, Spain

P. Ortiz
Departamento Sistemas Físicos, Químicos y Naturales – Universidad Pablo de Olavide, Seville, Spain

ABSTRACT: Architectural heritage is an important economic and cultural capital of European countries. A monument is more than just the construction itself, being part of the local identity and a source of memory of historical events. The concept of conservation of cultural heritage has evolved over the recent decades at the global level, in order to define multidisciplinary approaches to intervention in buildings, thus leading to their maximum preservation. Assessing the degradation condition of buildings over time is an essential issue to establish the necessary repairs and rehabilitation actions. In this sense, the maintenance of architectural heritage buildings requires methods, strategies and efficient plans. Within this context, and considering the investment required for the maintenance and repair of the built heritage, it is essential to define efficient tools that may be useful to planning an appropriate maintenance strategy for this kind of historical constructions. The aim of this paper is to present an innovative first approach of a new computerised tool (Art-Risk 3.0) for preventive conservation of heritage in urban centres based on models of artificial intelligence. This methodology is able to manage vulnerability, risks and functional service life of buildings in order to contribute to the preservation of built cultural heritage, helping owners, local, regional and national administrations to make decisions about conservation implemented by scientific criteria. The novelty of this challenge lies in its approach and results, a free software to evaluate decisions in regional policies, planning and management of heritage buildings, with a transversal development that includes urban, architectural, cultural heritage value, and the analysis of environmental, natural and socio-demographic situation around the monuments. This new model (Art-Risk 3.0) uses the fuzzy logic theory and the geographic information systems (GIS) for implements the environmental input variables and geological location of buildings located in the peninsular territory of Spain. The information obtained in this study is exceptionally relevant for the researchers and stakeholders responsible for the definition and implementation of maintenance programmes in building stocks. This analysis is extremely important in the implementation of maintenance programs in large building stocks. Art-Risk 3.0 model is one of the main results that will be obtained in the National Research Project (Art-Risk) funded by Ministerio de Economía y Competitividad (Spain) and Fondo Europeo de Desarrollo Regional (FEDER) [code: BIA2015-64878-R (MINECO/FEDER, UE)].

1 INTRODUCTION

The preventive conservation analysis focused on architectural heritage buildings is a novel procedure, in which the innovation should be oriented as a set of prioritised steps: i) restoration as a preservation event; and adding ii) a risks assessment approach over time (Prieto, 2019). Architectural heritage is at risk due to several external affections, e.g.: disasters, conflict, climate change, anthropic risks, and daily environmental affections, among many other factors. In this sense, new recommendations, guidelines and service life methodologies (Silva et al., 2016) should be developed to determine successful safeguard plans over cultural heritage sites in situations of emergency due to flooding, earthquakes, tornados, etc. Preventive strategies assessment varies according to the feasibility of operating on vulnerability and on the surrounding risk context. Thus, prevention encompasses a very mixed set of activities: some of them may concern the building, some others may occur in the surroundings, thus requiring an even more complex decision-making at an upper level, urban or even regional. In this sense, effective programmes are completely essential in order to achieve the preservation of cultural heritage for the present as well as for the future generations. In the context of cultural heritage conservation, policymakers in developed and developing nations are becoming more aware of the importance of identifying the best incentives mechanism or tool for the preservation of their cultural heritage in the historic areas.

2 RESEARCH AIM

The aim of this paper is to present a scheme of the future innovative tool Art-Risk 3, which has been developed within the Art-Risk project. This presentation has been done during the International Congress TechnoHeritage 2019 and a Workshop for its uses. The Ministry of Economy and Competitivity (Spain) and European Regional Development Fund (ERDF) have funded this project. The project presents a multidisciplinary team specialized in the protection and conservation of Heritage, from research institutions in Spain, Portugal, Italy, Chile and Colombia. The new tool developed Art-Risk 3 will be available (open access) to organizations and companies dedicated to the restoration and rehabilitation of heritage, aiding the implementation of scientific criteria and local-regional policies for the planning and management of maintenance actions, thus minimizing the risks of losses of cultural assets. This software has been developed using fuzzy logic, geographical information system (GIS) and Delphi method.

3 MATERIALS AND METHODS

3.1 *Description of the case studies*

Three parish churches of Seville city have been chosen for this initial probe of concept, which are shown in Figure 1. This particular geographical extension presents the climatic conditions: winters generally mild, a warm Mediterranean weather with annual average temperature around 18.0ºC; in summer the temperature can easily exceed 39-40ºC. These heritage constructions were built in the Mudejar-Gothic, Renaissance and Baroque styles between the 13th-18th centuries. Their architectural style was a unique Spanish artistic movement since it was influenced by both Islamic and Gothic Christian elements. These buildings are characterized by: a body of three naves with a timber roof (collar beam in the main nave) of Moorish origin and a vaulted Gothic apse (Figure 1).

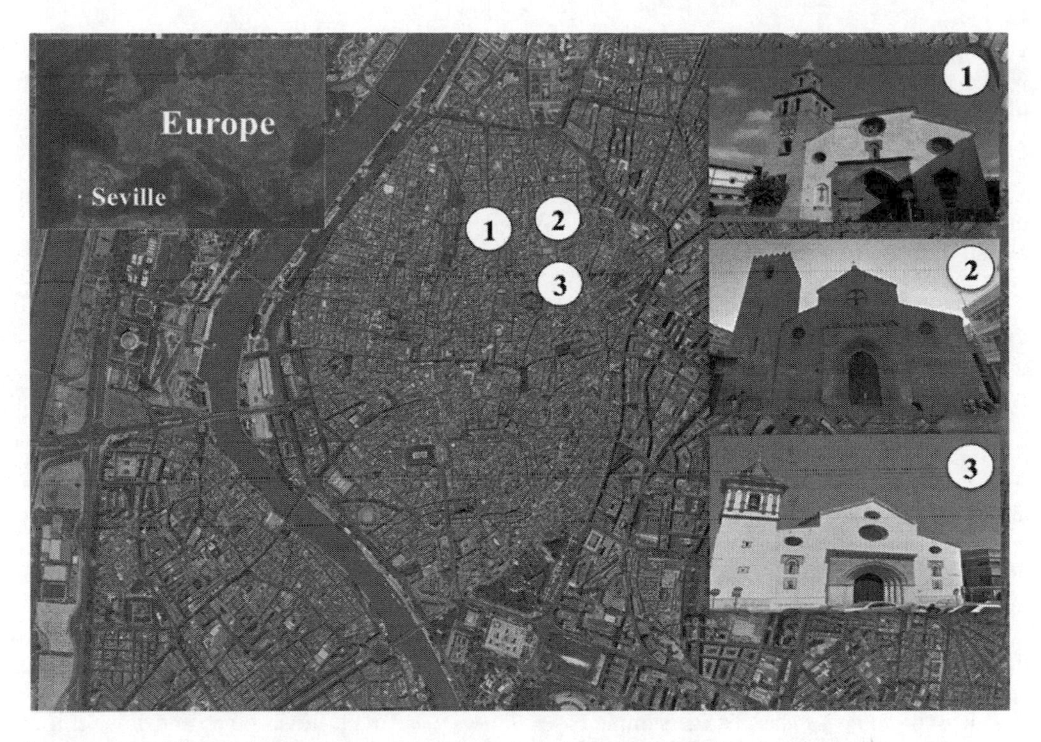

Figure 1. Location of the case studies in the historical city centre of Seville city (Spain)*.

**NOTE: 1. Parish church of Ominum Santorim (SE-OSM); 2. Parish church of Santa Marina (SE-MRN); 3. Parish church of San Román (SE-RMN)*

3.2 *Art-Risk 3.0 methodology*

In the real-life situations, human judgements are vague and cannot be translated to numerical values, since human decision-making are permanently linked with some degree of subjectivity. Fuzzy logic, introduced by L. A. Zadeh (1965), is stated as a powerful tool to be able to approach these kinds of uncertain situations. Fuzzy modelling using real measures of system variables is a tool that tolerates a control of nonlinear systems when there is no prior knowledge of the structure and dynamics system or when this is only partially known.

Performance of the methodology proposed is implemented in open-access software Xfuzzy, which can be executed on platforms containing the Java Runtime Environment. The last version of this system is named Xfuzzy3.5. The tool has been currently renovated (at the end of 2017) to include new functionalities and includes updated documentation of all tools. The Art-Risk 3.0 system designed to include environmental information, which can be provided by experts and users. Additionally, the model can be increased with the users' inputs and could be tested to upgrade in a continuous improvement sequence. This new methodology came from an extensive previous works developed from the University of Seville (Macías-Bernal et al., 2014), from University of Pablo de Olavide (Ortiz et al., 2016) and from Austral University of Chile (Prieto et al., 2019).

In general, fuzzy expert systems are planned in four stages: i) "fuzzification", in which input values, subject to certain imprecision and subjectivity, are represented by fuzzy sets. The Art-Risk 3.0 model is supported by 19 inputs (10 vulnerabilities and 9 risks) (Table 1). The three outputs' variables of the new fuzzy inference system are related with vulnerability, risks and functional service life; ii) knowledge base; iii) "inference" stage, in which fuzzy rules are defined such as e.g. propositional inference rules: IF "the drainage of water in the roof occurs slowly" AND "the constructive system is not completely adequate" THEN "the durability of the roofing system is low").; iv) and "defuzzification phase" phase. The Art-Risk 3.0 methods is supported by the 19 vulnerabilities and external risks variables related to the loss the functional performance of constructions, ranking the conservations actions which should be performed in the homogeneous set of parish churches examined.

4 RESULTS, DISCUSSION AND CONCLUSIONS

A practical application of the new fuzzy methodology (Art-Risk 3.0) was developed. This is a first partial prototype application of the full model. In terms of this initial effort, it was carried out on a total of three heritage buildings (parish churches) located in the historical city centre of Seville. Regarding the output's variables, the system can classify set of heritage buildings with similar constructive characteristics in based on the intrinsic vulnerabilities of the constructions –anthropic vulnerability and maintainability-, the external risk affection –natural, environmental, static-structural-, and the functional service life of the heritage buildings in which vulnerabilities and risk are involved.

In the set of case studies considered, the parish churches present a medium level of functional service life (Table 1 and Figure 1). It is ranged between 55.5% (case study 2 - parish church of Santa Marina) and 58.0% (case study 3 - parish church of San Román). Despite this, it is possible to classify the sample cases. One of the main advantages and novelties of this new system is that it also allows us to know the overall vulnerability of the building (considering the input variables AR1 - AR10) and the global risks (AR11 - AR19) to which the construction or set of constructions are subjected. Concerning, the case study 1 (SE-OSM) presents the greatest risks level (with 60.5%) and the case study 3 (SE-RMN) arranges the lowest external risk affection (46.0%). Considering the vulnerabilities associated to the set of cases studies analysed, case study 3 (SE-RMN) is around (40.0%) and case studies 1 (SE-OSM) and 2 (SE-SMR) are very close with 34.5% and 30.0% correspondingly (Table 1).

Art-Risk 3.0 presents the next innovative characteristics, which make it as a versatile tool for managing architectural heritage: i) it is based on international professional expert survey; ii) it is possible to apply in different environmental, economic, social and cultural context. The system is able to classify sets of heritage buildings (homogeneous constructive features), providing priorities of interventions and considering the buildings' vulnerability, considering the external risks affections and the functional performance as a whole integrated. This practical prototype application of the new model Art-Risk 3.0 helps in understanding how the system works and demonstrating the utility of the methodology in heritage parish churches in Southwest of Spain. More buildings are needed for the validation of the tool.

This kind of approach allows a further step in the development and improvement of efficient and effective new methodologies to support in the conservation of heritage buildings in regional approaches, through the ranking in terms of buildings priorities based on their overall vulnerability, risk and functional service life. These tools help in the development of preventive

Table 1. Practical application of the new system for analysing vulnerability, risk and functional service life in the three parish churches considered.

Inputs and outputs variables	ID	SE-OSM	SE-SMR	SE-RMN
Outputs' model	Vulnerability	30,0	34,5	40,0
	Risk	60,5	54,8	46,0
	Functional Service Life	56,0	55,5	58,0

maintenance programs of architectural heritage. Maintenance activities automatized can minimise the consumption of natural resources, allowing more rational management of future maintenance plans. This work is paramount, since it can be applied by diverse stakeholders (local end-users, communities, managers, architects, etc.) within the construction sector and of course can promote an effective safeguarding approach to heritage. This study can be extended to other kind of buildings and can also be adjusted to different environmental contexts.

ACKNOWLEDGEMENTS

This paper has been supported by and is based on the methodology developed by two Projects: RIVUPH, an Excellence Project of Junta de Andalucía (code HUM-6775), and Art-Risk, a RETOS project of Ministerio de Economía y Competitividad and Fondo Europeo de Desarrollo Regional (FEDER), [code: BIA2015-64878-R (MINECO/FEDER, UE)]. The authors acknowledge the support of the Natural and Anthropogenic Risks Research Center (RiNA) and Instituto de Arquitectura y Urbanismo, Universidad Austral de Chile.

REFERENCES

Macías-Bernal, J. M., Calama-Rodríguez, J. M., & Chávez-de Diego, M. J. 2014. Modelo de predicción de la vida útil de la edificación patrimonial a partir de la lógica difusa. Informes de la Construcción, 66(533), 006.

Ortiz, R., & Ortiz, P. 2016. Vulnerability index: A new approach for preventive conservation of monuments. International Journal of Architectural Heritage, 10(8),1078-1100.

Prieto, A. J., Macías-Bernal, J. M., Chávez, M. J., Alejandre, F. J., & Silva, A. 2019. Impact of Maintenance, Rehabilitation, and Other Interventions on Functionality of Heritage Buildings. Journal of Performance of Constructed Facilities, 33(2), 04019011.

Silva, A., De Brito, J., & Gaspar, P. L. 2016. Methodologies for service life prediction of buildings: with a focus on façade claddings. Springer.

Zadeh, L. A. 1965. Fuzzy sets. Information and control, 8(3),338-353.

Science and Digital Technology for Cultural Heritage – Ortiz Calderón et al. (Eds)
 ISBN 978-0-367-36368-0

The mechanical lamination of paper documents: A new review of the roller press and the hot vacuum press

L.Campins Fernández, T. Espejo Arias & M.R. Blanc García
Universidad de Granada, Granada, Spain

L.Crespo Arcá
Biblioteca Nacional de España, Madrid, Spain

N.Gómez Hernández
Instituto Nacional de Investigación y Tecnología Agraria y Alimentaria, Madrid, Spain

ABSTRACT: Mechanical lamination is a restoration treatment for deteriorated documents. It is considered an invasive procedure and it is barely recommended, despite this, and the fact that the machines used are outdated, it is still used in the restoration field today. The aim of this project is to broaden our knowledge on the effects of this treatment on the paper documents and artworks and reflect on its application, thus providing an updated review, which allows us to reflect on its suitability. The research begins with a literature review and continues with an experimental phase, based on the analysis of the physical and mechanical properties of four kinds of paper documents reinforced with Archibond tissue applied with both, roller press and hot vacuum press. In orden to compare the properties and establish the conclusions, the papers were analyzed before and after their lamination and artificial aging.

1 INTRODUCTION

This study is part of the research developed in recent years by the University of Granada focused on the study and evaluation of restoration treatments applied to graphic and documentary work. Mechanical lamination was, during a great part of the 20th century, the main restoration treatment used in libraries and archives throughout the world (McGath et al. 2015), and is still used nowadays. The aim of this study is to review and provide and updated approach to the treatment.

Mechanical lamination is based on the adhesion of an adhesive film or an adhesive plus tissue reinforcement, on one or both sides of a document, which provides the consistency necessary for its safe handling (Tacón Clavaín 2009). This is achieved through the use of synthetic thermoplastic adhesives, and the application of heat and pressure. The use of specific machinery allows the treatment to be carried out quickly, easily and on a large scale, but nowadays many authors warn about its disadvantages on the original support. It is claimed to induce physical, mechanical, chemical and aesthetic modifications on the supports (AIC. American Institute for Conservation n.d.), so it should only be applied in case of extreme necessity (Tacón Clavaín 2009). Finally, its widely spread, but not necessarily proved inconvenients have led to the disuse of the treatment and the advantages of it to be forgotten. Nowadays, manual lamination are recommended over mechanical lamination however, if there was a need to apply mechanical lamination, experts recommend the use of Archibond tissue along with the hot vaccum press (Tacón Clavaín 2009); nevertheless, roller presses are still conserved and used in some Spanish institutions.

1.1 *Historical notes*

Mechanical lamination was developed in the United States of America in the 1930s. The first mechanical laminations were done using steam heated hydraulic presses (Kathpalia 1973); in 1937, W.J. Barrow proposed a new system of mechanical lamination using a roller press machine which substituted the previous method and became the most popularly used treatment up till the 1980s. The roller press machine considerably reduced treatment time, improved adhesion and avoided the formation of air bubbles (Howard 1949). Indistinctly of which method was chosen, cellulose acetate was the most widespread adhesive film used either with or without additional reinforcement of tissue (McGath et al. 2015). In Europe, Polyethylene (Crespo and Viñas 1985), and to a lesser extent, Cellophane (Smith 1938), PVC and different polyacrylates (Feindt et al. 1981) were also used as adhesives. In Spain, from 1980 onwards, the hot vacuum press has been used as it supposedly reduces some of the negative effects of lamination such as flattening (Tacón Clavaín 2009). Regarding the use of adhesives, acrylic or vinylic resins are the preferred choice, and can be purchased in the form of a film or as a pre-laminated tissue (Muñoz Viñas 2010) where Archibond tissue is the most popularly known.

2 MATERIALS AND METHODS

Based on the literature review, an experimental methodology has been proposed. The materials and methods used are shown in Table 1. Four types of primary supports have been chosen: a sample paper (W), a 19th century artisan paper manuscript with iron gall inks (L) and two types of newsprint paper, one composed of chemical pulp (R1) and another of a mixture of chemical and mechanical pulp (R2), both dating from the beginning of the 20th century. Archibond tissue has been chosen as reinforcement and adhesive; this prelaminated tissue is made up of a 8,5-9,3 g/m^2 manila hemp tissue coated with a stable paraloid –polymethyl methacrylate and ethyl methacrylate– adhesive film which has a melting temperature between 70-80°C (Polyevart n.d.).

A Coinsa roller press and a Drytac HPG 220 hot vacuum press were used. The programmed temperature parameters were, in the roller press 95°C and in the hot vacuum press 110-115°C. The treatment time in the roller press was approximately 1 minute, at a pressure of 245-392 N/cm^2. For the hot vacuum press, due to the lack of unified criteria, two sample groups were prepared, the first was introduced in the machine for a total of 3 minutes whilst the second for a total of 5 minutes; in both cases the pressure used was -7.45 N/cm^2. Each paper sample was laminated on one of its flat surfaces with each of the three proposed methods. For every paper type, a non-laminated control sample was left for comparison.

Regarding the analysis method for the characterization of the control samples, fiber composition was analyzed in accordance to the UNE 57-021-92 standard using a Leitz optical microscope.

Next, all samples were analyzed using non-destructive methods of analysis: photographs of the surface of the samples were taken with a 4x magnification, with overhead and parallel

Table 1. Materials and methods.

Lamination presses*	Supports*	Methods of analysis	
Roller (R)	Watman (W)	Fiber composition	Grammage
Hot vacuum 3 min. (V-3)	Artisan paper (L)	Microscpic photography	Gurley permeability
Hot vacuum 5 min. (V-5)	Newsprint paper 1 (R1)	Optical properties	Tensile strength
	Newsprint paper 2 (R2)	Thickness	pH
	Archibond tissue (AT)		Artificial aging

*Next to lamination methods and materials their respective codes are shown.

light using an Olympus SZH10 Color View III with Cell-D 3.4 software. Determination of color coordinates in the CIELab space under C/2o lighting conditions (UNE-ISO 5631-1:2015), was done using a Lorentzen Wettre ELREPHO 070 spectrophotometer, and the differences in color (ΔE*) were calculated from the results. The thickness and grammage were obtained according to the UNE-EN ISO 534: 2012 and UNE-EN ISO 536: 2013 standards, for which a digital micrometer Mesmer-Buchel M372 and a precision balance Sauter K1200 were respectively used. For the measurement of Gurley permeability, the method proposed in the UNE-ISO 5636-5:2015 standard was followed using a Gurley porosimeter model 4190.

Destructive tests were then carried out. For the determination of tensile strength, tests were conducted in both directions of the fiber orientation of the primary support; the UNE-EN ISO 1924-2:2008 standard was modified, so that the prestressing of the sample was eliminated and the load was reduced to 100N. The pH of the samples was measured both at the surface and in aqueous extract according to the T529 om-88 and UNE-ISO 6588-1:2013 standards respectively, using in both cases a CRISON GLP22+ pHmeter. In aqueous extract, 0.5g of sample was used instead of the 2g stated by the standard norm, but the water-to-sample ratio was kept the same.

Finally, all samples were artificial aged in an ageing process lasting 144 hours which was carried out in an illuminated room at a temperature of 80°C and 60% relative humidity. After that, the non-destructive and destructive tests previously described were repeated.

3 RESULTS AND DISCUSSION

The results of the analysis of fibrous composition agrees with what was expected for the artisanal paper, made up of cotton, and for the newprint papers, composed of chemical and/or mechanical wood pulps; on the other hand, the Watman paper presents a much lower percentage of cotton than expected and contains a noticeable amount of chemical wood pulp.

The photographs show that the aesthetic properties are only modified on the laminated face; the samples treated with the roller press are more satin, while those treated with the hot vacuum press have a matte finish and present accumulations of adhesive, in these the veiling of the inks due to the tissue is also greater. The measurement of the optical properties shows an important color variation (Table 2) of the samples after ageing. The Archibond tissue, when measured independently, shows ΔE* > 2,5 and a significant reduction in brightness from 59.62% to 12.60% on the face containing the adhesive film; this suggests that a loss or migration of components occurred which indicate that Archibond tissue is in fact unstable.

One of the main evils attributed to mechanical lamination is the flattening of the primary support. Contrarily to these beliefs, the results of this study show that the thickness increases in all cases, although the increase is generally lower in the samples treated with the roller press (Figure 1). Likewise, larger reductions in paper thickness occur in more voluminous/porous papers such as Watman and the artisan paper, whereas flattening is not significant in industrial papers which have already been put through a calender. Regarding grammage, the combination of its results with those of thickness and artificial ageing, indicate that a hygroscopic modification occurs.

Table 2. ΔE* Variations.

Sample	ΔE*	ΔE*a	Sample	ΔE*	ΔE*a	Sample	ΔE*	ΔE*a	Sample	ΔE*	ΔE*a
W		3,71	R1		5,43	L		4,14	R2		4,05
RW	1,43	6,91	RR1	3,98	9,73	RL	1,20	6,89	RR2	1,57	5,06
VW-3	1,38	7,70	VR1-3	2,84	7,57	VL-3	3,03	6,02	VR2-3	1,22	5,69
VW-5	1,31	7,40	VR1-5	2,63	8,01	VL-5	2,71	6,10	VR2-5	1,30	5,32

Artificial aged (ΔE*a). ΔE* to de control sample, ΔE*a to the sample before its artificial aging.

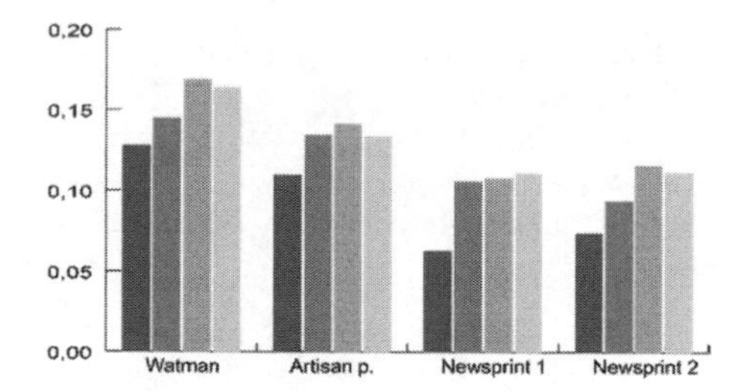

Figure 1. Increase of thickness.

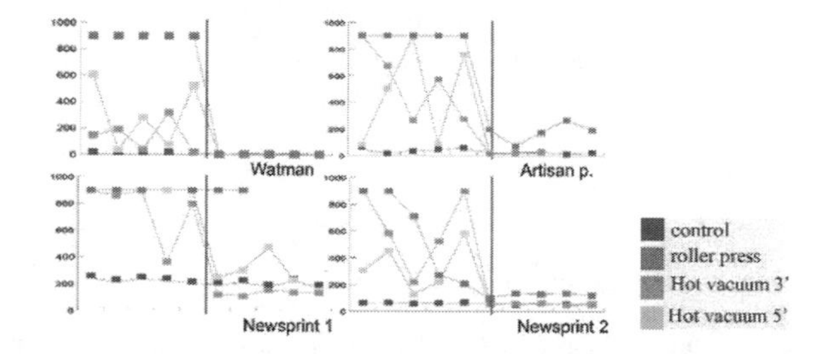

Figure 2. Gurley permeability before (left) and after (right) the artificial aging.

From the Gurley permeability results it is shown that Archibond tissue has an irregular distribution of the adhesive, resulting in laminations with accumulations of the adhesive, which cause an uneven reduction in the permeability of the primary support (Figure 2); this effect is minimized in the roller press. The drastic increase of the permeability of the laminated samples suggest, yet again, that a loss of the components making up the Archibond tissue occurs, which in turn induces the separation of the supports and in consequence the lamination is rendered useless.

The tensile strength results show that the lamination fulfills its main function of reinforcing the paper document independently of the method of application used. Also, according to the results of pH analysis, neither the use of high temperatures during treatment nor the use of a synthetic adhesive, cause or favor the acidification of the document.

4 CONCLUSIONS

Mechanical lamination with both laminators is a safe, fast and efficient method of treatment for document reinforcement and is appropriate when aqueous treatments are not possible. During the lamination of samples, we were able to see firsthand, that when using the roller press, it is not possible to control the parameters in a precise manner because of the antiquity of the machine.

With the proposed methodology we cannot determine whether one method is better than another. Both have advantages and disadvantages that must be assessed together with the nature and characteristics of the primary support. In general, the treatment with hot vacuum press is more suitable for matt and voluminous supports, although the veiling of the graph should be taken into consideration. On the other hand, the treatment with the roller press does not present disadvantages for bright, glossy and compact papers like those of newsprint. Additionally, we must consider that, with the materials used in this study, the variations in color have been greater than those desired in restoration treatments. Equally as important as choosing the method of treatment is the correct selection of materials to use. Variety of materials is scarce which often leads to the restorator choosing the wrong one. In this study,

Archibond tissue has proved to not be a suitable material for mechanical lamination due to the irregularity and the loss of its compounds.

Lastly, it is important to mention that the results of this study suggest that the hygroscopic behavior of the papers after treatment is modified, a phenomenon which has not been described previously and which requires further study. We also consider it is convenient to continue investigating the effects of temperature on the original support and to improve our understanding of the materials -adhesives and reinforcements- available for use in lamination.

ACKNOWLEDGEMENTS

The authors wish to express their gratitude to the project MAT2014-58650-P of the Universidad de Granada, and to the institutions that have participated in the study: the Biblioteca Nacional de España y el Instituto Nacional de Investigación y Tecnología Agraria y Alimentaria.

REFERENCES

AIC. American Institute for Conservation. n.d. "Lining - Wiki." Accessed February 26, 2019. http://www.conservation-wiki.com/wiki/Lining.

Crespo, Carmen, and Vicente Viñas. 1985. "La Preservación y Restauración de Documentos y Libros En Papel." Programa General de Información y UNISIST. Programa de Gestión de Documentos y Archivos (RAMP). París. http://unesdoc.unesco.org/images/0006/000635/063519so.pdf.

Feindt, W., W. Griebenow, A. Kallmann, and B. Werthmann. 1981. "La Laminación Con Polietileno," 388–94.

Howard, Nixon. 1949. "Lamination of Paper Documents with Cellulose Acetate Foil." ProQuest.

Kathpalia, Yash Pal. 1973. *Conservation and Restoration of Archive Materials. UNESCO*. París.

McGath, Molly, Sonja Jordan-Mowery, Mark Pollei, Steven Heslip, and John Baty. 2015. "Cellulose Acetate Lamination: A Literature Review and Survey of Paper-Based Collections in the United States." *Restaurator* 36 (4): 333–65. https://doi.org/10.1515/res-2015-0015.

Muñoz Viñas, Salvador. 2010. *La Restauración Del Papel*. Madrid: Tecnos.

Polyevart. n.d. "Archibond ® Sin Soporte." Accessed April 16, 2016. http://www.mecd.gob.es/cultura-mecd/areas-cultura/patrimonio/mc/polyevart/presentacion.html.

Smith, L. Herman. 1938. "Manuscript Repair in European Archives." *American Archivist* 1 (2): 51–77.

Tacón Clavaín, Javier. 2009. *La Restauración En Libros y Documentos : Técnicas de Intervención*. Ollero y Ramos.

Science and Digital Technology for Cultural Heritage – Ortiz Calderón et al. (Eds)
© 2020 Taylor & Francis Group, London, ISBN 978-0-367-36368-0

Morphology virtual reconstruction Reinaissance altarpiece from Antequera (Andalusian)

B. Prado-Campos
Departamento de Pintura, Facultad de Bellas Artes, Universidad de Sevilla, C/Laraña, Seville, Spain

A.J. Sánchez Fernández
Conservator-restorer, independent researcher, Cádiz, Spain

ABSTRACT: The Renaissance altarpiece of San Juan Bautista coming from Antequera (Málaga, Andalucia) (Fernández Rodríguez, 1943). It´s was located in the Church of San Zoilo. But for several reasons, it was dismantled and today it is dispersed in serveral locations and fragmented into pieces. The Movable Property Restauration Service of Antequera City Council is carrying out the Reclamation Project of the altarpice with the aim of retourning it to its original environment.

This work presents a virtual reconstruction of the altarpiece morphology. It is based on the search, location and documentation of existing altarpiece fragments. The main aim has been the morphology recovery and material reconstitution by virtual documentation. This initiative contributes to the recovery of the memory of the cultural heritage, helping the conservation-restauration project.

1 INTRODUCTION

The study main purpose of the altarpiece of San Juan Bautista was to document the existing parts to make a proposal to recover their original morphology, identifying possible missing fragments and the state of conservation of it. The physical materiality of the work, the dispersion and decontextualization of the parts made this objective difficult in a practical way. The reasons to dismantle and disperse to it was change of taste, interested to expose four tables in the museum and the deficient state of conservation the wall in the church location.

For this reason, virtual reconstruction and 3D modeling have been used. This work to contribute to the dissemination and knowledge of this cultural heritage, while applying a methodology for digital documentation with virtual reconstruction of altarpieces.

2 DESCRIPTION

The altarpiece is composed of structural elements (pilasters, columns, entablatures, niche and pediment) and ornamental elements (tables, sculpture and fantasy figures), a set of 14 pieces. It has 12 original tables, a table added painted black and another that has been lost (in the centre of the predella). There are 16 iconographic scenes and a sculpture. When it was dismantled, it was fragmented into different blocks: a single fragment for the predella, 4 tables, a sculpture, a niche, 6 columns and 2 entablatures for the first horizontal register. The second horizontal register is composed of a table and 2 fragments (3 tables, 1 square and 1 entablature in each fragment), and the attic is composed of a pediment and 2 fantasy figures.

3 METHODOLOGY

3.1 *Instruments*

A Sony D-50 digital camera (6.3 Mpixels with a maximum resolution of 3000 x 2000 pixels) with a Sony 18-55 mm lens were used for photographic documentation. Software: 123D Catch Autodesk for the 3D model; Blender for the 3D model. Hardware: Intel Core i7 computer, 2.8 GHz with 4GB of RAM, 750 GB hard drive and NVIDIA GeForce 720M graphics card.

3.2 *Photo shoots*

It has been done in the following steps: planning the photographs, Base/Height ratio range 1: 4, 60% overlap between images, not having lights and shadows too contrasted, the same camera setting was established and taking pictures from different angles and different elevations.

3.3 *Model creation*

The software use is 123 Autodesk Catch. Definition of reference points to scale them (introducing real measurements) and position them in space. And the software has tools for trimming the background.

Figure 1. Front and back virtual reconstruction. Right, altarpiece historical photography. Front and back planimetries. Right, layout of tables.

3.4 *Model editing*

The software use is blender. Virtual reintegration of lost volumes, creation of several light sources to avoid large contrasts and shadows, "orthographic view" option to obtain the orthophoto (Charquero Ballester, 2016), association of texture to volume for a photorealistic model and the quality of the final orthophoto is defined by two parameters: the resolution and the samplers.

4 DISCUSSION

The technology has advantages and disadvantages. The advantages are: it is a non-destructive technique, sensitive to testing working hypotheses and generating morphological data. We get an orthophotograph without perspective distortion. It is a low cost procedure in terms of time and budget and provides images and cartographies that support the intervention process documentation.

The disadvantages are: control of the stability of certain pieces to position correctly and be able to be photographed. The similarity of texture and color between the backgrounds and the piece to be recorded produced errors. As a solution, it was decided to arrange panels as backgrounds and include geometric reference elements. Patches without information appeared while obtaining the 3D model. To fill in them, it was necessary to shot pictures from different heights. Finally, difficulty managing large files in the editing software. As a solution, a "decimate" filter has been applied that reduces the polygons of the mesh

5 CONCLUSION

The parts of the altarpiece were in different rooms. This fact hindered 3D modeling since the parameters (light, distances, etc.) were not the same. However, the original configuration of the altarpiece of San Juan Bautista has been obtained, contextualizing all its parts. Two general orthophotos have been created (front and back). The rear view gives us an unusual picture that it will allow us to study the constructive system and the material history.

ACKNOWLEDGMENT

We want to thank the City Council of Antequera, the Culture and Heritage Area, City of Antequera Museum, Movable Property Restoration Center of the City Council of Antequera, and especially the conservator-restorer Dª Marisa Olmedo. She has given us the opportunity to know, document and ultimately "enjoy" a so relevant work.

REFERENCES

Fernández Rodríguez, J. M. 1943. *Las iglesias de Antequera*. Antequera: Publicaciones del centro de estudios andaluces.

Charquero Ballester, 2016. Práctica y usos de la fotogrametría digital en arqueología. Documentos de Arqueología y Patrimonio Histórico. *Revista del Máster Universitario en Arqueología Profesional y Gestión integral del Patrimonio, 1: 139–157.*

'Recomendaciones técnicas para la documentación geométrica de entidades patrimoniales', in IAPH http://www.iaph.es/export/sites/default/galerias/patrimonio-cultural/documentos/gestion-informacion/recomendaciones_tecnicas_documentacipm_geometrica.pdf (access 2018–06–05).

Mora García, R. T.; Céspedes López, Mª F.; Louis Cereceda, M. 2010. Fotogrametría y nube de puntos aplicado en la documentación del patrimonio construido. El caso de la torre de la Calahorra en Elche, in *X Congreso Internacional Expresión Gráfica aplicada a la Edificación*. Alicante: APEGA. Pp. 379–387.

Science and Digital Technology for Cultural Heritage – Ortiz Calderón et al. (Eds)
 ISBN 978-0-367-36368-0

Characterization of glass *tesserae* from the Mosque of Cordoba

A. Gomez-Moron
Andalusian Historical Heritage Institute (IAPH), Seville, Spain
Department of Physical, Chemical and Natural Systems, Pablo de Olavide University, Seville, Spain

N. Schibille
Centre National de la Recherche Scientifique/Université d'Orléans, Orléans, France

L.C. Alves
C2TN, IST/CTN, University of Lisbon, Lisbon, Portugal

A. Bouzas Abad
Andalusian Historical Heritage Institute (IAPH), Seville, Spain

P. Ortiz
Department of Physical, Chemical and Natural Systems, Pablo de Olavide University, Seville, Spain

M. Vilarigues
Research group VICARTE and Department of Conservation and Restoration, FCT-UNL, Campus de Caparica, Almada, Portugal

T. Palomar
Research group VICARTE, FCT-UNL, Campus de Caparica, Almada, Portugal

ABSTRACT: The *Maqsura* and the *Mihrab* are one of the most important locations of the Mosque of Cordoba (Spain), built during the expansion of the Mosque by the caliph Al-Hakam II (962- 965 AD). The main goal of this study was to identify the glass *tesserae* from the different historical restorations. The mosaics from the *Mihrab*, the *Sabat* chamber and ceiling mosaics presented different types of *tesserae* spread along the two doors and the ceiling attributed to the different campaigns. However, the mosaics from the *Bab Bayt al-Mal* chamber presented different glass compositions because it was completely removed and an exact replica was installed in 1916.

The chemical analysis detected a glass with high contents of PbO, mainly in the replacement *tesserae,* and in the green and yellow original *tesserae*. The original *tesserae* were classified into five different groups: *tesserae* with gold leaf, with inclusions, with bubbles, homogeneous opaque *tesserae* and red opaque *tesserae*.

1 INTRODUCTION

Built in 786 AD, the Great Mosque of Cordoba was significantly expanded during the caliphate of Al-Hakam II (962-965 AD). It was during this campaign that the three doors of the *Maqsura* and the *Mihrab* ceiling were decorated with glass *tesserae*. Ancient texts claim that a master mosaicist and the original *tesserae* came from the Byzantine Empire as a present from the Emperor Nikephoros II Phokas (Nieto Cumplido, 2007). In 1236, King Ferdinand III of Castile conquered Cordoba and the bishop of Osma consecrated the Mosque as the Cathedral of the Assumption of the Virgin Mary. In 1368, an altarpiece triptych was placed in

Table 1. Interventions in the *Maqsura* mosaics.

Mosaic	Interventions
Bab Bayt al-Mal door	– 1912-16 Ricardo Velázquez Bosco (J&M Maumejean Férres (Madrid)). A replica was made and installed in the door.
Mirhab door	– 1368 San Peter Chapel Altarpiece/in 1771-72 unmounted.
	– 1772 Balthasar Devreton consolidated the mosaics in the dome and employed repaintings on mortar for the mosaic intervention.
	– 1815 Patricio Furriel used coloured glasses+repainting and glasses+painted mortar for the mosaic intervention.
Sabat door	– 1815 Patricio Furriel restored the mosaics.

front of the *Mihrab* to hide it, while the mosaic on the *Bab Bayt al-Mal* chamber was covered with a canvas.

The *Sabat* chamber has since been used as the archive of the Cathedral. The removal of the triptych in the 18th century, and canvas at the beginning of the 20th century, revealed that the glass mosaics were severely damaged. The Cathedral thus ordered their comprehensive restoration. At least three different restoration campaigns are known from textual sources: in 1778 Devreton consolidated the mosaics in the dome, in 1815 P. Furriel restored the mosaics, and finally, R. Velázquez Bosco replaced the *Bab Bayt al-Mal* mosaics in 1912-1916 (Table 1).

In previous studies on glass tesserae were identified raw materials (Schibille *et al.*, 2012), glassmaking technique (Neri *et al.*, 2017) and weathering processes (Palomar *et al.*, 2011). The published studies of Islamic mosaics from relevant sites are limited (Moropoulou *et al.*, 2016) and this work will help to provide a large enough data set to allow a significant knowledge.

The aim of this work is to provide information about chemical composition of glass tesserae from Maqsura mosaics in order to distinguish original from historical interventions. This knowledge will be necessary to make a future restoration and conservation strategies of the Maqsura with guarantee.

2 MATERIALS AND METHODOLOGY

A total of 90 *tesserae* from *Maqsura* were analyzed: 14 samples from the *Bab Bayt al-Mal* door, 46 from the *Mihrab* door and dome, and 30 from the *Sabat* door. The samples were observed on a polished cross-section.

The characterization techniques used to study the samples have been the following: optical microscope studies were undertaken with a Leica model DM4000, scanning electron microscope observations were carried out with a Jeol JSM-5600LV microscope model with an energy dispersive X-ray spectrometer EDX INCA x-Sight Oxford model, the study of chromophores was identified by reflection spectrometer with fiber optic MAYA 200 PRO Oceans optics model and μ–Raman spectrometer LabRAM Horiba 300 Jobin Yvon model with Olympus microscope (785 nm laser) and, finally, the chemical analysis were undertaken with Laser Ablation Inductively Coupled Plasma Mass Spectrometer Thermofisher Element XR combined with a Resonetic UV laser microprobe (ArF 193 nm).

3 RESULTS

3.1 *Morphological characterization*

The morphological characterization (optical microscope and scanning electron microscope) allowed to distinguish between original *tesserae* and two different historical interventions (Davison, 2003). The original glass *tesserae* were classified according to their appearance into

Figure 1. Different original and intervention *tesserae* in cross-section.

five different groups: *tesserae* with gold leaf, with inclusions, blue *tesserae* with bubbles, homogeneous opaque *tesserae* and red opaque *tesserae* (Figure 1).

At least two different intervention *tesserae* could be identified, transparent glass *tesserae* with a painted layer on their surface and a group of glass *tesserae* with small and homogeneous inclusions (Figure 1).

3.2 *Base glass characteristics*

Multivariate statistical analysis (PCA) of ten base glass elements (B_2O_3, Na_2O, MgO, Al_2O_3, P_2O_5, Cl, K_2O, CaO, TiO_2, SrO) distinguished numerous base glass characteristics and clearly singles out the twentieth-century restorations, as well as, the high lead glasses (Figure 2). The twentieth-century samples from the *Bab Bayt al-Mal* door together with some individual samples from the *Mihrab* and *Sabat* doors were tightly clustered in the principal component space of PC1 and PC2 that account for over 60% of the total variance. They are characterized by high soda, very low chlorine and silica-related impurities that suggest the use of synthetic raw materials. The high lead glasses exhibit relatively high chlorine typical of lead glass from the Iberian Peninsula. The majority of samples show moderate levels of silica-related elements such as aluminum and titanium oxides, while the alkali contents (B_2O_3, Na_2O, MgO, K_2O) as well as CaO and SrO vary significantly, indicative of different fluxing agents.

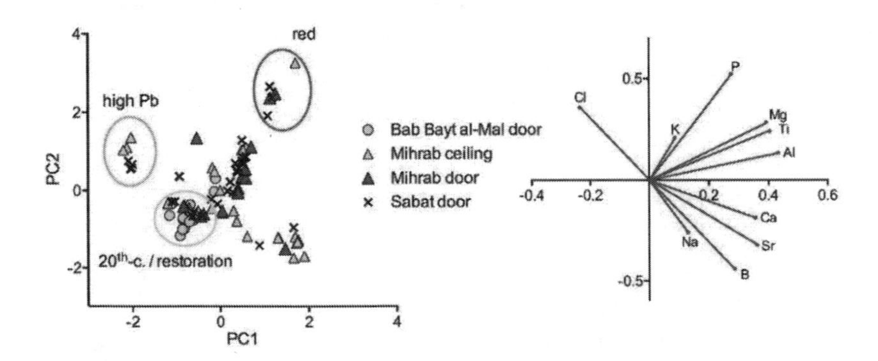

Figure 2. Multivariate statistical analysis of the base glass components.

Table 2. Chromophores of the glass *tesserae*.

Color	$Fe^{2+/3+}$	Co^{2+}	Cu^{2+}	Mn^{3+}	$Cu_2O+Cu°$
Yellow	✓				
Light grey		✓	✓		
Dark grey	✓	✓			
Grayish blue	✓	✓			
Light blue		✓	✓		
Blue		✓			
Light green			✓		
Turquoise			✓		
Mauve				✓	
Red					✓

Table 3. Opacifiers of the glass *tesserae*.

Location	Bab Bayt al-Mal door (20th century)	Mihrab and Sabat door	Mihrab and Sabat door (possible restorations)
$KAlSi_3O_8$	✓		
$Pb_2Sb_2O_7$	✓		
$Ca_2Sb_2O_7$	✓	✓	
SiO_2	✓	✓	✓
Fe_2O_3	✓		
Alkaline sulfate		✓	
SnO_2		✓	✓
$PbSn_{1-x}Si_xO_3$		✓	✓
$CaSiO_3$		✓	
$Mn^{2+}Mn^{3+}{}_6O_8(SiO_4)$		✓	

3.3 *Chromophores and opacifiers*

The most common chromophore found in the glasses was iron, which is an impurity of the glassmaking sand. Blue and grey *tesserae* contained cobalt with or without copper ions at different concentration. Mn^{3+} was detected in the purplish and burgundy samples, and nanoparticles of $Cu_2O+Cu°$ were observed in the red *tesserae* (Table 2).

The *Mihrab* and *Sabat* doors' opacifiers were SiO_2 (mainly quartz), $Ca_2Sb_2O_7$ and alkaline sulfate, the yellow and green *tesserae* showed $PbSn_{1-x}Si_xO_3$ crystals (Table 3). The twentieth-century *tesserae* from the *Bab Bayt al-Mal* door contained mainly $Pb_2Sb_2O_7$ and $Ca_2Sb_2O_7$, independently of the glass color (Table 3).

CONCLUSIONS

Based on the chemical composition and morphology we can distinguish between original *tesserae* and two different historical restorations. High contents of PbO were identified in the replacement *tesserae* from Maumejean (20th cent.), as well as the yellow and green original *tesserae*. The original *tesserae* were classified into five different groups: *tesserae* with gold leaf, blue with bubbles, high lead *tesserae* (green and yellow), homogeneous opaque *tesserae* without lead and red opaque *tesserae*. Two different restorations could be identified: painted layer on transparent glass *tesserae* from Patricio Furriel´s intervention (1815) and colored glass *tesserae* with inclusions and medium contents of lead from the intervention of Velázquez Bosco (1912, Maumejean glasses).

ACKNOWLEDGEMENTS

This work has been supported by the Project RIVUPH (HUM-6775), the Spanish Ministry of Economy and Competitiveness and the European Regional Development Fund (ref. BIA2015-64878-R), Art-Risk (code: BIA2015-64878-R (MINECO/FEDER, UE)), Portuguese Foundation for Science and Technology (Project UID/EAT/00729/2013) and the European Research Council (ERC) (grant agreement 647315 to NS).

REFERENCES

Davison, S. 2003. *Conservation and Restoration of Glass*. second edi. Edited by Butterworth-Heinemann. Oxford.

Moropoulou, A. *et al.* 2016. 'Analytical and technological examination of glass tesserae from Hagia Sophia', *Microchemical Journal*, 125, pp. 170–184. doi: https://doi.org/10.1016/j.microc.2015.11.020.

Neri, E. *et al.* 2017. 'Analyses of glass tesserae from Kilise Tepe: New insights into an early Byzantine production technology', *Journal of Archaeological Science: Reports*. Elsevier Ltd, 11, pp. 600–612. doi: 10.1016/j.jasrep.2016.12.036.

Nieto Cumplido, M. 2007. *La Catedral de Córdoba*. 2ª ed. Córdoba: Obra Social y Cultural CajaSur.

Palomar, T. *et al.* 2011. 'Patologías y estudio analítico de materiales procedentes de mosaicos de Carmona e Itálica', *Materiales de Construcción*, 61(304), pp. 629–636. doi: 10.3989/mc.2011.64310.

Schibille, N. *et al.* 2012. 'Chemical characterisation of glass mosaic tesserae from sixth-century Sagalassos (south-west Turkey): chronology and production techniques', *Journal of Archaeological Science*. Elsevier Ltd, 39(5), pp. 1480–1492. doi: 10.1016/j.jas.2012.01.020.

Science and Digital Technology for Cultural Heritage – Ortiz Calderón et al. (Eds)
 ISBN 978-0-367-36368-0

The Mudejar tower of the church "Sta. Mª de la Encarnación", Constantina (Seville). State of conservation and materials characterization

F.J. Alejandre Sánchez, F.J. Blasco López, V. Flores Alés, J.J. Martín del Río &
M. Torres González
Departamento de Construcciones Arquitectónicas II, Universidad de Sevilla, Seville, Spain

A.J. Correa
Dos studio, Tomares, Seville, Spain

ABSTRACT: The aim of this work is to describe the state of conservation of the tower of the church Sta. Mª de la Encarnación, Constantina (Seville), the characterization of its bricks, masonry and rendering mortars and establish the origin of their decay. The methodology consisted in determining the chemical composition by XRF, the contents of soluble salts and chlorides by ICP and ionic chromatography, the mineralogical composition by XRD, the study of physic-mechanical properties and finally the dating by ^{14}C of the rendering mortar. As a result of all the analyses developed it was determined that the main causes of decay were the heterogeneity of the quality of artisanal bricks and the presence of expansive efflorescence that damaged the structure through crystallization/dissolution cycles.

1 INTRODUCTION

The Parish Church of Sta. Mª de la Encarnación, whose architects were Martín de Gainza, Pedro Diaz and Vermondo Resta, was built in the 16th century and its tower was designed in 1569 by Hernán Ruiz II. This building was declared an Asset of Cultural Interest with an A degree of protection in 1983. The evident degradation of the materials from the north and east facades of the tower allowed to establish as hypothesis the existence of intrinsic and/or extrinsic (environmental) alteration factors. The intervention proposal for the restoration of the tower-facade of the church was presented in 2016. This proposal included previous studies of characterization through instrumental analytical techniques and physical-mechanical tests with the purpose of establishing the origin of the observed deterioration and proposing the most adequate treatments and compatible replacement materials. The programmed and developed actions during the intervention were focused on the cleaning, substitution and protection of the stone, mortars and bricks considering the importance of the results obtained from previous studies (Correa, 2016).

2 MATERIALS AND METHODOLOGY

Brick and mortar samples belonging to the church tower were studied to analyse their chemical, mineralogical, physical and mechanical properties. We also studied brick and mortar samples used in the intervention as replacement materials in order to evaluate the possible presence of anomalies and their compatibility with the original ones. In total 6 samples were analysed: unaltered brick and mortar (B1&M1), altered brick and mortar (B2&M2), replacement bricks and mortar (B3&M3). Major and minor chemical components and the trace elements were analysed by X–ray fluorescence (XRF) using a Panalitycal-Axios spectrometer. The

Figure 1. The Mudejar tower of Sta. Mª de la Encarnación church [(1)] Degradation in the North façade [(2)] B1 sample extraction [(3)] B2&M2 samples extraction [(4)].

content of soluble salts was studied according to UNE-EN 772-5:16 by preparing a suspension of 20gr of sample in 250ml of distilled water and was analysed by an atomic emission ICP (SPECTRO SectroBLUE). The content of chlorides was also studied by ionic chromatography (Metrohm 930 Compact ICFlexthe). The mineralogical composition was carried out by XRD (Bruker D8I 90) by means of the powder technique. Finally, the open porosity and apparent density were calculated through the vacuum method according to EN-1936:07 and compressive strength in bricks and mortars were determined according to UNE-EN 772-1:11

3 ANALYSIS OF RESULTS

3.1 *X-ray fluorescence*

The chemical composition is presented in Table 1. High SiO_2 contents assignable to the presence of quartz to decrease the plasticity of the clays, can be observed in the bricks. Phyllosilicates no decomposed and neoformation silicates were also detected. The raw material hardly contained carbonates due to the low content of CaO and MgO. The difference between the chemical composition of the original (B1-B2) and replacement brick (B3) was the higher content of S and Cl associated with the presence of soluble salts in the original samples.

The original mortars (M1-M2) had high contents of SiO_2 assignable to the presence of quartz and silicates from the sand (feldspars, micas and clays). The content of CaO and the loss on ignition at 900ºC can be attributed mainly to the calcium carbonate -$CaCO_3$- (See section 3.3) and the stoichiometric calculations from CaO indicated 51% (M1) and 44% (M2).

Comparing with reference lime mortars (Martín, 2018), it could be established that the maximum original lime dosage was between 1:2 and 1:1 (lime:sand ratio) by weight because the sand used in its preparation can contain naturally carbonated fractions (not all the carbonate content is always attributable to the carbonation of lime). Finally, the replacement grouting mortar was rich in CaO assignable to the portlandite -$Ca(OH)_2$- (still not carbonated) and also to the presence of dolomitic aggregate, considering the MgO content.

Table 1. X-ray fluorescence.

Sample	SiO_2 (%)	Al_2O_3 (%)	Fe_2O_3 (%)	MgO (%)	CaO (%)	Na_2O (%)	K_2O (%)	Cl (ppm)	S (ppm)	LOI (%)
B1	61.36	15.97	7.03	1.56	2.18	0.86	4.07	2683	223	3.10
B2	61.05	16.70	7.68	1.32	1.30	0.79	4.01	5610	891	3.50
B3	62.28	18.07	8.42	1.22	1.03	0.60	3.88	149	–	1.15
M1	28.87	6.97	3.02	1.38	28.74	0.64	2.06	2770	3837	24.54
M2	32.62	8.57	3.34	1.85	24.60	0.88	2.50	3773	1398	22.02
M3	2.82	1.02	0.94	17.75	35.65	0.08	0.11	134	5588	40.97

Table 2. Content of soluble ions and carbonates.

Sample	Grams per 1kg of bricks/mortar					
	Ca^{+2}	K^{+}	Mg^{+2}	Na^{+}	S	Cl^{-}
B1	0.33	0.21	0.019	0.20	0.03	0.25
B2	0.36	0.31	0.068	0.38	0.07	0.55
B3	0.02	0.01	0.001	0.01	0.00	0.01
M1	0.06	0.06	0.001	0.07	0.22	0.13
M2	0.18	0.13	0.010	0.18	0.11	0.24
M3	1.10	0.07	0.000	0.03	0.03	0.004

3.2 *Content of soluble ions extracted in the bricks and mortars*

The results obtained (Table 2) clearly indicate that there was a higher concentration of soluble cations in the altered bricks and mortars (B2 and M2) and the sulfur content (associable to sulphates $SO_4^{=}$) in B2 was more than twice that of the unaltered brick (B1). However, the salts present in the replacement brick and mortar (B3 and M3) were in very low concentrations.

3.3 *X-ray diffraction*

The minerals identified from the samples are shown in Figure 2. The raw material of the bricks was a non-limestone illitic clay with the addition of sand as degreaser to improve its workability. The presence of illite, a phase that disappears after 950°C, was detected in both types of bricks, especially in the altered ones. Traces of calcite, a mineral that decompose between 830°C and 870°C, can have a primary origin (calcite without decarbonate) or secondary, due to the quicklime -CaO- that has not reacted in the firing temperature. The shortage of calcite and dolomite in the raw material and the range of temperatures led to the formation of traces of high temperature silicates. Finally, another reaction takes place, consisting of the transformation of iron oxides (mainly goethite) into hematites -Fe_2O_3-, beginning at 850 °C. Thus, it is estimated a firing temperature between 850°-900°C that allows the structural illite clay not to collapse.

The mineralogy of the replacement brick (B3) was very similar to that of the original unaltered brick (B1). However, reflections of the illite had disappeared and there was a higher concentration of hematite, which means that the firing temperature was slightly higher (900°-950°C). M1 and M2 mortars were very similar; both were composed mainly by quartz and calcite. Notwithstanding, replacement mortar M3 was mainly composed by portlandite which had its origin in the dolomitic nature of the aggregate, since carbonation had not yet taken place.

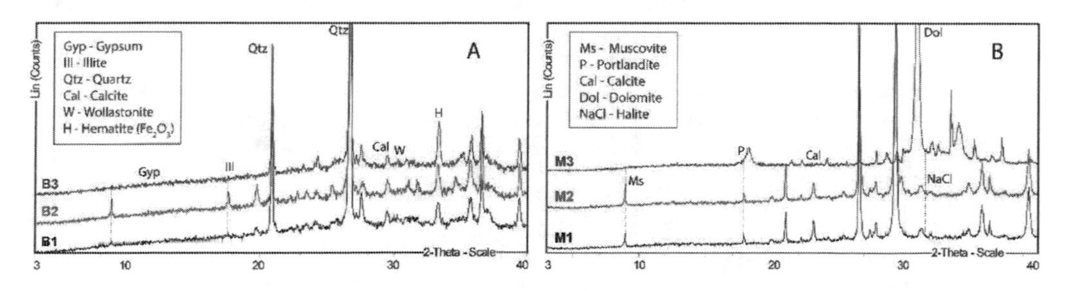

Figure 2. A) X-rays diffractograms of bricks (B1, B2&B3) and B) mortars (M1, M2&M3).

3.4 *Physical-mechanical properties*

The references for the open porosity of solid bricks show values from 35% to 37% (González, I. et al., 1998) and from 35% to 42% (Cultrone, G., 2001), therefore, the values obtained for the bricks (B1, B2 and B3) were within or below the common ranges considered. For lime mortars Cazalla et al. (2002) establish a porosity of 31-32% and Alejandre et al. (2004) from 40 to 50%, so that M1and M2 porosity was very high. Due to this high porosity the mortars showed a very low compressive strength that could not be tested because it was under the load detection limit of the strength testing machine (Table 3).

The result obtained for original brick (B1) was above the minimum acceptable values according to EN-772-1 ($5N/mm^2$). The replacement brick (B3) had similar value of compressive strength so it was defined as a suitable product in relation to its mechanical capacities (Table 3).

Table 3. Physical-mechanical properties.

Sample	D_r (gr/cm^3)	D_a (gr/cm^3)	P_o (%)	R_c (N/mm^2)
B1	2.70	1.78	33.87	11.01
B2	2.73	1.90	30.16	-
B3	2.74	1.81	33.92	10.73
M1	2.62	1.34	49.00	-
M2	2.63	1.35	48.80	-

3.5 *^{14}C-Dating*

Dating tests by ^{14}C indicated that currently the render mortars of the tower façade were probably from a restoration executed in the 17th century.

Table 4. ^{14}C - Dating.

Measured radiocarbon age (BP)	Calibrated Age 2σ (95%)	Probability
	[cal AD 1664: cal AD 1699]	17.59%
160 ± 25	[cal AD 1721: cal AD 1817]	53.64%
	[cal AD 1833: cal AD 1879]	10.10%
	[cal AD 1915: cal AD 1949]	18.64%

4 CONCLUSIONS

According to all the analyses developed it was determined that the main cause of decay was the presence of soluble sulphate salts combined with weathering conditions that generated expansive efflorescences through crystallization and dissolution cycles. The firing temperature heterogeneity was the main cause of the individual decay that bricks shown in the facade. With respect to the compatibility between the replacement materials and the original ones, according to the tests carried out, it could be considered that the selected bricks and the lime mortar were compatible, and their properties were appropriate for the requirements.

REFERENCES

Alejandre Sánchez, F.J., Tabales Rodríguez, M. A., Graciani García, A., Martín Del Río, J.J. 2004. El Real Monasterio de Santa Clara de Sevilla (España): estudio analítico de los tapiales, morteros

y ladrillos utilizados en diversas etapas constructivas. Actas VII Congreso Internacional de Rehabilitación del Patrimonio Arquitectónico y Edificación, CICOP, 35-38.
Cazalla, O, Rodríguez-Navarro, C, Cultrone, G., Sebastián, E.M. & De La Torre, M.J. 2002. The carbonation of lime mortrs: *The influence of aging of lime putty. Protection and Conservation of the Cultural Heritage of the Mediterranean Cities: 139-144.* Lisse:Balkema.
Correa Barrera, A.J. 2016. *Proyecto básico y de ejecución. Restauración del macizo mudéjar de la Iglesia Sta. Mª Encarnación.*
Cultrone, G., 2001. *Estudio mineralógico-petrográfico y físico-mecánico de ladrillos macizos para su aplicación en intervenciones del Patrimonio Histórico.* Tesis Doctoral, Universidad de Granada.
González, I., Galán, E., Miras, A. & Aparicio, P. 1998. New uses for brick-making clay materials from the Bailen area (southern Spain). *Clay Minerals 33: 453-465.*
Martín del Río, J.J., Flores Alés, V., Alejandre, F.J., Blasco López, F.J. 2018. New Method for Historic Rammed-earth Wall Characterization: The Almohade Ramparts of Malaga and Seville. *Studies in Conservation. Vol. 8. Núm. 1. Pag. 1-10.* DOI: 10.1080/00393630.2018.1544429.
UNE-EN 772-775: 2016. *Methods of test for masonry units - Part 5: Determination of the active soluble salts content of clay masonry units.*
UNE-EN 1936: 2007. *Natural stone test methods - Determination of real density and apparent density, and of total and open porosity.*
UNE-EN 772-771: 2011. *Methods of test for masonry units. Part 1: Determination of compressive strength.*

Science and Digital Technology for Cultural Heritage – Ortiz Calderón et al. (Eds)
© 2020 Taylor & Francis Group, London, ISBN 978-0-367-36368-0

A study of the durability of the inks serving to print contemporary graphic artwork

B. Sánchez-Marín, D. Campillo-García, T. Espejo-Arias & A. López-Montes
Department of Painting, Faculty of Fine Art University of Granada, Granada, Spain

M.R. Blanc García
Department of Analytical Chemistry, Faculty of Sciences University of Granada, Granada, Spain

ABSTRACT: Graphic artwork offers tremendous potential for artistic expression. The processes associated with artistic production have been greatly influenced by technological developments over recent decades, particularly regarding printing materials and techniques. The current study examines the durability of graphic artwork produced by contemporary media and printing materials by means of the analysis of a variety of inks. The study focuses on three techniques (offset, inkjet and laser printing) and identifies the elements that can lead the printed image to fade or suffer from alterations. This paper presents the main findings of the analyses and contributes to a better understanding of printing materials and techniques. Furthermore, this study paves the way for new research addressing the challenges of stability, conservation and durability of printed graphic artwork and identifies the better printing processes to apply in each case.

1 INTRODUCTION

Printed graphics is a resource with a great potential to express artwork. Yet there is a growing concern as to its conservation over time due to the nature of the media and the types of printing materials currently in use. The variety of these technologies that undergo deterioration from exposure generates concern as to their life span. This paper attempts to evaluate the performance, stability and durability of the three most current printing techniques (inkjet, laser and offset) by analysing the behaviour of their respective inks when serving as an alternative to treatments of chromatic reintegration and facsimiles. The stability of a printout must respond to a series of issues such as:

- Definition and incorrect identification of processes. The study gathered a series of printed works from museums, galleries and art fairs that apply these techniques and noted that there is in fact a great confusion and lack of information among artists as to how to define their production techniques.
- The study also identified a general lack of consensus as to how to define the different techniques and materials and a scarcity of specific information as to their components.
- The lack of understanding of the potential reactions of extrinsic factors that might affect a printout when displayed in different types of exhibitions (air contamination, humidity, temperature or exposure to light). The last example, exposure to light, often provokes a loss of quality of an image.

2 METHODOLOGY

The first aspect of this study's methodology was to decide upon which documents to analyse. There is in fact a vast assortment of options when printing and producing graphic art. These include a variety of techniques as well as they type of paper, ink, coatings, finishes, etc. The printers selected to produce the prints for this study are currently on the market.The three most common types currently serving in printing presses are offset, inkjet and laser. The study produced a total of 264 CMYK (yellow, cyan, magenta, black) samples printed on Somerset book soft white paper, a medium that is 100% cotton and devoid of glues, colours, and acidity.

The samples were subjected to four standardised tests: 1) resistance to ambient humidity (BELO illuminated Low Pressure Table), 2) resistance to immersion in water, 3) ageing in a weathering chamber (Solarbox 300rRH/xenon lamp with an irradiance of 300-800 nm) and 4) dry heat ageing (Beschikung-Loading Modell 100-800 Memmert ULE 600). A series of measurements were carried out before and after each test so as to contrast the potential differences. The measurements comprised dimension, weight and thickness, colour, pH level, and observations by optical microscope and photographs. The method put to use was followed by initiating a discussion of the findings so as to develop a system that allows to compare and certify the data as objectively as possible.

3 RESULTS AND DISCUSSION

The data gleaned from the study parameters suggest that there are no significant changes in dimension, weight and pH, beyond those due to the variability of the tools and measuring devices. A significant increase of thickness is apparent in all the prints during the environmental humidity and water immersion tests. There is little doubt that this dilatation is due to the capacity of absorption of the paper's cotton fibres.

The major findings of this study concern the optical variations and the perception of the colour attributes. These attributes were studied from the standpoint of a CIELab model with the axe L* representing the degree of luminosity, C* the chromatic value, and h* the value of tone. Figure 1 illustrates that an identical colour, in this case black, can react very differently depending on its printer and type of paper.

In the case of ageing in the weathering chamber the greatest alteration of the yellow, as illustrated in Figure 2, takes place among offset printouts after a period of 72 hours. The optical microscopic photograph illustrates this colourimetric alteration. The data listed in Figure 3, indicate the changes of luminosity (L*), chroma (C*) and angle of tone (h*).

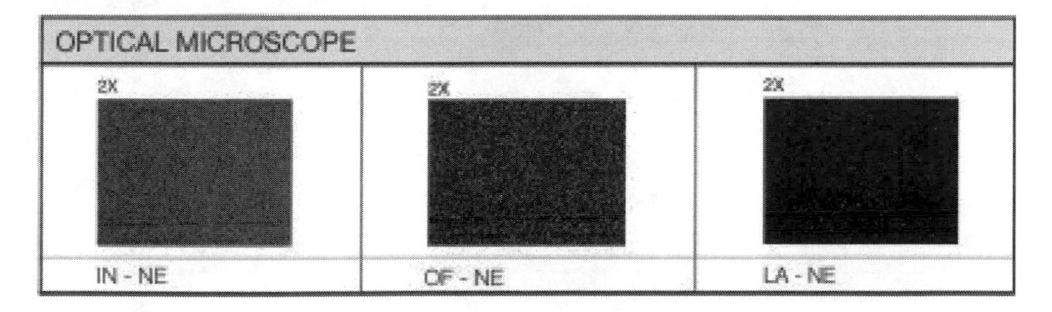

Figure 1. Optical microscopic view of the three types of printing using black ink.

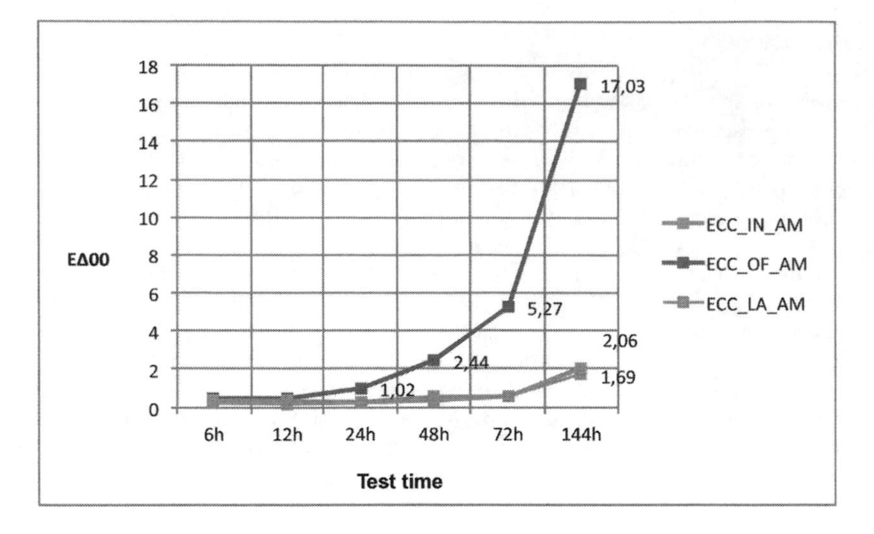

ECC_IN_AM = (Ageing Camera_Inkjet_Yellow)

ECC_OF_AM = (Ageing Camera_Offset_Yellow)

ECC_IN_AM = (Ageing Camera_Laser_Yellow)

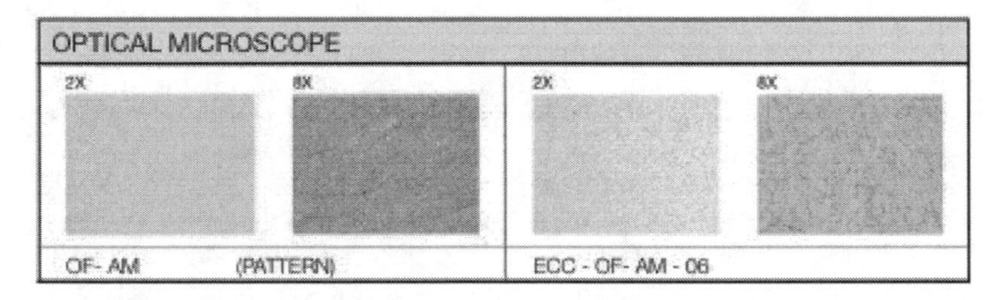

Figure 2. Top: graph indicating the total differences (E00) regarding the pattern over time of the samples subjected to ageing in a weathering chamber. Bottom: microscopic optical study.

STANDARDISED TEST		Ageing in a weathering chamber		
PRINTER		Offset		
COLOUR		Yellow		
INSTRUMENTATION				
Portable CM-2600d Spectrophotometers from Konica Minolta				
RESULTS				
Sample	ΔL*10	ΔC*ab,10	Δh*ab,10	ΔE00
ECC_OF_AM_01	0,25	-1,08	0,63	0,45
ECC_OF_AM_02	0,41	-0,52	0,69	0,48
ECC_OF_AM_03	0,56	-3,11	1,25	1,02
ECC_OF_AM_04	1,17	-8,89	2,27	2,44
ECC_OF_AM_05	1,86	-20,42	3,57	5,27
ECC_OF_AM_06	4,8	-56,65	7,54	17,03

Figure 3. Differences of luminosity (L*), chroma (C*) and angle of tone (h*) of an offset printout of a yellow colour.

4 CONCLUSIONS

As stated previously, resorting to printing techniques to produce works of art is very common today. The findings of this analysis indicate that digital printing systems, inkjet in particular, dominate the market. Moreover, the inkjet technique remains the most stable over time. The offset and laser methods, by contrast, are more likely to suffer optical variations. Yellow is the colour that is most prone to alteration. It practically completely disappears when aged in a weathering chamber.

Future research on this subject should conduct other trials such as image stability under different conditions of exposure to light (mostly UV radiation) and carry out fundamental tests such as chemical analyses of the ink.

Hopefully the current findings serve both as a starting point for future research and as a tool to develop further artistic projects. The findings will conceivably also lead to the perspective that printed images are living elements in constant evolution conditioned by their times and needs.

ACKNOWLEDGEMENTS

This study was carried out in the framework of the research project *New alternatives to understanding the materials and the processes of conservation and restoration of graphic work and documentary heritage* (MAT2014-58659-P) of the Spanish Ministry of Economy and Competitiveness. The authors would like to thank *Industrias Gráficas Sanmar* of Murcia (Spain), for its invaluable assistance and its facilities to print the samples.

REFERENCES

Antón, E. Muñoz, S. y Ruiz de Diego, S. 2016. Técnicas modernas de impresión en la creación artística contemporánea: Xerografía vs. Inkjet. *Conferencia llevado a cabo en Conservación de Arte Contemporáneo. 17ª Jornada. Museo Nacional Centro de Arte Reina Sofía.* Madrid, España.

Image Permanence Institute. 2009. *A consumer guide to Understanding Permanence Testing. Eastman Kodak Company*. Recuperado de https://www.imagepermanenceinstitute. org/webfm_send/311

López-Montes, A., Collado-Montero, F.J., Castillo, M.E.; Blanc, R.; Campillo, D., Espejo, T. 2017. Aging Analysis of a Color Facsimile Binding for the 14th Century Manuscript "Registro Notarial de Torres". Color Research and Application 42, (4), pp. 474–485. DOI:10.1002/col.22093 http://onlinelibrary.wiley.com/doi/10.1002/col.22093/full

UNE-EN ISO 16474–2, 2013. Método de exposición a fuentes luminosas de laboratorio. Madrid, Aenor.

UNE-EN ISO 11664–4, 2008. Colorimetría. Parte 4: Espacio colorimétrico L*a*b* CIE 1976. Madrid, Aenor.

UNE-EN ISO 12040. Tecnología gráfica. Estampas y tintas de estampación. Valoración del acelerador de luz usando un arco de luz filtrada xenón. Madrid, Aenor.

Wilhelm, H. 2006. *A 15-Year History of Digital Printing Technology and Print Permanence in the Evolution of Digital Fine Arte Photogaphy – From 1991 to 2006*. [Wilhelm Imaging Research]. Recuperado dehttp://wilhelmresearch.com/ist/WIR_IST_2006_09_HW.pdf

Science and Digital Technology for Cultural Heritage – Ortiz Calderón et al. (Eds)
© 2020 Taylor & Francis Group, London, ISBN 978-0-367-36368-0

Durability and stability study of *Debitus* grisailles

C. Machado & M. Vilarigues
Department of Conservation and Restoration & Research Unit VICARTE – Glass and Ceramics for the Arts, Faculty of Science and Technology, NOVA University of Lisbon, Caparica, Portugal

T. Palomar
Institute of Ceramic and Glass, Spanish National Research Council (ICV-CSIC), Madrid, Spain

ABSTRACT: During the restoration of stained-glass windows, to fill the grisailles losses is frequently used commercial *Debitus* grisailles. This article presents the study of durability and stability of five of these grisailles. Different alteration tests were carried out, samples were placed in high humidity chambers (80% and 95% RH) for 30 months and were submerged into distilled water for 3 months. The samples and degradation products were characterized by optical microscopy, μ-Raman spectroscopy and colorimetry. The pH of the water was measured during the experiment and the dissolved elements were analysed by inductively coupled plasma atomic emission. It was showed an alteration on the grisaille's colour and the formation of degradation products, that were identified as sulphates. Despite these alterations, these commercial grisailles presented a good durability and stability to be used in restoration works.

1 INTRODUCTION

Grisaille is a glass-based paint applied in the production of stained-glass panels, normally used for the creation of outlines and shadows. The grisailles are produced by mixing metals oxides with a ground lead-based glass. The obtained powder is mixed with a vehicle agent, such as vinegar or water, which will give the necessary plasticity to paint, and gum Arabic as a temporary binding agent. After fired at temperatures between 650-700°C, a thin layer of colourless glass with the metal oxides embedded is formed on the surface of the glass panel (Figure 1) (Schalm, 2000).

These paints were initially produced in the stained-glass workshops by the glass painters themselves (Machado *et al.*, 2019b); however, during the 19th century, glass paint manufacturers, like *Lacroix & Cie* founded in 1855 in Paris, made their products commercially available (Schalm *et al.,* 2003). This separation between the manufacturer and the user resulted in conservation problems that persist until nowadays, such as the incompatibility between the painting materials and the glass panels that can lead to the detachment of the painted layers (Machado *et al.* 2019a).

During the middle of the 20th century, *Lacroix & Cie* stopped the production because the demand was not enough to be rentable. However, at the end of the 20th century, the French section of the International Institute of Conservation (IIC) in partnership with the *Laboratoire de Recherche des Monuments Historiques* (LRMH), felt the need to create new formulations of glass paints to be used in conservation and restoration works. This research was trusted to Hervé Debitus, a conservator and specialist in painting on glass.

The *Debitus* grisailles started to become commercially available in 1991 after the publication of the paper "*Recherche pour une formulation nouvelle de grisailles*" (Debitus, 1991). These paints are sold until nowadays and are mostly used in the conservation and restoration of stained-glass windows, mainly for the chromatic reintegration during the process of filling losses.

Despite being one of the most used paints in restoration of stained-glass windows, no studies have been previously done to establish the compatibility, stability and durability of these

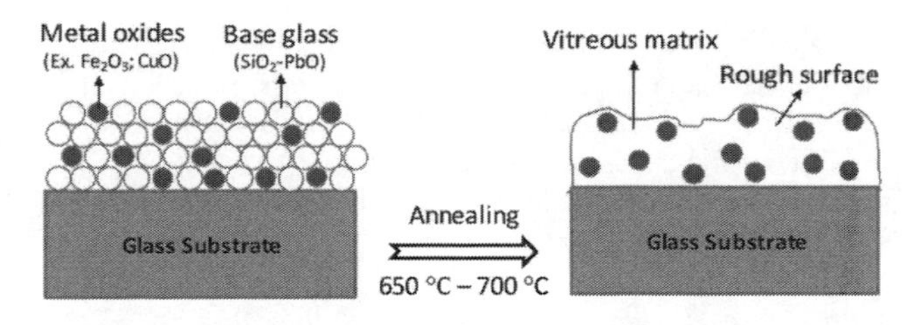

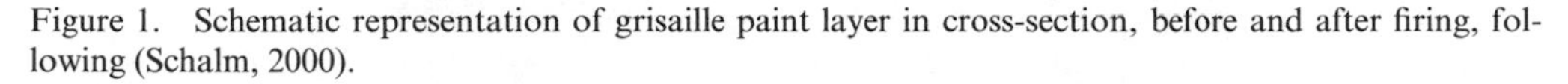

Figure 1. Schematic representation of grisaille paint layer in cross-section, before and after firing, following (Schalm, 2000).

materials. It is important, not only to understand the chemical and physical properties of the historical materials, but also the materials used in the conservation and restoration procedures to avoid future conservation problems in the historical windows.

In a previous study, the chemical, morphological and thermal characterization of these commercial materials was made (Machado, *et al.*, 2019a). It was observed a change in their original chemical composition from the theoretical formulation based in medieval treatises and in the manuscripts of *Lacroix & Cie*. Nowadays, the new formulation has a well-balanced ratio between the different components of the raw materials, a good homogeneity of the grisaille, a correct interdiffusion between these grisailles and the glass support, and a good thermal stability. These characteristics reveal a good-quality material, appropriate to be used in conservation and restoration works. However, in order to fully understand the stability and durability of these commercial grisailles, further studies needed to be done to assess the behaviour and the effect of ageing on these grisaille paints. Therefore, the main objective of this work was to assess the morphological and chemical durability and stability of these commercial grisailles.

2 METHODOLOGY

2.1 *Materials*

For this study, five of the most common grisailles were selected: one black (Noir Ordinaire), two browns (Brun XIII and Brun XVI) and two whites (Mousseline and Depoli Incolore). This choice was made by their representativeness within the colours found on historical grisailles.

Two alteration tests were developed: immersion in static conditions in distilled water for 3 months and exposure in high humidity chambers (95% and 80% RH) permanently closed at environmental temperature for 30 months.

The alteration tests were made on commercial powder before firing (Ø 0.15-35.50 μm) and painted samples (after firing). The painted grisailles were prepared with a mixture of water and gum arabic (less than 1 wt. %) and painted on 2x2 cm squares on the non-tinned side of float glasses. These samples were fired in an electric furnace (BARRACHA-Model E1) with a temperature ramp of 3 °C/min up to 680 °C, followed by a dwell of 30 minutes and a slow cooling. A fifth set was stored in normal environmental conditions for comparison.

2.2 *Analytical techniques*

The samples before and after the alteration tests were chemically, morphologically and chromatically characterized by optical microscopy, μ-Raman spectroscopy and colorimetry.

To observe the alteration in the morphology and the formation of degradation products the samples were observed by reflective light in an optical microscopy under plane and cross-polarised light. The microscope, an Axioplan 2 from Zeiss, is equipped with a halogen light HAL100 and digital camera (Nikon DMX1200F).

μ- Raman spectroscopy was used for the characterization of the metal oxides used as colouring agents and the degradation products. The spectrometer, LabRaman 300 Jobin Yvon, is equipped with a 17 mW HeNe laser operating at 632.8 nm. The laser beam was focused with a 50x Olympus objective lens. When need the laser power was controlled using neutral density filters 0.3 and 0.6, with a collection time of 10 seconds performing 15 scans.

To observe alterations in the colour of the samples a colorimetric study was done using UV-vis spectroscopy. The fibre optic spectrometer, an Avantes AvaSpec-2048, with a 300 lines/mm grating, and a range between the 200-1100 nm. The analyses were done with an integrating sphere (AVASPHERE-50-REFL) and 600 μm reflection probe (Avantes FC-UV600-2).

The aqueous solution was also analysed with a pH-meter along the 3 months. The equipment used was a Sartorius $DocupH_{meter}$, using an electrode with a KCl 3M solution.

To quantify the species lixiviated from the grisailles after the long-term contact with water, the aqueous solutions were analysed at the end of the alteration test by Inductively coupled plasma atomic emission spectroscopy (ICP-AES). A Horiba Jobin-Yvon Ultima model was used, equipped with a 40.68 MHz RF generator, Czerny-Turner monochromator with 1.00 m (sequential), and an AS500 autosampler.

3 RESULTS AND DISCUSSION

When comparing with the original samples (before alteration), a small change of colour was observed, as well as the formation of white crystal salts on their surface (Figure 2). The formation of these salts was favourable at high humidity chambers, mainly in the 80% RH chamber. The white crystal salts were formed due to the lixiviation of alkaline ions from the glass. (Vilarigues *et al.*, 2011)

The alteration of colour was confirmed with the colorimetric study. The colour change was more visible in the grisailles submerged in water than the ones from the high humidity chambers. The tendency of this alteration goes towards the yellow and red colours; however, the principal change occurs in the L* coordinate related to a whitening of the samples, mainly in the samples exposed at 95 % HR (Figure 3).

It was showed an increase of the pH that was stabilized after 10 days of experiment (Figure 4(a)). It was also observed an increase of the values of the alkaline and alkaline-earth elements in the ICP-AES analyses, mainly the calcium (Figure 4(b)). This calcium must come from the glass support because the grisailles does not have calcium in their chemical composition. This pH increase is higher in the painted samples than on the unpainted ones, and it can be related to the porosity of the grisaille, that can maintain the solution in permanent contact with the glass support favouring its punctual attack. (Palomar, 2017)

Hematite and cassiterite were identified by μ-Raman spectroscopy as the raw materials used in the production of the grisailles and responsible for the colour. The white crystal deposits formed during the alteration tests were identified as sulphates. Iron sulphate in the darker samples and sodium sulphate in the white ones. These products were formed due to the interaction of the lixiviated ions with the SO_2 present in the atmosphere ($SO_2 + H_2O + ½ O_2 \leftrightarrow SO_4^{2-} + 2H^+$).

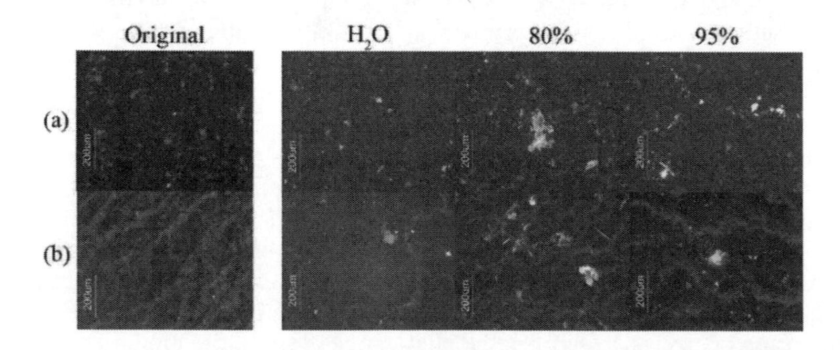

Figure 2. Optical microscopy image before and after alteration studies, with cross-polar light, in the painted samples (a) *Brun XVI* and (b) *Depoli Incolore*.

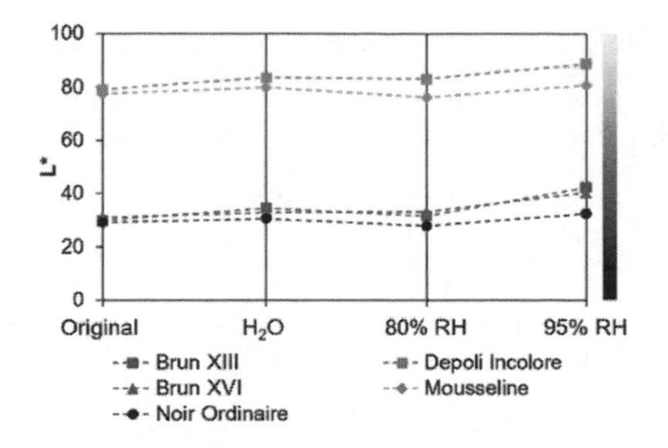

Figure 3. Variation of the L* coordinate according to the different alteration tests.

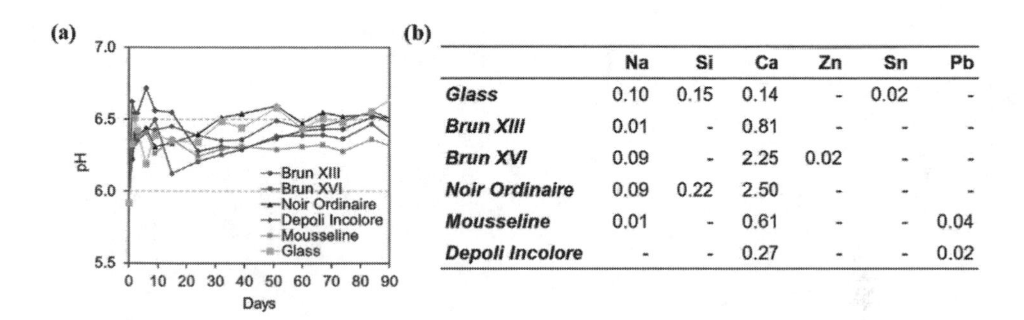

	Na	Si	Ca	Zn	Sn	Pb
Glass	0.10	0.15	0.14	-	0.02	-
Brun XIII	0.01	-	0.81	-	-	-
Brun XVI	0.09	-	2.25	0.02	-	-
Noir Ordinaire	0.09	0.22	2.50	-	-	-
Mousseline	0.01	-	0.61	-	-	0.04
Depoli Incolore	-	-	0.27	-	-	0.02

Figure 4. Aqueous solution analysis, (a) pH variation of painted samples along 90 days and (b) ICP-AES results (mg/l) of the solutions after the experiment.

4 CONCLUSIONS

It was possible to observe a small change of colour in the grisailles, as well as the formation of white crystal salts, identified as sulphates, on the surface. Regarding the solutions, it was observed an increase of the pH, related to the lixiviation of alkaline and alkaline-earth ions from the glass support. Despite these small changes, *Debitus* grisailles are morphological and chemical stable for being use in restoration works, since they are preferentially painted in the indoor glass surface to avoid their contact with the water.

ACKNOWLEDGEMENTS

The authors wish to thank the Fundação para a Ciência e Tecnologia de Portugal (project ref. UID/EAT/00729/2019, doctoral grant ref. PD/BD/136673/2018) and Fundación General CSIC (ComFuturo Programme).

REFERENCES

Debitus, H. 1991. Recherche pour une formulation nouvelle de grisailles. *Science et Technologie de la Conservation et de la Restauration des oeuvres d'art et du patrimoine* 2: 24–28.

Machado, C., Machado, A., Palomar, T., Alves, L. C., Vilarigues, M. 2019a.Debitus grisailles for stained-glass conservation: an analytical study. *Conservar Património*: In press.

Machado, C., Machado, A., Palomar, T. and Vilarigues, M. 2019b. Grisaille in Historical Written Sources. *Journal of Glass Studies*: In press.

Palomar, T. 2017. Chemical composition and alteration processes of glasses from the Cathedral of León (Spain). *Boletin de la Sociedad Española de Ceramica y Vidrio*, SECV 57(3): 101–111. doi: 10.1016/j.bsecv.2017.10.001.
Schalm, O. 2000. *Characterization of paint layers in stained-glass Windows: main causes of the degradation of nineteenth century grisaille paint layers*. Antwerpen: Universiteit Antwerpen.
Schalm, O., Janssens, K. and Caen, J. 2003. Characterization of the main causes of deterioration of grisaille paint layers in 19th century stained-glass windows by J.-B. Capronnier. *Spectrochimica Acta - Part B Atomic Spectroscopy* 58(4): 589–607. doi: 10.1016/S0584-8547(02)00282-3.
Vilarigues, M. *et al.* 2011. Corrosion of 15th and early 16th century stained glass from the monastery of Batalha studied with external ion beam. *Materials Characterization*. Elsevier Inc 62(2): 211–217. doi: 10.1016/j.matchar.2010.12.001.

Science and Digital Technology for Cultural Heritage – Ortiz Calderón et al. (Eds)
© 2020 Taylor & Francis Group, London, ISBN 978-0-367-36368-0

Removing insoluble salts from archaeological ceramics. Traditional methods under review

Á. Sáenz-Martínez & M. San Andrés
Painting and Conservation-Restoration Department, Faculty of Fine Arts, Complutense University of Madrid (UCM), Madrid, Spain

M. Álvarez de Buergo
Institute of Geosciences IGEO, Spanish Research Council and Complutense University of Madrid (CSIC-UCM), Madrid, Spain

I. Blasco
Painting and Conservation-Restoration Department, Faculty of Fine Arts, Complutense University of Madrid (UCM), Madrid, Spain

R. Fort
Institute of Geosciences IGEO, Spanish Research Council and Complutense University of Madrid (CSIC-UCM), Madrid, Spain

ABSTRACT: Although insoluble salts are one of the most common alteration forms that archaeological ceramics can present, the methods for their removal have not really changed within the past forty years. Furthermore, not many studies on the topic have been recently published. In this paper, a methodology to study the suitability of different traditional treatments for the removal of these salts, has been established. Ceramic mock-ups made out of a marketed red clay were fired up to 700ºC in laboratory conditions. Their composition and physical properties were studied. A validated method for the growth of calcium carbonate deposits was applied onto them. Afterwards, chemical tests based on three products and three application methods were studied. Finally, the mock-ups' characteristics were studied again to determine their changes due to the cleaning treatments. Results showed efficacy differences linked to the product and application method. In general, the products applied by immersion and cellulose pulp poultice were more effective than gel treatments, which were less effective with any of the three products tested.

1 INTRODUCTION AND AIMS

In archaeological sites, ceramics are the most abundant materials recovered due to their high physical and chemical resistance. However, they do not remain unalterable over time as they can present abrasions, breakings, material loss, stains and salts. Both soluble and insoluble salts are the most common and challenging alterations, as they appear in many pieces and can cause severe conservation problems. Their presence can imply a pressure increase inside the pores, which can lead to cracking, deformation, material loss and surface cover, and result into the decline of the artistic, social and historical values of these pieces.

The cleaning treatments applied for the removal of soluble salts are traditionally based on water baths. However, insoluble salts deposits present more hardness and adherence to the ceramic surfaces, which complicates their removal. Indeed, one of the most common solutions is to apply mixed treatments; once the chemical one has softened the deposits they are removed by mechanical means using scalpels, gravers and micro drillers, among others. Nevertheless, it is very difficult to control them and avoid any surface damage because chemical methods are based on products, such as acids, complex agents and ion exchange resins that react chemically with the insoluble deposits.

The use of hydrochloric, acetic and nitric acids at 10-20% vol. has been the most traditional treatment, especially for calcium deposits. However, some authors proved that ceramics might be damaged because of these treatments in terms of chemical and mineralogical composition, as well as physical properties, such as colour or roughness (Buys and Oakley, 1993; Fernández, 2003; Johnson and Iceland, 1995); nevertheless, they are still in use. Alternatively, chelating agents are able to form bounds with metallic ions to result in soluble salts. The most common of these agents derive from the ethylenediaminetetraacetic acid (EDTA) used at 5%-10% m/v (Pearson, 1987; Abd-Allah et al, 2010). More recently, the ion exchange resins have been applied with success (Casaletto et al., 2008; Crisci et al, 2010). However, they are not widely used. As applications methods, immersion for direct and poultices for indirect ones are the most common ones (Berducou, 1990). Although since the 1980s, gels and thickening agents have been explored as an alternative (Wolbers, 2000), they are not yet very common in archaeology conservation. Despite the removal of soluble and insoluble salts can imply serious risks for the conservation of ceramics, few specialized researchers have carried out the evaluation of cleaning methods (Casaletto et al., 2008; Johnson et al, 2011) and traditional treatments are still in use, although their applicability has not yet been established. All the commented above, led us to study the suitability of several treatments applied to remove insoluble salts from archaeological ceramics by following a methodology specifically designed.

2 MATERIALS AND METHODS

2.1 *Materials*

Thirty-six ceramic mock-ups of 5x5x1cm were manufactured in laboratory conditions with a red commercial clay (SIO-2® ARGILA). They were dried in a climatic chamber (25°C; 50% RH) for 72 hours. Afterwards, they were fired up to 120°C to evaporate the water inside the porous. Finally, they were fired again up to 700°C, reproducing archaeological ceramics' low firing temperatures.

2.2 *Methods*

Mineralogical and elemental composition. XRD data were collected on an X´Pert MPD equipment. Punctual elemental analysis was carried out with EDS x-ray X-Max microanalyzer.

Surface morphology. Roughness was measured with a TRACEit optical surface roughness tester. The measuring field was 5×5 mm and the resolution 500×500 pixels. Among the main roughness values (Ra, Rq and Rz), Ra (the arithmetic average of roughness' absolute values of the profile) was selected as the most suitable to compare the results before and after the cleaning tests performed. Besides, microphotographs were taken with a JEOL JSM 6335F scanning electron microscope and a stereoscopic microscope (SM) Leica MZ125 with digital camera Leica DC150.

Colour measurements. A Konika Minolta CM-2600d spectrophotometer was used to measure colour parameters (L^*, a^* and b^*) under D65 light source.

Physical properties. A density test following UNE-EN 1936:2007 was performed to determine their porosity, density and compactness index.

Growth of insoluble salt deposits. Three tests based on lime water were applied onto nine ceramic mock-ups: by immersion for 24 hours, by immersion for 24 hours and covered for 72 hours and with soil applied by brush and covered for 72 hours. All of them were put into a carbonation chamber for 3, 7 and 10 days simulating soil conditions at 20°C and 60% RH; and an atmosphere of 1600 ppm CO_2, boosting carbonation process. Another test with lime putty and soil directly applied onto three mock-ups was carried out. The ceramic pieces were put into carbonation chamber at the same conditions for 1, 4 and 7 days. The test performed with lime putty and soil directly applied and carbonated for 7 days allowed the growth of

Table 1. Chemical cleaning tests performed.

App. method / Product	Immersion (I)			Cellulose pulp poultice (CPP)			Vanzan NF-C gel (VG)		
	%	pH	Time (min)	%	pH	Time (min)	%	pH	Time (min)
Acetic acid	1	2.70	2	1	2.70	10	10	2.70	10
Nitric acid	1	1.00	2	1	1.00	10	1	1.16	15
EDTA tetrasodium salt	1	10.88	4	1	10.88	10	1	9.90	30

1-2 mm thickness deposits with enough hardness and adherence. Therefore, it was implemented to the ceramic mock-ups chosen for the cleaning tests.

Cleaning tests. Nine cleaning tests, gathered in Table 1, were carried out. According to the literature reviewed, a weak acid, a strong acid and a chelating agent were chosen as products. As application methods, one direct (immersion) and two indirect ones (poultice and gel) were selected. Japanese paper was used as barrier element for indirect methods. For the mechanical action it was decided to use wood swabs for being the least harmful. The neutralisation consisted in static baths of demineralised water, changed every 24 hours until conductivity measurements reached balance. The criteria followed were minimum product concentration and minimum chemical treatment time and constant pH.

3 RESULTS

In this section, the results of the studies carried out on the original ceramic mock-ups before (B) carbonation, and the results after the carbonation and cleaning treatments (A), are described.

Mineralogical and elemental composition. The main mineral components of ceramic mock-ups (B) were muscovite and montmorillonite. Hematite (Fe_2O_3), quartz (SiO_2), calcite ($CaCO_3$) and dolomite [$CaMg(CO_3)_2$] were also found. Elemental analysis (B) detected Al and Si as main elements and Ca, Fe, K and Mg as minors. After the cleaning treatments (A), Ca concentration increased in all the tests performed. The higher values were found in gel applications and reached the highest level with the chelating agent.

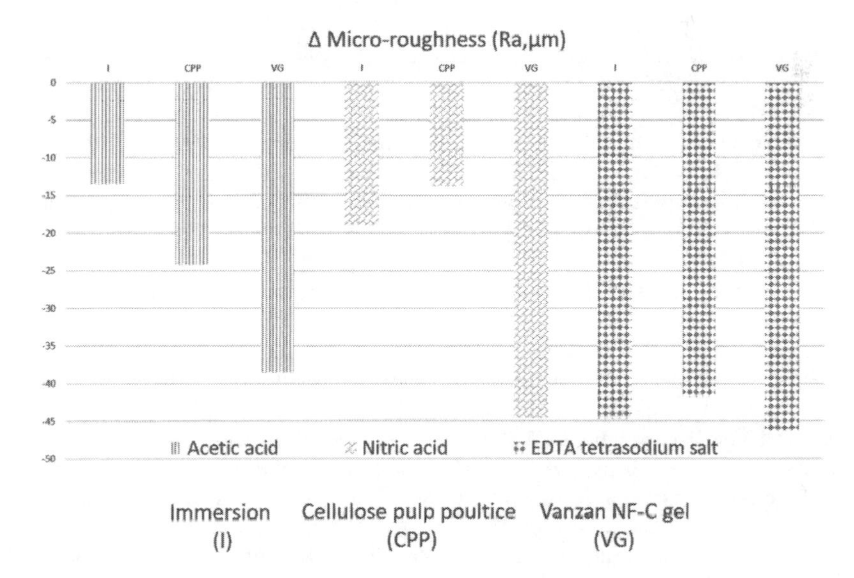

Figure 1. Decreases in micro-roughness after the cleaning treatments.

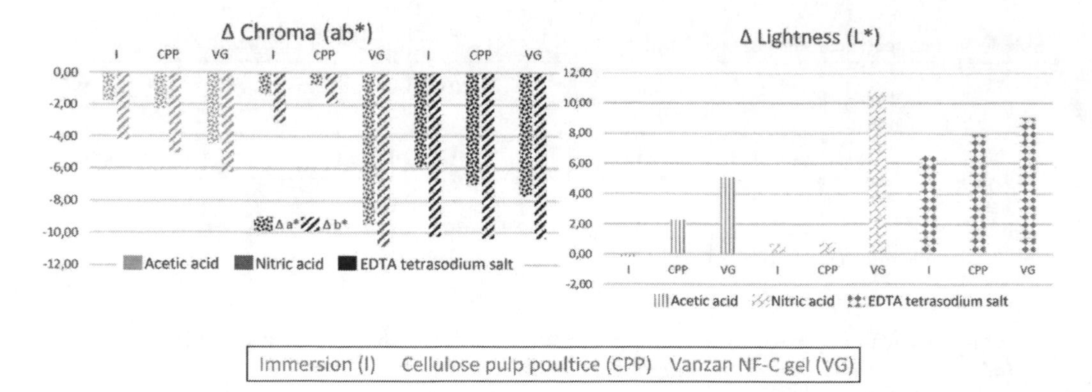

Figure 2. Variations in chroma (ab*) and lightness (L*) values after the cleaning treatments.

Table 2. Values of the total colour difference values after the cleaning treatments.

Product	Acetic acid			Nitric acid			EDTA tetrasodium salt		
App. method	I	CPP	VG	I	CPP	VG	I	CPP	VG
ΔE_{00}	2.6±1.5	3.8±2.1	6.1±1.4	2.0±0.6	1.6±1.4	12.7±2.2	8.7±1.0	9.8±2.6	10.6±1.2

Table 3. Average values of the physical properties studied before the cleaning treatments.

	Open porosity %	Real density gr/cm^3	Saturation %	Compactness index %
Average values	33.00±0.27	2.73±0.01	18.05±0.21	0.67±0.00

Surface morphology. Studies on mock-ups after the treatments (A) with SM detected incisions caused by mechanical action in all cases and residues of carbonate deposits, which were more abundant in the ceramics treated by indirect methods. With SEM it was possible to identify carbonate particles as granular and round shape aggregates of small size (<100nm) and to determine that surfaces were smoother, as carbonates had created a thin coating on them. This result is linked with micro-roughness ones, as Ra values decreased in all cases (13-46%). In acid treatments, a higher variation is determined in gel applications, meaning a higher carbonate content. However, in case of the chelating agent, Ra decreased more than in acid treatments but there were not significant differences among application methods (Figure 1).

Colour measurements. Chroma (ab*) and lightness (L*) values were studied. In all cases, these parameters differed within the application method used and cleaning agent, in the same way as roughness. In acid tests, higher L*ab* variations were found in gel application method. This is related to a higher carbonate content. In case of chelating agent tests, L*ab* variations were higher and similar among them, without notable differences among application methods (Figure 2).

The total colour difference was also established, and the results obtained go along with L*ab* variations, being the lightness the most influent parameter. Values are gathered in Table 2.

ΔE_{00} values under 1 (MokrzyckI and Tatol, 2011) are not noticeable to human eye, which would mean that the treatments have been effective as they have removed the carbonate deposits to the point that mock-ups ceramics colour is similar to the original state. In this case, all treatments are noticeable to human. Although nitric acid applied by cellulose pulp poultice showed the best results, according to ΔE_{00} values. This means that with the application of these treatments the ceramics seem very similar to their original appearance, before the growth of the deposits.

Physical properties. Before the cleaning treatments (B), porosity, density, water saturation and compactness index of all ceramic mock-ups showed a high homogeneity. Their values are gathered in the Table 3. A density test following UNE-EN 1936:2007 will be performed in the near future, in order to compare post and pre-treatments results.

4 CONCLUSIONS

According to the results, all the techniques applied allowed the study of the cleaning treatments' efficacy. Despite some amount of carbonate deposits are still present in all cases, treatments' efficacy depends on both the application method and the product.

In general, the treatments applied by immersion and cellulose pulp poultice were more effective, in terms of the removal of carbonate deposits, and in the length of the treatments, than those applied by gels.

The reason for this is probably the lack of time for the products to react with the deposits, as the gel releases the products in a very slow and controlled way.

In further research, application times will be extended to check whether the results change or not.

Besides, acid solutions were more effective by immersion and poultice application than by gel. The chelating agent was not as effective as acids, by any of the application methods.

Pending on the study of the mineralogical composition and physical properties after the treatments and according to experimental results already obtained, the treatments did not damage ceramic surfaces.

ACKNOWLEDGEMENTS

The authors gratefully acknowledge both research projects Geomateriales-2 (S2013/MIT-2914) and Top Heritage-CM (S2018/NMT-4372). They also thank Complutense University of Madrid and Banco Santander for Águeda Sáenz-Martínez's PhD scholarship (CT17/17-CT18/17).

REFERENCES

AENOR. 2007. *Métodos de ensayo para piedra natural. Determinación de la densidad real y aparente y de la porosidad abierta y total. Natural stone test methods. Determination of real density and apparent density, and of total and open porosity*. UNE-EN 1936: 2007. Madrid: AENOR.

AENOR. 2009a. *Conservación de edificios. Limpieza de elementos constructivos. Parte 5-1: Técnicas de limpieza química. Aplicación en forma de solución. Conservation of building. Cleaning of constructive elements. Part 5-1: Techniques of chemical cleaning. Solutions application.* UNE 41806-5-1:2009 IN. Madrid: AENOR.

AENOR. 2009b. *Conservación de edificios. Limpieza de elementos constructivos. Parte 5-2: Técnicas de limpieza química. Aplicación en forma de apósitos. Conservation of building. Cleaning of constructive elements. Part 5-2: Techniques of chemical cleaning. Dressing application.* UNE 41806-5-2:2009 IN. Madrid: AENOR.

Berducou, M. C. (Coord.) 1990. *La conservation en Archéologie. Méthodes et practique de la conservation.* Paris: Masson.

Buys, S. & Oakley, V. 1993. *The Conservation and Restoration of Ceramics.* Oxford: Butterworth-Heinemann Series in Conservation and Museology.

Casaletto, M.P., Ingo, G. M., Riccucci, C., de Caro, T., Bultrini, G., Fragalà, I., & Leoni, M. 2008. Chemical cleaning of encrustations on archaeological ceramic artefacts found in different Italian sites. *Applied Physics* 82: 35-42.

Crisci, G.M., La Russa, M.F., Macchione, M., Malagodi, M., Palermo, A.M. and Ruffolo, S.A. 2010. Study of archaeological underwater finds: deterioration and conservation. *Applied Physics* 100: 855-863.

Fernández, C. 2003. Las sales y su incidencia en la conservación de la cerámica arqueológica. *Monte Buciero* 9: 303-325.

Johnson, S. J. & Iceland, H. 1995. Identification of chemical and physical change during acid cleaning of ceramics. *Symposium – Materials Issues in Art and Archaeology IV* 352: 831-837.

MokrzyckI, W.S. & Tatol, M. 2011. Color difference ΔE - A survey. *Machine Graphics & Vision International Journal* 20 (4): 383-411.

Olive, J. & Pearson, C. 1975. The conservation of ceramics from marine archaeological sources. *Studies in Conservation* 20: 63-68.

Abd-Allah, R., al-Muheisen, Z and al-Howadi, S. 2010. Cleaning strategies of pottery objects excavated from Khirbet edh-Dharih and Hayyan al-Mushref, Jordan: four case studies. *Mediterranean Archaeology and Archaeometry*, 10-2: 97-110.

Pearson, C. 1987. Conservation of ceramics, glass and stone. In: Colin Pearson (ed.), *Conservation of Marine Archaeological Objects*. Oxford: Butterworth-Heinemann.

Wolbers, R. 2000. *Cleaning painted surfaces. Aqueous methods*. London: Archetype Publications.

Science and Digital Technology for Cultural Heritage – Ortiz Calderón et al. (Eds)
© 2020 Taylor & Francis Group, London, ISBN 978-0-367-36368-0

Development and optimization of an HPLC-DAD Q-TOF method for proteins identification in cultural goods

J. Romero Pastor, C. Imaz Villar, M. García Rodríguez, E. Parra & M. Martín Gil
Instituto del Patrimonio Cultural de España, Ministerio de Cultura y Deporte, Madrid, Spain

ABSTRACT: The identification of proteinaceous materials in cultural goods provides very valuable information such as the artistic technique, the presence of previous interventions, the influences from other artistic movements as well as the use or function of a specific archaeological vessel, etc. In order to clarify the source of proteins presented in cultural goods, we propose a proteomic approach to identify protein materials at low levels. We focused on the critical step of the preparation of samples and the analysis by a HPLC/DAD/MS/MS with tandem Quadrupole Time-Of-Flight detector (Q-TOF) which allows the accurate identification of the peptides and its proteinaceous natural source. The objective of the present work is to show the first benefits and results in the identification of proteinaceous materials, inter alia, egg yolk, milk, animal soft tissues or silk, in model and real samples from artworks.

1 INTRODUCTION

In cultural goods, proteins have been widely applied as binder (material holding together pigment particles in paint) in tempera paintings and as protein-based adhesive (substance used for sticking materials together) in a huge range of artworks (Mayer, 2005). One of the main difficulties to find a method for proteinaceous analysis is the small analyte amount present in a real sample and the complex matrix in which they are included. The common analytical methodologies focus on the amino acid composition after total hydrolysis do not give information of biological origins of the protein or chemical modifications when aging (Garnier, 2016; Tokarski, 2006; Whitcher, 2010). Different analytical techniques have been used to get a better understanding of the proteinaceous media, such as chromatography–mass spectrometry (GC-MS), matrix-assisted laser desorption/ionisation time-of-flight mass spectrometry (MALDI-TOF-MS), high performance liquid chromatography with diode-array detection (HPLC-DAD) and HPLC with tandem mass spectrometry (HPLC/MS/MS) (Crhova, 2014; Fremount, 2010; Lluveras-Tenorio, 2017; Cuervo, 2017). In this work, peptide mass fingerprinting and peptide sequencing using tandem mass spectrometry with Agilent MassHunter BioConfirm software and UniProt Database () provide a more precise and allows the characterization of exact amino acid sequences of peptides, providing thus the accurate identification of proteins. It allows also the identification of protein modifications.

2 MATERIALS AND METHODS

2.1 *Samples*

Pure proteins were achieved from Sigma Aldrich, i.e. casein (ref. C6780-1G), collagen (ref. 5608-100G) or albumin (ref.5503-1G), and fresh silk pattern selected from IPCE Archives in order to detect proteinaceous components. In addition, real samples were also chosen from IPCE Archives, particularly, Mariano Fortuny´s tunic, dated in 1910, belong to Museo Nacional del Traje-Centro de Investigación del Patrimonio Etnológico (Madrid, Spain) and San Luis de Tolosa"/ Painting on wood panel (XV century) belongs to the altarpiece of Capilla de Santiago, Catedral de Toledo (Toledo, Castilla La Mancha)

Figure 1a. Imagen of the pink silk tunic created by Fortuny in 1910, the piece shows geometric patterns in gold. Figure 1b: Imagen of San Luis de Tolosa.

2.2 *Sample preparation: extraction, tryptic digestion and peptides purification*

Samples were taken using a sharp scalpel and were placed in an Eppendorf tube where bicarbonate buffer were added. For extraction of proteins before hydrolysis, samples were heated during 20 min at 40 °C and were sonicated twice during 10 min. They were centrifugated and the supernatant were transferred to another Eppendorf (Figure 2). After extraction a tryptic hydrolysis were carried out. For peptides purification before injection Thermo Scientific Pierce C18 Spin Columns were used.

2.3 *HPLC-Q-TOF analysis*

All samples were analysed by LC-MS (Q-TOF) system from Agilent Technologies (Palo Alto, CA, USA). The final liquid chromatography conditions were performed using an AdvanceBio Peptide Map (2.1 mm internal diameter, 100 mm length, 2.7 µm film thickness) at 40ºC and a flow of 0.5 mL/min. Mobile phases were ultrapurified water (eluent A) and acetonitrile (solvent B), both containing 0.1% of formic acid. The gradient pattern was finally established as follows: from 0.3 to 3min, to 2% B; from 3 to 20min, to 50% B, from 20 to 30 minutes, to 75% B; from 30 to 35 minutes, to 100% B; from 35 to 38min, to 100% B and from 38.1 to 40 minutes, to 2% B. The injection volume was 20 µL.

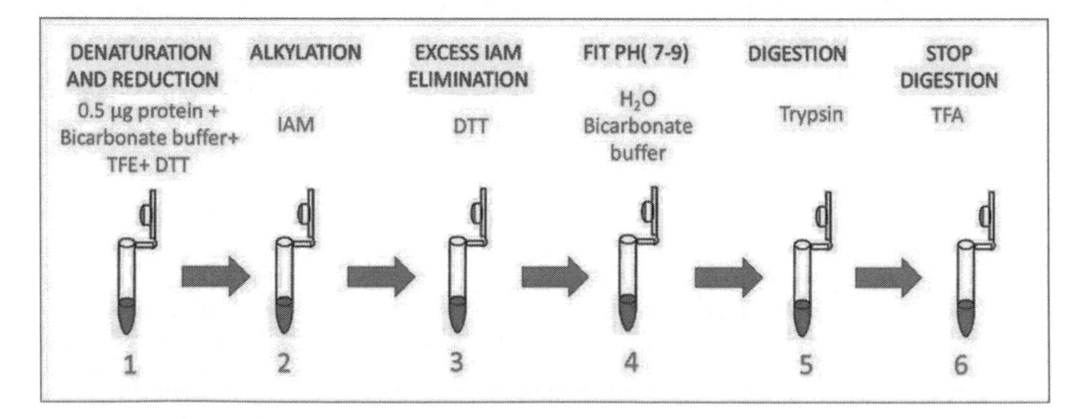

Figure 2. Procedure of enzymatic hydrolysis (DTT:dithiothreitol; TFE:tetrafluoroethylene; IAM:iodoacetamide; TFA: trifluoroacetic acid.).

2.4 *Data analysis*

All data treatment were studied with an Agilent MassHunter BioConfirm software and UnitProt database. Data obtained from LC–MS/MS were searched for against publicly available UNIPROT databases. In the case of LC–MS/MS, sequence data from at least two specific peptide fragment (they will be named in further discussion as the Uniprot reference names, e. g. P04653), the matching databases are sufficient for unambiguous identification of a protein (Crhova, 2014).

3 RESULTS AND DISCUSSION

The proposal method using proteomic analysis in cultural goods showed the possibility of identification at low levels its proteinaceous natural source. Different assays were carried out in order to optimize sample preparation, chromatographic conditions and Q-TOF parameters. Reference materials were used to establish detection limit of assays, ranging to 20 ng/μl.

The Figure 3a shows extracted biomolecules (MS/MS) of 500 ng of casein compared to Uniprot Database (P04653) and its identification using Agilent MassHunter BioConfirm software. The results allowed to detect enough peptides in order to be compared to Uniprot database.In Figure 3b the results of the analysis showed the presence of some specific fragments for the characterization of Fibroin light chain associated to silk..

REAL SAMPLES. The method was applied to real textile sample from a Fortuny´s tunic, in order to study the presence of a possible proteinaceous adhesive on the surface of a textile. The sample was analysed using the above mentioned methodology and the results showed the presence of some specific fragments for the characterization of the presence of ovalbumin

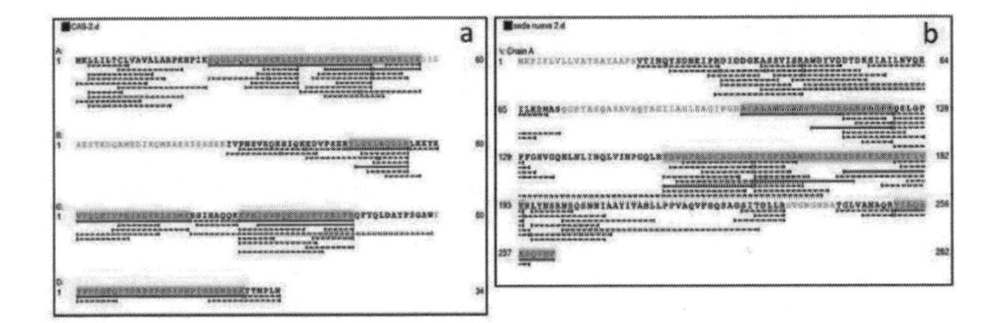

Figure 3a. Protein sequence of a digested Sigma Aldrich Alpha-S1-casein (P04653). Figure 3b: Protein sequence of a digested fibroin light chain associated to silk. (P21828).

Figure 4. a) Image of optical microscopy of the Fortuny´s textile sample; and b) two ovalbumin peptides identified (in green colour) on the textile surface from gallus gallus (Chicken) Uniprot reference (P01012), being compared to blank sample and mobile phase (in red and black colours, respectively) in order to confirm a total absence of impurities and contaminations.

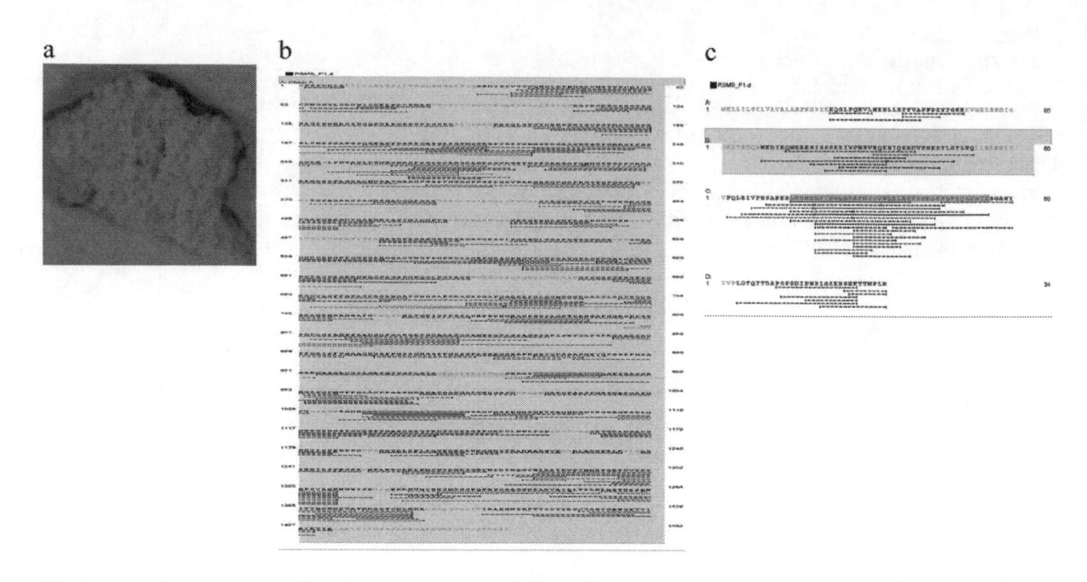

Figure 5. a) Image of optical microscopy of the San Luis de Tolosa sample; b) collagen peptides identified (in green colour) from bovine collagen Uniprot reference (P02453), c) casein peptides identified (in green colour) from casein-bovine alpha a.

from white egg of chicken. The identified protein could be associated to an adhesive between metal powder and textile.

It also was analysed a microsample of the black zone taking from San Luis de Tolosa in order to study the nature of the ground layer.

4 CONCLUSIONS

The present work shows the development of an analytical procedure focused on the identification of the natural proteinaceous source in cultural goods using HPLC/DAD/MS/MS Q TOF. An specific methodology based on an optimized extraction/purification steps and analytical conditions allowed us to obtain sensitivities of 20 ng/μl. Trypsin digestion protocol remains completely consolidated as the preferred method for the identification of the animal proteins, even in presence of aging processes. In fact this is so because the methodology allows in ideal cases to sequence the protein or proteins present, in order to realize the changes suffered by primary structures of the macromolecules. In particular, the presence of ovalbumin, from white egg, were identified as adhesive in a Fortuny´s textile, promoting a better knowledge of Mariano Fortuny and his work in fashion and textiles.

REFERENCES

Crhova M., Kuckova S., Santrucek J., Hynek R. 2014. Peptide mass mapping as an effective tool for historical mortar analysis. *Constr Build Mater* 50: 219–225.

Cuervo D., Loli C., Fernández-Álvarez M., Muñoz G., Carreras D. 2017. Determination of doping peptides via solid-phase microelution and accurate-mass quadrupole time-of-flight LC-MS. *J Chromatogr B An alyt Technol Biomed Life Sci.*1065-1066:134–144.

Fremout W., Dhaenens M, Saverwyns S, Sanyova J, Vandenabeele P, Deforce D, Moens L. 2010. Tryptic peptide analysis of protein binders in works of art by liquid chromatography-tandem mass spectrometry. *Anal Chim Acta.* 658(2):156–162.

Garnier D., Tokarski R. 2016. Proteins in Art, Archaeology, and Paleontology: From Detection to Identification. *Chem. Rev.* 116(1): 2–79.

Lluveras-Tenorio A., Vinciguerra R., Galano E., Blaensdorf C., Emmerling E., Colombini M.P., Birolo L., Bonaduce I. GC/MS and proteomics to unravel the painting history of the lost Giant Buddhas of Bāmiyān (Afghanistan). Plos One 12(4): e0172990.

Mayer R. 2005. Materiales y tecnicas del arte. Tursen-Hermann Blume.

Tokarski C., Martin E., Rolando C., Cren-Olivé C. 2006. Identification of Proteins in Renaissance Paintings by Proteomics. *Anal. Chem.* 78(5): 1494–1502.

Whitcher Kansa, S., Howard S., Campbell S., Thomas-Oates J., Collins M. 2010. Distinguishing between archaeological sheep and goat bones using a single collagen peptide. *J. Arch Sc.*, 37: 13–20.

Science and Digital Technology for Cultural Heritage – Ortiz Calderón et al. (Eds)
© 2020 Taylor & Francis Group, London, ISBN 978-0-367-36368-0

Voids and masonry behind finishing mortars detected by infrared thermography and ultrasonic velocity measurements

R. Bustamante
Escuela Técnica Superior de Arquitectura, Universidad Politécnica de Madrid, Madrid, Spain

R. Otero
Proyecto Geomateriales, Madrid, Spain

ABSTRACT: Infrared thermography (IRT) and ultrasonic velocity (USV) measurements were combined in an experimental study to evaluate their effectiveness when detecting holes and type of masonry behind 15-mm thick finishing mortar. The voids were detected in three prototypes using IRT activated by solar radiation or artificial heat source; masonry was also approximately identified. USV semi-direct measurements gave lower values around the holes. However, differences in mortars' stiffness affect the results obtained in USV measurements, and this was confirmed by tests on samples made with the same mixtures of the finishing mortars.

1 INTRODUCTION

The condition of buildings can often be assessed visually to determine the type of masonry and its discontinuities. But hidden gaps can result from (a) functional changes, e.g., no-longer used niches, doors, windows, passages, arches, etc., (b) after-execution works masonry, or (c) use for passing pipes, downspouts, conduits, etc. Non-destructive techniques are an option for studies prior to intervention (Binda et.al. 2000). Ground Penetrating Radar (GPR) in masonry with two leaves, connected or not, has proved effective to explore internal structure, but IRT is more effective detecting air gaps due to separations of plaster by only 2 mm, compared to the minimum 8-mm spaces that GPR requires (Cotic et al. 2014). Benefits from combining these techniques have been demonstrated for assessing old structures (Kilic, 2015) and for evaluating treatments of materials (Avdelidis et al. 2004). IRT allows the mapping of blind spaces as well as detecting seismic damage (Bisegna et al. 2014), which can be a hybrid technique if combined with sunlight (Sfarra, et al. 2016). IRT shows that cement-mortar coatings have higher surface temperatures than lime-mortar, which is warmer than plaster-mortar (Palaia et al. 2009). Combination of active IRT with digital holographic speckle pattern interferometry offers insight into the bulk of materials by surface illumination of marquetry and wall-painting type of artworks: defects become visible with both techniques (Tornari et al. 2016). Also, the potential of stimulated IRT to detect cracks in metals has been studied (Bodnar, 2014). Characterization by USV has been applied to masonry, concrete, stone and wood. It is often used in historical buildings along with other techniques such as IRT (Grinzato, et al. 2004). It has been shown that, in homogenous materials, radar and ultrasound tomography can recognize voids of size $12\times11\times24 cm^3$ minimum (Wendrich et al. 2006). Juxtaposition of IRT, USV measurements and laboratory analyses was found useful to assess strength of timber (Kandemir-Yucel et al. 2007).

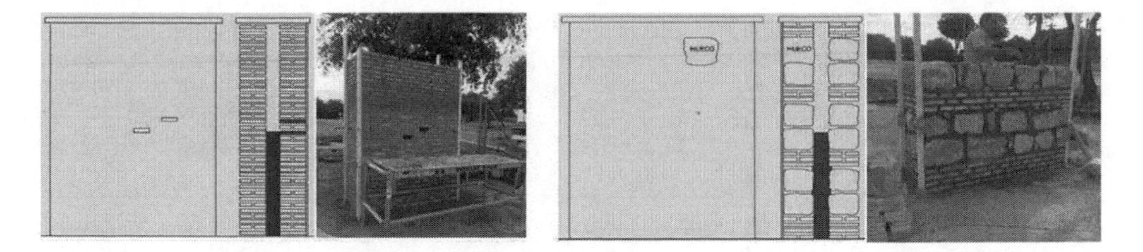

Figure 1. Voids location, wall section, and construction process in outdoor Prototypes 1 (left) and 2 (right).

2 METHODOLOGY

2.1 *Prototypes construction*

Two outdoor prototypes 2-m high×1.50-m wide×0.60-m thick were built at a UPM campus. Prototype 1 is a cavity wall of 2 brick leaves, 23-cm thick each. Prototype 2 is a cavity wall of 2 leaves each with three sections of limestone ashlars (2 rows each) alternating with four sections of brick (4 rows each) (Figure 1). In both prototypes, a 15-mm thick render of lime-Portland cement mix covered one side of the wall: south of Prototype 1 and north of Prototype 2.

Prototype 3 was built in the laboratory to simulate an indoor surface. It consists of two brick leaves separated by an 11-cm empty space, length 70 cm, height 50cm, and width 60cm. USV measurements were made in three conditions of the cavity: empty, half filled with 35-mm size gravel, and completely filled. The 15-mm thick finishing mortar is half gypsum, half aerial lime.

2.2 *Implemented test techniques*

IRT measures thermal radiation and displays surface temperatures using a colour scale. IRT was taken after solar heating of outdoor Prototypes 1 and 2. Measurements were made with a camera Extech Instruments FLIR TIK, MO 297, detector resolution 1024×768 pixels, thermic sensibility (NETD) < 25 mK (0,025°C), special resolution (IFOV), 1.23 mrad with lens 45°, interval of lecture temperature -40°C to 150°C. IRT of Prototype 3 was taken before and after heating with an Osram lamp P2/7 1000W R7s (64741) EMK 189mm, power 1000 W, luminous flux 2500 lm, colour temperature 3200 K. The camera was 2m from the prototype with focus 60cm from the heated surface.

USV measures acoustic properties of materials by recording the propagation of ultrasonic waves in elastic media. The square of the velocity of propagation times the bulk density (ρ_b) equals the dynamic modulus of elasticity (E_{dyn}). A TICO Ultrasonic Instrument ZI 10006 of Proceq Testing Instruments, 54 kHz, was used. Measurement interval was 15 to 6550 μs with

Table 1. Physical and thermal properties of finishing mortars. ρ_b: bulk density; λ: thermal conductivity; *a*: thermal diffusivity; *b*: thermal effusivity; CS: compression strength; USV: ultrasonic velocity; E_{dyn}: dynamic modulus of elasticity.

Mortars	ρb [kg/m³]	λ (1) [W/m.K]	*a* [x10⁻⁹m²/s]	*b* [W.s/m²K]	CS (MPa)	USV m/s	E_{dyn} [GPa]
Portland cement-aerial lime	1837	1.30	708	1550	5.7	2455	10.98
Gypsum plastering	1275	0.80	627	1010	3.75	2152	5.90
Aerial lime	1665	1.00	601	1290	0.5	940	2.60

(1) Technical Building Code of Spain.

resolution of 0.1 μs. Also, prismatic samples 40×40×160mm^3 of the 3 finishing mortars were prepared to determine their physical and mechanical properties (Table 1) and to compare the USV measurements in laboratory with those obtained *in situ*. Mortars were: 1:3 mixture of Portland cement-aerial lime with silica sand; 0.8 ratio water/gypsum; and 1:1.5 mixture of aerial lime with silica sand.

3 RESULTS

3.1 *Thermographic measurements*

Thermograms of outdoor Prototypes 1 and 2 were taken in a very dry, hot day: 37.4ºC at 13:00. On the south side of Prototype 1 the two holes 23x5cm^2 behind the finishing mortar are clearly identified (geolocation 196º S, Figure 2, left). At the same time on the north side of Prototype 2, the space of a missing limestone ashlar around 25×25cm^2 behind the same type of finishing mortar is visible (geolocation 317ºNO, Figure 2, right). Emissivity 0.95.

In the left half of Prototype 3, without a brick on the middle and plastered with gypsum mortar (Figure 3.a), IRT detected the hole of 12×5cm2 behind the plastered surface, at indoor temperature 25°C, relative humidity 29%, prototype surface temperature 33.21°C, and emissivity 0.93 (Figure 3.b). After 2 hours of artificial heating the binding masonry became visible (left of Figure 3.c).

A brick was then removed from the right half, and this half was plastered with lime mortar (Figure 3.d). In the IRT test, the ambient temperature was lower, 20ºC, and it was not possible to detect the hole since the prototype surface temperature was fairly uniform: average 23.49°C (Figure 3.e). After 1 hour of artificial heating, which increased temperatures, the binding

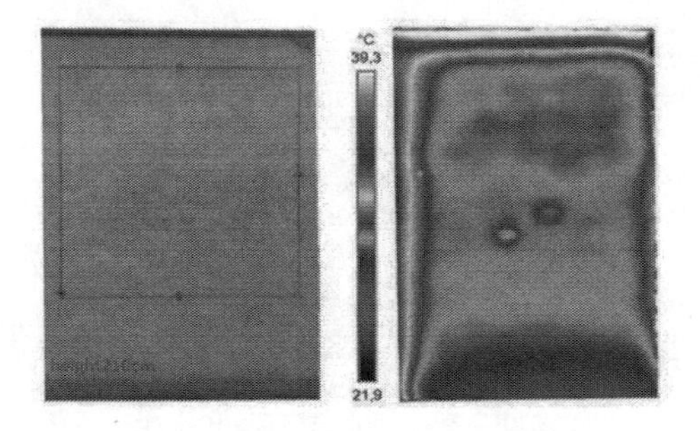

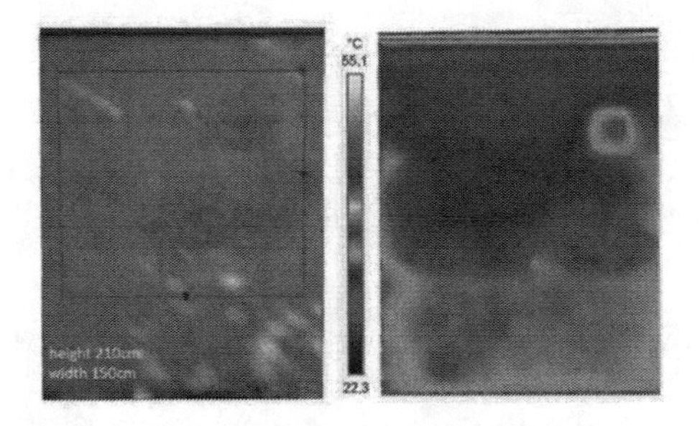

Figure 2. IRT detection of voids behind finishing mortar: in Prototype 1 (left) and in Prototype 2 (right).

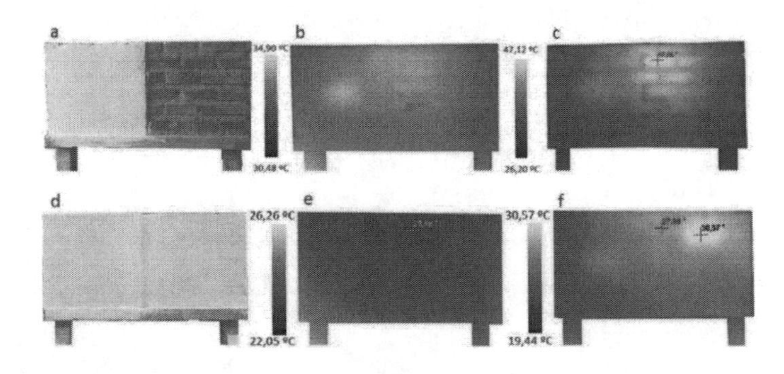

Figure 3. Voids, masonry detected by IRT (Prototype 3).

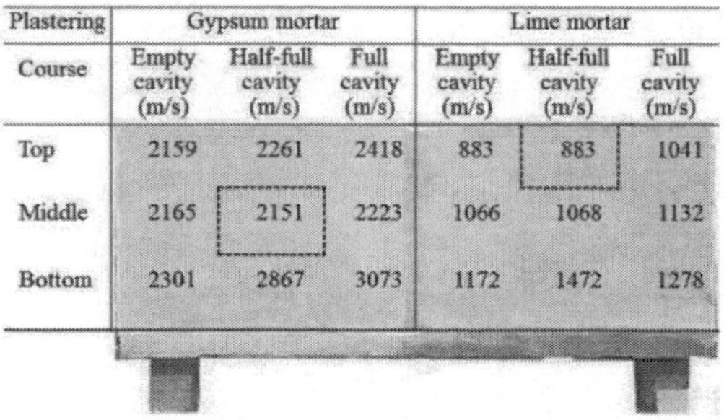

Plastering	Gypsum mortar			Lime mortar		
Course	Empty cavity (m/s)	Half-full cavity (m/s)	Full cavity (m/s)	Empty cavity (m/s)	Half-full cavity (m/s)	Full cavity (m/s)
Top	2159	2261	2418	883	883	1041
Middle	2165	2151	2223	1066	1068	1132
Bottom	2301	2867	3073	1172	1472	1278

Figure 4. Holes (boxes in dotted line) detected by USV measurements (Prototype 3).

could be seen on gypsum and lime plastering, and also the hole under the lime plastering (Figure 3.f). After 5 hours, the void in the lime mortar stayed visible but the brick rows were no longer defined.

3.2 *Ultrasonic velocity measurements*

In Prototype 1, lowest values at 1.10m and 1.20m heights by semi-direct USV measurement (13-cm sender-to-receiver distance) coincided with the masonry holes. The average 1807m/s obtained *in situ* is lower than 2455m/s obtained in laboratory samples. In Prototype 2, with the same mortar, 1503 m/s measured at 1.60m height is close to missing limestone ashlar, and 40% lower than results obtained in mortar samples. Lower values by semi-direct measurement in Prototype 3 were obtained coinciding with the holes indicated in the boxes with dotted lines in Figure 4. Generally, lower values occur in the holes of the top row and with empty cavity. The average interval of 2180±31 m/s (middle row) and 2279±107 m/s (top row) in the gypsum plaster is very close to 2152 m/s obtained in samples. And USV 1089±31 m/s (middle row) and 936±74 (top row) also approach the 940 m/s obtained in the lime samples. It was observed, however, that highest values in the bottom row are produced by backscattering of ultrasonic waves towards the prototype base of concrete reinforced with steel bars.

3.3 *Comparison of samples/in situ ultrasonic velocity measurements*

The lowest result in the USV measurement of finishing mortar of Prototype 2 in comparison with Prototype 1 could be due to irregular thickness of mortar applied on non-flat surfaces of limestone ashlars and variety of joints masonry. USV results in gypsum and lime plastering are close to results obtained in test samples since the plaster has a uniform thickness in Prototype 3. Dynamic modules of elasticity of the tested mortars are approximately: E_{dyn} (Portland cement-aerial lime) = 2 E_{dyn} (gypsum) or 4 E_{dyn} (aerial lime). That is, USV test results are also affected by the mortars' composition: hard silica sand grains and soft calcite matrix in the aerial lime, harder grains in the lime-Portland cement, and homogeneous mass of gypsum plaster.

4 CONCLUSIONS

Missing blocks behind finishing mortars can be detected by IRT complemented by semi-direct USV measurements. The masonry courses, the thickness of the stone ashlars, bricks and binding joints can be characterized by IRT. Lower values of USV measurements were recorded near the holes in the masonry of three prototypes and they were affected by the stiffness (E_{dyn}) of the finishing mortar. However, variations of USV values must be considered in the context of the surface of the wall under study, since constructive aspects can alter *in situ* results: dampness, concrete reinforcements with bars or connecting pieces of the cavity wall leaves.

ACKNOWLEDGEMENT

This work was carried out within the *Proyecto Geomateriales* on non-destructive testing funded by the Comunidad de Madrid (2014-2018).

REFERENCES

Avdelidis, N. P. & Moropoulou, A. 2004. Applications of infrared thermography for the investigation of historic structures. Journal of Cultural Heritage 5:119–127.

Binda, L., Saisi, A., Tiraboschi, C. 2000. Investigation procedures for the diagnosis of historic masonries, Construction and Building Materials 14:199–233.

Bisegna, F., Ambrosini, D., Paoletti, D., Sfarra, S. & Gugliermetti, F. 2014. A qualitative method for combining thermal imprints to emerging weak points of ancient wall structures by passive infrared thermography – A case study. Journal of Cultural Heritage, 15: 199–202.

Bodnar, J.L. 2014. Crack detection by stimulated infrared thermography, Eur. Phys. J. Appl. Phys. 65: 31001 https://doi.org/10.1051/epjap/2014130332.

Cotic, P., Jaglicic, Z. & Bosiljkov, V. 2014. Validation of non-destructive characterization of the structure and seismic damage propagation of plaster and texture in multi-leaf stone masonry walls of cultural-artistic value. Journal of Cultural Heritage, 15: 490–498.

Kandemir-Yucel, A., Tavukcuoglu, A., Caner-Saltik, E.N. 2007. In situ assessment of structural timber elements of a historic building by infrared thermography and ultrasonic velocity, Infrared Physics & Technology, Volume 49, 3: 243–248.

Kilic, G. 2015. Using advanced NDT for historic buildings: Towards an integrated multidiscipli-nary health assessment strategy. Journal of Cultural Heritage, 16: 526–535.

Palaia, L., Monfort, J., Navarro, P., Sánchez, R., Gil, L., Álvarez, A., López, V., Tormo, S. 2009. Infrared thermograph image analysis for the identification of masonry coating in historic buildings, in relation to several samples prepared as patterns. Structural Studies, Repairs and Maintenance of Heritage Architecture X, 373–382, vol 109 WIT Press.

Sfarra, S., Marcucci, E., Ambrosini D. & Paoletti, P. 2016. Infrared exploration of the architectural heritage: from passive infrared thermography to hybrid infrared thermography (HIRT) approach. Materiales de Construcción, 66 [323], July–September 2016, e094.

Wendrich, A., Trela, C., Krause, M., Maierhofer, C., Efner, U. & Wöstmann, J. 2006. Location of Voids in Masonry Structures by Using Radar and Ultrasonic Traveltime Tomography. ECNDT– Tu.3.2.5. Berlin 25–29.

Tornari, V., Andrianakis, M., Hatzigiannakis, K., Kosma, K., Detalle, V., Bourguignon, E., Giovannacci, D., Brissaud, D. 2016. Complementarity of digital holographic speckle pattern interferometry and simulated infrared thermography for Cultural Heritage structural diagnostic research, International Journal of Engineering Research & Science (IJOER), Vol-2, Issue-11, November- 2016.

Science and Digital Technology for Cultural Heritage – Ortiz Calderón et al. (Eds)
© 2020 Taylor & Francis Group, London, ISBN 978-0-367-36368-0

Development of restoration mortars to improve functionality at low temperatures

L. Valdeón-Menéndez, A. Rojo-Álvarez, F.J. Mateos-Redondo & J. Castro Bárcena
Gea asesoría geológica, Llanera, Asturias, Spain

ABSTRACT: One of the main problems presented by traditional lime mortars, usually used in restoration of historical building heritage, is that their degree of carbonation (set processes) slows down significantly or is canceled out under low temperature environmental conditions (<5-10°C). The research undertaken focused on understanding how different types of industrial or even traditional additives can improve the carbonation processes in traditional lime mortars in order to obtain better functionality at low temperatures. Different instrumental technics have also proven their different effectiveness to detect changes in the carbonation processes that are transformed over time at different temperatures. The addition of small quantities of stable chemicals, such as acrylic resins or antifreeze fluids, also inert particles, has provided very promising results in the way that they increase the quality of lime mortars when carbonating up to around 1 degree Celsius of ambient temperature. Below the freezing point no conclusive results have been obtained, although the results have opened new variables for new lines of research that can bring new mortars to the materials conservation industry.

The carbonation quality of most traditionally manufactured lime-sand mortars is impaired when these processes act at low ambient temperatures, below 10°C, fact that have been reflected from roman times mortars research to current traditional mortars manufacturers (Sánchez-Moral, S. 2004. Gamarra Campuzano A. García Mulero M.J. 2012. Arizzi, A., Viles, H and Cultrone, G. 2012). Since industrial Portland mortar types are clearly rejected in historical heritage conservation works, traditional lime-sand mortars are potentiated to be used in restoration projects all over Europe, taken into account that working operations and carbonation processes are strongly limited during severe cold weather with long periods of time below or near the freezing point. Having said that, the main purpose of the research is to find additive components which can help traditional lime mortars to improve their functionality at low temperatures.

For this, the research follows two parallel ways, in one hand by assessing the quality of a variety of mortars while they forge at standard and low temperatures but also by checking the ability of some instrumental and analytical techniques to properly detect the quality properties involved. Several mortars have been prepared by adding a variety of additives to a traditional manufactured lime mortar, checking through different techniques the quality of carbonation.

1 METHODOLOGY

The traditional mortar mix dosage measured in volume is 3 part of 0.2 mm pure silica sand by 1 part of CL90 lime powder, without additives. Once the mix have been prepared the mortar probes go to the climatic chambers to carbonate and being tested.

The temperatures selected: T1 = -18° ± 3°C; T2 = 1 ± 0.5°C; T3 = 8°C ± 3; T4 = 20°C.

*Corresponding author: correo@geaasesoriageologica.com

The additives: Additions to the mix start at some lower percentages 2.5% or 3% and has been increased to 5% and 6% depending on the quality of the results obtained. Often the behavior caused by different concentrations has been compared to foresee the quality tendencies of the resulting mortars. The additives are:

- Chemically active substances: Water soluble thermoplastic polymer (Aquazol by CTS); Water soluble acrylic resin (PRIMAL by CTS).
- Substances and particles that lower the freezing point: Antifreeze ethylene glycol; Virgin beeswax (not soluble in water) (CTS).
- Inert particles: Nanoparticles of silica (TECNAN); Particles of glass fiber (Pintures Sagrada Familia).

Test chamber conditions: 2-6 months carbonation times under the four different temperatures. Various analysis and physical quality parameters were obtained, from the first mixtures specifically mineral composition with X-Ray Difractometry (XRD) and Scanning Electron Microscopy (SEM), porosimetry with Hg porosimeter, electrical conductivity with a conductivity meter probe, pH values with peachometer and ultrasonic velocity with a PUNDIT instrument associated to 1 GH sensors.

2 RESULTS

2.1 *Validity of parameters and instrumental techniques*

While checking the carbonation processes under different techniques, the representativeness of the values serves to classify the effectiveness of each technique.

pH and electrical conductivity show a reduction with carbonation time, but so tiny that we consider is not representative enough to assess this processes (Figure 1). Porometry evaluated by means of Hg porosimetry shows also very tiny variations although some times may inform about the so tiny binder particle variations, usually also too subtle.

DRX mineralogy can inform about the progressive inverse transformation between portlandite and calcite in this case only when clear mineralogical changes has actually been produced at 1°C and 25°C temperatures (Figure 2).

SEM microscopy helps to monitor tiny transformations in cristal development of the binder while changing (Figure 3).

Longitudinal ultrasonic velocity is the property which better show from a mechanical point of view the progressive quality transformation of mortars probes with carbonation time. They are sensitive enough to detect compactation of the probes, even at near the freezing temperature (1-2°C), but measurements taken to probes below the freezing point become not useful since they really does not get stay rigid when defrosted.

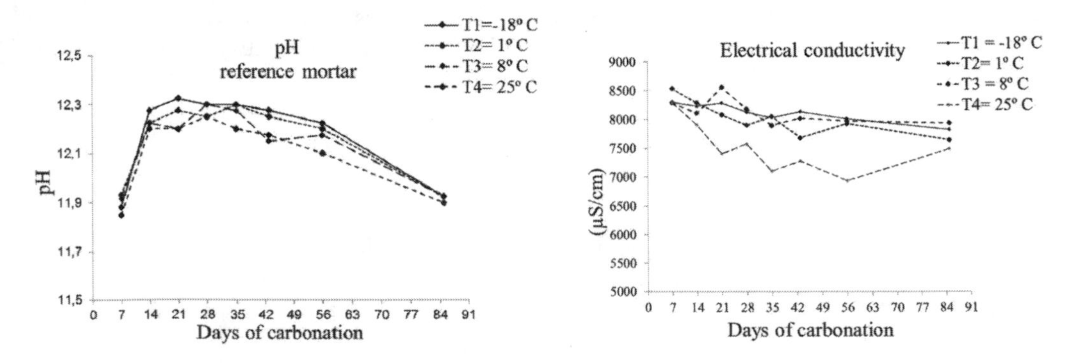

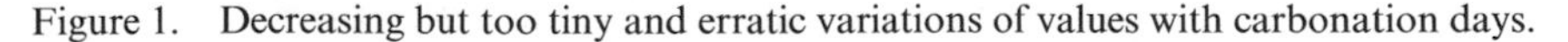
Figure 1. Decreasing but too tiny and erratic variations of values with carbonation days.

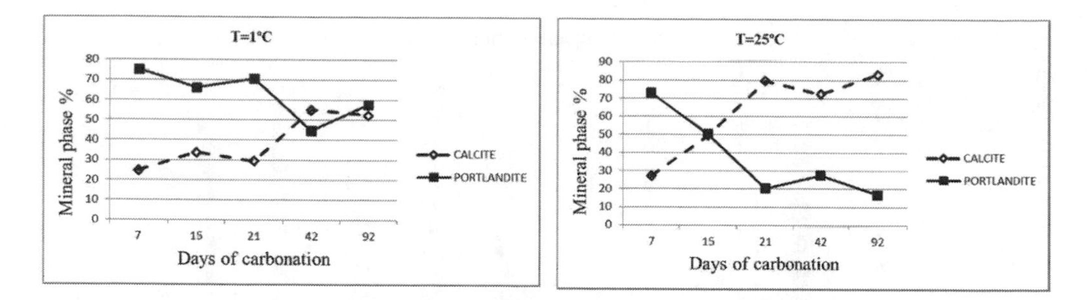

Figure 2. While portlandite decrease, calcite grows up at 25°C, even at 1°C carbonation temperature (left) (relative calcite/portlandite DRX proportions).

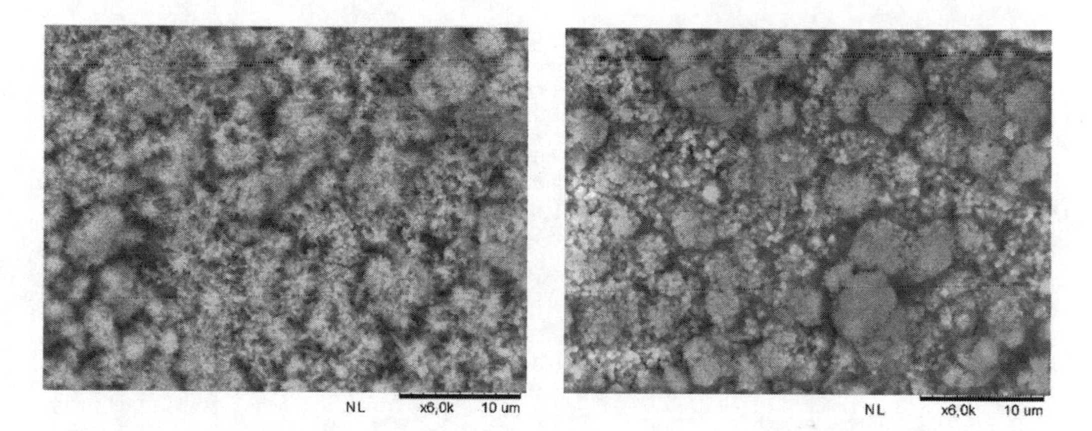

Figure 3. Binder growing calcite crystals from star-like crystal formations while carbonation processes progress (right picture).

2.2 *Quality assessment of low temperature mortars dosed with several additives*

After being trying carbonation processes we have seen the all the probes carbonating under -18°C and 8°C temperatures show no useful results using any additive. At -18°C frozen water inside the porous system do not allow CO_2 to combine with Ca anymore. At 8 or 10°C there are a constant amount of liquid water within the mix and the climatic chamber that permanently keep the mortar probes too wet and soft. The excess of water inside the pore system prevents the exchange of gases which is essential to launch the processes of carbonation.

But, substantial improvements were found in the quality of the mortar probes with some addtives when mortars evolve at 1°C (Figure 4). The ones show in the figure increase the Vp values of lime binder mortar type without additives when carbonating at 1°C.

The better mechanical consistency is achieved with the addiction of 3% acrylic resin, ethylene glycol (5%). Nanoparticles (5%) shows also acceptable results.

It is possible that polymerized resin and nanoparticles cause better Vp values due to their physical although microscopic presence in the mix, but ethylene glycol is chemically integrated into the mortar mixture. In any case the mortars have been mechanically more compact and manageable with these additives.

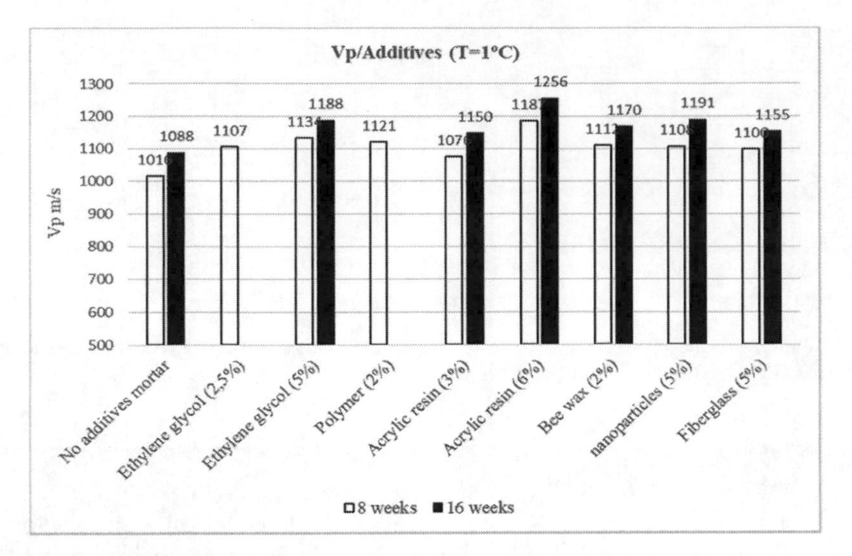

Figure 4. Some additives added to lime mortars increase the intergranular mechanical coherence measured through ultrasonic velocity (Vp). The best values are obtained for the acrylic resin (6%), ethylene glycol (6%) and nanoparticles (5%).

ACKNOWLEDGMENT

European Union; PCTI Asturias; IDEPA Gobierno del Principado de Asturias.

REFERENCES

Arizzi, A., Viles, H. and Cultrone, G. 2012. "Experimental testing of the durability of lime-based mortars used for rendering historic buildings". Construcion and Building Materials 28, 807–818.

Gamarra Campuzano A*. García Mulero M.J.*Gamarra & García, conservación y restauración, S.L.P° Albers, n° 63 08198 St. Cugat del Vallès. 2012: Las cales hidraulicas en la conservación y musealizacion de yacimientos arqueológicos. En II JORNADAS FICALFórum Ibérico de la Cal, FICAL, Laboratori de Materials EPSEB,Grup de Recerca GICITED, UPC.

Sánchez-Moral, S., García-Guinea, J., Luque, L., González-Martín, R. y López-Arce, P. 2003. "Cinética de carbonatación de morteros experimentales de cal de tipo romano". Materiales de Construcción 54, 275.

Science and Digital Technology for Cultural Heritage – Ortiz Calderón et al. (Eds)
 ISBN 978-0-367-36368-0

Electrochemical and microstructural study of La-based protective coatings on metal substrates

A. Doménech-Carbó
Depto. Química Analítica, Universidad de Valencia, Valencia, Spain

J. Peña-Poza, F. Agua, M. García-Heras & M.A. Villegas
Instituto de Historia, CCHS-CSIC, Madrid, Spain

ABSTRACT: The conservation of metallic items in museums and other Cultural Heritage sites need direct protection, e.g. by means of coatings able to delay the advance of corrosion. This work describes the electrochemical study (voltammetry of immobilized particles) of a set of metal slabs of copper, bronze, lead and steel protected with $La(NO_3)_3$- or $La(AcO)_3$- silica based sol-gel coatings subjected to different induced corrosion tests aimed at detecting changes in the metal probes. Microstructural features of coated and uncoated areas after the ageing treatments were studied by field emission scanning electron microscopy. The results pointed out the doped coatings ability to behave as an effective protective strategy in the case of copper, bronze and lead substrates, acting as a barrier effect and/or by means of a self-healing mechanism. However, steel samples need thicker coatings since small pits were detected on the coated surface.

1 INTRODUCTION

Lanthanum-based conversion coatings have been prepared for metals and alloys of technological interest; however, with the unique exception of brass (Fan et al. 2011), there is no disposal of studies on protective coatings on historical metals, in particular those widely represented in artifacts and works of art: copper, brass, lead and iron, whose conservation offers considerable difficulties. Remarkably, testing the performance of protective coatings on historical metal objects has the inherent additional difficulty of the unavailability, by obvious reasons of conservation of the objects integrity, of several analytical techniques for working on historical pieces. The present objectives are: 1) testing the protective properties of $La(NO_3)_3$- or $La(AcO)_3$-silica based sol-gel coatings on copper, bronze, lead and steel probes, and 2) introducing the voltammetry of immobilized particles (VIMP) as a novel methodology for studying such conversion coatings and self-repairing features. It consists (Scholz et al. 2015) on the record of the voltammetric response of a sparingly solid sample attached to an inert electrode in contact with a suitable electrolyte. Since the amount of sample can be restricted at the microgram-nanogram level, the VIMP is particularly applicable in the study of Cultural Heritage.

2 EXPERIMENTAL

The model metal substrates used were copper, bronze, lead and steel slabs from 1.0 to 1.5 mm in thickness, and about 20x30 mm in size. The sol-gel formulation was calculated to obtain a theoretical mol composition of $1La_2O_3{\cdot}99SiO_2$, starting from tetraethoxysilane (TEOS) and as dopants lanthanum acetate ($La(AcO)_3$) and lanthanum nitrate ($La(NO_3)_3$), used separately. The coatings deposition was carried out by dipping at withdrawal rate of 2 mm s^{-1}, 25 °C and

35 % relative humidity. The partial thermal densification was accomplished in a forced air stove at 60 °C for 72 h. The average coatings thickness was about 220±15 nm. Resistance of the coatings was tested by accelerated ageing cycles in a climatic chamber Dycometal model CCK-81 (ISO-9142 standard), and in a Kesternich chamber Dycometal model VCK-300 under SO_2 atmosphere (DIN-50018 standard). Additionally, samples were exposed to an atmosphere of organic acids (mixture of HCOOH and CH_3COOH), since they are usually present in the emissions inside museum showcases. Electrochemical experiments were undertaken in 0.25 M HAc/NaAc, pH 4.75 and 0.10 M H_2SO_4 aqueous solutions using a conventional three-electrode cell and a CH 660I potentiostatic device. Potential were referred to the AgCl (3 M NaCl)/Ag reference electrode. Voltammetric experiments were carried out at sample-modified paraffin-impregnated graphite electrodes of 3 mm diameter (Alpino Maxim, CH type) using VIMP protocols (Scholz et al. 2015). Microstructural and textural features of the coatings were observed by field emission scanning electron microscopy (FESEM) with cold-cathode Hitachi S4800 equipment. Slabs were sputter-coated with graphite on a Jeol JEE 4b vaporizer to make them conductive and observed under 7-15 kV acceleration voltages. A selection of results is the following.

3 RESULTS AND DISCUSSION

3.1 *Copper*

Figure 1 compares the square wave voltammograms from the copper slab coated area with those recorded for two (brown and greenish) uncoated areas of the same slab, after ageing under the Kesternich chamber. A cathodic peak appears at –0.15 V vs. Ag/AgCl (C_{Cu1}), attributable (Doménech-Carbó et al. 2010) to the reduction of cuprite (Cu_2O), forming the primary patina, to copper metal. The signal C_{Cu1} is followed by a broad wave at ca. –0.75 V vs. Ag/AgCl (C_{ox}), corresponding to the reduction of dissolved oxygen. In the uncoated areas the cathodic peak C_{Cu1} is enhanced relative to that recorded for coated areas. This feature denotes the possible presence of 'green' copper corrosion products from malachite ($Cu_2CO_3(OH)_2$), atacamite ($Cu_2Cl(OH)_3$) and/or brochantite ($Cu_4SO_4(OH)_6$), whose reduction to Cu° occurs at a potential almost identical to that of cuprite. Atacamite is discarded since no Cl^- is expected in the test.

The microstructure observed by FESEM (Figure 2a) shows a precipitate emerging from the cracked coating plus microcrystals formed between the metal surface and the coating. Estimated average size of the precipitate grains is ~20 μm, while for the

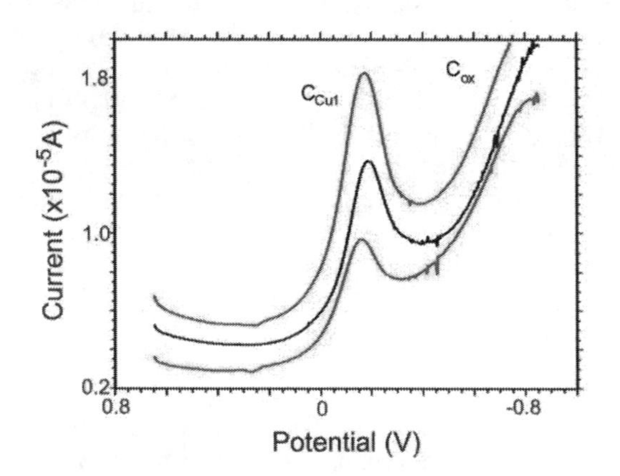

Figure 1. Square wave voltammograms of coated (with $La(NO_3)_3$ doped silica) area of the copper slab (red line); uncoated brown area (black line); uncoated green area (green line), after ageing under the Kesternich chamber.

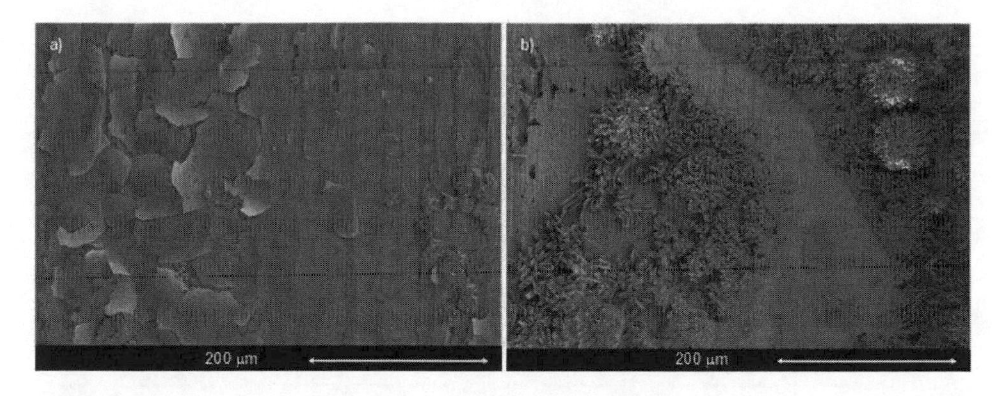

Figure 2. FESEM images of the copper slab after ageing under the Kesternich chamber conditions: a) area coated with $La(NO_3)_3$ doped silica, b) uncoated area.

microcrystals is <5 μm. These microcrystals are compatible with those studied by Zhang et al. (2009). The precipitate of the uncoated area (Figure 2b) could be attributed to the quick formation of a corrosion layer, composed by malachite (humid carbonation) and brochantite (acid attack), provided by the H_2SO_4 generated during the ageing under the Kesternich chamber. This suggests that, even though the sol-gel coating was microcracked during ageing, the early generation of cuprite, induced by the lanthanum dopant, could seal the cracks as a repairing barrier.

3.2 *Bronze*

The voltammetric behavior of bronze slabs after ageing under the Kesternich chamber is similar to that of copper. After the treatment under the organic acids environment, on the coated area (coated with $La(AcO)_3$ doped silica) the voltammogram indicates the presence of cuprite, while upon the uncoated area small green pits appeared. These could only be attributed to malachite, since neither Cl^- ions nor SO_4^{2-} ions are expected during the organic acids treatment to induce green species such as atacamite or brochantite. In the FESEM images of the coated area after the organic acids treatment, the main feature is the layer preservation, which could be due to the obvious improved chemical resistance of bronze compared to copper. On the coated area small grains of low crystallinity cuprite are detected. On the uncoated area a roughly microstructure with a continuous precipitate is observed, suggesting that a secondary cuprite patina has been formed.

3.3 *Lead*

In the voltammograms of lead samples (Figure 3a) cathodic peaks at 1.0, 0.6 and -0.8 V appear. The two first signals (marked Pb(IV)) are attributable to the reduction of different Pb(IV) forms, associated to plattnerite (PbO_2), whereas the more prominent peak at -0.80 V (marked Pb(II)) corresponds to the reduction of litharge (PbO) to metal lead.

The signal at 1.0 V can be assigned to an "impermeable" crystalline PbO_2 form, while the signal at 0.6 V could be attributed to a "porous" less crystalline (non-stoichiometric) PbO_2. In the anodic scan voltammogram (Figure 3b), the oxidative dissolution of the deposit of lead metal previously formed occurs at -0.60 V (marked Pb) accompanied by minor peaks attributable to the formation of lead deposits of different grain size. In comparing the voltammograms for the parent uncoated areas (Figure 3c and 3d), the relevant feature is the lowering of the Pb(IV) peak at 1.0 V and the broadening of the oxidative dissolution of lead. The protective coating has apparently increased the "over-oxidation" of lead to form a significant

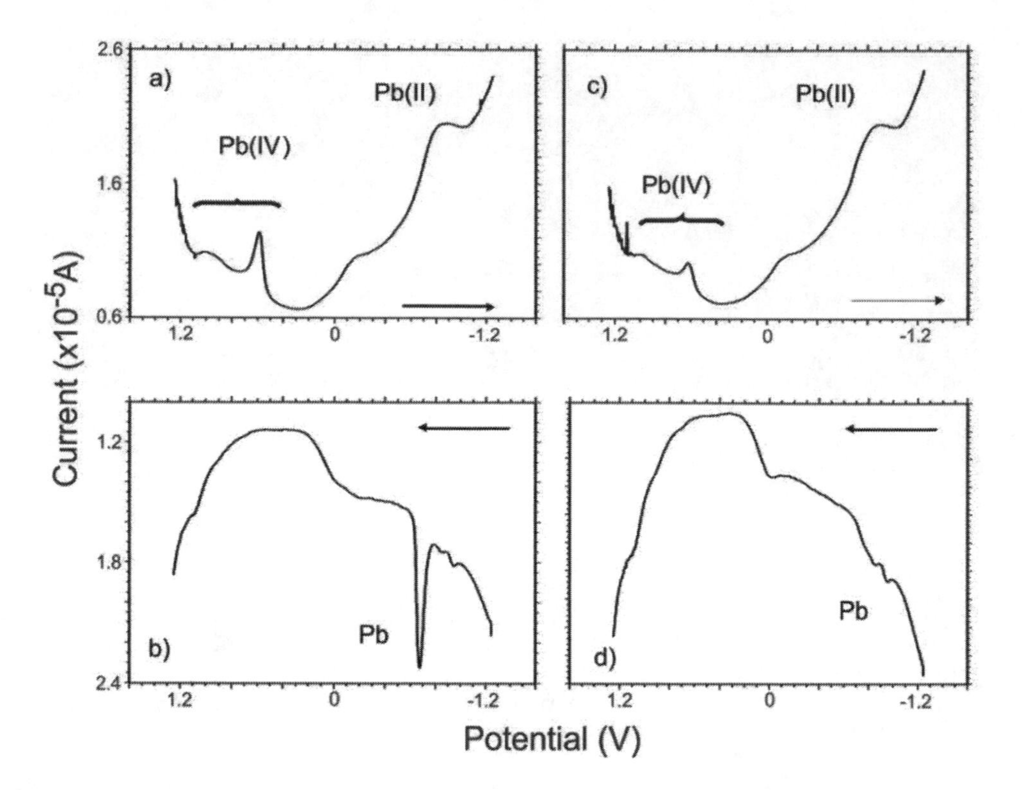

Figure 3. Square wave voltammograms of: a) and b) lead sample coated with $La(NO_3)_3$ doped silica and aged under the Kesternich chamber; c) and d) uncoated area of the same slab. Potential scan initiated at a) and c) 1.25 V in the negative direction; b) and d) -1.05 V in the positive direction.

amount of PbO_2 forms (Figure 3a). The microstructure of the coated area after the Kesternich chamber treatment shows good resistance against the ageing, and small precipitates appeared with incipient crystallinity (probably elongated plattnerite crystals ~1 μm). The parent uncoated and aged area showed a profuse crystallization in the shape of lamellae, probably due to litharge and/or to another Pb(II) compound such as anglesite ($PbSO_4$). The presence of cerussite ($PbCO_3$) and/or hydrocerussite ($Pb_3(CO_3)_2(OH)_2$) cannot be discarded, since they use to crystallize in the lamellae shape.

3.4 *Steel*

The voltammogram of the steel sample coated with $La(AcO)_3$ doped silica, weathered under the climatic chamber conditions, on the uncorroded coated area is dominated by the reductive dissolution of crystalline hematite (Fe_2O_3) at -0.40 V, with a shoulder at ca. -0.30 V attributable to a low crystalline hydrated hematite, and a wave at ca. -0.15 V, which correspond to crystalline and hydrated goethite forms α-FeO(OH). On the slightly corroded area of the aged coated sample, the hematite signals vanish and the voltammogram was dominated by the reduction of crystalline goethite, whereas on the uncoated aged area corrosion pits and a rust layer was detected corresponding to the reduction of hydrated goethite. Concerning microstructure, the coated and aged area shows heterogeneous precipitates distributed as pits that, according to the voltammograms, are attributed to hematite and/or goethite (crystalline and/or hydrated). The uncoated and aged area presents a corrosion layer with pits and abundant heterogeneous precipitates related to hydrated goethite.

4 CONCLUSIONS

Copper: on coated areas more or less crystalline and hydrated cuprite is formed; on uncoated areas cuprite, malachite and brochantite appeared. The lanthanum doped coating behaves as a repairing barrier providing a fine sealing cuprite precipitate that hinders corrosion. Bronze: on coated areas cuprite crystals are formed, while malachite appeared in the greenish patina of uncoated areas. A combined protective mechanism took place: some barrier effect due to the resistant coating, and generation of a fine precipitate enhancing the barrier effect and/or sealing the coating microcracks. Lead: on coated areas a self-repairing microcrystallization occurs, while on uncoated areas a common patination/passivation layer is formed. The lanthanum role can be interpreted in terms of the inhibition of the cathodic corrosion counterpart as the result of the probable local precipitation of lanthanum oxides or hydroxides. Steel: the corrosive tests determine the replacement of crystalline hematite on the coated non corroded areas by hydrated hematite and goethite in partially corroded areas of the coating. On uncoated areas the corrosion layer shows pits and heterogeneous precipitates of hydrated goethite. Thicker coatings are needed in steel to provide an effective barrier against corrosion.

ACKNOWLEDGEMENTS

Financial support of MINECO/FEDER MAT2015-65445-C2-2-R and TOP Heritage ref. P2018/NMT-4372 (CAM-EU). Professional support of TechnoHeritage Network on Science and Technology for the Conservation of Cultural Heritage.

REFERENCES

Doménech-Carbó, A., Doménech-Carbó, M.T., Martínez-Lázaro, I. 2010. Layer-by-layer identification of copper alteration products in metallic works of art using the voltammetry of microparticles approach. *Analytica Chimica Acta* 610: 1–9.

Fan, H., Li, S., Zhao, Z., Wang, H., Shi, Z. 2011. Improving the formation and protective properties of La-conversion coatings on brass by use of La_2O_3 nanoparticle incorporation with electrodeposition. *Corrosion Science* 53: 3821–3831.

Scholz, F., Schröder, U., Gulaboski, R., Doménech-Carbó, A. 2015. *Electrochemistry of Immobilized Particles and Droplets*. Berlin-Heidelberg: Springer International Publishing, 2nd edition.

Zhang, D.F., Zhang, H., Guo, L. et al. 2009. Delicate control of crystallographic facet-oriented Cu_2O nanocrystals and the correlated adsorption ability. *Journal of Materials Chemistry* 19: 5220–5225.

Science and Digital Technology for Cultural Heritage – Ortiz Calderón et al. (Eds)
© 2020 Taylor & Francis Group, London, ISBN 978-0-367-36368-0

Red resins, lapis lazuli and vermilion: High symbolism luxury pigments identified in a *Bifaz* from the late fifteenth century preserved in the Camp del Turia (Valencia)

E. Alba Pagán & M.L. Vázquez de Ágredos-Pascual
Art History Department, Universitat de València, Valencia, Spain

M. Miquel Juan
Art History Department, Universidad Complutense de Madrid, Madrid, Spain

L. Rojo Iranzo
Art History Department, Universitat de València, Valencia, Spain

ABSTRACT: The use of red resins to represent the blood of Christ began to be frequent in the Middle Ages. These additives of organic origin had a viscous texture and a color that resembled the vital liquid, and that contributed to give more realism to the Passion of Christ. The study of the pictorial film of a *Bifaz* of the late fifteenth century from Camp del Turia (Valencia) has been developed in the *Laboratory of Analysis and Diagnosis of Art Work* of the University of Valencia, and the *Science Park* of the same institution, using techniques such as Fourier Transform Infrared Spectroscopy (FTIR) and Gas Chromatography-Mass Spectrometry (GC-MS). Both techniques have revealed the employ of a resin colored for this purpose, and other materials of great importance and tradition in the representation of the sacred images of Christianity, as is the case of the blue pigment that was used in the veil of the Virgin: the expensive lapis lazuli. This mineral, from the pigment was made, is a complex sulfur-containing sodium aluminum silicate $(Na,Ca)_8(AlSiO_4)_6(SO_4,S,Cl)_2$. The cost of some of these raw materials, given their distant origin, allows relating the artistic production of this icon with a high color economy, and a strong religious symbolism.

1 THE *BIFAZ* OF THE CHURCH OF *SANT JAUME APOSTOL* (LA POBLA DE VALLBONA, VALENCIA). THE FIRST NEWS IN HISTORICAL SOURCES

The medieval *Bifaz* of the church of *Sant Jaume Apostol* at La Pobla de Vallbona (Figure 1a-b) is an almost unique treasure, representing the true face of the Christ and the Virgin “de los Dolores”. It is a monstrance, with a structure of ornate and gilded wood, commonly used in private prayers, testimony of a tradition rooted in the ancient Crown of Aragon since the XVth century, which was especially focused on the devotion to the image of the Virgin. This territory was the family patrimony of King Martí l’Humà and María de Luna, and on her death in 1406 was inherited by her son Martin of Sicily, and from 1409 by his bastard son Federic, of whom Alfons el Magnànim was tutor. The king integrated them to the Royal Patrimony, passing in 1436 to his brother Enric and later to the infant Enric Fortuna. This *Bifaz* is related to the splendorous medieval past of the church, when it still depended on the prestigious *Cartuja de Portacoeli* in Serra (Valencia). From that same time are the mural paintings in Linear Gothic style (Figure 2) and the splendid table (coming from the altarpiece) of Sant Jaume, work of Joan Reixach that has great resemblance to the *Monstrance-Bifaz*. Its close relationship with the altarpiece of Reixach that is preserved in Pego (Alicante) is also evident. The first written notice of the *Bifaz* goes back to 1759. We owe It to the general visitor at the time of Archbishop Andrés Mayoral, who in his inventory of the church spoke of an image of “Jesus and Mary

Figure 1a-b. The *Bifaz* conserved at the Church of *San Jaume Apostol*. Figure 1a (to the left): the side painted with the *Virgen de los Dolores*. Figure 1b (to the center) the side with the face *of Christ (Veronica)*. Figure 2. (to the right) medieval mural paintings in lineal Ghotic style conserved at the Church of *Sant Jaume Apostol (detail)*.

painted in a square with a foot for the Supplications" (*Church Archives*, Visit 1759).This is an image of devotion or *Andachtsbild*, which in Valencia had a fundamental role in the renovation of the International Gothic style towards new forms imported from Nordic countries, but in this case stands out for presenting the true image of Christ or Veronica, or in other words: the face of Christ impregnated in the cloth during the Passion. The presence of *vero ikon* in Valencian lands is linked to the image of the Virgen de los Dolores, sometimes with the Ecce Homo on the reverse, as in the Parish of Madonna of the Assumption in Pego (Alicante) and the Archbishop's Palace in Valencia, and *The Virgin* of the National Gallery (NG1335), possibly attributed to the Valencian school of Reixach. The physicochemical study of the colors used in this work of art gave coverage to questions of great relevance to the historical-artistic and cultural level. In particular, there were four objectives that we proposed with this research: (1) to carry out the physicochemical characterization of the *Bifaz*, (2) to provide information on the possible chronology of this important piece of art, (3) to establish relationships with other similar works of art made between the Middle Ages and the first Renaissance, and (4) to contribute to the symbolic-cultural study of the *Bifaz* starting with the scientific results obtained on its materiality.

2 MATERIALS AND METHODS

A total of 12 samples from the Bifaz were taken (Table 1). The physicochemical analysis techniques that were used in the study were as follows:

(a) *Optical microscopy (LM): we use* a Leica DMR optical microscope with an incident/transmitted light system and polarization system in both cases.
(b) *Scanning Electronic Microscopy combined with energy dispersive X-ray (SEM-EDX)*: The sample has been analyzed using a Hitachi Variable Pressure Scanning Electron Microscope (VP-SEM) model S-3400N, equipped with a Bruker dispersed X-ray energy spectrometer model XFlash® with a silicon drift droplet detector (SDD), with Dura-Beryllium window (8 μm) and energy resolution of 125 eV @5.9 keV. The analysis was performed

Table 1. Coloring materials identified in the Bifaz conserved at the Church of Sant Jaume Apostol.

Sample	Colour	Coloring material identified	Chemical formula
		VERGIN SIDE OF THE *BIFAZ*	
BZ-1	$White_1$	Lime White	$CaCO_3$
BZ-2	$White_2$	Lead White	$(PbCO_3)_2 \cdot Pb(OH)_2$
BZ-3	Black	Carbon black	Carbon Black
BZ-4	Red	Hematite	Fe_2O_3
BZ-5	Brown	Ochre	Fe-based compounds
BZ-6	Blue	Natural Lapis Lazuli	$(Na,Ca)_8(AlSiO_4)_6(SO_4,S,Cl)_2$
		CHRIST SIDE OF THE *BIFAZ*	
BZ-7	$White_1$	Lime White	$CaCO_3$
BZ-8	$White_2$	Lead White	$(PbCO_3)_2 \cdot Pb(OH)_2$
BZ-9	Black	Carbon black	Carbon Black
BZ-10	Red_1	Hematite	Fe_2O_3
BZ-11	Red_2	Red resin + Vermilion	Red resin + HgS
BZ-12	Brown	Ochre	Fe-based compounds

using 20 kV, a working distance of 10 mm, and pressure inside the chamber of the chamber of 60 Pa.

(c) *RAMAN Spectroscopy*. Raman analyses were performed using a HORIBA Jobin Yvon iHR320 spectrometer with a Peltier-cooled CCD for detection and 785 and 532nm doubled YAG laser as excitation. A 50x magnification LWD objective was used to focus the laser on the sample and collect the scattered light. The measurements were performed using a laser power between 10 and 30mW, an integration time of 120s and up to 5 spectral accumulations.

(d) *Fourier Transform Infrared spectroscopy (FTIR)*. Fourier Transform Infrared Spectroscopy analyses of potential organic binders were obtained with a Thermo Nicolet Nexus spectrophotometer. Spectra were acquired between 4000 and $400cm^{-1}$, 64 scans with $4cm^{-1}$ of resolution. FTIR coupled with ATR (Attenuated Total Reflectance) was also performed in the same analytical conditions.

(e) *Gas chromatography-mass spectrometry* (GC-MS): a technique specializing in the characterization of organic substances. The gas chromatograph used in the characterization of the samples was an Agilent 6890N (Agilent Technologies, Palo Alto, CA, USA). The chromatograph is docked with an HP 5973 mass detector. The column used was an HP-5MS-5% phenyl and 95% polydimethylsiloxane). The oven temperature program was 60-220 °C, with an increase of 1 °C/min; the temperature was maintained at 220 °C for 3 min. The injector temperature was at 250 °C. The injection volume was 1 μ L (95:5), with an inlet pressure of 7.96 psi. The carrier gas was He. The interface temperature was 280 °C. For the mass detector, the ionization temperature was set at 230 °C. The GC-MS database (NIST Library version 2002) was used for the possible identification of the organic components.

3 RESULTS AND DISCUSSION

The paint support is wood of probably local origin. Chromatographic analyses identified the use of a protein-type organic binder to link the pigment particles to each other (and to these with the support). The painter prepared the wood of both sides with a thin layer of gypsum $CaSO_4 \cdot 2H_2O$ and a protein-nature additive (possibly egg yolk). This composition corresponds to one of the most traditional in the icons and altarpieces of the Middle Ages. Also the colors that the painter employed for the sacred images on both sides of the *Bifaz* are characteristic of the XIIIth and XIVth centuries. The face of the Virgin was painted with carbon black, ochre earth, hematite Fe_2O_3, white lime ($CaCO_3$) and white lead $(PbCO_3)_2 \cdot Pb(OH)_2$. This last

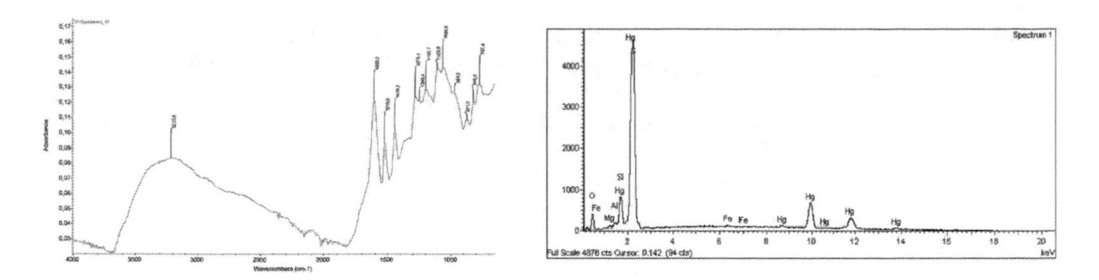

Figure 3a-b. Spectrums by FTIR and EDX of the two components mixed by the painter to prepare the coloring material used to represent the blood of Christ: a red resin (to the left: spectrum by FTIR) and vermilion (to the right: spectrum by EDX).

pigment was also used in the veil that covers his head, whose dark blue was identified as lapis lazuli. On the other hand, the painter employed in the face of Christ was made carbon black, ocher, hematite, white of lime and white of lead. A mixture of carbon black and ochre served him to obtain the colour dark brown of the hair. Finally, the red lines that come out of the head of Christ, and that must refer to the blood of Passion, were made with a mixture of one red resinmixed with vermilion (HgS) (Figure 3a-b).

In the Middle Ages it was very common to use these resins to represent the blood of Christ, especially the so-called Blood of Drago (*Dracaena cinnabari*). This is a dark red dye whose origin is located in East Asia. These red resins, of great symbolism, were also widely used in medieval times for the decoration of manuscripts and to prepare coloring varnishes. Between the thirteenth and fifteenth centuries the red resin more similar to the Blood of Drago that was imported to the Iberian Peninsula for different artistic and medicinal purposes came from the Canary Islands (*Dracaena draco*) (Calvo 1997:202). The use of this type of red resins, therefore, to represent the blood of Christ in the *Bifaz* of the Church of *Sant Jaume Apostol* represents a choice of high religious significance. The importance of this organic color was reinforced with that of vermilion which also had a high symbolism in the Middle Ages. In this sense, in the earliest version of the *Mappae Clavícula*, dating from about 800 A.C., there are two descriptions for making artificial cinnabar (or vermilion). According with Thompson (1933): "Artificial vermilion was the most important medieval addition to the palette of classical antiquity and rules for its manufacture must have been in great demand for some centuries after its discovery" (Gettens *et al.* 1993:161-162). These same authors indicate that "the manufacture of vermilion was a novelty in the eighth century, a commonplace in the fourteenth" (*Ibidem*, 162). For its part, lapis lazuli, in this period was probably imported from Afganistan into Europe mainly by way of Venice, the principal port for trade with the East. The high cost of this imported raw material and the long laborious process of extraction explain that the value of lapis lazuli pigment in the Middle Ages was comparable to that of gold (Plesters 2002:38-39).

4 CONCLUSIONS

The physicochemical results obtained with the scientific study of the *Bifaz* of the Church of *Sant Jaume Apostol* (Pobla de Vallbona, Valencia) offer interesting information about the high quality of the piece. At least two of the coloring matters identified were imported: the lapis lazuli and the vermilion. While the source of origin of the first was the Middle East through Venice, the raw material to prepare the vermilion would possibly proceed from the most important quarry in Spain from Antiquity: Almaden, which are still the most relevant source of mercury in the world. Although the source of origin closest of the resin named *Dracaena draco* was the Canary Islands, it is possible that the origin substance identified to represent the blood of Christ (and mixed with the vermilion) corresponds to another red resin of local origin and properties very similar to the additive obtained from *Dracaena* species. The complex manufacturing process of these three coloring matters and the high knowledge that the painter should have in their respective use -including the possible technical incompatibilities that would have meant a rapid deterioration of the *Bifaz*- also refer to the high technical quality of this small piece and the great skill of his artist, possibly very close to Joan's Reixach workshop (or Joan Reixach himself). This last conclusion is also possible thanks to the historical and stylistic study of this work of art.

BIBLIOGRAPHY

Alba, E. 2011. "Arquitectura y Artes figurativas. La iglesia de San Jaime Apostol. Reforma y transformación: aditamentos barrocos y renovación de la iglesia". In: *La pobla de Vallbona: historia, geografía y arte*. Valencia: Universidad de Valencia.

Alba, E.; Vázquez de Ágredos, M.L.; Griñena, P. 2019. "Colour analysis and mural archeology for a restoration and conservation project for the walls of the church of Sant Jaume in la Pobla de Vallbona (valencia). Project for the restoration of this medieval baroque church", In: *11th European Symposium on Religious Art, Restoration & Conservation (Proceeding book)*, pp. 108–112 (M.L. Vázquez de Ágredos, I. Rusu, Cl. Pelosi, L. Lanteri, A. Lo Monaco & N. Apostolescu, Eds.). Torino: Lexis Compagnia Editoriale.

Calvo, A. 1996. *Conservacion y Restauración. De la A a la Z*. Barcelona: Ediciones Del Serbal.

Gettens, R.J. 2002. "Vermilion and Cinnabar". In: *Artist's Pigments: A Handbook of their History and Characteristics*, vol.2, pp.37–65 (A. Roy, Ed.). London: Archetype Publications.

Miquel, M. 2013. "¡Oh, dolor que recitar ni estimar se puede! La contemplación de la piedad valenciana medieval a través de los textos devocionales". In: *Anuario de la Historia de la Iglesia*, 22, pp. 291–315. Pamplona: Universidad de Navarra.

Plesters, J. 2002. "Ultramarine Blue, Natural and Artificial". In: *Artist's Pigments: A Handbook of their History and Characteristics*, vol.2, pp.37–65 (A. Roy, Ed.). London: Archetype Publications.

Science and Digital Technology for Cultural Heritage – Ortiz Calderón et al. (Eds)
© 2020 Taylor & Francis Group, London, ISBN 978-0-367-36368-0

Evaluation of silver nanoparticles effectiveness as biocide by multi-spectral imaging

J. Becerra & A.P. Zaderenko
Department of Physic, Chemic and Naturals Systems, Universidad Pablo de Olavide, Seville, Spain

I. Karapanagiotis
Department of Management and Conservation of Ecclesiastical Cultural Heritage Objects, University Ecclesiastical Academy of Thessaloniki, Thessaloniki, Greece

P. Ortiz
Department of Physic, Chemic and Naturals Systems, Universidad Pablo de Olavide, Seville, Spain

ABSTRACT: A biocidal treatment was developed using silver nanoparticles which were stabilized with tetraorthosilicate (TEOS). The composite material was tested as inhibitor of biofilm growth on limestones. Samples were subjected to artificially accelerated weathering conditions for 60 days. The chromatic change (ΔE^*) of the stone surfaces were measured. Moreover, a complementary method was designed, that allows the quantification of the biofilm growth, using multi-spectral images at different wavelengths. In particular, segmentation of homogeneous regions (biofilm and stone) was carried out, to quantify the surface area which was affected and altered.

The capability of the devised method to study the effectiveness of biocidal treatment was assessed.

1 INTRODUCTION

Biodeterioration is a common damage in historical buildings. Restorers have been using different routes to prevent biodeterioration. These are usually chemical treatments which have some disadvantages such as, for instance, short-term effectiveness, high toxicity and incompatibility with the original substrate (Nugari and Salvadori, 2003).

Recent advances in nanotechnology and nanomaterials offer the possibilities to apply new treatments for the protection of Cultural Heritage materials. In this regard, silver nanoparticles (NPs) have shown excellent biocidal properties (Banach and Pulit-Prociak, 2016) which are related to multifactorial processes such as deterioration of cell wall or inhibition of protein synthesis by the microorganisms (Nowicka-Krawczyk et al., 2017). The biocidal properties are related to the morphological characteristics of silver NPs, particularly with their size and shape (Raza et al., 2016). Silver, combined with other NPs (e.g. titanium oxide) have been recently suggested as biodeterioration inhibitors (Becerra et al., 2018).

2 MATERIALS AND METHODS

2.1 *Synthesis*

Silver NPs stabilized with tetraorthosilicate (AgTEOS) were synthetized according to Becerra et al. (2019) as briefly described next. Silver nitrate ($AgNO_3$) was purchased from Panreac, TEOS from Acros and sodium borohydride ($NaBH_4$) from Sigma-Aldrich. All other chemicals were reagent grade. Water was purified using a Milli-Q reagent grade water system from Millipore. Briefly, an aqueous solution of $NaBH_4$ (15 mL, 0.12 mmol) is added to an aqueous solution of $AgNO_3$ (15 mL, 0.015 mmol) under magnetic stirring in ice bath. Then, a solution of 33 µL of TEOS in 20 mL of ethanol is added. The resulting suspension is stirred for 1 hour in ice bath.

AgTEOS nanoparticles were characterized by dynamic light scattering (DLS; Zetatrac Analyzer, Microtrac, USA) and UV-visible (UV-Vis) spectrophotometry (Ocean Optics) equipped with an HR4000 detector (Dunedin, FL, USA).

2.2 *Treatment application on limestone*

The effectiveness of AgTEOS NPs was evaluated for the protection of limestone slabs which were obtained from Estepa quarry (Sevilla, Spain). The selected substrate is a white oolitic limestone which has a total porosity of 13%, an average pore radius ranging within 0.022-7.87 µm and has been widely used in historical buildings in southern Spain.

Four doses of 100 µL of AgTEOS NPs suspension (concentration: 0.03 mg/mL) were brushed on the stone samples. The samples were left to dry at room temperature for three days.

The treated samples were analyzed by colorimetry (PCE-CSM 2), optical microscopy (Zarbeco MiScope MP2) and scanning electron microscopy (SEM, JEOL JSM-6510). The color changes (ΔE^*) were calculated according to the equation $\Delta E^*=(\Delta L^2+\Delta a^{*2}+\Delta b^{*2})^{1/2}$, where ΔL, Δa^* and Δb^* are the color variations caused by the treatments and are defined by the CIELAB scale.

2.3 *Accelerated microalgal growth assay*

A greenhouse was designed and built to keep the relative humidity and temperature at 90±5% and 28±2°C, respectively, and to apply a natural day-night cycle with average sunlight of 14 h.

Biofilms were cultivated/produced on samples which were extracted from facades of historical buildings of Seville. A dose of 100 µL of microalgal culture (OD750nm: 0.1) was spread onto stone surface, and the samples were placed in a platform with a constant source of capillary absorbed water. The duration of the assay was 60 days.

The treatment was evaluated by colorimetry, optical microscopy and multispectral imaging (multispectral camera MuSIS HS). The images were captured at wavelengths of 420, 520, 620 and 720 nm. Biofilm extension quantification (E) was measured using Image-Pro Plus 6 software. The images were converted to greyscale and, based on the spectral study of the greyscale histogram, a threshold analysis was carried out for the binarization of the images. The limit between altered and non-altered zones was established at the value of 110. Zones with values higher than 110 were classified as altered zones. Consequently, the biofilm extension was calculated as the percentage of pixels measured withitn the altered zone versus the total number of pixels.

3 RESULTS AND DISCUSSION

3.1 *Synthesis*

The AgTEOS suspension showed high stability due to the small size of the NPs and their zeta potential (125±19 mV), which is higher than the value reported by Koutsoukos et al. (2006) for stable NPs suspensions (|30|). The average hydrodynamic diameter was 5±1 nm, which is in agreement with the interpretation of the peak silver surface plasmon made by Raza et al. (2016). In this case it was placed at 385 nm, meaning a NPs size less than 20 nm.

3.2 *Treatment application on limestone*

The ΔE^* caused by the treatment on the limestone sample was of 6.8±0.4. This induced color change is perceived by the human eye (Figure 2.) but it is only slightly higher than the maximum value recommended (Pinho et al., 2015) for conservation treatments of the Cultural Heritage ($\Delta E^*<5$).

The comparison between untreated and treated limestone surfaces at different magnifications (Figure 1) showed that the AgTEOS NPs did not form superficial layer on limestone and did not block the pores of the stone. This is due to the small size of the NPs which can penetrate the limestone pores.

3.3 *Accelerated microalgal growth assay*

Aesthetic changes caused by algal fouling was visible on untreated samples after accelerated microalgal growth (Figure 2.A), whereas on treated samples the formation of algal fouling was prevented (Figure 2.B). In particular, ΔE^* on untreated samples was 13.9. However, on samples treated with AgTEOS, ΔE^* was only 3.1 corresponding to a color change which is hardly perceived by the human eye.

Figure 3 shows images captured at different wavelengths. The extension of the biofouling measured at λ=520 nm was 40% in the case of untreated samples. The difference in the extension of biofouling in relation with the wavelengths might be due to both aged or dominant microorganism in algal colonies (Mehrubeoglu et al., 2013). In the case of

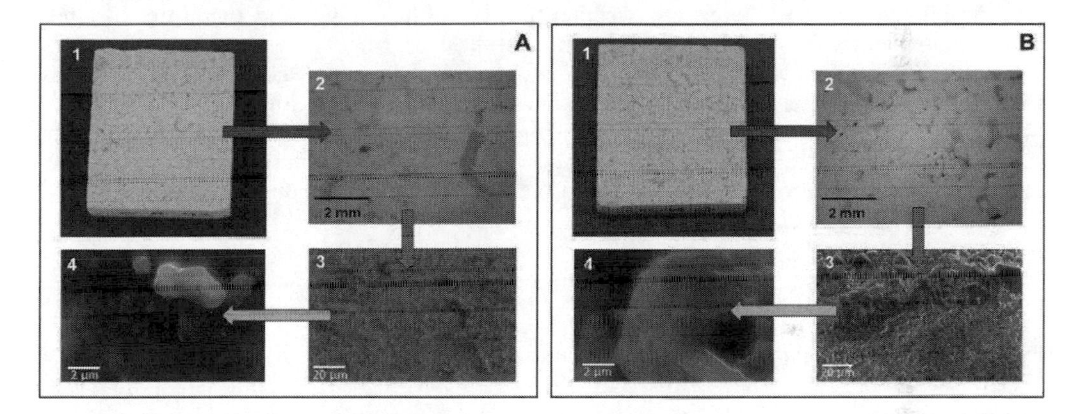

Figure 1. Images of untreated (A) and treated (B) samples collected with a camera (1), optical microscopy (2) and SEM (3-4), thus corresponding to increasing magnifications.

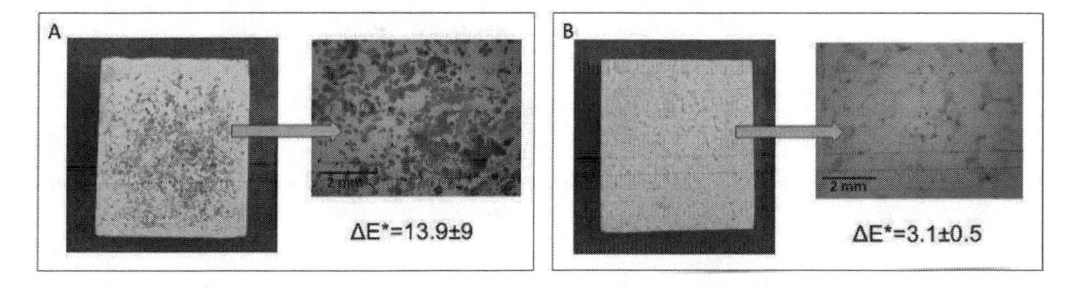

Figure 2. Images of untreated (A) and treated (B) samples after accelerated microalgal growth assay and corresponding color changes (ΔE^*).

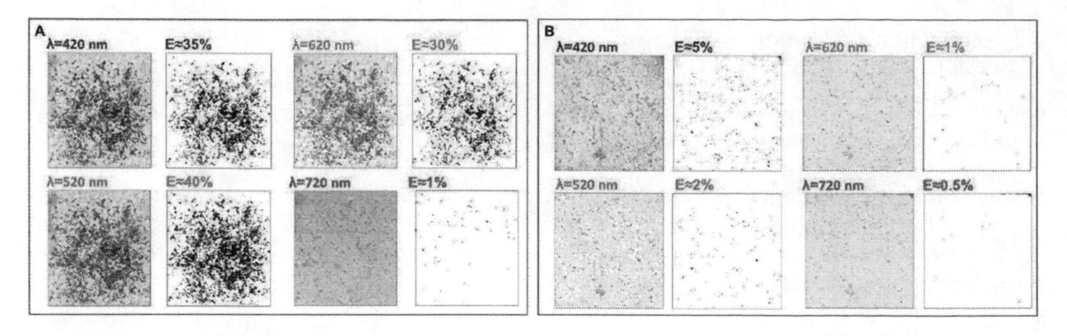

Figure 3. Images of untreated (A) and treated (B) samples captured with multispectral camera at different wavelengths (λ). Images after digital analysis used for biofilm extension quantification (E) are included.

treated samples, the values of E were related to the limestone pore system and some disperse algal pustules.

4 CONCLUSIONS

Suspension of AgTEOS NPs is a stable colloidal system. NPs correspond to a small hydrodynamic diameter which promotes their penetration through the limestone porous system.

The chromatic change caused by the NPs ($\Delta E^*=6.8$) is perceptible by the human eye. However, the NPs appeared to be effective in preventing biofilm growth and therefore their use can be recommended for biocide treatment of limestones.

Biofouling extension was measured using multispectral imaging, associated with digital image analysis. The differences observed among different wavelengths are related to the wavelength absorption by the (fresh and aged) microorganisms present in the biofouling. The image of the untreated samples at λ=520 nm shows high value of biofilm extension. At λ=720, the biofilm does not absorb the radiation and therefore the corresponding image cannot provide any useful information.

ACKNOWLEDGEMENTS

Ministerio de Economía y Competitividad and FEDER (Art-Risk BIA2015-64878-R and CTQ2013-48396-P). Research teams TEP-199 and FQM-319 from Junta de Andalucía. J. Becerra is grateful to the Ministerio de Educación, Cultura y Deporte for his pre-doctoral fellowship (FPU14/05348), to the University Pablo de Olavide for funding his pre-doctoral research stay (Ref. AE-18-1) and the University Ecclesiastical Academy of Thessaloniki for his stay as visiting researcher.

REFERENCES

Banach, M., Pulit-Prociak, J., 2016. Chapter 12 – Synthesis, characteristics, and biocidal activity of silver nanoparticles, Fabrication and Self-Assembly of Nanobiomaterials. Elsevier Inc. https://doi.org/10.1016/B978-0-323-41533-0.00012-X.

Becerra, J., Ortiz, P., Zaderenko, A.P., Karapanagiotis, I., 2019. Assessment of nanoparticles/nanocomposites to inhibit micro-algal fouling on limestone facades. Build. Res. Inf. In press. https://doi.org/10.1080/09613218.2019.1609233.

Becerra, J., Zaderenko, A.P., Sayagués, M.J., Ortiz, R., Ortiz, P., 2018. Synergy achieved in silver-TiO2 nanocomposites for the inhibition of biofouling on limestone. Build. Environ. 141, 80–90. https://doi.org/10.1016/j.buildenv.2018.05.020.

Koutsoukos, P.K., Klepetsanis, P.G., Spanos, N., 2006. Calculation of Zeta-Potentials from Electrokinetic Data, in: Somasundaran, P. (Ed.), Encyclopedia of Surface and Colloid Science. Taylor & Francis Group, New York, pp. 1097–1113. https://doi.org/10.1081/E-ESCS-120000059.

Mehrubeoglu, M., Teng, M.Y., Zimba, P. V., 2013. Resolving mixed algal species in hyperspectral images. Sensors (Switzerland) 14, 1–21. https://doi.org/10.3390/s140100001.

Nowicka-Krawczyk, P., Zelazna-Wieczorek, J., Koźlecki, T., 2017. Silver nanoparticles as a control agent against facades coated by aerial algae — A model study of Apatococcus lobatus (green algae). PLoS One 12, 1–14. https://doi.org/10.1371/journal.pone.0183276.

Nugari, M.P., Salvadori, O., 2003. Biocides and Treatment of Stone: Limitations and Future Prospects, in: Koestler, R.J., Koestler, V.H., Charola, A.E., Nieto-Fernandez, F.ed (Eds.), Art, Biology, and Conservation: Biodeterioration of Works of Art. The Metropolitan Museum of Art, New York, pp. 518–535.

Pinho, L., Rojas, M., Mosquera, M.J., 2015. Ag–SiO2–TiO2 nanocomposite coatings with enhanced photoactivity for self-cleaning application on building materials. Appl. Catal. B Environ. 178, 144–154. https://doi.org/10.1016/j.apcatb.2014.10.002.

Raza, M., Kanwal, Z., Rauf, A., Sabri, A., Riaz, S., Naseem, S., 2016. Size- and Shape-Dependent Antibacterial Studies of Silver Nanoparticles Synthesized by Wet Chemical Routes. Nanomaterials 6, 74. https://doi.org/10.3390/nano6040074.

Science and Digital Technology for Cultural Heritage – Ortiz Calderón et al. (Eds)
 ISBN 978-0-367-36368-0

Study of solvents and their implication in the in-depth penetration of nanolimes in different limestones

J. Becerra, T. Núñez, P. Ortiz & A.P. Zaderenko
Department of Physic, Chemic and Naturals Systems, Universidad Pablo de Olavide, Seville, Spain

ABSTRACT: This research is focused on the improvement of nanolimes ($Ca(OH)_2$ nanoparticles) as consolidant for limestones. For that, the $Ca(OH)_2$ nanoparticles were doped with zinc oxide quantum dots (ZnO QDs) which have fluorescence under UV light.

The selection of an optimal mixture of solvents based on their surface tension and evaporation rate improved the penetration depth of the treatment, decreasing the chromatic change on the stone surface. Furthermore, the effectiveness of the treatment was also related to the physical properties of stones as higher in-depth consolidation was achieving when the limestone substrate has a higher porosity and higher medium porous size.

1 INTRODUCTION

The degradation of our Cultural Heritage is due to different intrinsic and extrinsic factors that cause loss of mechanic and aesthetic properties of the building materials. Between the most common weathering form are powdering, erosion, alveolization and loss of materials (Ortiz and Ortiz, 2016). For that, the application of consolidation treatments is a common practice in the restoration of stones and mortars of historical buildings.

The search of new consolidants more compatible and effective in limestone has favored the develop of new investigation, as the consolidant based on nanolimes. The $Ca(OH)_2$ nanoparticles (NPs) are hexagonal platelets of portlandite with a size about 100 nm (Borsoi et al., 2018), so this treatment has the same advantages of typical limewater (compatibility and durability) together a nanometer size that favor the deeper penetration into the stone pore system (Daniele et al., 2008). Nevertheless, recent studies have showed that their penetration on these materials is lower than expected, because of they achieve only a pre-consolidation in the surface of the material. It is due to their accumulation near the surface (Borsoi et al., 2017), that decreases their penetration depth into construction materials and, consequently, generates a white haze and a low effectiveness (Jang and Matero, 2018).

2 MATERIALS AND METHODS

2.1 *Consolidant*

$Ca(OH)_2$ NPs doped with ZnO QDs have been synthetized according to Becerra et al. (2019). This nanocomposite has been previously characterized and it is composed by portlandite platelets with spherical particles of ZnO posed on their surface. The ZnO QDs have an average size of 8 ± 1 nm, without any noticeable increment in the average size of the $Ca(OH)_2$ NPs.

The nanocomposite suspension was centrifugated at 5000 rpm during 10 min. The pellets with the $Ca(OH)_2$ NPs with ZnO QDs was resuspended in the selected solvents or mixture of solvent at the concentration of 1.6 g/L.

2.2 *Selection of solvents*

The properties of the solvent have a high influence in the penetration depth of the nanoparticles into the stone pore (Borsoi et al., 2016). In this case, we have studied solvents and solvent mixtures that favor the penetration depth of the nanocomposite into stone pore and have a low evaporation rate to avoid the back migration of the treatment during the solvent evaporation. Table 1 shows the selected solvents used in this assay and their main properties. Additionally, the following mixtures of solvents were also studied: octanol/isopropanol (concentrations; 1:1, 1:3, 3:1) and octanol/acetone (concentration; 1:1, 1:3).

Contact angles of selected solvents and solvent mixtures were analyzed on Jerez de la Frontera limestone surfaces to assess a good penetration into stone pore. The contact angle was measured taking photograps with high magnification of the drops and using the software ImageJ® to calculate the contact angle.

2.3 *Limestone samples*

Different limestone from historic quarries in Seville, Cádiz and Alicante (Spain) have been used to assess the effectiveness of nanolimes with ZnO QDs as consolidant. Table 2 shows the selected quarries and the main properties of the limestones, previously characterized by Guerrero (1990). Slabs of 2.5x2.5x0.5 cm were employed in the assays.

2.4 *Application methods*

The optimal application method was studied on Jerez de la Frontera limestone samples. For that, three doses of 200 μL was brushed or spread with a pipette on the sample surface. The dried of the samples was made at room temperature (24 ± 2°C) and medium humidity (60%).

The color changes after treatment (ΔE*) were measured by colorimetry (PCE-CSM 2) and calculated according to the equation $\Delta E^*=(\Delta L^2+\Delta a^{*2}+\Delta b^{*2})^{1/2}$, where ΔL, Δa* and Δb* are the color variations caused by the treatments defined by the CIELAB color-system.

2.5 *Assessment the treatment as consolidant for Cultural Heritage*

Three doses of 200 μL of treatment was spread on stone sample surfaces with a pipette. The samples where dried at room temperature (24 ± 2 °C) during 7 days.

Table 1. Properties of selected solvents for this assay (Wypych, 2014).

Solvent	Viscosity at 25°C (mPas)	Surface tension at 20°C (mN m-1)	Boiling point (°C)	Evaporation rate (butyl acetate=1)
Acetone	0.303	22.68	56.1	6.60
Isopropanol	2.100	21.40	82.2	2.90
Octanol	7.363	26.92	204.0	0.01
Water	0.895	71.97	99.61	-

Table 2. Quarries and main properties of the limestones.

Quarry	Sample	Pore diameter (μm)	Open porosity (%)
El Puerto S. María (Cádiz)	calcarenite	10-100	17 ± 3
Utrera (Sevilla)	biosparite	0.1 – 1	8 ± 1
Jerez de la Frontera (Cádiz)	biosparite	0,1-25	8 ± 4
Novelda	biosparite	0.01 – 0.5	5 ± 2

The color changes were calculated according to previous method. The consolidation was analyzed by peeling test, using Scotch Cristal tape (3M) with 10 repetitions over the same location. The percentage of consolidation was calculated according to the equation $\%C=((TRM_{untreated}-TRM_{treated})/TRM_{untreated})*100$, where TRM is the total removed material in the untreated ($TRM_{untreated}$) and treated ($TRM_{treated}$) samples. The graphical representation of the results was made according to Drdácký et al. (2015). Finally, the penetration depth was measured under UV light (λ=254 nm) in the cross-section of the samples.

3 RESULTS

3.1 *Selection of solvents*

Table 3 shows the results of the contact angles measures on stone surfaces. As it was expected, the solvents with highest surface tension had the highest contact angles, such as water (71.97 mN m^{-1}; 46°) or octanol (26.92 mN m^{-1}; 17°). Taking into account both contact angles and evaporation rates, the best result was obtained with the mixture of octanol and acetone (1:1) because of the contact angle is low (10°) what allows a fast penetration of the treatment on the stone pore, and a high retention by the presence of octanol (evaporation rate=0.01). For that, this mixture of solvents was selected for next studies.

3.2 *Application methods*

Three different methods were employed to assess the treatments. The color changes generated after deposition of the treatment on the stone surface was used to select the optimal application method. The application systems caused similar color change, always less than 5 and, consequently, admissible for Cultural Heritage. In this sense, the least ΔE^* was obtained by brushing ($\Delta E^*=4.3 \pm 3.0$) and this was the application method chosen for next assays.

3.3 *Assessment the treatment as consolidant for Cultural Heritage*

The treatment was assessed in four limestones with different average pore size and open porosity (Table 2). The treatment is applicable in all the stone with a $\Delta E^*<5$ (Figure 1A), except to Utrera limestone ($\Delta E^*=7.8$). The color changes were higher in the yellowish limestones. The %C was very similar in limestones (between 10-20%), decreasing the remove materials specially at the beginning of the peeling test sequence. The non-lineal tendency in the treated samples were more near to the horizontal line because of the reduction of removed materials (Figure 1B). The higher penetration depth was achieved in Puerto de Santa María limestone (Figure 1C). This limestone has the highest average pore size and porosity, what allowed the better penetration on the treatment by the stone pore. In other stones, the treatment only penetrated a few microns, remaining on the sample surfaces, and because of that this application methods are not recommended.

Table 3. Results of contact angle assay.

Solvent	Water	Oct	Iso	Oct//Iso 1:1	Oct/Iso 1:3	Oct/Iso 3:1	Oct/Acet 1:1	Oct/Acet 1:3
Contac angle (°)	46 ± 0.9	417 ± 0.9	414 ± 0.2	12 ± 1.6	14 ± 0.2	17 ± 1.1	10 ± 3.3	9 ± 2.5

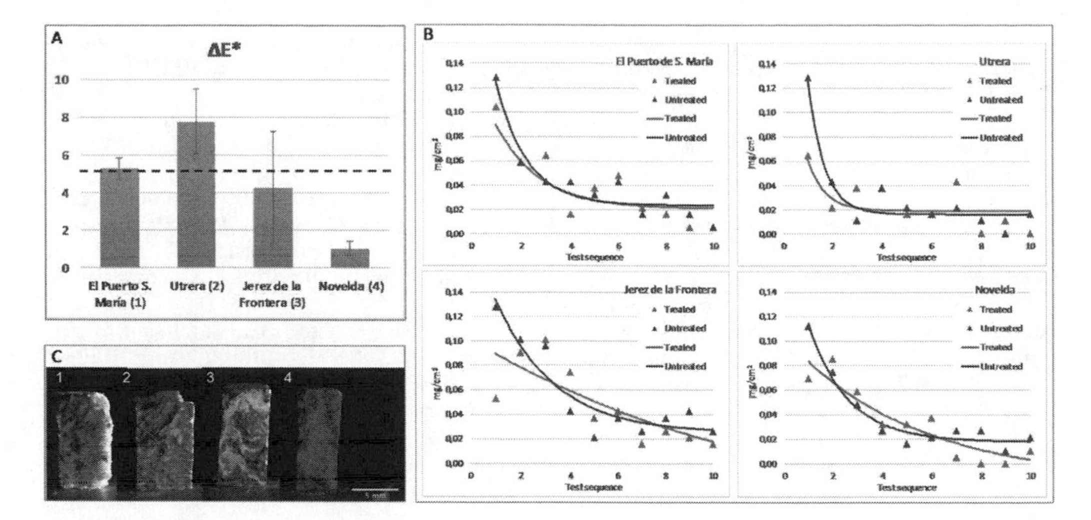

Figure 1. A color changes after treatment application. B. Consolidation treatment (symbols) and non-linear regression (lines) achieved in the peeling test assay. C. Photographs under UV radiation of the cross-section samples.

4 CONCLUSIONS

A proof of concept to improve the consolidation of limestones was carried out with $Ca(OH)_2$ NPs doped with ZnO QDs. Octanol-acetone (1:1) was chosen after the evaluation of different solvent mixtures because it showed a low contact angle and had a low evaporation rate that helped the penetration of treatment into the limestones. The best application method was brushing as it produced the lower color change.

The limestone properties have a great influence in the effectiveness of this consolidant based of $Ca(OH)_2$ NPs. The limestone with a high total porosity and a high medium pore size showed the highest penetration depth, although the effectiveness (peeling test) was lower than other limestones (only 10%). In fact, Puerto de Santa María results showed that it is necessary repeat applications of treatments due to its porosity. The treatment improved the mechanical properties of the other stones according to peeling test results. The color changes were acceptable for Cultural Heritage, except the slabs from Utrera with $\Delta E^* \approx 8$.

ACKNOWLEDGEMENTS

Ministerio de Economía y Competitividad and FEDER (Art-Risk BIA2015-64878-R and CTQ2013-48396-P). Ministerio de Educación, Cultura y Deporte (FPU14/05348). Research teams TEP-199 and FQM-319 from Junta de Andalucía.

REFERENCES

Becerra, J., Ortiz, P., Martín, J.M., Zaderenko, A.P., 2019. Nanolimes doped with quantum dots for stone consolidation assessment. Constr. Build. Mater. 199, 581–593. https://doi.org/10.1016/j.conbuildmat.2018.12.077.

Borsoi, G., Lubelli, B., van Hees, R., Veiga, R., Santos Silva, A., 2018. Application Protocol for the Consolidation of Calcareous Substrates by the Use of Nanolimes: From Laboratory Research to Practice. Restor. Build. Monum. 22, 99–109. https://doi.org/10.1515/rbm-2016-0008.

Borsoi, G., Lubelli, B., van Hees, R., Veiga, R., Santos Silva, A., 2017. Evaluation of the effectiveness and compatibility of nanolime consolidants with improved properties. Constr. Build. Mater. 142, 385–394. https://doi.org/10.1016/j.conbuildmat.2017.03.097.

Borsoi, G., Lubelli, B., van Hees, R., Veiga, R., Silva, A.S., 2016. Optimization of nanolime solvent for the consolidation of coarse porous limestone. Appl. Phys. A 122, 846. https://doi.org/10.1007/s00339-016-0382-3.
Daniele, V., Taglieri, G., Quaresima, R., 2008. The nanolimes in Cultural Heritage conservation: Characterisation and analysis of the carbonatation process. J. Cult. Herit. 9, 294–301. https://doi.org/10.1016/j.culher.2007.10.007.
Drdácký, M., Lesák, J., Niedoba, K., Valach, J., 2015. Peeling tests for assessing the cohesion and consolidation characteristics of mortar and render surfaces. Mater. Struct. Constr. 48, 1947–1963. https://doi.org/10.1617/s11527-014-0285-8.
Guerrero, M.A., 1990. Diagnóstico del estado de alteración de la piedra del Palacio Consistorial de Sevilla. Causas y mecanismos. Universidad de Sevilla (Spain).
Jang, J., Matero, F.G., 2018. Performance evaluation of commercial nanolime as a consolidant for friable lime-based plaster. J. Am. Inst. Conserv. 57, 95–111. https://doi.org/10.1080/01971360.2018.1486126.
Ortiz, R., Ortiz, P., 2016. Vulnerability Index: A New Approach for Preventive Conservation of Monuments. Int. J. Archit. Herit. 10, 1078–1100. https://doi.org/10.1080/15583058.2016.1186758.
Wypych, A., 2014. Databook of Solvents. ChemTec Publishing, Toronto.

Science and Digital Technology for Cultural Heritage – Ortiz Calderón et al. (Eds)
© 2020 Taylor & Francis Group, London, ISBN 978-0-367-36368-0

Supports used for mounting of large format photograph and digital prints. Analytical and structural characterization

E. Navarro, M.San Andrés, L. Castelo, R. Chércoles & J.M. de la Roja
Facultad de Bellas Artes. Departamento de Pintura y Conservación-Restauración. Universidad Complutense de Madrid, Madrid, Spain

1 INTRODUCTION

In the market there is a wide variety of mounting supports which can be face or back mounting and, not only serve to support photograph images and digital prints but also to make them presentable. All of them are semi rigid support, without framing, that have been replacing the techniques of traditional matting and framing. This type of mounting improves the aesthetic and the quality of the image. In many cases the current demands on their permanence respond satis- factorily in relation to the temporality for which they were originally created. However, when these works become part of museum collections, questions about their long-term behavior arise. In fact, the types of supports that are used by laboratories and photographers are very diverse, but there is no guarantee that they comply with the necessary conservation standards (Pénichon *et al.*, 2002; Pénichon and Jürgens, 2005; Zorn and Dobrusskin, 2011).

There are several materials involved in this type of assembly. Those that constitute the pho- tographic/print image (PI), i.e. the paper and the image, and the own support on which the image is mounted through an adhesive. As for face-mounting system with a poly (methylmethacrylate) (PMMA), two methods can be established: face-mounting with Pressure-Sensitive Adhesive (PSA) and face-mounting with sili- cone. This last method was patented as Diasec® system, and it was introduced in the 70s of last century by the Swiss Heinz Sovilla-Bruhlhardt, and since the 90s has spread throughout Europe and the US (Smith, 2012; Herrera, 2014). In this assembly, the PI is attached to the PMMA support using an electric press equipped with two pressing rollers. The process is shown in Figure 2. First, a primer is applied to the PMMA and to the image side of PI. When primer has evaporated silicone adhesive is uniformly dispensed at one end of the PMMA forming a thin layer. Then the image is placed on the upper roller of the electric press and the PMMA on the lower roller. Pressure is applied, and the image is perfectly adhered to the PMMA. Once the process is finished, the silicone is left to cure for 48 hours. (Pénichon and Jürgens, 2001).

In relation with back-mounting supports used are frequently multilayer composite materials which have a compact or foamed core around which, layers of other materials are placed and symmetrically distributed in the upper and lower plane; the number and nature of these layers varies from one support to another. On one side of the support, the PSA is attached directly by pressure and optionally heating (60°C); this adhesive is protected by a sheet of siliconized paper which is removed just before the PI mounting and this last process is again achieved by pressing (Figure 2).

In this paper, a multianalytical approach to the study of five different supports is pre- sented. The main objective has been to determine their structures and composition. For that, different analytical techniques have been used in order to identify in detail the nature of polymeric material used, and some additives pre- sent, such as plasticizers, fillers, stabilizers, and so on. Adhesives used in the mounting process have also been studied.

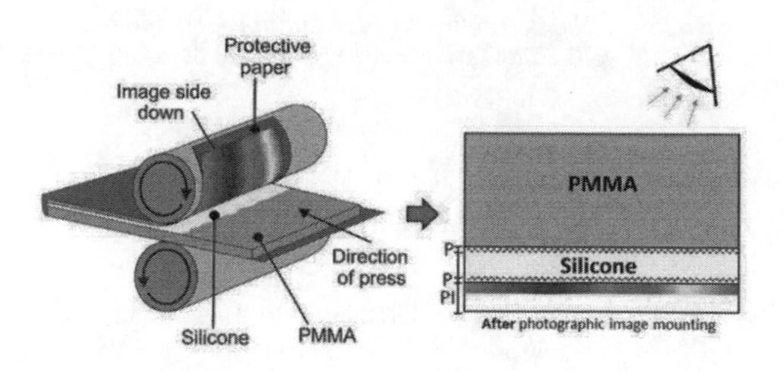

Figure 1. Face-mounting with PMMA and silicone. Inspired by an article (Pénichon and Jürgens, 2001).

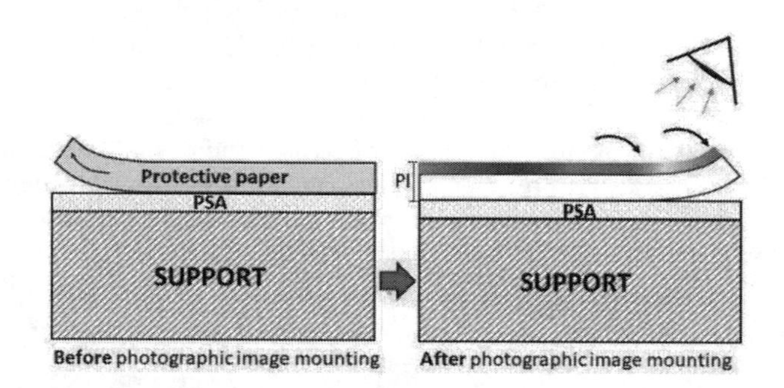

Figure 2. Back-mounting with pressure-sensitive adhesive (PSA).

From these results, in a second part will be study the long-term behavior of these materials in order to establish guidelines for their conservation-restauration of photograh/print image.

2 METHODOLOGY

2.1 *Materials*

Five different materials used in face and back mounting photographs and digital print have been selected. All of them are widely used in the field of fine arts. Four of them corresponding to the back-mounting with PSA system; these supports are commercialized as *Forex, Dibond, Kapafix*, and cellulosic support. The fifth support studied is a PMMA obtained by casting pro- cess, which is used in face-mounting with silicone or PSA. They are also studied adhesives used in the two types of mounting process. The PSA, which is supplied as a double-sided adhesive with a protective siliconized paper sheet on each face, and the adhesive, silicone sealant. The primer applied in this last system has also been studied.

2.2 *Methods*

Structural study: This study includes the identification of layers (number and thickness) and its morphology for each support. The techniques used have been: stereoscopic microscopy (SM), light microscopy (LM) and scanning electron microscopy (SEM).

Compositional study: The layers of each support have been analyzed to identify organic and in- organic components. The analytical techniques used have been: FTIR-ATR spectroscopy and Py-GC-MS to identify polymeric matrix and additives; SEM-EDX to identify fillers present.

The adhesives have been analyzed by FTIR-ATR spectroscopy and Py-GC-MS.

3 RESULTS AND DISCUSSION

Dibond and *Kapafix* supports are multilayer composite materials which display a very complex structure with layers of different thickness and composition. In Table 1 are shown results corresponding to the study by SM, SEM-EDX, FTIR-ATR spectroscopy and Py-GC-MS. Both supports present a very thick foamed core symmetrically rounded of several layers. *Dibond* sup- port is constituted for 5 layers. The black foam core (layer 3) is composed by mixture of HDPE (poly-meric matrix), plasticizers: phthalates, lubricants: fatty acids and bulking agents: CaCO3 and carbon black. Around this foamed core, there are four layers whose symmetrical distribu- tion and thickness, as is indicated in image corresponding to this support in Table 1. Layer 1 and 5 are of HDPE (polymeric matrix) mixed with plasticizers (DEGDB), lubricants: (fatty acids) and bulking agent (TiO2). Layers 2 and 4 are metallic sheet of aluminum.

Kapafix support presents a very much more complex structure with a total of 11 layers which have very different thickness and composition. The foam core (layer 6) presents an open cell morphology. Its main component is a polyester-urethane (polymeric matrix) additive with: plas- ticizers (phthalates), lubricants (fatty acids) and bulking agent (CaCO3). On this foamed core are disposed five layers on one side, and another five on the other side. The more external layers (1 and 11) are made of HDPE additive with lubricants (fatty acids) and bulking agent (TiO2). Lay- ers 2 and 10 are a very thin aluminum sheet. Layers 3, 5, 7 and 9 corres- pond to an adhesive identified as poly (buthylacrylate) (PBA) with plasticizers (phthalates: DBP, DEHP) and lubri- cants (fatty acids). Layers 4 and 8 are cardboards of cellulose and both contains CaCO3 as filler.

As regards *Forex*, PMMA and cellulosic supports, the structure is much simple. They present only one layer, however, all of them are composite materials (Table 2). The most complex one is Forex support. This is a foam material which polymeric matrix is a mixture or a copolymer of PVC-PMMA. Other components identified have been plasticizers phatalates (DHEP and DBP), lubricants (fatty acids and their esters,) and bulking agents (CaCO3, TiO2).

PMMA support is used with silicone or PSA face-mounting system. Because in this case the PI is observed through support, this has to be necessarily transparent. This characteristic is asso- ciated to PMMA obtained by casting process. As expected the main component is PMMA (pol- ymeric matrix); other components present are: plasticizers (phthalates: DHEP), lubricants: fatty acids and their esters and an anti-UV substance (Tinuvin-P).

Finally, cellulose support contains cellulose as main component. Other substances detected have been: plasticizers (DEGDB), lubricants (fatty acids) and bulking agent (CaCO3).

The adhesive used in the PSA face-mounting system is a poly (butyl acrylate) plasticized with phthalates (DBP and DEHP). In the face mounting system with silicone are used a primer and sealant adhesive. The former is a mixture of an organic compound (2-propanol) and poly (dimethyl siloxane) (PDMS), and the sealant adhesive also is PDMS.

Table 1. Structure and composition of multilayer supports. Dibond and Kapafix.

Support (Image SM/Detail SEM)	Layer	Analysis		
		SEM-EDX	FTIR-ATR [bands (cm^{-1}) assign- ments]	Py-GC-MS (m/z) * (assignments)
Dibond 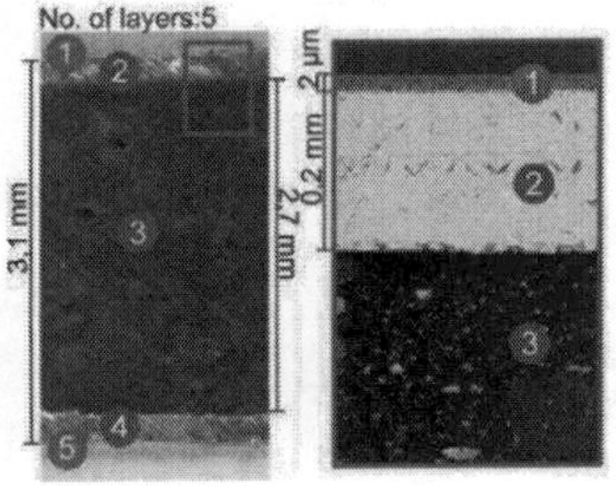	①, ⑤	C, **Ti**, Al	**PE:** 2928,2855,1470, 727 and 715 **Plasticizers?**: 1714	**DEGDB**: 51, 77, 105, 149 **FA**:(**16:0**): 41,60,73,97,129,213, 256; (**18:0**): 43,73,97,129,185,241,285
	②,④	C, **Al**, Fe		
	③	C, **Ca**	**HDPE:** 2914,2897,1472,1463,730, /717/ **CaCO3**: 1424,874, 711 **Plasticizers?**: 1735	**DEHP**: 43,71,97,111,149,167,279 **DOP**: 43,70,112,149,167,261,279 **FA**:(**16:0**):73,213,256; (**18:0**):73, 241, 285
Kapafix 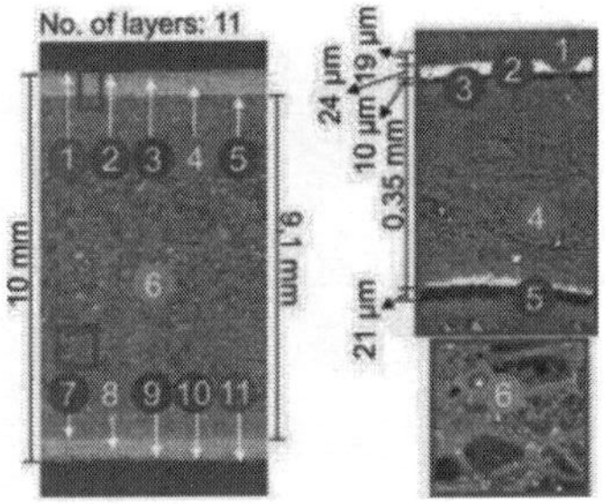	①, ⑪	C, **Ti**, Al	**HDPE:** 2914,2847,1471,1462,729 and 718 **Plasticizers?:** 1712	**FA:** (**16:0**): 41,60,73,97, 129, 213 256; (**18:0**): 43,73, 97, 129, 185, 241,285
	②, ⑩	C, **Al**, Fe		
	③, ⑤, ⑦, ⑨	C	**PBA**: 2992,2949,2927,2850, 1721,1455,1397,1378,1267,1237,1190, 1157,1026,985,965,954,858,839, 750, 502, and 436	**DBP**:41,59,104,149,167,205,223 **DEHP**: 57,106,149,167,222, 279 **FA**:(**16:0**):43,73,97,129,185,213 256;(**18:0**):43,73,97,129,185,241 284
	④, ⑧	C, **Ca**	**Cel**: 3335,3278,2915,2848, 1646, 1633, 1465, 1336, 1423, 1369, 1261, 1315, 1236, 1205, 1159, 1102, 1052, 1027, 1001, 985, 897 and 659	**FA:** (**14:0**): 29, 55, 73, 97, 185 and 228; (**15:0**): 43, 60, 73, 129, 199 and 242

		CaCO3:1424, 872 and 717	
⑥	C, **Ca**	**PESU**: 3306,2967,2875,2925,2856, 2277,1701,1594,1508,1306,1455, 1341,1437,1374,1410,1374,1219, 1068,1042,1016, 851,812 and 763 **CaCO3**:1424, 872 and 717	**DEP**: 59, 76, 105, 149, 177, 222 **DBP**: 41,59104,149,167,205,223 **DEHP**: 57,106,149,167,222,279 **FA**: (**16:0**) 55,73,97,129,213,256; (**18:0**): 60,73,97,129,185,241,284

* Only are indicated m/z value corresponding to additives which are part of polymeric matrix.
Cell: cellulose; DBP: dibuthyl phthalate; DEHP: bis (ethylhexyl) phthalate; DEP: diethyl phathalate; PBA: Poly(buthylacrylate); DEGDB: Diethyl- ene glycol dibenzoate; DOP: Dioctyl phthalate; FA: fatty acids; PE: polyethylene; HDPE: High-density polyethylene; PESU: poly(ester urethane) (14:0): myristic acid; (15:0): Pentadecanoic acid; (16:0): Palmitic acid; (18:0): Stearic acid

Table 2. Structure and composition of monolayer supports. Forex, Cellulosic support and PMMA.

Support (Image SM)/Detail SEM)	Analysis		
	SEM- EDX	FTIR-ATR [bands (cm^{-1}) assign- ments	Py-GC-MS (m/z) * (assignments)
Forex	C, **Cl**, **Ca**, **Ti**	**PVC:** 2956,1327,1240,678, 609 **PMMA**: 1731,1191,1151 **CaCO3:** 1425,875,715 **Plasticizers?**:2358,2336,1731, 1018,846	**PMMA:** 41,69,100 **DHEP**: 41, 57, 149, 167, 242, 279 **DBP:** 43, 71, 149, 167, 202, 229, 270 **FA: (16:0) and 18:0**: 43, 60, 73,157, 185, 213, 256; 43, 73, 97, 129, 241, 285 **Hexadecyl and stearyl palmitate**: 57, 83, 97, 111, 224, 257; 57, 83, 111, 207, 257, 285
Cellulosic support	C, **Ca**	**Cell:** 3333,3275,2899,2852, 1640 1652-1635, 1451, 1426, 1335, 1369, 1314, 1241, 1203, 1157, 1106, 1052, 1026, 998, 983, 898, 661 **CaCO3**: 1424,872,702	**DEGDB:** 51, 77, 105, 149 **FA**: (**14:0**): 29, 55, 73, 97, 185, 228, (**15:0**): 43, 60, 73, 129, 199, 242 (**16:0**): 43, 73, 97, 129, 185, 213, 256; (**18:0**): 43, 73, 97, 129, 185, 241, 284
Poly(metyl methacrylate)	C	**PMMA:** 2992,2949,2927,2850, 1721,1479,1434,1385,1363,1267, 1189,1140,1062,985,965,839,750, 481	**DHEP**: 32, 57, 70, 83, 149, 167, 279 **FA**: (**16:0**): 43, 60, 73, 97, 129, 213, 256, (**18:0**): 43, 60, 73, 97, 185, 241, 285 **MS**: 43, 69, 74, 143, 255, 267, 298 **Tinuvin-P**: 65, 78, 93, 154, 168, 196, 225

* Only are indicated m/z value corresponding to additives and comonomers which are part of polymeric matrix..DBP: dibuthyl phthalate; DEHP: bis (ethylhexyl) phthalate; PMMA: Poly methyl methacrylate; DEGDB: Diethylene glycol dibenzoate; MS: Methyl stearate, FA: fatty acids; (14:0): myristic acid; (15:0): Pentadecanoic acid; (16:0): Palmitic acid; (18:0): Stearic acid.

4 CONCLUSIONS

The analyses carried out on the photographic support have allowed characterizing the structure of supports selected (number and thickness of layers) and their compositions (polymeric matrix, organic additives belonging to the group of phthalates and fatty acids, and inorganic ad- ditives like CaCO3, TiO2 and carbon black). In addition, we have characterized the main adhe- sives used in mounting process (PSA and PDMS).

The nature of these compounds can affect to the long-term behavior of these supports, specially plasticizers and lubricants. It is also very important to be in mind its effects on the photo- graph and digital print supported which can be affected by the degradation products generated, such as VOCs and acidic compounds.

ACKNOWLEDGEMENTS

Projects HUM2015-68680-P and Top Heritage-CM (S2018/NMT-4372). PhD scholarship (BES-2016-078292) given to E. Navarro financed by the MINECO (Spain). Authors thanks to CNME, CAI of UCM and IPCE Laboratory. Also to: Muñoz y Kramer Mounting Studio and IMAGENDECOR Photographic laboratory

REFERENCES

Herrera, R. 2014. La conservación de fotografía contemporánea. Nuevos retos y problemas. *15th Jornada de Arte Contemporáneo*: 81–96.

Pénichon, S & Jürgens, M. 2001. Two finishing techniques for contemporary photographs. *Topics in Pho- tographic Preservation* 9: 85–96.

Pénichon, S; Jürgens, M. & Murray, A. 2002. Light and dark stability of laminated and face-mounted photographs: a preliminary investigation. *studies in conservation* 47(3): 154–159.

Pénichon, S; Jürgens, M. 2005. "Plastic lamination and face mounting of contemporary photographs". Constance McCabe (Ed.), *Coatings on Photograhs: Materials, Techniques, and Conservation*. Washington, DC: American Institute for Conservation: 218–233.

Smith, M. 2012. Face-mounting techniques for contemporary photographs and digital images. *7th AICCM Book*, Paper and Photographic Materials Symposium: 75–81.

Zorn, S. & Dobrusskin, S. 2011. 'Diasec' and other finishing techniques. Investigation of accelerated light fading. *Studies in Conservation* 56: 257–266.

Science and Digital Technology for Cultural Heritage – Ortiz Calderón et al. (Eds)
© 2020 Taylor & Francis Group, London, ISBN 978-0-367-36368-0

The use of Nanorestore Gel® in the cleaning treatments of easel paintings

C. Pernía Escobar, J.M. de la Roja & M. San Andrés
Painting and Conservation-Restoration Department, Faculty of Fine Arts, Complutense University of Madrid, Madrid, Spain

ABSTRACT: Gelled media of a very varied nature are being used in recent years for the cleaning of artworks. In this study, a practical analysis of the product known as Nanorestore Gel®, a chemical hydrogel designed for the cleaning of artistic works, has been carried out. For that, an experimental protocol has been developed to control the cleaning efficiency and the secondary effects of its action on the chosen pictorial surface (oil and acrylic paint). In order to achieve so, prototypes with specific characteristics has been made to be submitted to experimental tests. In this first practical approach, three gels have been tested: Dry, Maximum Dry and Extra Dry. Attending to the cleaning results obtained, the Dry gel presents the best behavior.

1 INTRODUCTION

Cleaning treatments have been the subject of multiple investigations related to the products used, methods of application and cleaning effectiveness. This issue is evident in the middle of the last century and is still topical due to the variety of different products that are continuously appearing in the market, which are the result of the development of new materials and, a priori, may be of interest in this area. For years, organic solvents and aqueous media based on the use of surfactants, buffer solutions, chelating agents and enzyme preparations have been used, among others (Wolbers, 2000; Cremonesi and Signorini, 2016). In addition, gelled media of a very varied nature are being used in the last years and chemical gels with interesting characteristics have been introduced more recently (Baglioni and Chelazzi, 2013; Baglioni et al., 2014).

In this paper, a practical analysis of the product known as Nanorestore Gel® is shown. This is a chemical hydrogel designed for the cleaning of artistic works, whose peculiar polymeric structure provides retention characteristics that avoid or reduce the risk of penetration and retention of the cleaning medium inside the painted layers. The purpose of this work is to develop an experimental protocol that makes it possible to control the cleaning efficiency and the secondary effects of its action on pictorial surface of easel painting. To carry out this purpose, several prototypes with specific characteristics about nature of pictorial layer, varnishes and dirt have been submitted to cleaning tests with different variants of Nanorestore Gel® (Dry, Extra Dry and Max Dry). Other studies made with these gels are for cleaning remains of adhesives and dirt in works of art on paper (Domingues, et al., 2013).

In a first approach, to determine the cleaning efficiency, colorimetric and gloss measurements have been made, beside stereoscopic microscopy analysis. The results obtained have made possible to establish the viability of the experimental design and assess the interest of this type of gels.

2 METHOD AND MATERIALS

2.1 *Materials*

Nanorestore Gels® are transparent chemical hydrogels, made up of a semi-interpenetrating network that is the result of the reaction of poly (2-hydroxyethyl methacrylate) (pHEMA) and poly (vinylpyrrolidone) (PVP) or povidone. The cross-linked structure gives the gel a specific maximum load of fluid retention. Three different types of gels are marketed (Gel Dry, Extra Dry and Max Dry): each one has a specific capacity to retain the solvent medium that has been chosen to clean the pictorial surface. These gels can be loaded with different solvents (CSGI, 2019).

In order to carry out the cleaning treatment with the Nanorestore Gels®, two models have been prepared. These models were made on DM supports with canvas cloth, applying different pictorial techniques, the first with white acrylic paint and the second with white oil (three patterns in each of them). Different protection layers were applied to each pattern: Winsor & Newton varnish (Ketone Resin), Titan varnish (acrylic resin) and Regal Varnish Gloss (low molecular weight aliphatic resin). The preparation process of both models and their subsequent varnishing has followed a procedure similar to that used in an easel painting. Subsequently, a layer of dirt (Wolbers dirt) has been applied on these already prepared specimens (Figures 1a and 1c). On these models the three gels selected were tested (Figure 1b).

2.2 *Methods*

2.2.1 *Cleaning procedure*

After preliminary tests carried out with the gels to determine the loading solvent, an aqueous solution composed of 70% water and 30% Ethanol was chosen, and a cleaning protocol was established. The gels were introduced in glass jars with the loading solution; after the preliminary tests, an exposure time of 5 minutes was chosen. During this time the gel was covered with a Mylar® plastic and a glass to provide a little weight and pressure; finally, the gel was removed and the area was rinsed with a dry swab.

2.2.2 *Cleaning evaluation*

Colors measurements. In order to perform the colorimetric measurements, a Konica Minolta CM 2600d spectrophotometer was used, with a range of 400nm-700nm and an interval of 10nm. Reflection optical geometry (8°). Diameter of measurement area of 3 mm. The measurements have been expressed in CIELAB coordinates (L*a*b*) assuming an illuminating CIE D65 pattern and observer pattern CIE 1964 10°. The values of the CIELAB coordinates used correspond to the averages of at least five measurements. The chromatic values have been obtained through the software CM-S100w 1.91.0002 SpectraMagic and processed in spreadsheet. These measurements have been carried out in the following stages of the testing process: patterns before their soiling, areas stained with Wolbers dirt and, finally, measurements have been made on these same areas after cleaning. From each of them, 5 colour measurements have been taken and the average value has been calculated. To determine the colour variations experienced (ΔE_{00}), the values L, a and b have been taken into account before applying the dirt on the canvas and after cleaning, and the CIE2000 formula has been applied (Luo and Rigg, 2001: 340-350):

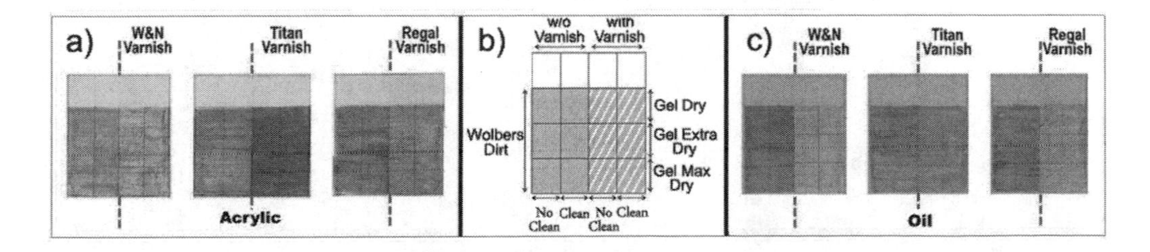

Figure 1. Scheme of the different patterns realized in order to test cleaning efficiency: acrylic model (a) oil model (c); explanatory diagram of the process (b).

$$\Delta E_{00} = \sqrt{\left(\frac{\Delta L'}{K_L S_L}\right)^2 + \left(\frac{\Delta C'}{K_C S_C}\right)^2 + \left(\frac{\Delta H'}{K_H S_H}\right)^2 + R_T\left(\frac{\Delta C'}{K_C S_C}\right)\left(\frac{\Delta H'}{K_H S_H}\right)}$$
$$\text{where } K_L = 1; K_C = 1; K_H = 1$$

Eq. 1. CIE2000 Formula

As a general interpretation of data obtained: the lower the value of ΔE_{00}, the more effective the cleaning will be. However, this value may have certain inaccuracies, since in the application of equation 1 the initial amount of dirt in the samples is not taken into account. For this reason, and to make an assessment that also includes this parameter, the cleaning index (IL) has been calculated (Redsven et al., 2003) applying equation 2:

$$\%\ Cleaning\ Index = \frac{(L^{*}cleaned - L^{*}soiled)}{(L^{*}unsoiled - L^{*}soiled)}\ x\ 100$$

Eq. 2. Cleaning Index Formula

Taking into account the variables considered in the application of this equation, the % Cleaning Index allows to assess in a more realistic way the cleaning efficiency of the gels tested. Likewise, it is deduced that the higher the value of the cleaning index, the more efficient the used gel is.

Gloss measurements. To determine the effect of the mechanical action of cleaning, gloss measurements have been made. For this, a TRIO NK 20/60/85º glossmeter was used, and three measurements were made, this number being sufficient, since the dispersion of values was minimal

3 RESULTS

In Figures 2 and 3 are shown the results obtained. At first glance (Figure 2a) significant differences can be appreciated which are confirmed with the colorimetric measurements made.

Regarding cleaning, the ΔE_{00} values are in accordance with the % cleaning index reached (Figures 2b and 2b). In the acrylic models ΔE_{00} is generally greater than in oil ones, which highlights, as explained in equation 2, a minor cleaning effectiveness of the gels on acrylic painting. Moreover, take into account these results gels tested have an overall better cleaning power in varnished surfaces, and they generally better behave in oil painting than in acrylic one, where the results are quite heterogeneous. Overall, the gel that gives better results in both acrylic and oil paint is the Dry.

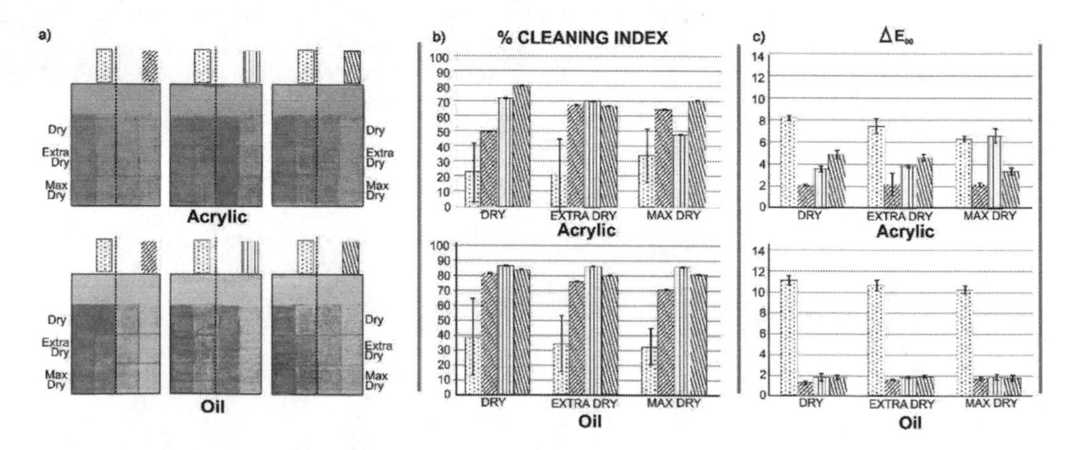

Figure 2. Cleaning results obtained with the different gels tested (Dry, Extra Dry and Max Dry): a) Mocks up after the cleaning process; b) and c) colorimetric measurement results.

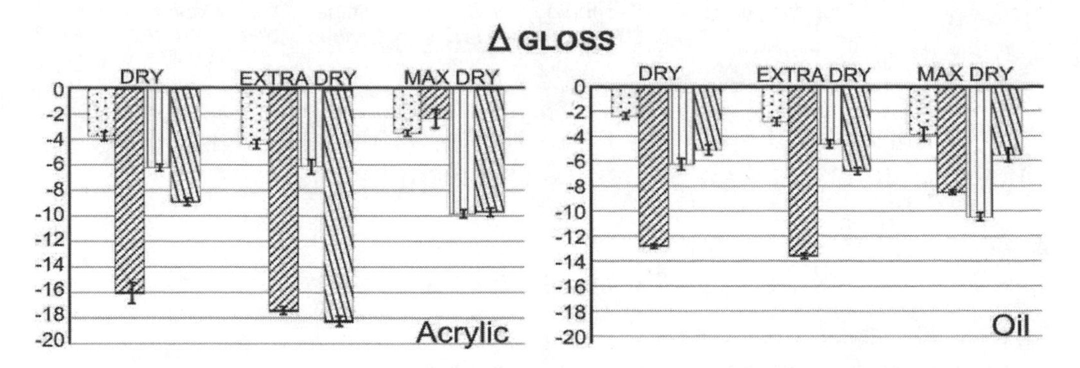

Figure 3. Gloss changes after the cleaning process.

Finally, it's important to analyse the gloss values of the samples (Figure 3). In all cases, as consequence of cleaning process, a decrease of gloss is observed. But this effect is more significant on acrylic patterns than in the oil ones.

In the unvarnished surfaces the variation values are minor and this can be related to dirt still present on painted surface.

4 CONCLUSIONS

As result of this experimental study, the gel that gives better results in both acrylic and oil paint is the Gel Dry. This is probably due to the fact that the Gel Dry is characterized by an interpenetrating network of polymers a little more open than the Max Dry and Extra Dry, so it absorbs and subsequently deposits more solvent in contact with the surface, thus facilitating the cleaning action. The results of this study are configured as a first practical approach to the study of the effectiveness of cleaning with Nanorestore gels, so these remain pending to be contrasted with further, more detailed and specific analyses on patterns with different characteristics and considering additional scientific results.

ACKNOWLEDGEMENTS

This research has been supported by the Project HUM2015-68680-P and Fondos Estructurales (FSE and FEDER) and the Project Top Heritage-CM (S2018/NMT-4372).

REFERENCES

Baglioni, P. & Chelazzi, D. 2013. Nanoscience for the Conservation of Works of Art. Royal Society of Chemistry.

Baglioni, P., Chelazzi, D. & Giorgi, R. 2014. Nanotechnologies in the Conservation of Cultural Heritage: A Compendium of Materials and Techniques. Springer.

Cremonesi, P. & Signorini, E. 2016. Un approccio alla pulitura dei dipinti mobili. Padova, Italia: Il Prato.

CSGI, 2019. Nanorestore Gel® Dry. Technical Sheet.

Domingues, J., et al. (2013). Innovative hydrogels based on semi-interpenetrating p(HEMA)/PVP networks for the cleaning of water-sensitive cultural heritage artifacts. Langmuir, 29(8), 2746–2755.

Luo, M. R. & Rigg, G. C., "The Development of the CIE 2000 Colours-Difference Formula: CIEDE2000", Color Research and Application, 5 (26), 2001, 340–350.

Redsven, I., Kuisma, R., Laitala, L., Pesonen-Leinonen, E., Mahlberg, R., Kymäläinen, HR., Hautala M. & Sjöberg, A-M. 2003. Application of a proposed standard for testing soiling and cleanability of resilient floor coverings. Tenside Surfactants Detergents, 40 (6), 346–352.

Wolbers, R. 2000. Cleaning painted surfaces. Aqueous methods. London: Archetype Publications.

Science and Digital Technology for Cultural Heritage – Ortiz Calderón et al. (Eds)
© 2020 Taylor & Francis Group, London, ISBN 978-0-367-36368-0

Poly-materials contemporary artworks by Sante Monachesi and Lucio Fontana: Study of materials and techniques

N. Macro
Physical and Chemical Sciences Department, University of L'Aquila, L'Aquila, Italy

M. Ioele
Istituto Superiore per la Conservazione ed il Restauro, Rome, Italy

M. Lazzari
Centre for Research in Biological Chemistry and Molecular Materials (CIQUS), University of Santiago de Compostela, Santiago de Compostela, Spain

ABSTRACT: Working with contemporary art could be very challenging, both for restorer and for conservation scientists, considering the consistent variability of the materials used by artists. In this paper, we present two examples of Italian contemporary artworks as paradigms of this variability of materials and techniques. To perform the best conservation intervention possible, it is mandatory to deeply study how was the artwork realized, moreover considering the short lifespan of modern materials like synthetic polymers. The artworks, coming from the "Scuola di Restauro ENAIP Botticino" (Brescia, Italy), were restored in 2018. To better understand the specific features of each artwork and the different degradation patterns we chose to apply a simple analytical approach. After a preliminary observation via optical microscopy, the samples were analyzed by FTIR spectroscopy, to characterize the organic components. Scanning Electron Microscopy coupled with Energy Dispersive Spectroscopy (SEM-EDS) was useful to study the inorganic fractions and the structural morphology.

1 INTRODUCTION

Numerous contemporary and modern Art collections nowadays include objects partially or completely realized in plastics (Quye & Williamson 1999). Many plastic polymers may show signs of extended degradation, often loosing color and shape, after few decades (Wypych 2018), even when they aren't exposed at extreme environmental conditions, but just to the normal ones found in museums and galleries (Lazzari et al.2011). Collecting and caring for plastic objects are today priorities for museums and conservators, and so are the necessities of studying their degradation patterns and to develop new strategies for their conservation. The lack of comprehension these fields may lead to inaccurate or even wrong conservative interventions which can affect, often irremediably, the artworks.

In this paper we briefly present results on two Italian contemporary works of art (Figure 1) partially realized using synthetic polymers. Our purpose was to study the materials chosen by the artists and to assess the conservative conditions of the artworks in order to help the conservators with the not easy task of restoring contemporary art. Our approach involved the sampling of small fragments from the artwork. First the samples were all observed through optical microscopy, then analyzed via Fourier transformed infrared spectroscopy (micro FTIR) and when mandatory to study the stratigraphy with SEM-EDS.

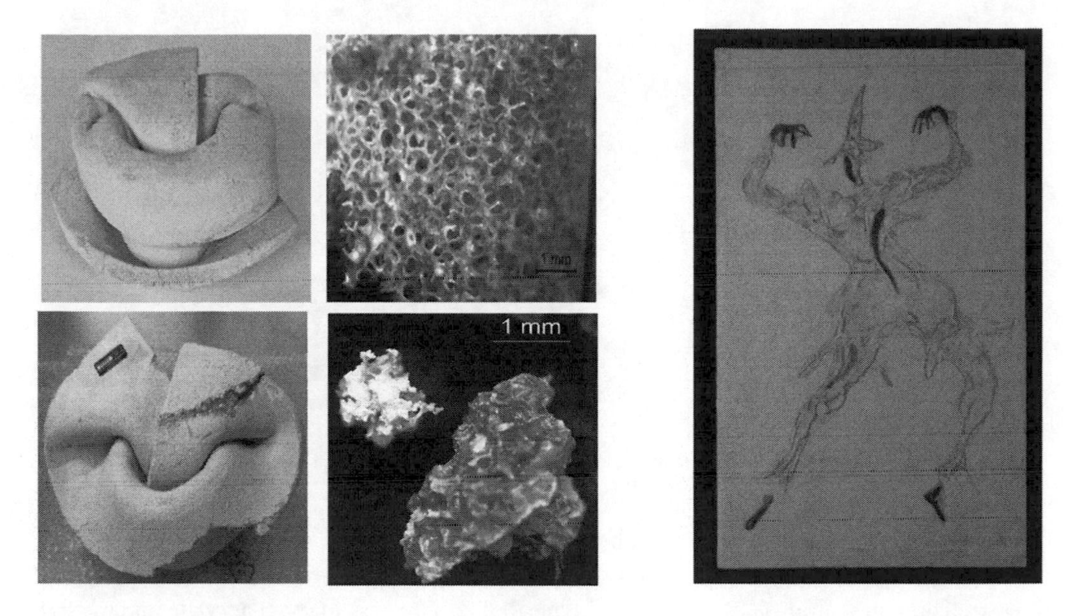

Figure 1. Artworks studied in this work: Composizione Astratta by Sante Monachesi, 1966 (left) and detail of its nucleus (upper center) and of the adhesive on the lower portions (lower center). Arlecchino by Lucio Fontana, 1951 (right).

2 MATERIALS AND METHODS

Micro-FTIR analysis were performed using a Thermo Scientific™ Nicolet™ iN™10 Infrared Microscope equipped with an MCT detector, liquid nitrogen cooled. The measurements were performed in transmission on diamond cell, in the region 4000–650 cm^{-1} at a resolution of 4 cm^{-1}, for 32 scans. SEM-EDS analysis were executed with a Zeiss EVO 60, equipped with a micro probe OXFORD Inca Pentaflex EDS for semi-quantitative analysis. The sample was analyzed in low pressure, in resin block cross section. All samples were previously observed using a microscope Leica M125, equipped with a Leica DFC 420C camera.

3 RESULTS AND DISCUSSION

3.1 *Sante Monachesi, Composizione Astratta*

Composizione Astratta (literally "Abstract Composition"), was realized in 1966 by the Futurist Italian artist Sante Monachesi (Macerata, 10 January 1910 – Rome, 28 February 1991). Monachesi is famous for his action-sculptures, mostly made of Perspex (polymethylmethacrylate) and rubber foam (polyurethane). The nucleus of the artwork is made of rubber foam, covered by a layer of white plaster which was showing evident signs of degradation, such as darkened surface and extended cracks.

FTIR analysis on the external layer (Figure 2) allowed to recognize the presence of calcium carbonate ($CaCO_3$) by the sharp absorptions at 2514, 1803, 875 and 712 cm^{-1} and the broad one at 1466 cm^{-1} (Learner & Gallery 1998).

The absorptions at 1733 cm^{-1} (C=O stretching), 1245 and 1025 cm^{-1} (C-O stretching) are typical of polyvinyl acetate (PVA) (Lobo & Bonilla 2003). According to these evidences, the artist plausibly mixed together grinded calcium carbonate with PVA to obtain a plaster with which he covered the internal nucleus.

The qualitative characterization of the rubber foam core of the artwork through infrared spectroscopy allowed to recognize (Figure 2) vibrational modes of C-H stretching (2988 and

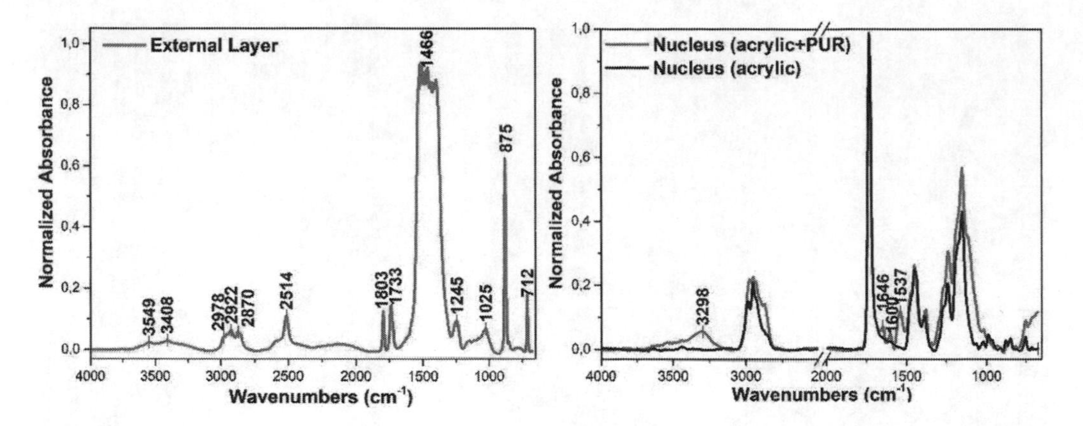

Figure 2. FTIR spectrum of the external layer (left). Comparison between spectra from different depths of the internal core of Composizione Astratta (right).

2951 cm^{-1}) and bending (1448 and 1387 cm^{-1}), carbonyl stretching at 1729 cm^{-1}, C-O stretching in the range 1300-1100 cm^{-1} typical of copolymers of ethyl acrylate and methyl methacrylate (Chiantore & Lazzari 1996). A previous conservation intervention, performed to fix in place a detached part, left on the lower portion of the artwork a darkened area. The infrared spectrum shows features typical of proteins such as collagen, present in animal glues such as rabbit-skin glue or bone glue.

3.2 *Lucio Fontana, Arlecchino*

Lucio Fontana (Rosario, 19 February 1899-Comabbio, 7 September 1968) sculptor, ceramist, painter, founder of Spatialism. Extremely creative with matter, space and figurativeness, famous for his renowned cut canvas, experimented with ceramic and mural painting too.

Arlecchino, one of the only Fontana's mural paintings, was victim of a bad restoration intervention, of which we don't have any track. The artwork, removed from his original support, was placed in a frame, probably to make the artwork movable. The area surrounding the human figure was filled with an ivory-colored plaster, probably altering the original depth wanted by the artist. The whole surface of the artwork appeared glossy probably due to the presence of a finish layer.

During the restoration some small samples of the artwork were collected in order to study the materials and to help the conservator to distinguish between Fontana's original design and the previous restoration contributions.

The first sample analyzed came from the most external surface.

FTIR analysis showed this layer contained calcium carbonate, recognizable by the absorptions at 2516, 1798, 1425, 877 and 712 cm^{-1} (Andersen & Brecevic 1991) and barium sulphate (signals at 1187, 1125 and 1084 cm^{-1}) (Learner & Gallery 1998). The appearance of signals linkable with an organic substance (2930, 2857 and 1730 cm^{-1}), required further investigation, so a cotton swab soaked in acetone was rubbed onto the surface of the artwork.

FTIR analysis of the extract confirmed the presence, on the artwork, of a layer of polyvinyl acetate (PVA) (Lobo & Bonilla 2003). A second sample was collected on the left lower corner, where plausibly the previous restoration didn't modify the original material. The FTIR analysis showed the presence of gypsum ($CaSO_4 \cdot 2H_2O$).

The EDS compositional maps (Figure 4) confirmed the artwork is composed by two layers of different plasters. The upper layer is 0.15 mm thick and contains titanium, barium and zinc (probably as white pigments) and calcium associated with sulphur most likely as gypsum. Inclusions, conceivably of calcium carbonate, are mostly elongated and sub-rounded. The

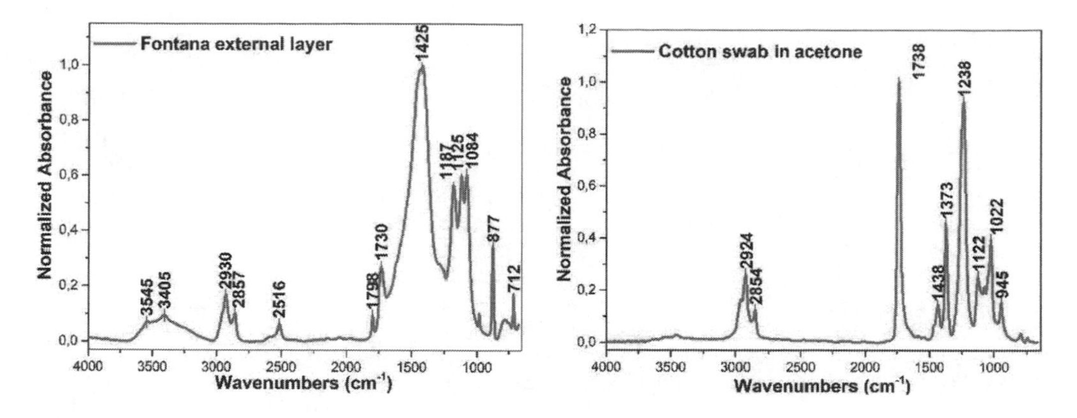

Figure 3. FTIR spectra of the external sample (left) and of the cotton swab (right).

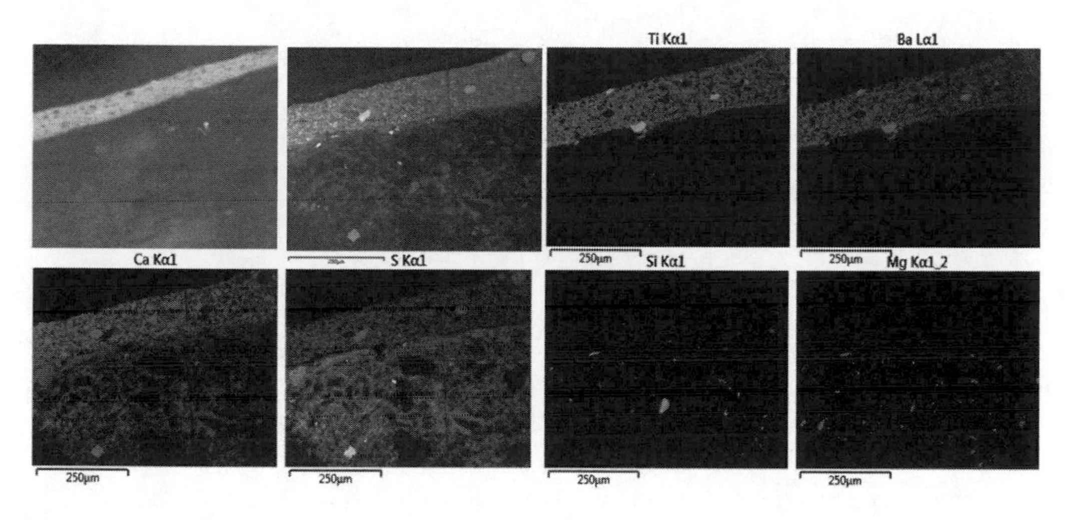

Figure 4. Optical microscopy picture of the cross section (left) and SEM image with EDS maps.

lower layer is 1.8 mm thick; the inclusions can be both angular (mostly formed by calcium sulphate) and elongated (containing silicon and aluminum, likely in form of oxides).

4 CONCLUSIONS

Modern materials are often susceptible to fast weathering; conservator's tasks, dealing with such complicated media, is surely not an easy one. The lack of information about material's degradation patterns and realization techniques may lead to wrong interventions which may cause the loss of the original artist's design and the artwork's reading. To study two Italian contemporary artworks, both victims of incorrect restorations, we applied a simple analytical approach which provided a detailed picture of compositions and state of preservation of the works of art. Micro-FTIR was used to investigate the organic components while SEM-EDS contributed characterizing the inorganic ones. Both the analytical techniques proved to be useful in detecting restoration components, not initially parts of the artist's project.

REFERENCES

Andersen, F. A. & Brecevic, L. 1991. Infrared spectra of amorphous and crystalline calcium carbonate. *Acta Chem. Scand*, 45(10), 1018–1024.

Chiantore, O. & Lazzari, M. 1996. Characterization of acrylic resins. *International Journal of Polymer Analysis and Characterization*, 2(4), 395–408.

Lazzari, M., Ledo-Suárez, A., López, T., Scalarone, D., & López-Quintela, M. A. 2011. Plastic matters: an analytical procedure to evaluate the degradability of contemporary works of art. *Analytical and bioanalytical chemistry*, 399(9), 2939–2948.

Learner, T., & Gallery, T. (1998). The use of a diamond cell for the FTIR characterisation of paints and varnishes available to twentieth century artists. *In Postprints: IRUG2 Meeting* (pp. 7–20).

Lobo, H., & Bonilla, J. V. (Eds.).2003. Handbook of plastics analysis (Vol. 68). *Crc Press.*

Quye, A., & Williamson, C. 1999. Plastics: collecting and conserving. *NMS Publishing Ltd.*

Wypych, G. (2018). Handbook of material weathering. *Elsevier*.

Science and Digital Technology for Cultural Heritage – Ortiz Calderón et al. (Eds)
© 2020 Taylor & Francis Group, London, ISBN 978-0-367-36368-0

Some applications of a novel surface enhanced Raman spectroscopy (SERS)-based strategy for the detection of art materials

M. Lazzari & D. Reggio
Centre for Research in Biological Chemistry and Molecular Materials (CIQUS), and Department of Physical Chemistry, Faculty of Chemistry, University of Santiago de Compostela, Santiago de Compostela, Spain

ABSTRACT: This paper shows recent results focused on the validation of a new nanodestructive and non-invasive sampling methodology based on novel SERS substrates for the detection of degradation markers and dyes from different types of museum artefacts. The procedure allows achieving detection limits much lower than those reached so far, opening the way to the use of SERS as an innovative tool for both the compositional analysis and the monitoring of ageing of museum artefacts.

1 INTRODUCTION

Although Raman spectroscopy has long being used for the detection of art materials, e.g. for the identification of pigments and dyes, its inherent disadvantages such as a large sample requirement, poor sensitivity, poor selectivity, and the inability to identify conclusively many compounds, strongly limited its potential and wider applicability. After the significant analytical advancement disclosed by *in situ* Raman spectroscopy, the breakthrough of surface-enhanced Raman spectroscopy (SERS) is further showing the potentials of the technique. Signal enhancement by several orders of magnitude is mainly due to the interaction of the Raman transition moment of the adsorbed molecules with the light-induced electromagnetic fields on the appropriate metal surface or substrate. Additionally, SERS has the advantage of quenching the fluorescence, thus allowing to reach even lower limits of detection with respect to traditional Raman spectroscopy (Vandenabeele et al. 2007). In the cultural heritage field, micro-destructive methods are generally preferred to destructive ones, due to the limited amounts of material required and SERS-based methodologies are emerging as the most promising applications (Casadio et al. 2016). In particular, SERS has been extensively used for the identification of pigments in paintings and the detection of natural and synthetic dyes in complex mixtures, and more in general for the identification of trace compounds in archaeological objects (Pozzi & Leona 2016). Silver nanoparticles, in the form of aqueous colloids, are traditionally applied to enhance the Raman signal, and analysis can be performed on a microsample taken from the object or directly *in situ*.

In this publication, we summarize the results obtained within a European Commission project focused on the application of nanomaterials and nanotechnologies to the conservation and protection of artworks, with a special emphasis for contemporary art (Gomez et al. 2018b). A novel nanodestructive and non-invasive methodology, based on a recently developed Al-coated SERS substrate (Gomez & Lazzari 2014, Gomez et al. 2018a), was applied to solve differently analytical challenges, from the detection of degradation markers from the surface of polymeric museum artefacts (Gomez et al. 2019a) or linseed oil films (Gomez et al. 2019b) to the identification of dyes in Japanese woodblock prints and different pen drawings. We also showed that the successful application of the procedure depends on the fine-tuning of an unconventional sampling approach, easily adaptable to the diverse analytical challenges usually faced in heritage science.

2 EXPERIMENTAL

Substrates were fabricated by ultraviolet nanoimprint lithography using a commercial nanostructured surface consisting of a square lattice of inverted pyramidal pits as a master mould. Details of the fabrication procedure are reported in Gomez & Lazzari 2012, and Gomez et al. 2018b). Silicone strip samplers were fabricated by casting and thermal curing of a liquid prepolymer under vacuum at 80°C for 2 h. Raman spectra were collected either in a Spectrometer Renishaw InVia Flex, equipped with two continuous wave lasers emitting at 785 nm and 514 nm or in a Witec Alpha 300R+ Confocal Microscope, equipped with three continuous wave lasers at 488, 532 and 785 nm, with singles and double grating options. All the SERS measurements were performed with a 10 s accumulation and 0.1-5% laser power, i.e. between 0.09 mW for 532 nm laser and 9 mW for 785 nm laser.

3 RESULTS AND DISCUSSION

Independently on the analytical target, either an artistic polymer surface, an oil painting or a woodblock print, the first step of the procedure shown in Figure 1a-c consists in the gentle pressure of a silicone strip sampler onto the surface of interest for 10 sec. Target molecules get physisorbed onto the strip and then dissolved in ca. 20 µl of an adequate solvent and transferred onto appositely prepared SERS-active Al-coated 3D structures (Figure 1d). The substrates have an average SERS enhancement factor of 10^9 for a 514 nm excitation wavelength, as determined for rhodamine 6G (Gomez et al. 2018b). The upper molecular weight limit of the extracted molecules is around 2 kDa, with a lower molecular weight cutoff of just few hundreds of Daltons.

In the case of aged polymeric surfaces, the silicone strip sampling aims to extract molecular markers of the degradation process, thus providing experimental evidence not achievable otherwise, even in conditions in which traditional spectroscopic techniques or pyrolysis-GC /MS have not enough sensitivity, and offering at the same time invaluable information on the mechanism of degradation of polymers (Lazzari et al. 2011, Rodriguez-Mella et al. 2014). The sensitivity of the approach was demonstrated by analyzing oligomeric degradation markers resulting from the artificial photo-ageing of model polymers commonly used in artworks, namely ABS, PVAc and polyisoprene (PI), and the natural ageing of contemporary artworks from the CGAC of Santiago de Compostela realized in natural PI. SERS analysis of surface products sampled either from the artworks (Figure 2a) or the corresponding material remainders available in the artist's studio, demonstrates the formation of small molecules essentially

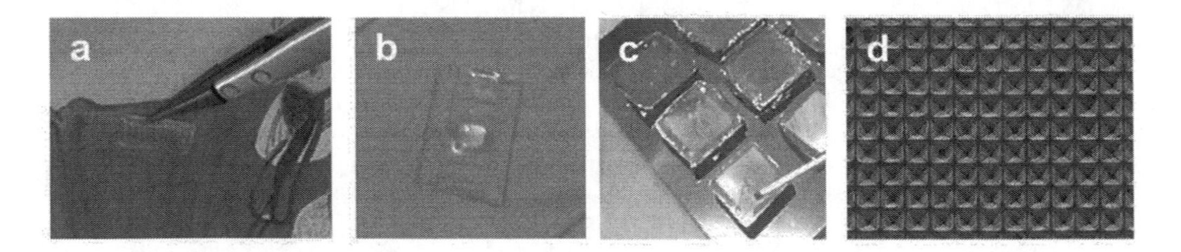

Figure 1. Photographs corresponding to the different steps of the sampling procedure, consisting in the application of a silicone strip sampler onto an artistic surface (a); followed by dissolution of the target molecules in the opportune solvent (b) and transfer (c) onto the Al-coated SERS substrate (d; SEM image of inverse pyramids 1.8x1.8x1 µm size).

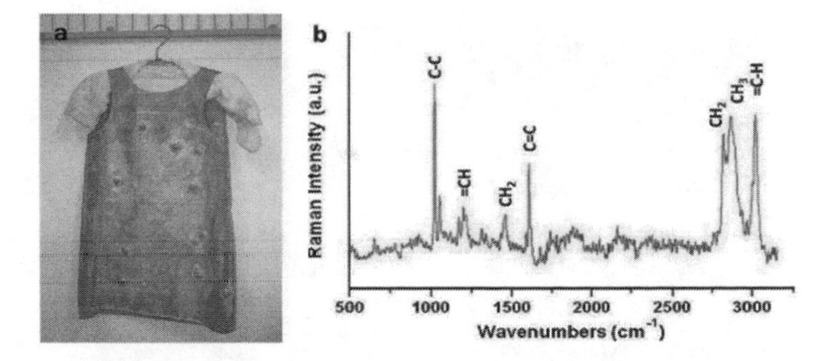

Figure 2. Polyisoprene-based artwork by Andrés Pinal from the collection of the Galician Center of Contemporary Art (a: *O traxe de Pepa, 1996*), and SERS spectrum at 514 nm excitation wavelength of surface products from the artwork, performed in 2018 (b).

consisting in isoprene oligomers, indicating that extensive deterioration is taking place at molecular level (Gomez et al. 2019a). An example is shown in Figure 2b.

Linseed oil is another organic material widely used in modern and especially traditional paintings, which drying and degradation mechanisms are well-known (Lazzari & Chiantore 1999, Scalarone et al. 2001). Considering that other drying oils and organic medium follow similar ageing pathways, linseed oil seemed an excellent model for a wider research on SERS for organic materials degradation in artworks. As an example of the investigation, Raman spectra of fresh linseed oil and touch-dry films collected from 3D Al-coated SERS substrates immediately after dropping around 20 ng of oil, and 24 h later are shown in Figure 3a and 3b, respectively, whereas spectra such as the one shown in Figure 3c could be obtained by micro-extraction from a silicone strip sampler applied onto photo-aged linseed oil films, herewith considered as behavioral model of oil paintings. By comparing the SERS spectrum of the fresh oil with that after 24 h drying, the expected decrease of the bands associated with C=C double bonds is clearly visible and correspond to the macroscopical formation of a touch-dry film. On the other hand, the structure of the molecules sampled from photo-aged films may be related with those formed by the typical auto-oxidation mechanism (Lazzari & Chiantore 1999) corresponding to low molecular weight carboxylic species released from the dried linseed oil network. The detection of these species from the surface of oil paintings may be

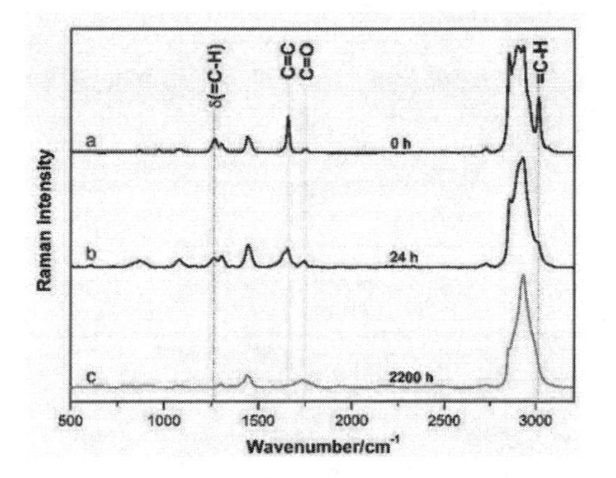

Figure 3. SERS spectra at 514 nm excitation wavelength of raw (a), touch-dry linseed oil (b) and molecules removed with a silicone strip sampler, mostly fatty acids fragments, from a 2200 h photo-aged film (c) onto the SERS substrate.

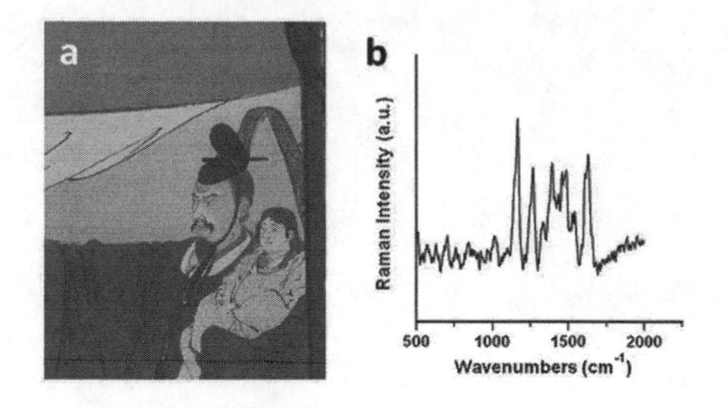

Figure 4. Detail a Japanese woodblock print (a), and SERS spectrum at 785 nm excitation wavelength of the molecules sampled from the red area, corresponding, e.g., to the man`s kimono visible in Figure 4a (b).

considered as a diagnostic marker of the ageing process, which should be further proved on real oil painting.

In addition, in order to validate the methodology, the sampling approach and the substrate were tested to face more traditional analytical issue, such as the identification of dyes. The procedure was applied on a Japanese woodblock print dated to the beginning of the XX century. No visible signs of material removal were reported after sampling; yet Raman spectra collected allowed discriminating the different coloring media. Different azo dyes, Prussian blue and carbon black, and also inorganic pigments, e.g. arsenic sulfide and mercury sulfide (vermilion), were easily detected. As an example, the SERS spectrum of the red dye sampled as shown in Figure 1a, is reported in Figure 4b. The spectral features match those of reference acid red 73, a common dis-azo β-naphthol colorant. Further details of these detections will be reported in a forthcoming publication.

4 CONCLUSIONS AND FINAL CONSIDERATIONS

We have demonstrated the applicability of Al-coated SERS substrates in combination with a nanodestructive, and therefore almost non-invasive, sampling for the detection of degradation markers released at the surface of polymeric materials and other organic materials, e.g. drying oils. In addition, we also showed the approach permits to detect dyes from paper surfaces, without leaving any visible signs as an effect of sampling. This enormous potential is being further explored to expand the knowledge on the detection of organic materials in artworks and museum objects.

ACKNOWLEDGMENTS

The European Commission (Project NANORESTART: H2020-NMP-21-2014/646063 and European Regional Development Fund) and the Xunta de Galicia (Centro singular de investigación de Galicia accreditation 2016–2019, ED431G/09, and GPC ED431B 2018/16) supported this work. The authors are also in debt with Antonio Mirabile, a paper conservator working in Paris, for providing the Japanese woodblock print, and Thais López Morán, conservator at the Centro Galego de Arte Contemporáneo of Santiago de Compostela.

REFERENCES

Casadio, F., Daher, C. & Bellot-Gurlet, L. 2016. Raman spectroscopy of Cultural Heritage materials: overview of applications and new frontiers in Instrumentation, sampling modalities, and data processing. *Topics in Current Chemistry*: 374, 62.

Gómez, M. & Lazzari, M. 2012. PFPE-based materials for the fabrication of micro- and nano-optical components. *Microelectronics Engineering* 97: 208–211.

Gómez, M. & Lazzari, M. 2014. Reliable and cheap SERS active substrates. *Materials Today* 17: 358–359.

Gómez, M., Kadkhodazadeh, S. & Lazzari, M. 2018. Surface enhanced Raman scattering (SERS) in the visible range on scalable aluminum-coated platforms. *Chemical Communications* 54: 10638–10641.

Gómez, M., Reggio, D., Lazzari, M., Rodríguez-Arias, I. & López-Quintela, M.A. 2018. Nanotechnologies for contemporary art conservation: Some applications on plastics. In *Conserving Cultural Heritage*. Mosquera & Almoraima Gil Eds. CRC Press, 181–183.

Gómez, M., Reggio, D. & Lazzari, M. 2019. Detection of degradation markers from polymer surfaces by a novel SERS-based strategy. *Talanta* 191: 156–161.

Gómez, M., Reggio, D. & Lazzari, M. 2019. Linseed oil as a model system for SERS detection of degradation products in artworks. *Journal of Raman Spectroscopy* 50: 242–249.

Lazzari, M. & Chiantore O. 1999. Drying and oxidative degradation of linseed oil. *Polymer Degradation and Stability* 65: 303–313.

Lazzari, M., Ledo-Suárez, A., López, T., Scalarone, D. & López-Quintela, M.A. 2011. Plastic matters: an analytical procedure to evaluate the degradability of contemporary works of art. *Analytical and Bioanalytical Chemistry* 399: 2939–2948.

Pozzi, F., Leona, M. 2016. Surface-enhanced Raman spectroscopy in art and archaeology. *Journal of Raman Spectroscopy* 47: 67–77.

Rodriguez-Mella, Y., López-Morán, T., López-Quintela, M.A. & Lazzari M. 2014. Durability of an industrial epoxy vinyl ester resin used for the fabrication of a contemporary art sculpture. *Polymer Degradation and Stability* 107: 277–284.

Scalarone, D., Lazzari M. & Chiantore O. 2001. Thermally assisted hydrolysis and methylation-pyrolysis-gas chromatography/mass spectrometry of light-aged linseed oil. *Journal of Analytical and Applied Pyrolysis* 58: 503–512.

Vandenabeele, P., Edwards, H. G. M. & Moens, L. 2007. A decade of Raman spectroscopy in art and archaeology. Chemical Reviews 107: 675–686.

Science and Digital Technology for Cultural Heritage – Ortiz Calderón et al. (Eds)
© 2020 Taylor & Francis Group, London, ISBN 978-0-367-36368-0

Non-destructive techniques applied to *in situ* study of Maqsura at Cordoba cathedral (Spain)

A. Gómez-Moron
Andalusian Historical Heritage Institute (IAPH), Seville, Spain
Department of Physical, Chemical and Natural Systems, Pablo de Olavide University (UPO), Seville, Spain

P. Ortiz, R. Ortiz, J. Becerra & R. Gómez-Cañada
Department of Physical, Chemical and Natural Systems, Pablo de Olavide University (UPO), Seville, Spain

R. Radvan, A. Chelmus, L. Ratoiu, L. Ghervase, I.M. Cortea, C. Constantin & L. Angheluta
National Institute for Research and Development in Optoelectronics, Măgurele, Romania

ABSTRACT: The Maqsura and its Mihrab is one of the most outstanding locations of the Mosque-Cathedral of Cordoba built during the reign of Caliph Al-Hackam II (962-965 AD). The ART4ART mobile laboratory has allowed to study materials, state of conservation and structure characteristics at Maqsura without sampling. 3D digitization, elementary and molecular physicochemical analysis, Laser Doppler Vibrometer (LDV) and Ground Penetration Radar (GPR) were employed to study mosaics, mural paintings roof and subsoil. 3D digitization using photogrammetry offers the possibility to digitally manipulate the models and observe small details on the surface, but also to analyze the monument degradation in time, by subsequent 3D models of the same areas. GPR and LDV techniques are noninvasive methods to investigate internal structure of the walls or their state of preservation (cracks, detachments). Chemical analysis was used for material identification *in situ*.

This multidisciplinary approach based on non-destructives techniques allows us to study Maqsura materials and its state of conservation in situ and real-time without sampling assay, what makes this approach as a good protocol for the diagnosis of the Cultural Heritage.

1 INTRODUCTION

The Mosque-Cathedral of Cordoba is located in the South of Spain, region of Andalusia, and it is a World Heritage site since 1984.The site was originally a Catholic Basilica, but when Muslims conquered Spain in 711 A.D., this church was demolished (784 A.D.) and was built the Great Mosque of Cordoba. The Maqsura and Mihrab were built during the ampliation of Mosque by Umayyad Caliph Al-Hakam II between 962 and 965 A.D. When the Roman Catholic church was installed again by Catholic Monarchs in 1236 A.D., architectural structures and the Caliphal decoration of Maqsura were preserved.

Maqsura consists of three adjacent rooms decorated with colorful glass mosaics and mural paintings (Figure 1). The mosaics were made by tesserae brought from Constantinopla by al-Hakam II and have suffered three interventions: Devreton reformed the domes in 1772,

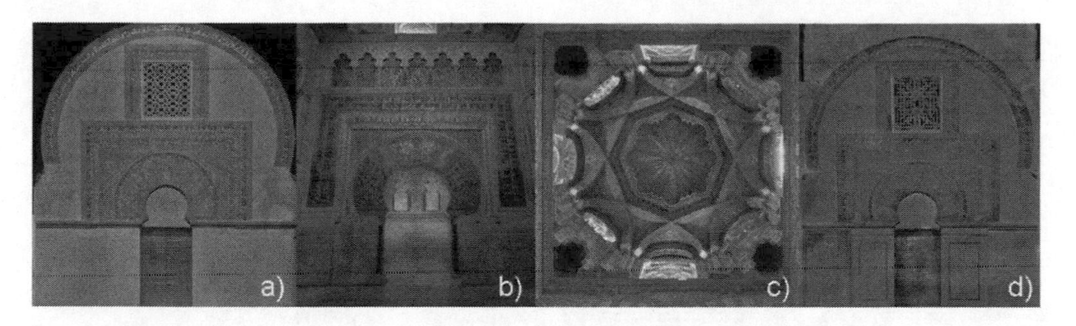

Figure 1. a) Bab Bayt al-Mal door b) Mihrab door c) Mihrab ceiling d) Sabat door.

P. Furriel restored the mosaics in 1815-1918, and finally R. Velázquez Bosco restored the mosaics again in 1912-1916 (Nieto Cumplido 1998).

The mobile laboratory ART4ART was employed to study and diagnose onsite and without sampling the different materials of Maqsura and their state of conservation.

2 METHODOLOGY

ART4ART is a mobile laboratory equipped with several non-invasive techniques. The facilities are specially designed as a heritage ambulance following the three principles of good diagnosis practice: non-contact, non–destructive and fast-response analysis (Angheluta Laurentia et al. 2015, Angheluta Laurentia et al. 2019).

A multiapproach diagnosis of mosaic, roof and subsoils were planned with the following techniques:

a) Photogrammetry for accurate digital reconstruction of surfaces with a Nikon D810 camera (60mm) at 36 Megapixels resolution and Agisoft Photoscan software.
b) Laser doppler vibrometer (LDV) for the identification of hidden defects. The laser beam from the LDV was directed at the surface of interest and the vibration amplitude and frequency from 30MHz to 30 GHz are extracted from the Doppler shift of the laser beam frequency due to the motion of the surface. Sine excitation was used in LDV technique to measure the deflection shape at 5,1kHz due to a good signal-to-noise ratio and the possibility to perform a scan in a short time. The measured magnitude is ranged between 0 and 25 mm/s. Burst Chirp was also used in LDV technique, it is a fast swept sine signal followed by a zero signal. The measured magnitude is ranged between 0 and 60 mm/s.
c) Ground penetrating radar (GPR) is a technique that uses electromagnetic pulses that are directed into the ground and records the signal that is reflected from buried objects, so it is useful to detect undersoil cavities or buried objects. Two antennas with different frequency were used in the GPR investigations 800 MHz antenna with a travel time of the electromagnetic pulse at 50 ns and 500 MHz antenna at 100 ns.
d) Portable X-ray fluorescence spectrometry (PXRF) was employed for glass tesserae chemical characterization without the requirement of sampling (Davison 2003). The equipment is a TRACER III-SD Bruker with a Rh tube, working at 40kV and 10.60 mA with 10 s of acquisition time per spectrum (Cortea 2019).

3 RESULTS

3.1 *Photogrammetry*

Four zones (three from the Mihrab dome and the one from the Mihrab mosaic wall) were scanned for accurate digital reconstruction of their surfaces and their 3D/Digital elevation

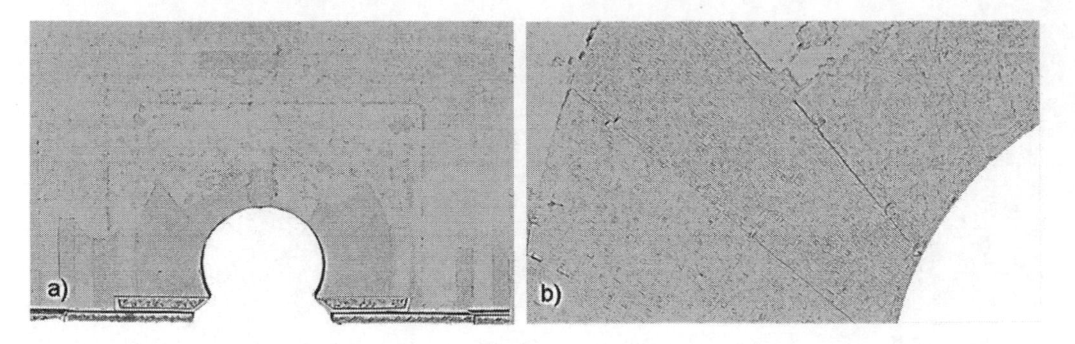

Figure 2. a) General photogrammetry of Mihrab mosaic wall. b) Detailed view of an arch with restorations.

Model. The most interesting images show the details of the 3D model of Mihrab mosaic wall where it can distinguish different interventions (Figure 2).

3.2 *Laser doppler vibrometer*

LDV was used in two modes. The results using the sine excitation function showed low magnitudes distributed on the surface with only some small areas that revel a different vibration intensity (40-60 mm/s) (Figure 3a). This means that the surface conditions on mosaics are similar without important crack. In the case of the Burst Chirp excitation function, the vibration magnitude was a bit higher due to the function characteristics (Figure 3b). Red small areas (approx. 60 mm/s) showed irregularities that might be due to interventions made in that area or *tesserae* missing.

3.3 *Ground Penetrating Radar (GPR)*

GPR was carried out on the soil of Bab Bayt al-Mal door (Area 1) and Sabat door (Area 2). Area 1 showed a strong reflection, observed at both 500 and 800 MHz radargrams, that may be due to two graves (approximatively at 40 cm depth from the bottom of a grave). The Area 2 showed linear reflections at the depth of 0,3 meters. The results also indicate a change of propagation, the medium become inhomogeneous from the surface till the depth of 0,6 meters. A low dielectric permittivity medium was detected between the depth of 0,6 and 1,5 meters due to the low intensity reflections, that may be a small crypt (Figure 4).

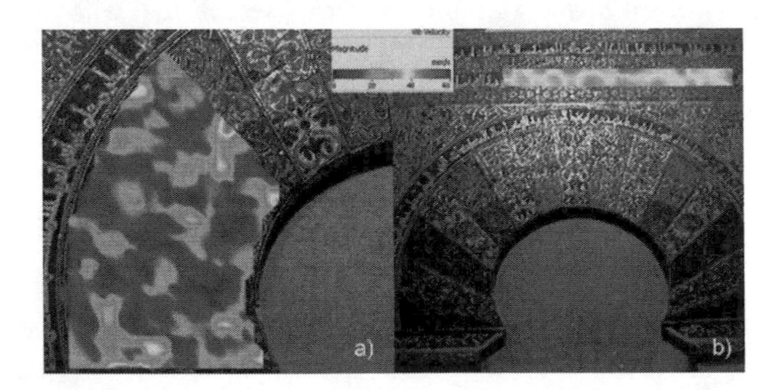

Figure 3. a) Sine function results of different zones of Sabat door with similar surface conditions b) Burst Chirp function results of different zones of Sabat door, red zones are irregularities.

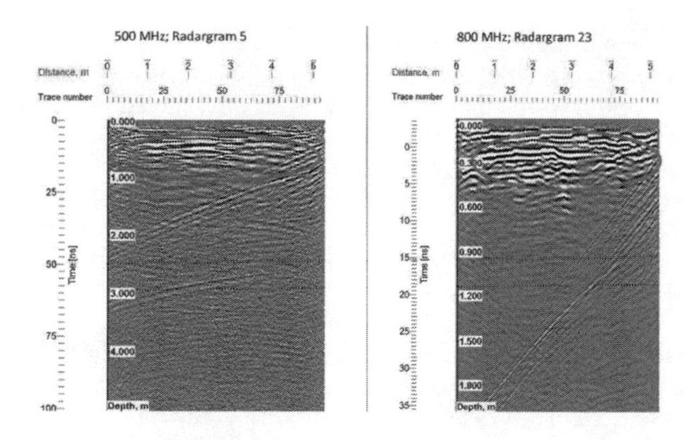

Figure 4. a) 500 MHz Radargrams of the Sabat door that identifies a possible small crypt (red) b) 800 MHz Radargrams of the Sabat door, possible crypt in red.

Table 1. XRF analysis areas and main elements identified.

Area description	*Sample*	*Detected elements**
Golden	*1*	*Pb (ma), Ca (mi), Bi, Fe, Cu, Sn (tr)*
	2	*Ca, Fe, Sr, Mn (ma), K, Pb, Si, Au (mi), Zr, Ti, S, Cu, Cr, Zn (tr)*
	4	*Pb (ma), Fe, Au (mi), S, Ca, Sr, Cu, Mn (tr)*
	25	*Mn, Ca, Fe (ma), Sr, Si, Pb, Au (mi), K, Cu, Rb, Ti, S, Zn (tr)*
	26	*Ca, Fe, Mn, Au (ma), Sr, Pb, Si, K, Rb (mi), Ti, Cu, Zn (tr)*
Blue	*5*	*Pb (ma), Fe, Au, S (mi), Ca, Cu, Sr, Bi (tr)*
	6	*Ca, Fe (ma), Cu, Si, Sr, Mn, K, S, Ti (tr)*
	11	*Ca, Fe, Mn, Sr (ma), Si, Cu, Pb, K, Au (mi), Zn, Rb, S (tr)*
	13	*Pb, Ca, Fe (ma), Cu, Si, Co, Sr, S, Mn (mi), Bi, Ti, Zr, Zn, K (tr)*
	15	*Cu, Ca, Fe (ma), Si, Pb, As, Sr, Mn (mi), Ti, Zr, S (tr)*
	17	*Ca, Fe (ma), Cu, Si, Pb, Sr, Mn (mi), K, Zr, Ti, Zn, S, Sn (tr)*
	14	*Ca, Fe (ma), Mn, Cu, Si, Pb, Sr (mi), K, Ti, Zr, S (tr)*
Violet	*19*	*Ca, Mn, Fe, Sr (ma), Si, Pb, K, Zr (mi), Cu, Ti, Rb, S, Al (tr)*
	21	*Pb (ma), Mn, Sn, Ca (mi), Fe, S, Sr, Si, BI (tr)*
Turquoise	*12*	*Ca, Fe, Cu (ma), Pb, Si, Mn, Sr (mi), K, Zr, Ti, S, Rb (tr)*
Olive		*Pb (ma), Ca (mi), Cu, Fe, S, Bi (tr)*
Green	*3*	*Pb (ma), Ca (mi), Fe, Bi, Cu, Sn (tr)*
	10	*Cu (ma), Ca, Fe (mi), Si, Sr, As, Mn (tr)*
	20	*Cu (ma), Fe, Ca, Mn (mi), Si, Sr, Pb, As (tr)*
Red	*7*	*Fe, Cu, Ca, Sr (ma), Pb, Mn, Si, K, S, Ti, Zn, Sn (tr)*
	16	*Fe, Cu, Ca, Sr, Pb (ma), Mn, Si, K, Zr (mi), Zn, Ti, Sn, Rb, As, S (tr)*
	24	*Fe, Cu, Ca (ma), Sr, Mn, Pb, Si, K (mi), Ti, Zn, Au, Rb, S (tr)*
White	*8*	*Pb (ma), Ca, S (mi), Fe, Bi, Sr (tr)*
	23	*Pb (ma), Ca (mi), S, Cu, Fe, Bi, Sr (tr)*
Black	*9*	*Ca, Fe, Mn, Cu, Sr (ma), K, Zr, Pb, S (mi), Ti, Rb (tr)*
	22	*Ca, Mn, Fe, Sr (ma), Si, K, Cu, Pb, Zr (mi), Ti, Rb, Zn, S (tr)*
Integration area	*27, 28*	*Pb (ma), Fe (mi), S, Ca, Bi, Sn (tr)(only 28 Sr (mi), Cu, Mn, Zr, Ti, Si (tr))*

* major (ma), minor (mi), trace elements (tr)

3.4 *Portable X-ray Fluorescence Spectroscopy (XRF)*

XRF is a non-invasive elemental analysis technique that allowed to analyze 28 areas with glass *tesserae* without sampling and in a fast way (Table 1). The blue and green *tesserae* contained Co and/or Cu. Manganese was found mainly in black and violet areas. Lead was mostly in areas with repainting, probably painted when Furriel restored the mosaics in 1815-1918. Green *tesserae* contained lead and tin as chromophore or opacifying compounds. The golden areas showed gold. Arsenic lines was noticed in dark green and turquoise blue and it was correlated with high intensities of Cu lines, which may point out to the use of emerald Green ($Cu(CH_3COO)_2$ 3 $Cu(AsO_2)_2$).On the other hand, red areas were characterized by a higher content of Fe, probably due to the use of red iron oxide (Fe_2O_3).

4 CONCLUSIONS

The mobile laboratory for *in situ* investigation has allowed us to study Maqsura materials and its state of conservation in real-time without sampling assay or preparation. This multidisciplinary study has helped to increase the knowledge about Maqsura and to evaluate the state of conservation following the three principles of good diagnosis practice: non-contact, non–destructive and fast-response analysis.

Glass *tesserae* have been analyzed to know principal chromophores, opacifying and bulk glass composition. GPR studies showed that there could be two graves and a crypt. LDV allowed us to detect intervened areas.

ACKNOWLEDGEMENTS

This paper has been supported and based on the Methodology developed by Art-Risk, a RETOS Project of Ministerio de Economía y competitividad and Fondo Europeo de Desarrollo Regional (FEDER), (BIA2015-64878-R (MINECO/FEDER, UE)) and the research group PAI-TEP-199 from Junta de Andalucía.

REFERENCES

Angheluta Laurentiu-Marian; Ene Dragos Valentin 2015. An Interdisciplinary Field Campaign For Modern Investigation And Monitoring in Preservation And Restoration, International Journal of Conservation Science, Vol. 6.

Angheluta, Laurentiu, Radvan, R 2019. Multi- and Inter- Disciplinary Approaches in the Scientific Documentation and Monitoring of Cultural Heritage Assets. Vol. Fascile II.

Cortea, I., Luminița Ghervase, Ovidiu Țentea, Anca Constantina Pârău, and Roxana Rădvan. 2019. First Analytical Study on Second-Century Wall Paintings from Ulpia Traiana Sarmizegetusa: Insights on the Materials and Painting Technique. International Journal of Architectural Heritage, January. Taylor & Francis.

Davison, S. 2003 Conservation and Restoration of Glass, Butterworth-Heinemann in Conservation and Museology, Butterworth-Heinemann.

Nieto Cumplido, M. 1998. Capítulo VI. Ampliación de al-Hakam II en La Catedral de Córdoba, Caja Sur.

Science and Digital Technology for Cultural Heritage – Ortiz Calderón et al. (Eds)
© 2020 Taylor & Francis Group, London, ISBN 978-0-367-36368-0

In situ study by XRF and LDV of mural paintings in Magdalena church (Seville, Spain)

A. Tirado-Hernández, J. Becerra, R. Ortiz & P. Ortiz
Department of Physical, Chemical and Natural Systems, Pablo de Olavide University, Seville, Spain

A. Gómez-Morón
Andalusian Historical Heritage Institute, Seville, Spain
Department of Physical, Chemical and Natural Systems, Pablo de Olavide University, Seville, Spain

L. Ghervase, I. Cortea, A. Chelmus & R. Radvan
National Institute for Research and Development in Optoelectronics, Măgurele, Romania

ABSTRACT: The characterization and diagnosis of mural paintings located in Santa María Magdalena Church (Seville, Spain) were studied by X-ray fluorescence (XRF) and Laser Doppler Vibrometer (LDV). Laser Doppler vibrometer (LDV) is a non-destructive and non-contact technique based on recording the answer of materials to different sound waves in order to evaluate deformations, cracks, blisters, detachments and other damages on painting walls of the chorus. XRF analyses discovered the main materials used by Lucas Valdés in the frescoes of 18th century. The main weathering forms found in the temple after visual inspections are cracks and lacunas, However, LDV could not detect cracks at the ratio volume/distance employed. Further studies must be carried out to improve the use of this methodology in mural paintings.

1 INTRODUCTION

Mural paintings of Santa María de la Magdalena Church (Seville, Spain), an example of Sevillian baroque architecture, were studied by X-ray fluorescence (XRF) and Laser Doppler vibrometer (LDV). At the top of the lateral walls of this transept are located two 18th century mural paintings by Lucas Valdés (Figure 1): *The triumphal entry of San Fernando in Seville*, located on the right side of the transept, which represents the entrance of this king in the city accompanied by the Virgin and a large ecclesiastical retinue; and *Auto de Fe in the times of San Fernando*, located on the left side of the transept. It represents an Auto of Faith in which Dominican friars appear and along with San Fernando that transports firewood for the bonfire.

XRF analyses were carried in situ for identification of the pigments employed by Valdes so was carried out on the paintings located on the top of the side walls of the transept. LDV is a non-destructive and non-contact technique used in order to evaluate damages on walls. LDV was carried out on the chorus painting walls. The Figure 2 indicates the analyses zones in the church for each technique.

2 METHODOLOGY

2.1 *X-ray fluorescence (XRF)*

The development of hand-held XRF equipment was a great advance in the field of restoration, allowing in situ investigations (Radvan et al, 2016; Mantler & Schreiner, 2000). XRF was

Figure 1. *Auto de Fe in the times of San Fernando*, located on the left side of the transept (left) and *The triumphal entry of San Fernando in Seville*, located on the right side of the transept (right).

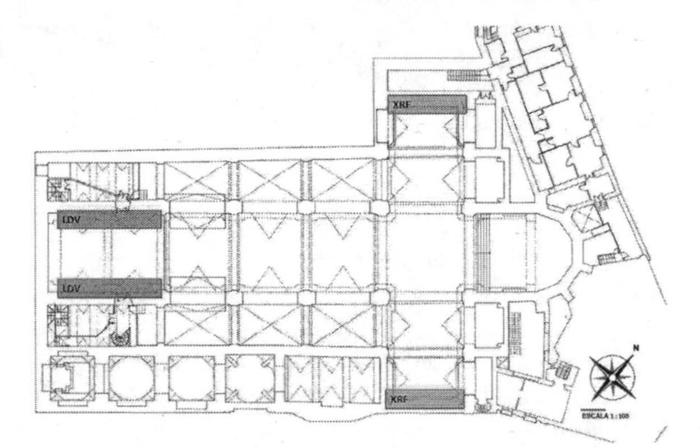

Figure 2. In situ XRF and LDV analyses zones in the church.

carried out on paintings located on the top of the side walls of the transept. 53 spectra were acquired (29 areas on the right side and 24 areas on the left side), on various chromatic areas. The analysis has been carried out thanks to Tracer III-SD XRF analyser (conditions: 40 kV, 10.60 μA). Identification of elements was performed through Bayesian deconvolution with ARTAX 7.4.0.0 software. Microsoft Office Suite 2016 was used for the element distribution. Data used for elemental distribution were normalized with respect to the Rayleigh scattering peak of Rh.

2.2 *In situ study of weathering forms*

Prior to the use of non-destructive techniques, a visual inspection of the main weathering forms has been carried out according to ICOMOS classification (ICOMOS-ISCS, 2008).

2.3 *Laser Doppler Vibrometer (LDV)*

LDV is a non-destructive and non-contact technique based on recording the answer of materials to different sound waves with a view to evaluate damage on walls such as deformations, cracks, blisters and detachments (Longo et al. 2010). The application of LDV to the diagnosis of monuments is based on the principle that it is possible to evaluate the structural state of an object from the examination of its vibrations (Esposito, 2005).

The laser beam is directed at a surface, and the vibration amplitude and frequency are extracted from the Doppler shift of the laser beam frequency due to the motion of the surface. The Doppler Effect is the apparent frequency change noticed by an observer when the source of a wave is moving relative to the observer. This frequency shift is proportional to the velocity of relative movement between the observer and the wave source. When two coherent light beams combine, the resulting intensity has a component that is related to the difference in path lengths of the two beams (Tabatabai et al. 2013). LDV measure the velocity of a point on a surface along the direction of the laser. A crack changes the vibration characteristics of structure.

Four measurements were carried out in the chorus, in order to evaluate the results in comparison with the in-situ observations of the weathering forms. Studied object were excited using a common loudspeaker.

3 RESULTS

3.1 *X-ray fluorescence (XRF)*

The Table 1 shows the major (ma), minor (mi) and trace elements (tr) identified for the main chromatic areas investigated.

As expected, intense calcium lines were registered on all analyzed areas. Taking into account the general high content of sulfur and iron also identified, we can assume that the mural substrate comprises gypsum plaster and clay.

In terms of the color palette, the presence of Co, Ni, As, K and Bi within the blue areas indicates the use of smalt. For the red areas, registered data suggests the use of multiple red pigments: red iron oxide (L9), vermilion (L10), and red lead (R16) - this last one degraded into brown. For the green areas a copper based pigments was used, while for the yellow ones the use of yellow ochre can be suggested based on the high content of iron, along with registered traces of potassium, silica and manganese. Violet hues might have been obtained by mixing smalt and red ochre. The use of burnt umber could also be inferred taking into account that manganese lines appear with higher intensities within the brown and black areas. For the white, a calcium based pigment was used (calcium carbonate).

For the metallic decorations of the mural paintings gold leaf was identified. As indicated by the high lead content (L8), the gold leaf was most probably applied on top of a preparatory layer based on lead white.

Some of the areas on the right side have more intense zinc, titanium and barium lines, probably associated with pigments used in more recent restoration.

Table 1. Summary results of the XRF analysis carried out on the left (L) and right (R) side of mural paintings of Magdalena Church.

Area description	Analyzed area	Identified elements
Golden	L8	Pb, Au (ma), Ca, Fe, As, Cu (mi), Cl, Sr, Hg, Mn, Ba, Bi (tr)
Blue	L14	Ca, As (ma), Fe, Co, Ni, Sr, K, Bi (mi), Si, Zn, Sr, Pb, Mn, Cu, Cl (tr)
Violet	L21	Ca, Fe (ma), As, Co, S, Ni, K (mi), Bi, Sr, zn, Mn, Cu, Cl (tr)
Green	R4	Cu, Ca (ma), Fe, S (mi), Zn, Pb, As, Sr, Co, Cl, K (tr)
Dark red	L9	Fe, Ca, Pb (ma), Hg, S, Co (mi), Sr, Si, Mn, K, Cl (tr)
Red	L10	Hg, Pb, Ca (ma), Fe, S, As (mi), Sr, Au, Cl, Co (tr)
Brown	L17	Ca, Fe (ma), S, Pb (mi), Sr, Si, K Co, Cl (tr)
	R9	Ca, Fe (ma), S, Pb (mi), Sr, Mn, K, Co (tr)
	R16	Pb, Ca, Fe (ma), As, S, Mn (mi), Bi, Sr, Hg, Ti, Cl, Co, Cu (tr)
Yellow-ochre	L6	Ca, Fe (ma), S (mi), K, Zn, Ba, Co, Zn, Sr, Si, Ti, Mn, Cu (tr)
	R27	Ca, Fe (ma), S, Sr, Ti, Zn, Ba (mi), K, Si, Mn, Co (tr)
White	L5	Ca (ma), Fe, S (mi), K, Sr, Co, Cl (tr)
	R1	Ca (ma), Ti, S, Fe (mi), Pb, Sr, K, Ba, Zn (tr)
Black	L23	Ca (ma), Fe, S (mi), K, Sr, Co, Cl (tr)
	R6	Ca (ma), Fe, S (mi), Sr, Pb, K, Mn, Co (tr)

This in situ and non-destructive analysis allows to reduce the sampling for a good diagnosis practice in Cultural Heritage.

3.2 *In situ study of weathering forms*

In the interior of the church, loss of material in the cornices and loss of painting areas in frescoes were detected, caused fundamentally by dampness and capillarity. Efflorescences usually appeared above the base and concretions were located in the vaults of the Epistle Side. With a very high frequency, fractures and cracks (Figure 3) were detected in the vaults and arches of the side aisles and in the central nave (Tirado et al. 2018).

3.3 *Laser doppler vibrometer (LDV)*

The main weathering forms found in the church are lacunas and cracks according to the results of visual inspection. However, regarding the results of LDV, cracks were not detected while some small differences on surface are recorded as no significant according to the results of a Burst Chirp function, a fast swept sine fallowed by a zero signal (Figure 4). The reason could be that the speaker volume used to excite the surface was too low compared with distance between the wall/ceiling and the speaker and that did not make the surface to vibrate or that the crack is only on surface. Further studies where the volume of the speaker is higher must be carried out to find out the optimal conditions to cause vibration of the surface in order to obtain better results.

Figure 3. Crack in the roof of the central nave of Santa Magdalena Church.

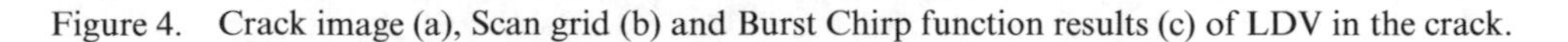

Figure 4. Crack image (a), Scan grid (b) and Burst Chirp function results (c) of LDV in the crack.

4 CONCLUSIONS

XRF is an in-situ and non-destructive technique that was found useful for mural painting characterization and diagnosis before restorations. XRF results allowed us to determinate the composition of the pigments used in the frescoes and to reduce the future sampling according to good diagnosis practice in Cultural Heritage. After visual inspection, efflorescences, lacunas, concretions, loss of material and overall fractures and cracks were detected. LDV could not detect cracks at the ratio volume/distance employed, although it is proved that it is a very useful technique in diagnosis in Cultural Heritage. Further studies must be carried out in order to improve its application on mural paintings.

ACKNOWLEDGEMENTS

This paper has been supported by ART-RISK, a retos project of Ministerio de Economía y Competitividad and Fondo Europeo de Desarrollo Regional (FEDER), (code: BIA2015-64878-R (MINECO/FEDER, UE)) and the research team PAI-TEP-199 from Junta de Andalucía. Infrastructure assured by Program OPTRONI & INOVA OPTIMA.

REFERENCES

Esposito, E. 2005. Recent developments and applications of laser doppler vibrometry to cultural heritage conservation. *Proceedings of Conference on Lasers and Electro-Optics*-Europe, IEEE, 673.

ICOMOS-ISCS. 2008. Illustrated glossary on stone deterioration patterns.

Longo, R., Vanlanduit, S., Vanherzeele, J. & Guillaume, P. 2010. A method for crack sizing using Laser Doppler Vibrometer measurements of Surface Acoustic Waves. *Ultrasonics*. 50 (1), 76–80.

Mantler, M. & Schreiner, M. 2000. X-ray Fluorescence Spectrometry in Art and Archaeology. *X-Ray Spectrometry*, 29, 3–17.

Radvan R, Bors, C, Ghervase, L. 2016. Portable X-Ray fluorescence investigation of certain bronze beads of hoard Tărtăria i and their specific corrosion. *Romaina Journal of Physics*. 61, (9-10), 1530–1538.

Tabatabai, H., Oliver, D., Rohrbaugh, J. & Papadopoulos, C. 2013. Novel Applications of Laser Doppler Vibration Measurements to Medical Imaging. *Sens Imaging*, 14(1-2): 13–28.

Tirado-Hernández, A., Ortiz, R. & Ortiz P. 2018. 3D laser scanning applied to diagnosis in vaults. *Conserving Cultural Heritage* – Mosquera & Almoraima Gil (Eds). Taylor & Francis Group, London, 145–149.

Science and Digital Technology for Cultural Heritage – Ortiz Calderón et al. (Eds)
© 2020 Taylor & Francis Group, London, ISBN 978-0-367-36368-0

Non-destructive techniques applied to the study and diagnosis of ceramic and glazed terracotta tombs in Omnium Sanctorum church (Seville, Spain)

A. Tirado-Hernández, J. Becerra, R. Ortiz & P. Ortiz
Department of Physical, Chemical and Natural Systems, Pablo de Olavide University, Seville, Spain

A. Gomez-Moron
Andalusian Historical Heritage Institute, Seville, Spain
Department of Physical, Chemical and Natural Systems, Pablo de Olavide University, Seville, Spain

L. Ghervase, I. Cortea, L. Angheluta & R. Radvan
National Institute for Research and Development in Optoelectronics, Măgurele, Romania

ABSTRACT: The purpose of this work was to carry out the diagnosis and vulnerability analysis of the ceramic and glazed terracotta tombs of Omnium Sanctorum church (Seville, Spain) by non-destructive techniques such as X-Ray fluorescence (XRF) for chemical analyses of tiles, and 3D digitization to identify the main weathering forms. Efflorescence, detachment, cracks and loss of glazed materials were detected by 3D digitization and compared with the results obtained by in situ inspection, while XRF analysis discovered the composition of tiles without sampling. The combination of these non-destructive techniques allowed to know the composition of the materials and to establish that dampness by capillarity may be the main cause of weathering.

1 INTRODUCTION

Omnium Sanctorum (14th-century church of Seville, Spain) contains an arcosolium constituted by two arches decorated with tiles of the 16th, 18th and 20th centuries, and with two terracotta sculptures attributed to Mercadante de Bretaña, the tombs of Sánchez-Dalp Dukes. The glazed ceramic pieces have different decorative motifs: laceries, floral ornaments, heraldic shields, figures of angels and saints (Santo Domingo and San Ramón Nonato). The Figure 1 shows the area of study in the church and its location in the temple.

The activity and results presented here are part of a conservation status monitoring plan and summed up this phase for analyses that can reveal a great deal of information without involving overly sophisticated methods. It is worth noting that the proposed working method pursues the exclusive use of non-invasive methods, without inducing changes on the surface or in depth.

2 METHODOLOGY

2.1 *X-ray fluorescence (XRF)*

The development of hand-held x-ray fluorescence (XRF) equipment was a major breakthrough in the field of art, allowing in situ investigations of elemental composition without sampling (Radvan et al, 2016; Fosters et al., 2011). 13 and 18 XRF spectra were acquired in the ceramic and glazed terracotta works, respectively. The analyses have been carried out

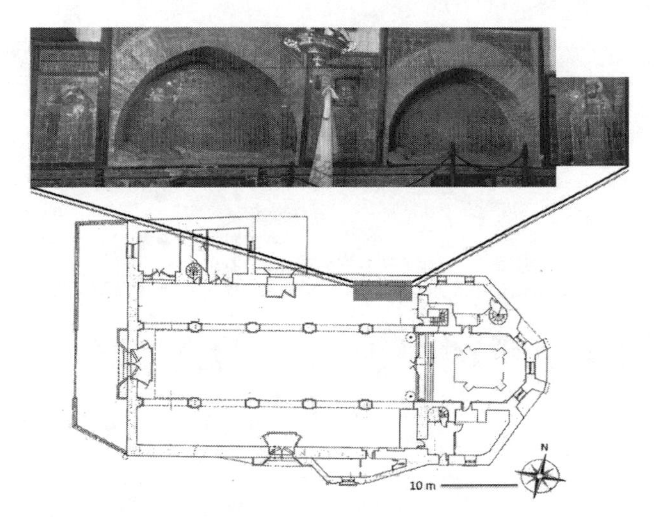

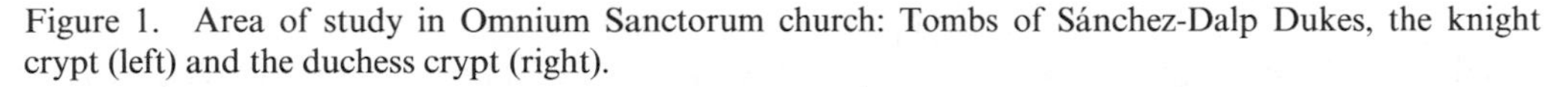

Figure 1. Area of study in Omnium Sanctorum church: Tombs of Sánchez-Dalp Dukes, the knight crypt (left) and the duchess crypt (right).

according to the different chromatic areas that have been distinguished at naked eye. A TRACER III-SD equipment was used for the analysis. Identification of elements was performed through Bayesian deconvolution. Data used for elemental distribution were normalized with respect to the Rayleigh scattering peak of Rh.

2.2 *In situ study of weathering forms*

A first approach to the vulnerability of the area of study has been made through a visual inspection of weathering forms according to ICOMOS classification (ICOMOS-ISCS, 2008)

2.3 *3D digitization*

In the preservation and conservation of Cultural Heritage assets, 3D digitization has a very important role, considering that can deliver a digital model of the subject having a 3D replica. (Angheluta & Radvan, 2017), thanks to which details of the object of study that can not be observed with the naked eye can be noticed. 3D images were taken thanks to a NIKON D810 (60 mm) camera. A total number of 2911 images have been taken to reconstruct the 3D models of the crypts and the arcosolium tiles, covering an area larger than 7.7 m^2.

3 RESULTS

3.1 *X-ray fluorescence (XRF)*

The Table 1 summaries the main chemical elements detected in the materials used in ceramics and glazed terracotta by XRF. The chemical elements in greater proportion for each chromatic variable are highlighted in bold.

The composition of the terracotta has a high Fe content, according to other Sevillian monuments, because of the high concentration of iron in Sevillian clays and because of this pieces were usually cooked at low temperature (700-800 ºC). Quartz was added to these clays in order to achieve more consistency. They seem similar to other clays studied in the city such as those of the Covent of Santa Paula (Gómez-Morón, 2016). Further studies as cross section and X-Ray Diffraction (XRD) are recommended.

Table 1. Most relevant elemental presence of the main elements, for all analyzed area.

Analyzed area	Tiles	Glazed terracotta	Observations
Terracotta	-	**Ca, Fe, S, Pb, Si,** Cr, K, Mn, Rb, Sr, Ti, Zn, Zr	To make painted or polychrome terracotta objects, a preparatory layer of plaster (Pb, S) is applied traditionally before decorating it. The other elements belong to terracotta.
Golden	-	**Pb**, Bi, Fe, Si, Ca	It could be lustre.
Black	**Pb, Mn, Fe**	**Pb, Cu, Fe,** Mn, Rb, Zn, Co	The mixture of Mn, Co, Cu and Fe was used to achieve black color.
Blue	**Pb, Fe, Co, As, Ni**	**Pb,** Fe, Co	The main source of cobalt used in smalt preparation in Middle Age was smaltite and later on, erytthrite and cobaltite. In glazed works, Cobalt blue might be the most plausible pigment.
Green	**Pb, Cu, Fe, As,** Zn	**Pb, Cu, Fe,** Bi, Ca	The most probable pigment is malachite.
Violet	-	**Pb,** Bi, Si, Cu, Co	Violet color could be due to the combination of Co and Bi.
Blue-Green	-	**Pb**, Cu, Bi, Fe, Ca	Cu is responsible for the color.
Yellow	**Pb, Zn, Fe,** Sb, Si	-	Possible use of lead antimony yellow (known since the 16-14th century BC). The highest presence of Zn is interesting, considering that zinc was used to achieve a brighter yellow color since 14th century.
Dark- Brown	**Pb, Fe,** Mn, Ca	-	The highest manganese lines observed could possibly indicate manganese dioxide or manganese brown.
Orange	**Pb, Fe**, Sb, Sn, Bi	-	Orange color is due to the use of lead antimony yellow and ochre.
White	**Pb,** Fe, Si, K	-	White lead is the most possible pigment.

The addition of Zn in yellow and green to enhance the color in ceramic, it was not detected in the terracotta except in black color.

XFR results allow to establish the main chromophores located in the arcosolium, such as cobalt for blue colors or copper for green colors. The presence of lead may be due to the use of certain pigments, such as yellow antimonite lead, the existence of a preparatory white layer of lead white, as in the cover of Santa Paula, or that it was added as an opacifier of pigments.

This technique allows to make a map of composition and to reduce the sampling on Cultural Heritage. However, samples of cross section of ceramic and terracotta should be taken to clarify its composition and to know if there is a preparatory layer of lead white under the glaze.

3.2 *In situ study of weathering forms*

The main weathering forms observed by naked eye were fissures, loss of material, efflorescence, chromatic alteration, displacement and sanding in both crypts and in the tiles. However, the magnitude of these weathering forms was different between both study areas. The chromatic alteration was much greater in the tiles, while the presence of efflorescence and disintegration is greater in the tombs.

3.3 *3D digitization*

3D digitization allowed the localization and quantification of the main weathering forms found by in situ study, such as efflorescence, detachment, disintegration, discoloration, lacunas, cracks and loss of glazed materials. This 3D model has a high quality that allows to detect weathering forms in more details and to find some damages that are not

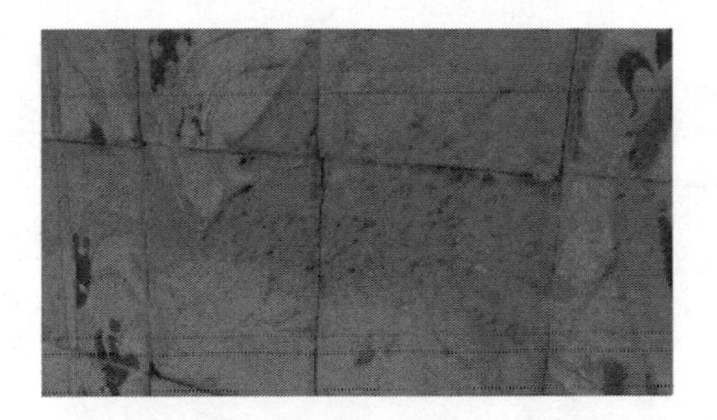

Figure 2. Detail of disintegration in tiles, identify by 3D digitization, not perceptible to the naked eye.

Figure 3. Disintegration located in hidden part of the sculpture (left) and efflorescences (right) identified by 3D digitization in the Duke crypt.

perceptible at a glance or in a photo taken with a standard camera such as disintegration on tiles (Figure 2).

3D model can be directed in all directions, allowing the access to hidden areas such as the left part of the sculptures, like the face of the Duke crypt (Figure 3); or to areas that due to its orientation, can not be observed correctly such as efflorescences on the edge of the arch of the crypts (Figure 3). In addition, it is possible to zoom in it to better appreciate the damages and the magnitude and real extension of weathering forms. The Figure 4 shows an image of a crack in the Duke tomb taken with a standard camera (left) and the same crack obtained by the 3D model (right). Details of the crack, as well as the volumes, proportions and colors can be better appreciated than in a standard image.

Efflorescences, detachment and cracks may be caused by dampness by capillarity, detected by visual inspection, so it would be necessary a restoration based on the elimination of dampness and salts and the consolidation of the possible losses of glazes, according to the international criteria of restoration.

Figure 4. Standard image (left) and 3D model image (right) of cracks and loss material of the Duke crypt.

4 CONCLUSIONS

The combination of these non-destructive techniques allowed to know the composition of the tiles and glazed terracotta of the arcosolium and to carry out a diagnosis of the main weathering forms such as efflorescence, detachment, disintegration, discoloration, lacunas, cracks and loss of glazed materials, with a high level of detail and accuracy, without causing any damage to the works of art. The use of these techniques is recommended in order to reduce the sampling and to improve the diagnosis according to a good laboratory practice in Cultural Heritage.

ACKNOWLEDGEMENTS

This paper has been supported by Art-Risk, a RETOS Project of Ministerio de Economía y Competitividad and Fondo Europeo de Desarrollo Regional (FEDER), (code: BIA2015-64878-R (MINECO/FEDER, UE)) and the research team PAI-TEP-199 from Junta de Andalucía. Infrastructure assured by Program OPTRONI & INOVA OPTIMA.

REFERENCES

Angheluta, L. & Radvan R. 2017. 3D digitization of an antique decorative textile artifact using photogrammetry. *Romanian reports in physics*. 69, 801.

Fosters, N., Grave, P., Vickery, N. & Kealhofer, L. Non-destructive analysis using pxrf: methodology and application to archaeological ceramics. *X-Ray Spectrometry*, 40 (5), 389–398.

Gómez-Morón, A. 2016. Caracterización química de los materiales empleados en la Portada de Santa Paula (Sevilla). IAPH. Consejería de Cultura.

ICOMOS-ISCS. 2008. Illustrated glossary on stone deterioration patterns.

Radvan R, Bors, C, Ghervase, L. 2016. Portable X-Ray fluorescence investigation of certain bronze beads of hoard Tărtăria i and their specific corrosion. *Romania Journal of Physics*. 61, (9-10), 1530–1538.

Science and Digital Technology for Cultural Heritage – Ortiz Calderón et al. (Eds)
© 2020 Taylor & Francis Group, London, ISBN 978-0-367-36368-0

In situ application of a consolidant on the Roman theatre of Cádiz

G.M.C. Gemelli, M.J. Mosquera, M. Galán, A. Pelaez, J.M. Perez &
M.L.A. Gil Montero*
TEP-243 Nanomaterial Group, Departamento de Química Física, Universidad de Cádiz, Campus de Puerto Real, Cádiz, Spain

ABSTRACT: The Cadiz Theatre represents an archaeological site of special cultural interest as it is considered to be the oldest and the largest of the Iberian Peninsula. Its proximity to the sea and to one of the main streets of the city promotes the acceleration of the degradation phenomena.

Thus, the objective of the work has been the application of a consolidant to minimize the action of the atmospheric agents. Prior to the *in situ* application, preliminary laboratory tests have been performed including: XRD, XRF, colorimetric, hydric and mechanical analysis, in order to characterize the stone substrate. The same analysis were repeated, after the application, in order to evaluate the compatibility and the effectiveness as consolidant product.

In accordance with the laboratory results, the consolidant treatment was applied *in situ* on the Theatre. The *in situ* analysis highlighted an increase in the grains cohesion as well as a significant reduction of the water absorption.

1 INTRODUCTION

The Theatre of Cádiz is an example of Roman architecture remaining in Cadiz, a coastal city placed in the south of Spain. It is one of the largest and oldest theatres in the country, constructed by Lucius Cornelius Balbus the Younger, in the last part of the 1st century BC (Pachón *et al.* 2015). In 1980, the remains of the Roman Theatre were accidentally discovered under the "*Populo*" district (Arévalo González *et al.* 2012), prompting a great interest about the archaeological site, with the aim of recovering and restoring its remains. The Roman Theatre was built in "*ostionera*", a local stone largely used in the architecture of the historical quarter of Cádiz.

The old town is located in an area constantly surrounded by high traffic density, which leads to elevated pollution levels by carbon dioxide, soot and other components that enhance the chemical alteration of the monument constituent materials. In addition, its location close to the sea and the exposure to the predominant east winds entails a high degree of humidity (between 70 and 90% RH), as well as a saline environment that accelerates its mechanical alteration. Furthermore, the rough surface of the material promotes both the accumulation of rainwater and the formation of soil deposits that facilitate the growth of microorganisms and vascular plants (Figure 1). Due to the risk to the materials posed by the factors stated above, it is advisable to apply a consolidant treatment in order to mitigate further degradation processes.

For this study, an innovative nanostructured consolidant, previously developed by our research group (TEP-243), was applied. Specifically, alkoxysilane-based product was synthetized through a fast and cost-effective route. This process is based on an inverse micelle mechanism promoted by n-octylamine to obtain crack-free mesoporous materials (Facio *et al.* 2017).

As it is well known, alkoxysilane products polymerize *in situ* inside the pore structure of the disintegrating stone, through a classic sol-gel process, and significantly increase the cohesion

*Corresponding author: almoraima.gil@uca.es

Figure 1. Location of the Cádiz Roman Theatre.

of the material (Wheeler 2005). Furthermore, the low viscosity of alkoxysilane facilitates its penetration into the porous structure of the stone and avoids the use of organic solvents, toxic for the environment and for the operators (Illescas and Mosquera 2012).

Before the *in situ* application, preliminary tests have been made on laboratory samples in order to evaluate the compatibility and the performance of the treatment, considering the Compatibility Indicators (CI) proposed by Delgado Rodriguez and Grossi (Rodrigues and Grossi 2007).

2 MATERIALS AND METHODS

2.1 *Stone Characterization*

Mineralogical characterization and chemical analysis were performed using XRD and XRF analysis. In accordance to European Standards, water absorption by capillarity and by total immersion were assessment to study the hydric properties of the stone in question.

2.2 *Synthesis and Application procedure*

The synthesis route was as follow: an ethoxysilane (TES40) was mixed with hydrophilic fumed silica (OX50), under ultrasonic agitation (60 W cm-3 for 10 minutes), in the presence of aqueous dispersion of a surfactant (n-octylamine) (Facio and Mosquera 2013). The application of the nanostructured product was performed by brushing. After 2 months, weight remained stable on all of the treated samples and the following tests were assessment.

2.2.1 *Assessment of Compatibility and Effectiveness*

To verify the compatibility of the treatments with the stone in question, change in colour and reduction in water vapour permeability were evaluated following the Compatibility Indicators (CI) (Rodrigues and Grossi 2007). In accordance with the obtained results, the effectiveness of the consolidant was subsequently evaluated. Therefore, capillarity and total immersion absorption tests were repeated, in order to study the behaviour of the treatment against water. Furthermore, a Drilling Resistance Measurement System (DRMS from SINT Technology) was used to evaluate the penetration depth and the improvement in mechanical resistance of the treatment. Conforming to the obtained laboratory results, the nanostructured consolidant product was applied *in situ* on an angular rectangular block of the Roman Theatre. Before the *in situ* application, the salts were removed by five consecutive applications of a commercial cellulose pulp poultice soaked in water (Egartner and Sass 2016) in order to promote the consolidant penetration. The reduction of the water-soluble salts concentration in the block were monitored measuring the water conductivity of the pulp poultice after each application. *In situ* consolidant evaluation was performed by measuring the reduction of water absorption and the increase in grain cohesion by using the sponge test method (UNI 11432:2011) and the peeling test (Drdácký *et al.* 2012).

3 RESULT AND DISCUSSION

In agreement with the XRD and XRF analysis (see Figure 2 and Table 1), the Ostionera stone is a biocalcarenite composed mainly of CaCO3 (79%) and minority of SiO2 (17%).

The physical parameters recorded from the laboratory samples, before and after the treatment are include in Table 2. The results highlighting that the treatment could be considered compatible for *Ostionera* stone according to CI for the application of consolidant product as conservative treatment.

The increase in mechanical resistance, evaluated using the DRMS test is evidenced by the average profiles of drilling resistance on both treated and untreated samples (Figure 3). The

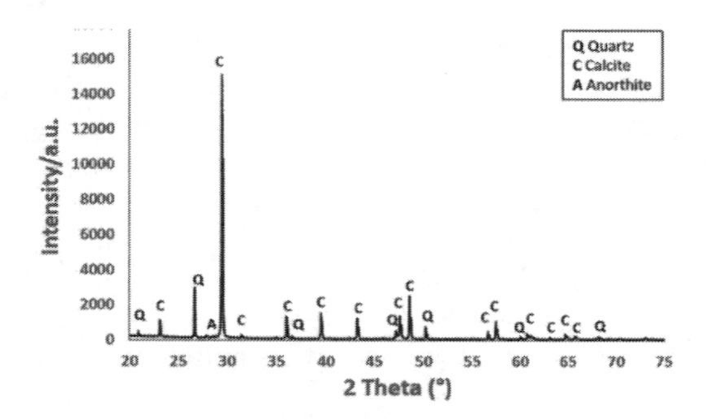

Figure 2. Mineralogical analysis of the Ostionera stone.

Table 1. Mainly elements contents of the Ostionera stone.

Formula	CaO	SiO_2	Fe_2O_3	Al_2O_3	MgO	K_2O	Na_2O	P_2O_5
Concentration (%)	78.96	16.9	1.52	0.65	0.305	0.242	0.13	0.1

Table 2. Physical parameters of the Ostionera stone measured before and after the treatment.

Property	Before treatment	After treatment	Reduction (%)
ΔE*	-	1.48 ± 0.4	
Open porosity (%)	27.80 ± 1.3	9.19 ± 0.5	18.6
WAC* (%w/w)	13.19 ± 2.1	1.03 ± 0.4	91.9
WAI** (%w/w)	15.73 ± 0.8	6.70 ± 0.3	57.4
WVD*** ($D \cdot 10^{-6}$ ($m^2 \cdot s^{-1}$))	4.53 ± 0.7	4.24 ± 1.3	6.3

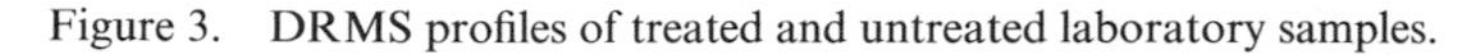

* Water absorption by capillarity; ** Water absorption by total immersion; *** Water vapor diffusivity.

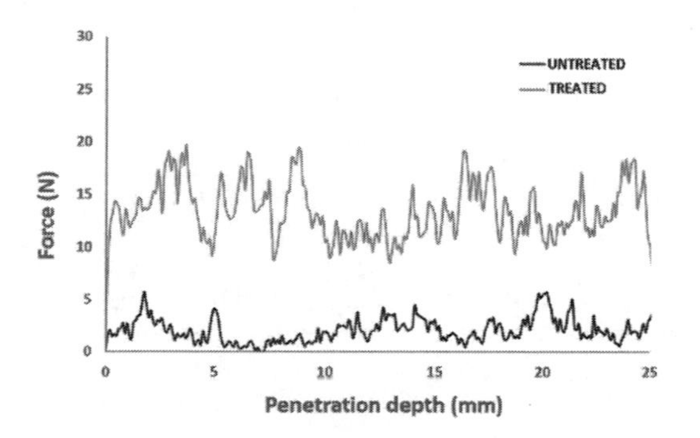

Figure 3. DRMS profiles of treated and untreated laboratory samples.

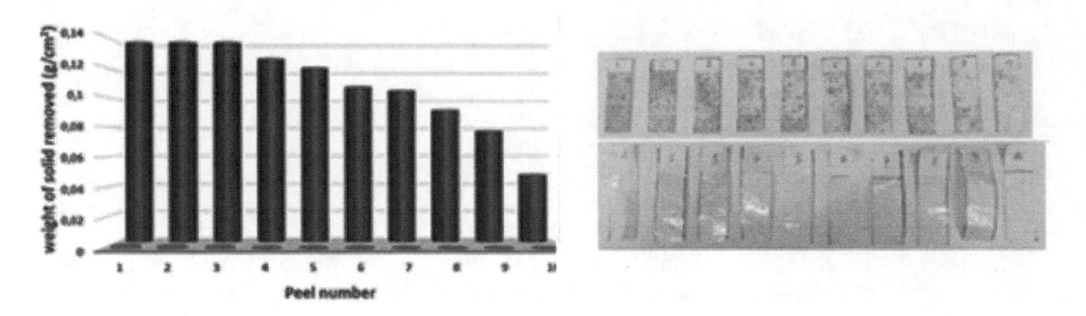

Figure 4. *In situ* Scotch Tape Test.

drilling resistance test confirms a significant improvement of the mechanical resistance up to a depth of 25 mm.

In line with the laboratory tests, the *in situ* application confirms the compatibility and effectiveness of the treatment. Specifically, the ΔE^* value indicate a barely perceptible color change ($\Delta E^* = 2.26$) and the obtained values from the contact sponge method confirms the reduction in water uptake around 70%. Moreover, the peeling test demonstrate an increase in grain cohesion and superficial resistance as can be seen in Figure 4.

4 CONCLUSIONS

This study aims to demonstrate the importance of laboratory evaluation of conservation products useful for preliminary compatibility assessment with the stone substrate. Moreover, adequate consolidant treatment could extend the monument lifetime, allowing to transmit it to the future generations. The *in situ* analysis confirms an increase in the grains cohesion and their adhesion to the substrate, as well as a significant reduction of the water absorption highlighted mitigate the environmental degradation phenomena. The long-term performance of the consolidant treatment is currently under monitoring.

REFERENCES

Arévalo González, A., Bernal Casasola, D., Sanchez, V., Bustamante-Álvarez, M., Yanes Bustamante, A., and Cobo Fernández, A., 2012. *El proyecto Theatrum Balbi. Hacia una dinamización del conocimiento y de la valorización del teatro romano de Cádiz*. I Congreso Internacional 'El patrimonio cultural y natural como motor de desarrollo: investigación e innovación'. Jaén: Universidad Internacional de Andalucía.

Drdácký, M., Lesák, J., Rescic, S., Slízková, Z., Tiano, P., Valach, J., Drdácký, M., Lesák, J., Rescic, S., Slížková, Z., Tiano, P., and Valach, J., 2012. Standardization of peeling tests for assessing the cohesion and consolidation characteristics of historic stone surfaces. *Materials and Structures (Dordrecht, Netherlands)*, 45 (4), 505–520.

Egartner, I. and Sass, O., 2016. Using paper pulp poultices in the field and laboratory to analyse salt distribution in building limestones. *Heritage Science*, 4 (1), 1–13.

Facio, D.S., Luna, M., and Mosquera, M.J., 2017. Facile preparation of mesoporous silica monoliths by an inverse micelle mechanism. *Microporous and Mesoporous Materials*, 247, 166–176.

Facio, D.S. and Mosquera, M.J., 2013. Simple Strategy for Producing Superhydrophobic Nanocomposite Coatings In situ on a Building Substrate. *ACS Applied Materials & Interfaces*, 5 (15), 7517–7526.

Illescas, J.F. and Mosquera, M.J., 2012. Producing surfactant-synthesized nanomaterials in situ on a building substrate, without volatile organic compounds. *ACS Applied Materials and Interfaces*, 4 (8), 4259–4269.

Pachón, P., Compán, V., Rodríguez-Mayorga, E., and Sáez, A., 2015. Control of structural intervention in the area of the Roman Theatre of Cadiz (Spain) by using non-destructive techniques. *Construction and Building Materials*, 101, 572–583.
Rodrigues, J.D. and Grossi, A., 2007. Indicators and ratings for the compatibility assessment of conservation actions. *Journal of Cultural Heritage*, 8 (1), 32–43.
Wheeler, G., 2005. *Alkoxysilanes and the Consolidation of Stone*. Journal of the American Institute for Conservation.

Science and Digital Technology for Cultural Heritage – Ortiz Calderón et al. (Eds)
© 2020 Taylor & Francis Group, London, ISBN 978-0-367-36368-0

Rain simulation device to test durability of building materials

M.L.A. Gil, L.A.M. Carrascosa, A. Gonzalez, M.J. Mosquera & M. Galán
TEP-243 Nanomaterial Group, Departamento de Química Física, Universidad de Cádiz, Campus de Puerto Real, Cádiz, Spain

A. Morgado-Estevez & J. Vilaverde-Ramallo
Departamento Ingeniería en Automática, Electrónica, Arquitectura y Redes de Computadores, Universidad de Cádiz, Cádiz, Spain

M. Palomo-Duarte
Departamento de Ingeniería Informática, Universidad de Cádiz, Cádiz, Spain

ABSTRACT: Water is the most aggressive agent of decay which continuouslyattacks building materials. For this reason, the research on hydrophobic and superhydrophobic materials has increased in the last years. These treatments present a series of drawbacks, such as low-adhesion to the substrates and subsequently, low durability. In the present work we have developed an automatic rain test simulation, based on an Arduino microcontroller to automatize the process. The simulator includes temperature, flow rate, and pH sensors as main variables of rain water. The combination of these variables allows to use different cycles of assays. This portable device permits the evaluation the durability of the materials testing the chemical stability of the coatings against water contact with a continuous water dropping.

In order to test the new device, different building materials were treated with two superhydrophobic products and a commercial product and one developed in our research group (UCA). After each essay, the dynamic and static contact angle were evaluated. The UCA product maintains for all the materials the conditions of superhydrophobicity.

1 INTRODUCTION

Water is the most aggressive agent of decay, which continuously attacks building materials through different mechanisms, such as salt crystallization, acid rain or ice swelling. For this reason, the research on hydrophobic and superhydrophobic materials (with hydrophobic and repellent properties), has increased in interest in the last years (Liu 2013). Most of the currently available treatments are coatings that produce functionalized surfaces, therefore the persistence of these properties for a long time in particular environmental conditions, is a critical factor. These treatments present a series of drawbacks, mainly low-adhesion to the substrates and subsequently, low durability. Among others, a critical issue in the specific case of building materials, is the loss of the chemical functionalities of the surface under the rainwater and acid rainwater attack (Barati Darbanda 2018). This irreversible loss reduces the materials lifetime and increases the costs involved in its maintenance and repair. In spite of this fact, the number of works published studying the durability of hydrophobic and superhydrophobic treatments on building materials is significantly scarce, however, there are works about durability test of superhydrofobic treatments on materials such as wood (Panek 2017) bamboo (Jingpeng 2015) and carbon steel (Wang 2016). The durability assays utilized in these works are water spraying, immersing in sulphuric acid rain and exposure to UV/water condensation cycles. These tests demand their own infrastructure and have an expensive maintenance. In this work, we have developed a simple, portable and automatic device to simulate in the laboratory the effects of rainwater and acid rainwater in order to evaluate the durability of the coatings. The simulator allows optimizing products in the laboratory before the expensive and laborious application and evaluation *in situ*.

2 METHODOLOGY

The rain simulation is a portable device that was thought to be used in any place with little requirements, just requiring power supply and a water drain. It can be hanged on the door of a laboratory cabinet above the sink, using the tap as a water supplier for the test. The device allows programing cycles combining three variables the water flow, pH and temperature in real time. The operation of the device is controlled via an Arduino microcontroller that runs a specifically developed software program, that can be downloaded from https://github.com/nekuneko/SimuRain.

2.1 *Components of the rain simulation device*

As can be seen in Figure 1, the rain simulation device has seven components: (1) two electro valves that open and closes the circuit, regulating the flow of water from the tap (6), which provides water pressure and/or acid solution. The acid or alkali solution is stored in a tank (7) and allows to adjust the pH of the rain. There are three devices to make the measures: a flowmeter (3) that measures the volume of water per unit of time. A pH sensor with a range from 3 to 8, (4) to measure the acidity of the solutions used for the test. A temperature sensor, (5) to measure the water outlet temperature. Finally, a shower head allows the distribution of the water over the sample.

All these sensors are controlled by Arduino module microcontroller (2), protected in a waterproof box. The Arduino module includes a USB port to connect with a computer,

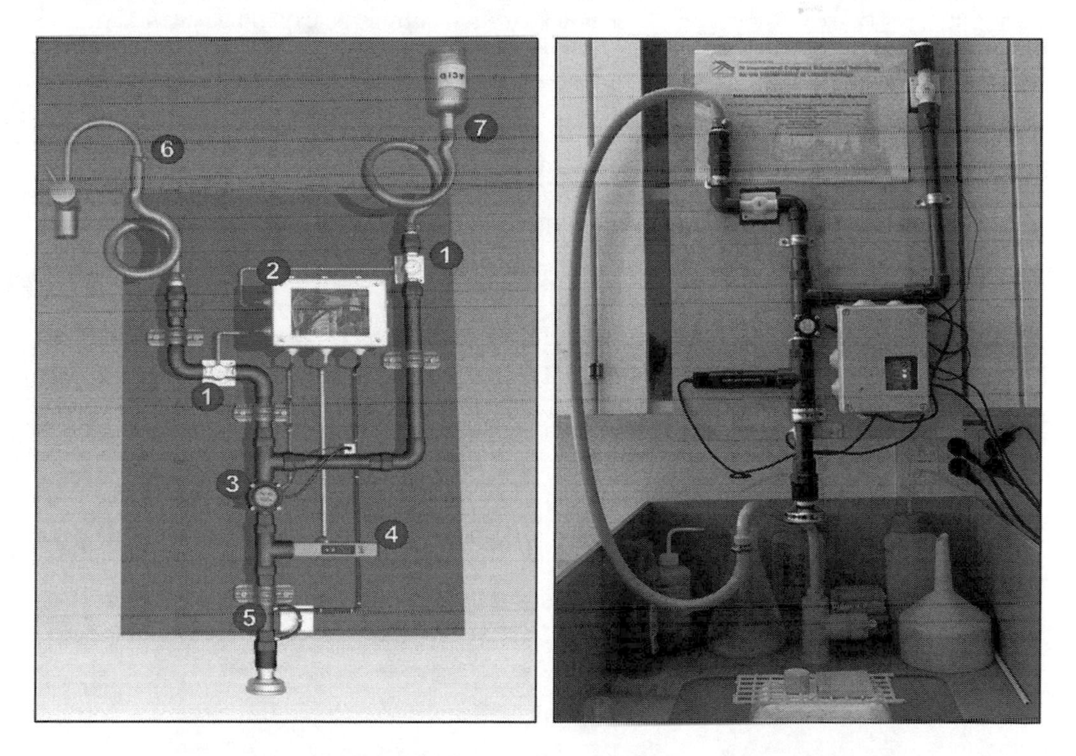

Figure 1. Left: Rain simulator scheme (1) Electro valves, (2) Microcontroller, (3) Flowmeter, (4) pHmeter, (5) Thermometer, (6) Water inlet, (7) Acid inlet. Right: Photograph of the rain simulator device.

a real-time clock, a micro SD card reader to save the data a reset button that reset the system and a set of LEDs to shows in real time the state of the device according to the following code: a red LED indicates that the power input is open. A yellow1 LED that open the electro-valve 1 and allow the water input. A yellow2 LED that open the electro-valve 2 for the acid solution input. A green LED that informs that the process is being executed and finally a blue LED that open the sensors of flow, temperature or pH.

2.2 *Durability evaluation*

To carry out a durability test,two superhydrophobic products were applied by brush,: a commercial product and a synthesized product (UCA), by our research group (Facio 2013, Carrascosa 2016) on four different materials: Concrete, granite, wood, and sandstone. For the test, four cycles of 300, 900, 1800 and 2000L/m^2 with a water flow of 45mL/s were performed at pH 7, after every test samples were dried for 24 h in an oven at 100°C and was carried out a study of the dynamic and static contact angle values with a commercial video-based, software-controlled contact angle analyzer, model OCA 15plus, from Dataphysics Instruments.

3 RESULTS AND DISCUSSION

3.1 *Data obtained from the rain simulator*

The data obtained from the rain simulator is saved in real time on a micro SD card, in a CSV text file that can be easily processed by spreadsheet software. As an example, Figure 2 shows the results obtained in an experiment carried out under heavy rain conditions, that takes 7,5 min, saving data every 2 s. In the experiment, the flow took three different values: 55mL/min during the first 6 minutes, 133mL/min during the next 48 seconds and finally 200mL/min during the last 34 seconds. The total accumulated flow during the experiment was 15 L, the pH was changed from 8.00 to 7.60 and the temperature from 22°C to 20°C.

3.2 *Roughness performed after the treatments*

For verifying that the surfaces have developed roughness with the treatment, Atomic Force Microscopy (AFM) images have been taken. Results (Figure 3) showed that in general all of the coatings produced roughness on all the surfaces.

3.3 *Contact angle and hysteresis*

After the rainwater simulator test, the durability was evaluated measuring the static angle (formed by the intersection of the solid, liquid, and solid-gas interface) and dynamic contact angle (the lowest horizontal angle that the droplet requires to roll on the surface), and the hysteresis which is the difference between the advancing and receding angles. As we can see in

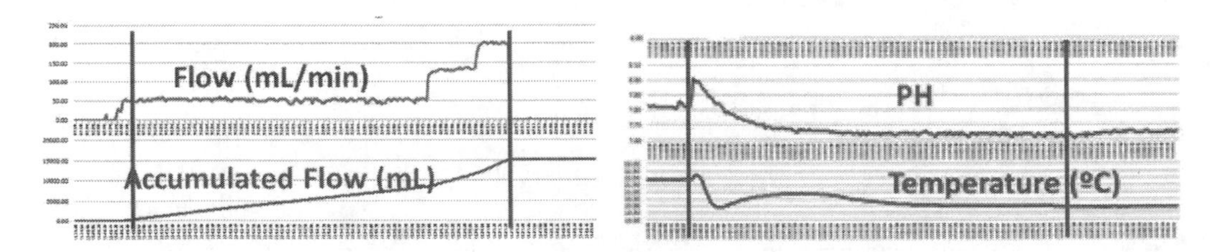

Figure 2. Output data, the red lines point out the start and finish of the test.

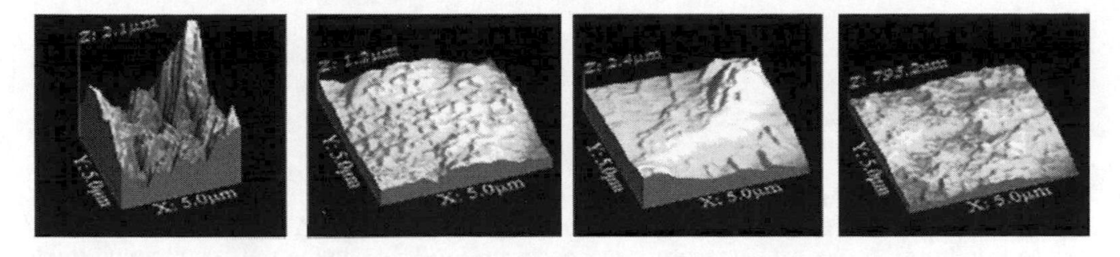

Figure 3. AFM images from left to right concrete, wood, granite, and sandstone.

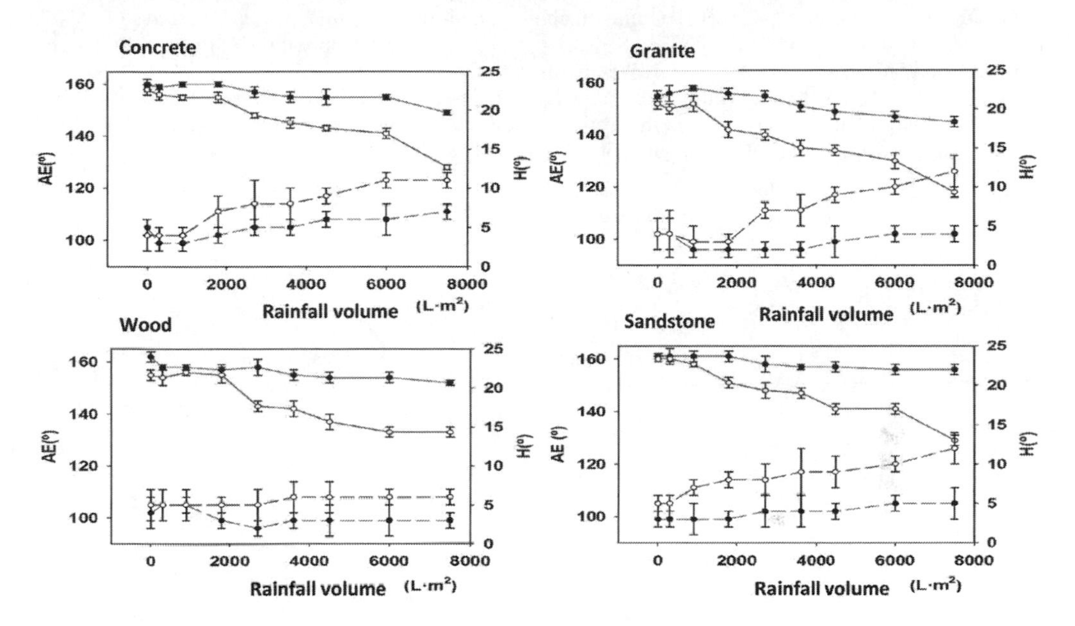

Figure 4. CA and H obtained after the rainwater simulator test. The blue line and dotted blue line indicate static angle and hysteresis of a commercial product. Redline and dotted red line indicate static angle and hysteresis of UCA product.

Figure 4 a decrease of the static contact angle has been obtained for all the materials tested. The UCA product obtains, for all the materials, a higher resistance of the coating under conditions of continued rain, maintaining static angles of 150° and hysteresis below 10° conditions of superhydrophobicity). Additionally, hysteresis decreases as the receding angle decreased.

4 CONCLUSIONS

A portable simulator rain device has been designed to perform durability measures in different materials. It combines three experimental variables and allows to store the data in a micro SD card using a CSV text file that can be easily processed by spreadsheet software. To test the results of the device, we evaluated the durability of two superhydrophobic products in four different materials. The results obtained confirm the utility of the simulator, which is capable to test the durability of the products and material studied. The decrease of the static contact angle and hysteresis are caused by surface chemistry changes and the loss of surface nanostructure due to the erosion caused by the strong attack of the rain.

REFERENCES

Barati Darband, Gh. Aliofkhazraei, M. Khorsand, S. Sokhanvar, S. Kabolia, A. Science and Engineering of Superhydrophobic Surfaces: Review of Corrosion Resistance,Chemical and Mechanical Stability *Arabian Journal of Chemistry* https://doi.org/10.1016/j.arabjc.2018.01.013

Carrascosa, L. A. M. Facio, D. S. Mosquera, M. J. 2016. Producing Superhydrophobic Roof Tiles. Nanotechnology 27(9): 095604.

Jingpeng L. Huanhuan Z. Qingfeng S. Shenjie H. Bitao F. Qiufang Y. Chenye Y. and Chunde J. Fabrication of superhydrophobic bamboo timber based on an anatase TiO_2 film for acid rain protection and flame retardancy *RSC Advances* (5): 62265–62272.

Liu, Y. Liu, J. Li S. Liu, J. Han, Z. Ren, L. 2013. Biomimetic superhydrophobic surface of high adhesion fabricated with micronano binary structure on aluminum alloy. *ACS Appl. Mater. Interfaces* (5): 8907.

Pánek, M. Oberhofnerová, E. Zeidler, A. Šedivka, P. Efficacy of Hydrophobic Coatings in Protecting Oak Wood Surfaces during AcceleratedWeathering Coatings (7): 172.

Wang, L. Yang, Z. Zhenhua L. Tao Shenga, Y.M.Hu. De-Quan Y. A study of the mechanical and chemical durability of ultra-ever dry superhydrophobic coating on low carbon steel surface Colloids and Surfaces A: Physicochemical and Engineering Aspects. (497): 16–27.

Science and Digital Technology for Cultural Heritage – Ortiz Calderón et al. (Eds)
© 2020 Taylor & Francis Group, London, ISBN 978-0-367-36368-0

Evaluation of cleaning, consolidation and adhesion processes of the Portada dels Apóstols

D. Juanes, G. Contreras & M. Domenech
Valencian Institute for Conservation, Restoration and Research of Cultural Heritage, Valencia, Spain

L. Ferrazza
Scientific Conservator Freelance, Castellón, Spain

M.T. Pastor
MACVAC Museum, Vilafamés, Spain

ABSTRACT: The Portada dels Apostols of the Church of Santa María la Mayor in Morella is one of the few examples that has survived of polychromed façades in our country. The quantity and quality of the polychrome, and the environmental conditions in which it is, represent a challenge for its restoration and conservation.

Our previous studies of materials used in the façade showed a disturbing state of conservation that affected the physical and chemical nature of the materials, with the presence of a superficial crust of compact and resistant dirt, giving rise to a very heterogeneous and complex structure. Faced with this situation, the suitability and efficacy of cleaning, consolidation and adhesion treatments of the polychromy were being evaluated in situ and in the laboratory in order to obtain a suitable and optimized intervention methodology. The results have provided knowledge of the materials used and the execution technique, and a valuable knowledge of how to conserve and restore similar works elsewhere.

1 INTRODUCTION

The Fachada dels Apostols is one of the few remaining examples of polychrome facades in Europe, so its study and conservation was a priority objective (Figure 1). The state of conservation, the environment and the deterioration suffered by the Fachada dels Apostols is incomparable in Europe. The search for similar cases was unsuccessful, so this project was developed for its study and conservation. The research was divided in three phases: 1) previous studies in order to identify what materials were present in the façade and its state of conservation, 2) cleaning and consolidation test in laboratory and in situ for developing a specific intervention protocol, and 3) the design of the intervention process. To carry out the project, there were national and European entities that are renowned for their work and knowledge in heritage conservation, which have participated in advising on intervention processes (Pérez García et al., 2017).

The results of this project are being used successfully in the restoration work currently being carried out on the façade. In addition, the uniqueness of the work stands out at a European level and constitutes a start point of further studies and conservation processes of this kind of cultural heritage in the rest of Europe (Rivas López, 2008).

Figure 1. Fachada dels Apostols of the Collegiate Church of Santa Maria la Mayor in Morella.

2 METHODOLOGY AND TECHNIQUES

2.1 *First phase. Previous studies*

In order to identify the materials used in the Fachada dels Apóstols, the execution technique, the possible previous interventions and its state of conservation, an exhaustive study was carried out using a non-invasive methodology that maintains artistic historical values unchanged. The analysis methodology followed a sequence that began with general analyses without sampling such us thermographic study and georadar, continued with punctual analyses without sampling (EDXRF) and finally, a selective sampling was carried out. The micro samples were analysed by optical microscope, SEM-EDX and FTIR (Juanes et al., 2018).

2.2 *Second phase. Cleaning and consolidation test*

The main objective was developing a specific intervention protocol. In situ and laboratory studies were carried out on the viability of pre-consolidation systems, cleaning and consolidation of polychromies, designed on the basis of the results obtained from materials, execution technique and the general types of alteration identified.

2.2.1 *Laboratory consolidation tests.*

The objective was to carry out a comparative study of the behaviour and stability against ageing of different polymers separately and their possible suitability and applicability. Six natural and synthetic organic consolidants have been tested: rabbit glue, sturgeon glue, Acryl 33 Acryl ME, Lascaux Medium for Consolidation, Aquazol 200, Aquazol 500, mixtures of rabbit glue and Plextol B500, mixtures of sturgeon glue and Plextol B500 and mixtures of Aquazol 200 and Acryl ME (Pastor Valls and Juanes Barber, 2017).

Homogeneous films of the consolidating products were prepared and then subjected to a climatic test setting humidity and temperature conditions that simulated the environment of the paintings (Figure 2). Subsequently, a study was carried out on the variation of mechanical behaviour, pH, brightness and colour.

2.2.2 *In situ consolidation and cleaning tests.*

Different cleaning methods were tested depending on the type of dirt (mineralized crusts, dust, etc.): Laser cleaning, micro abrasion, mechanical methods and chemical methods. On the

Figure 2. Thermo-hygrometric laboratory test for the evaluation of different glues for use in the consolidation process.

other hand, the 6 natural and synthetic organic consolidants that had been previously studied in the laboratory were tested.

3 RESULTS

3.1 *Materials and making process*

The results of the analysis showed that a fossil limestone was used as the support material. On the stone there is a polychrome structure and interventions with the presence of different layers of colour on a ground layer made of white lead. The making process in terms of technical complexity and finishes is similar to its equivalents in easel painting (Figure 3). The most recent polychromy is oil, although conclusive results have not yet been obtained. Finally, the oldest pigments have been identified and those used in different interventions (lead white, yellow ochre, red ochre, azurite, smalt blue, green copper, calcium carbonate, red lead, vermilion, carbon black and organic black). Analyses also showed that the polychromy was rich in gilded elements and decorations. Remains of gold leaf have been found in investments, crowns, jewellery, wings, hair, stars, etc.

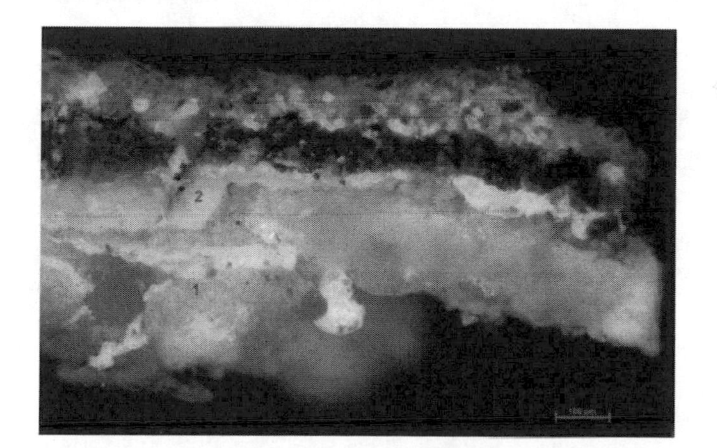

Figure 3. Microsample cross section from the tympanum (1) Support. Calcite mortar. (2) White, thin and irregular layer of lead white. (3) Blue layer, composed by azurite and with presence of white lead and earth pigment. (4). Blue layer, azurite and with the presence of gypsum and possibly salts. (5). Blue layer, very fine esmalt blue.

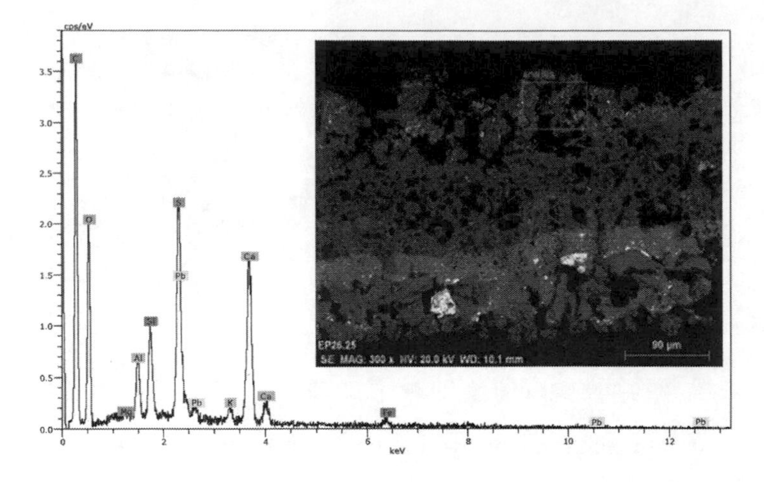

Figure 4. SEM-EDX of the surface crust. Gypsum deposits and silicates were detected.

3.2 *State of conservation*

Studies show that the main damage is the presence of a layer of dirt, crusts and surface dust on all surfaces (Figure 4.). A change in colour was detected with the passage from blue to green located on the external part of the starred background of the tympanum, which affects the iconographic reading and could correspond to a change in the pigments due to their alteration by the presence of chlorine (azurite: passage from blue to green), or as a change in coatings (yellowing).

The presence of disintegrated layers with lack of cohesion could be related to the alteration and/or loss of the binder, both due to its exposure to external factors of alteration and due to the action of cleaning products in old interventions.

3.3 *Conservation methodology.*

The laboratory and "in situ" tests of the 6 consolidating products resulted in the sturgeon glue being the most appropriate according to its behaviour. Once the tests of materials and processes had been completed, an intervention protocol was drawn up for the polychrome which essentially consists of the following steps:

1. Dry surface cleaning with suitable means (brush, aspiration, etc.) as far as possible.
2. Swab cleaning with water and alcohol at 50-50 (%) in order to remove as much crust and surface dirt as possible.
3. Impregnation of the area with water and alcohol at 50-50 (%).
4. Pre-consolidation with sturgeon glue and occasionally rabbit glue.
5. Removal of the superficial crust with suitable means.
6. Consolidation with sturgeon glue of the exposed polychrome layer.

4 CONCLUSION

In conclusion, the work carried out in these three phases has made it possible to know the materials used and the execution technique, which will provide invaluable information to art historians. On the other hand, the analysis of the state of conservation, its special circumstances due to being outside in extreme climatic conditions, and the fact that there were no similar examples, motivated an extensive study of materials and processes that could be used in their conservation and restoration. These have provided valuable knowledge of how to conserve and restore similar works elsewhere.

ACKNOWLEDGEMENT

We would like to show the acknowledge for their collaboration to National Institute for Research and Development in Optoelectronics (INOE 2000, Romania), the Instituto Andaluz del Patrimonio Historico (IAPH, Spain), the Centre for the Restoration of Artefacts of Catalonia (CRBMC, Spain), the Department of Physical, Chemical and Natural Systems of the Pablo de Olavide University (UPO, Spain) and Conservation Technologies Division of Elengroup (El.En. S.p.A., Italy). Finally, we would to show the acknowledge to Mrs. Carmen Amoraga General Director of Culture and Heritage. Department of Education, Research, Culture and Sport, Generalitat Valenciana (Valencian Government, Spain)

REFERENCES

Juanes, D., Ferrazza, L., Pastor, M. & Contreras, G., 2018. Alteration of the polychromies in stone on the outside: the Façade dels Apostols of the Archpriestly basilica of Santa Maria de Morella, in: *II Colóquio Investigações Em Conservação Do Património. 28–29 September 2018*. Lisbon.

Pastor Valls, M. & Juanes Barber, D., 2017. Consolidación y adhesión de pintura vinílica: estudio de la viabilidad y comportamiento frente al envejecimiento de diversos polímeros, in: *La Ciencia y El Arte VI. Ciencias Experimentales y Conservación del Patrimonio*. Madrid, pp. 304–321.

Pérez García, P., García, A. & Ferrazza, L., 2017. Aportación de los Estudios Científicos a la Restauración del Pórtico de la Gloriade la Catedral de Santiago de Compostela: Análisis de las Alteraciones, in: *La Ciencia y Arte VI. Ciencias Experimentales y Conservación del Patrimonio*. Ministerio de Educación, Cultura y Deporte, Madrid, pp. 134–152.

Rivas López, J., 2008. *Policromías sobre piedra en el contexto de la Europa Medieval: aspectos históricos y tecnológicos*. Universidad Complutense de Madrid, Servicio de Publicaciones.

Science and Digital Technology for Cultural Heritage – Ortiz Calderón et al. (Eds)
© 2020 Taylor & Francis Group, London, ISBN 978-0-367-36368-0

Evaluation of the penetration depth of nano-biocide treatments by LIBS

J. Becerra
Department of Physic, Chemic and Naturals Systems, Universidad Pablo de Olavide, Seville, Spain

M.P. Mateo
Departamento de Ingeniería Naval e Industrial, Laser Applications Laboratory, Universidade da Coruña, Ferrol, Spain

G. Nicolás
Departamento de Ingeniería Naval e Industrial, Laser Applications Laboratory, Universidade da Coruña, Ferrol, Spain

P. Ortiz
Department of Physic, Chemic and Naturals Systems, Universidad Pablo de Olavide, Seville, Spain

ABSTRACT: In this research, two nanocomposites were studied as potential biocides for contemporary and historical buildings. These treatments are based on silver and titanium dioxide nanoparticles synthesized by a bottom-up method with sodium borohydride as reduction agent.

The treatments were tested on limestone from Novelda quarry (Alicante). This construction material was employed during the restoration of the Town hall of Seville (Spain). The inhibition of the biofouling growth on limestone was demonstrated previously by our research group, but their long-term effectiveness depends on the in-depth penetration of the nanocomposites. Because of that, the characterization of the penetration depth of the deposited nanocomposites in the limestones was carried out by Laser-Induced Breakdown Spectroscopy (LIBS) depth-profiling. LIBS showed successful results due to its high sensitivity to detect and locate the nanoparticles despite the small amount of nanoparticle treatment. Moreover, LIBS could be applied *in situ* without sampling for restoration control.

1 INTRODUCTION

The development of new treatments based on metallic nanoparticles for the conservation of our Cultural Heritage has generated new needs related to the characterization *in situ* and the evaluation of the effectiveness. In the case of silver nanoparticles and nanocomposites, whose biocidal properties have been confirmed previously (Becerra et al., 2019b, 2018), the low concentration impeded their detection with common analytical techniques, such as SEM-EDX, due to the low sensitivity to detect trace elements of these techniques (Kearton and Mattley, 2008).

The use of a spectroscopy technique based on laser ablation (LIBS) can eliminate this inconvenience. LIBS is a minimally invasive technique that provides the elemental composition of the sample when this is irradiated by a pulsed laser beam. Furthermore, this technique does not require previous sample preparation, can be used *in situ* and allows to achieve depth-profiling without additional equipment. The first proofs of concept have successfully measured the penetration depth of both, silver nanoparticles (Mateo et al., 2019) and nanocomposites (Becerra et al., 2019a), although any comparison between treatments has been carried out yet.

2 MATERIALS AND METHODS

2.1 *Synthesis*

Two biocidal treatments were studied in this research, silver nanoparticles and silver/TiO_2 nanocomposites, both stabilized with trisodium citrate (AgCit and AgCit/TiO_2, respectively). The synthesis of these nanomaterials was carried out according to Becerra Luna et al. (2018).

The treatments were characterized by UV-Visible (UV-Vis) spectrophotometry (Ocean Optics spectrometer equipped with an HR4000 detector, Dunedin, FL, USA) and Dynamic Light Scattering (DLS; Zetatrac Analyzer, Microtrac, USA).

2.2 *Application of treatments*

The treatments were applied on limestone from Novelda. This limestone has been widely used in the built of historical and contemporary buildings (Fort et al., 2002) and is a biosperite with a low average pore size (0.01-0.5 μm) and a 5% of open porosity (Guerrero, 1990). Two doses of 200 mL of aqueous suspension of nanoparticles at the same silver concentration (0.015 mg/mL) were applied on each stone samples. The limestone samples were dried during three days at room temperature before the assays.

The samples were analyzed by colorimetry (PCE-CSM 2) and color changes (ΔE^*) were calculated according to the equation $\Delta E^*=(\Delta L^2+\Delta a^{*2}+\Delta b^{*2})^{1/2}$, where ΔL, Δa^* and Δb^* are the color variations caused by the treatments defined by the CIELAB color-system.

2.3 *LIBS analysis*

A Q-Switched Nd:YAG laser beam (Brilliant, Quantel, 5 ns pulse width, 27 mJ pulse energy) was used in this assay, working at the second harmonic wavelength of 532 nm. The laser beam was focused on the limestone samples by a plano-convex quartz lens to achieve the correct energy density to ablate the sample and produce the plasma. The light emitted was guided by a quartz optical fiber to the entrance slit of an Echelle spectrograph (Mechelle, Andor), where the plasma light was dispersed and the spectral region from 200 to 850 nm detected in an ICCD camera (iStar, Andor). The data acquisition conditions were a time delay of 2.5 μs and an integration time of 10 μs. The assay was carried out in air under atmospheric pressure.

3 RESULTS AND DISCUSSION

3.1 *Synthesis*

The silver nanoparticles stabilized with citrate showed a good stability to be applied as biocide for stone. The colloidal stability decreased in the case of AgCit/TiO_2 nanocomposite (Table 1), although its value (|25| mV) is near to the limit established by Koutsoukos et al. (2006) for stable NPs suspensions (|30| mV). Moreover, the presence of TiO_2 increased the hydrodynamic diameter from 36 nm in the case of AgCit nanoparticles to 72 nm for AgCit/TiO_2 nanocomposites, and the size dispersions of particles was higher with a width band in the UV-Vis spectrum.

Table 1. Characterization of the biocidal.

Biocide	Hydrodynamic diameter (nm)	Zeta potential (mV)	UV-Vis spectroscopy peaks (nm)
AgCit	36±8	-63±3	390 (narrow band)
AgCit/TiO_2	72±18	-24.8±0.3	390 (width band)

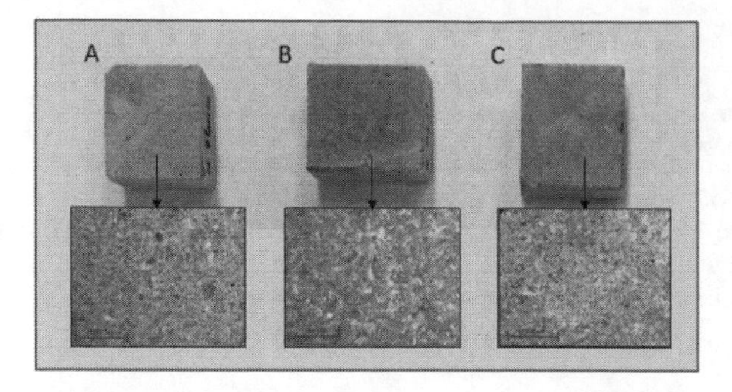

Figure 1. Images of untreated (A) and treated samples with AgCit nanoparticles (B) and AgCit/TiO_2 nanocomposite (C). Inset, a magnification of the area of each limestone sample.

3.2 *Application of the treatment*

Figure 1. shows untreated and treated samples. In the case of samples treated with AgCit nanoparticles, the aesthetical changes were not visible (Figure 1.B), while in the samples treated with AgCit/TiO_2 nanocomposites was possible to see a brownish haze on the surface (Figure 1.C). These perceptions are in agreement with the ΔE^* caused by the treatments. AgCit/TiO_2 nanocomposites caused the higher ΔE^* (11.6) due to the deposition of the nanocomposite on the limestone surface, although it was not visible the formation of nanocomposite aggregates at macroscopic scale. The ΔE^* was greater than the allowed for Cultural Heritage ($\Delta E^*<5$), and this treatment only was recommended to non-visible areas. AgCit nanoparticles could be used on monuments because of the ΔE^* was 1.2. The smaller hydrodynamic diameter of the nanoparticles favored their penetration through the stone pore, avoiding nanoparticles accumulations on the stone surface.

3.3 *LIBS analysis*

The emission lines of Ag (328 and 338 nm) and of Ti (306, 326, 327, 337, 353, 368, 376, 451, 461 and 521 nm), characteristic of the nanobiocidals, were detected by LIBS in the treated samples. The detection of Ag peak around 338 nm was used to compare the penetration of the two treatments. In this sense, several ablations were made in the same place (Figure 2) to study the evolution of Ag intensity with the number of in-depth laser pulses. A decay trend of

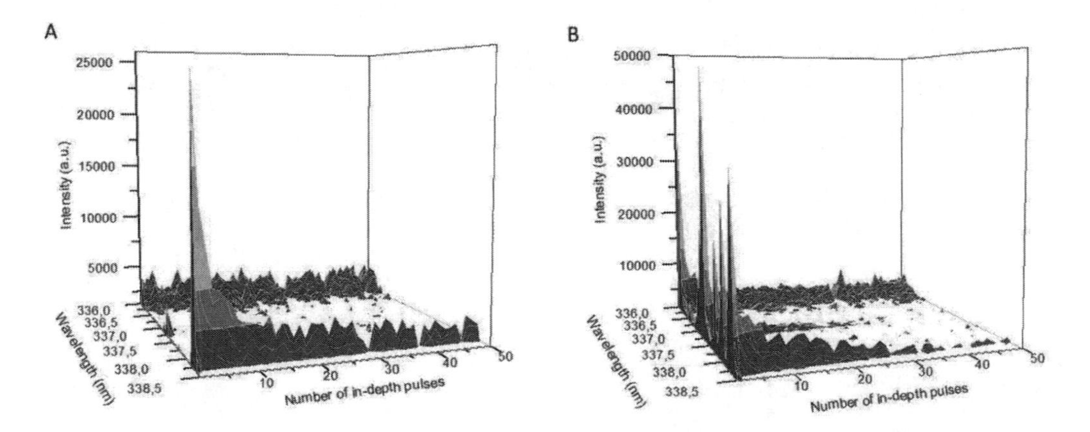

Figure 2. Evolution of the spectral signal with the number of in-depth pulses to analyze the penetration of AgCit nanoparticles (A) and of AgCit/TiO_2 nanocomposites (B) in the limestone from Novelda.

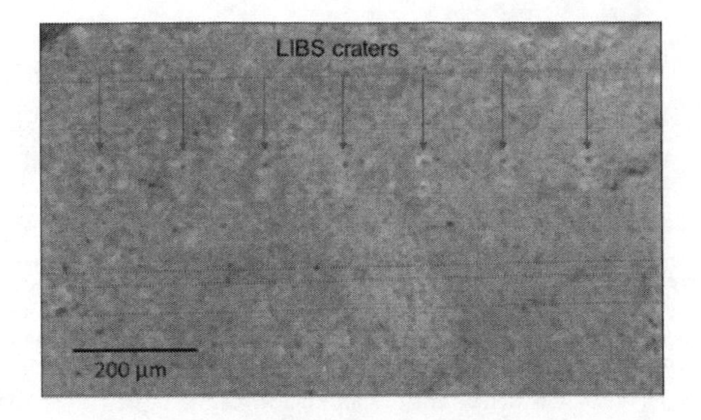

Figure 3. Craters generated by LIBS analyses on the limestone surface.

Ag signal was found in agreement with the diffusion of the nanoparticles and nanocomposites into the limestone matrix. In addition, the depth of different craters generated by laser ablation in the samples under study was measured by optical microscopy, resulting in an average ablation-rate of 6 µm/pulse. This number was used to convert the number of laser pulses from which the Ag signal was negligible into depth to estimate the penetration depth of the nanotreatments. AgCit/TiO_2 achieved a penetration range of 156-228 µm, while the silver nanoparticles penetrated more than 270 µm in this type of limestone. As shown in Figure 3, LIBS is a minimally invasive technique, but it can perform the comparison of the penetration depth of different biocidal treatments.

CONCLUSIONS

The penetration depth of two biocidal treatments based on silver nanoparticles applied on Novelda limestone has been studied based on the main emission line of Ag at 338 nm in the acquired emission-spectra. In addition, depth profiles were generated by plotting the evolution of the Ag intensity when the number of in-depth laser pulses was increased. The calculated penetration depth of the biocidal treatments after conversion of laser pulses into depth was 246-276 µm for AgCit and 156-228 µm for AgCit/TiO_2. The lower penetration depth of AgCit/TiO_2 was due to its higher hydrodynamic diameter that blocks the diffusion of the nanocomposite in the limestone.

In summary, the capability of LIBS has been demonstrated for the comparison of the penetration depth of biocidal treatments based on nanocomposites. The results evidence the potential of LIBS as an analytical technique in the restoration field related to nanoparticle treatments.

ACKNOWLEDGEMENT

Ministerio de Economía y Competitividad and FEDER (Art-Risk BIA2015-64878-R and CTQ2013-48396-P). Research teams TEP-199 and FQM-319 from Junta de Andalucía. J. Becerra is grateful to the Ministerio de Educación, Cultura y Deporte for his pre-doctoral fellowship (FPU14/05348).

REFERENCES

Becerra, J., Mateo, M., Ortiz, P., Nicolás, G., Zaderenko, A.P., 2019a. Evaluation of the applicability of nano-biocide treatments on limestones used in cultural heritage. J. Cult. Herit. In press. https://doi.org/10.1016/j.culher.2019.02.010.

Becerra, J., Ortiz, P., Zaderenko, A.P., Karapanagiotis, I., 2019b. Assessment of nanoparticles/nanocomposites to inhibit micro-algal fouling on limestone facades. Build. Res. Inf. In press. https://doi.org/10.1080/09613218.2019.1609233.

Becerra, J., Zaderenko, A.P., Sayagués, M.J., Ortiz, R., Ortiz, P., 2018. Synergy achieved in silver-TiO2 nanocomposites for the inhibition of biofouling on limestone. Build. Environ. 141, 80–90. https://doi.org/10.1016/j.buildenv.2018.05.020.

Becerra Luna, J., Zaderenko Partida, A.P., Ortiz Calderón, P., 2018. Biocide Treatments on Limestones Based on Silver Nanocomposites, in: 5th International Conference YOCOCU 2016, Youth in Conservation of Cultural Heritage. Museo Nacional Centro de Arte Reina Sofía, Madrid, pp. 294–297.

Fort, R., Bernabéu, A., García del Cura, M.A., López de Azcona, M.C., Ordóñez, S., Mingarro, F., 2002. Novelda Stone: widely used within the Spanish architectural heritage. Mater. construcción 52, 19–32. https://doi.org/10.3989/mc.2002.v52.i266.332.

Guerrero, M.A., 1990. Diagnóstico del estado de alteración de la piedra del Palacio Consistorial de Sevilla. Causas y mecanismos. Universidad de Sevilla (Spain).

Kearton, B., Mattley, Y., 2008. Laser-induced breakdown spectroscopy: Sparking new applications. Nat. Photonics 2, 537–540. https://doi.org/10.1038/nphoton.2008.173.

Koutsoukos, P.K., Klepetsanis, P.G., Spanos, N., 2006. Calculation of Zeta-Potentials from Electrokinetic Data, in: Somasundaran, P. (Ed.), Encyclopedia of Surface and Colloid Science. Taylor & Francis Group, New York, pp. 1097–1113. https://doi.org/10.1081/E-ESCS-120000059

Mateo, M., Becerra, J., Zaderenko, A.P., Ortiz, P., Nicolás, G., 2019. Laser-induced breakdown spectroscopy applied to the evaluation of penetration depth of bactericidal treatments based on silver nanoparticles in limestones. Spectrochim. Acta Part B At. Spectrosc. 152, 44–51. https://doi.org/10.1016/j.sab.2018.11.010.

Vulnerability assessment: Agents and Mechanisms of Decay

(Physical, Chemical and Biological)

Science and Digital Technology for Cultural Heritage – Ortiz Calderón et al. (Eds)
© 2020 Taylor & Francis Group, London, ISBN 978-0-367-36368-0

Modelling of the conservation state of gypsum-based plasters in the Real Alcazar of Seville

A. Silva
CERIS, IST - University of Lisbon, Lisbon, Portugal

M.T. Freire
CERIS and National Laboratory for Civil Engineering, Lisbon, Portugal

F.J. Blasco-López, F.J. Alejandre Sánchez & V. Flores Ales
Department Architectural Constructions II, Universidad de Sevilla, Seville, Spain

A. Calero
University of Granada, Granada, Spain

ABSTRACT: Plasterworks are gypsum-based decorative elements characteristic of Muslim art, providing a relevant historical and patrimonial heritage. Its natural ageing, associated with the action of degradation agents led to the appearance of anomalies, requiring the adoption of adequate and regular maintenance strategies. The *Real Alcazar of Seville (Spain)*, presents an extensive area of decorative plasters, and the establishment of a hierarchy of preventive conservation can be a very difficult task. In this study, a model to evaluate the overall degradation condition of plasterworks is proposed, based on a numerical index, called severity of degradation (S_w), which contemplates: i) the area affected by the different anomalies that can occur; ii) their severity, through the adoption of various degradation levels; iii) and the relative importance of each anomaly. The evaluation of the degradation condition is performed through visual inspections and through the application of non-destructive tests and image analysis.

1 INTRODUCTION

Plasterworks are gypsum-based decorative elements characteristic of Muslim art, executed on walls, arches and vaults by Muslim masters in Spain, over more than nine centuries. During this period of time, various artistic styles have been created, providing a relevant historical and patrimonial heritage to the buildings in which these plasterworks are included. Their natural ageing, associated with the incidence of degradation agents and mechanisms, naturally led to the appearance of several anomalies (Figure 1), which can occur in the plasterwork and in its fastening systems. In severe situations, the adherence of these plasterworks to the substrate is compromised, leading to the detachment of fragments, thus jeopardizing the users' safety, with serious irrecoverable losses to the building's heritage value.

In this sense, the plasterworks require the adoption of adequate and regular maintenance strategies. Nevertheless, many of the Spanish monuments, such as the Real Alcázar of Seville, present an extensive area of decorative plasters, and therefore, the establishment of a hierarchy of preventive conservation and intervention actions can be a difficult task. In this study, a model to evaluate the overall degradation condition of decorative plasterworks is proposed. For this purpose, a set of plasterworks in the Real Alcázar, corresponding to periods from XIV to XVI centuries is analysed. The evaluation of the degradation condition of the plasterworks under

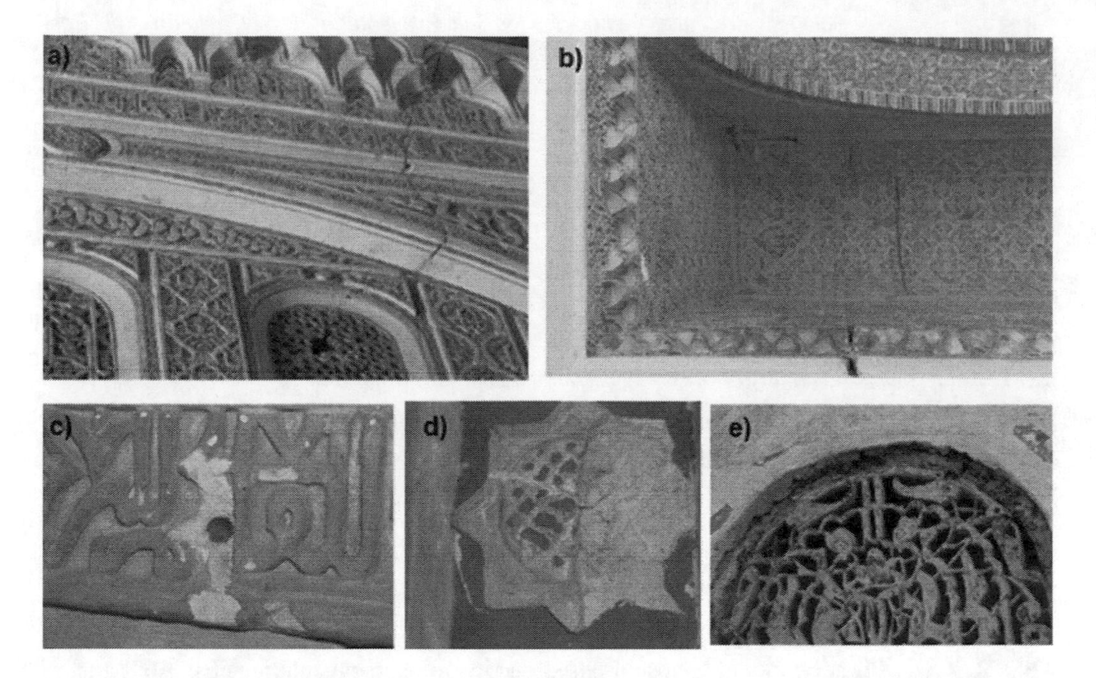

Figure 1. Examples of anomalies in the plasterworks of the Real Alcazar: a) Hall of the Ambassadors, with severe cracking and detachment of material; b) crack detection in plasterworks at *Patio de las Doncellas*; c) damage caused by rusted metal fixing; d) detachment; e) wear by water action.

analysis is performed through visual inspections and through the application of non-destructive tests (e.g. surface hardness, humidity) and image analysis (photogrammetry and thermography).

2 METHODOLOGY

In this study, a mathematical model is proposed, to describe the degradation of decorative plasterworks, based on previous studies related with the service life prediction of façade claddings in Portugal (Silva et al., 2016). For this purpose, a numerical index, called severity of degradation (S_w) - equation (1) -, is used, in order to quantify the overall degradation of a plasterwork, considering the anomalies detected, their severity and the area of the plasterwork affected by each one. Different anomalies have different levels of severity that need to be quantified, considering their different effects in the degradation condition of the plasterwork, the propensity to cause new anomalies or conditions of progression of the degradation phenomenon, and the aesthetic and safety impacts.

$$S_w = \frac{\sum(A_n \times k_a \times k_{a.n})}{A \times \sum k} \quad (1)$$

Where S_w represents the severity of degradation of the plasterworks, in %, k_n is the multiplying factor of anomaly "n", as a function of their degradation level, within the range K = {0, 1, 2, 3, 4}, $k_{a,n}$ is the weighting factor corresponding to the relative weight of the anomaly "n", A_n is the area of the plasterwork affected by an anomaly "n", in m^2, A is the façade area, in m^2, and Σk is the multiplying factor corresponding to the highest degradation level of a plasterwork of area A.

For the application of the degradation model proposed, the anomalies that may occur in the decorative plasterworks must be identified. Table 1 presents the proposed classification

Table 1. Classification of anomalies in plasterwork and the diagnosis methods proposed.

Type of anomaly	Anomaly	Diagnosis methods
Aesthetics and visual	A-E1 Presence of moisture	Visual inspection, hygrometrer and thermography
	A-E2 Accumulation of dirt	Visual inspection
	A-E3 Stains or color change	Visual inspection
	A-E4 Biological colonization/ biodeterioration	Visual inspection and chemical analysis in laboratory
	A-E5 Parasitic vegetation	Visual inspection
	A-E6 Efflorescences/Cryptoflorescences	Visual inspection and chemical anlaysis in laboratory
Loss of	A-A1 Adhesion failure/Detachment/ blistering	Visual inspection, percussion test with adherence Hammer, hygrometer, ultrasound and thermography
Loss of integrity	A-I1 Cracking	Inspection and measurement of the cracks width
	A-I2 Loss of material, loss of cohesion, Disintegration	Visual inspection and measurement in depth of the plaster´s alteration
	A-I13 Impact and other mechanical actions	Visual inspection
	A-I14 Pulverulence of the plasterwork	Measurement of the surface hardness
	A-I15 Pulverulence of the polychromy	Solubility test

system, for the anomalies that may occur in the decorative plasterworks in the Real Alcazar of Seville. In this study, two groups of anomalies are considered: i) evolutive anomalies, whose intensity and severity tend to increase over time (e.g. anomalies due to the presence of damp); ii) anomalies caused by inadequate maintenance actions or conservation treatments (e.g. limescale residues, resin stains, reintegration with Portland cement, among others), which are discrete events, but also compromise the conservation state of the decorative plasterworks. Naturally, different anomalies have different levels of severity. For instance, visual or aesthetic anomalies, such as stains or color changes, present less serious consequences when compared with the detachment of plaster elements from the lining or ceiling. In addition, the detachment of plaster elements presents different levels of severity, depending on whether it jeopardizes the users' safety or leads to the irreparable loss of those elements. Therefore, the global degradation rate of the decorative plasterworks (S_w) considers different levels of degradation (k_n) and takes into account different weighting factors ($k_{a,n}$) that allow considering the different effects of the different anomalies, their extension in the coating, their propensity to cause new anomalies or conditions of progression of the degradation phenomenon, and their aesthetic and safety impacts.

The conservation state of the decorative plasterworks is evaluated through *in situ* surveys, using an inspection form, which allows collecting information related with the materials' characteristics, the environmental exposure conditions, the use and maintenance conditions and the anomalies observed (their extension, location in the façade and severity). Moreover, during the fieldwork survey, some samples can be collected for more detailed analysis in laboratory.

3 APPLICATION OF THE METHODOLOGY TO A CASE STUDY

The proposed methodology was applied to a case study, located in the Patio de las Doncellas. The analysis of this case study encompasses the visual identification of the anomalies in situ, with the compilation of photographs, complemented by measurements and a collection of schematic drawings of the location of the anomalies. The characterization of the plasterwork was carried out, aided by photogrammetry and image processing applications. Also, computer-aided design (CAD) was used to map the observed anomalies, in order to quantify the areas affected by each type. Figures 2 and 3 show the mapping of the anomalies observed in the case study analyzed. Some of them are easily identified in situ, such as loss of integrity and aesthetic anomalies. Nevertheless, other anomalies are impossible to quantify only by visual analysis, as it is the case of the detachment of the plasterwork from the substrate. In this case, the visual

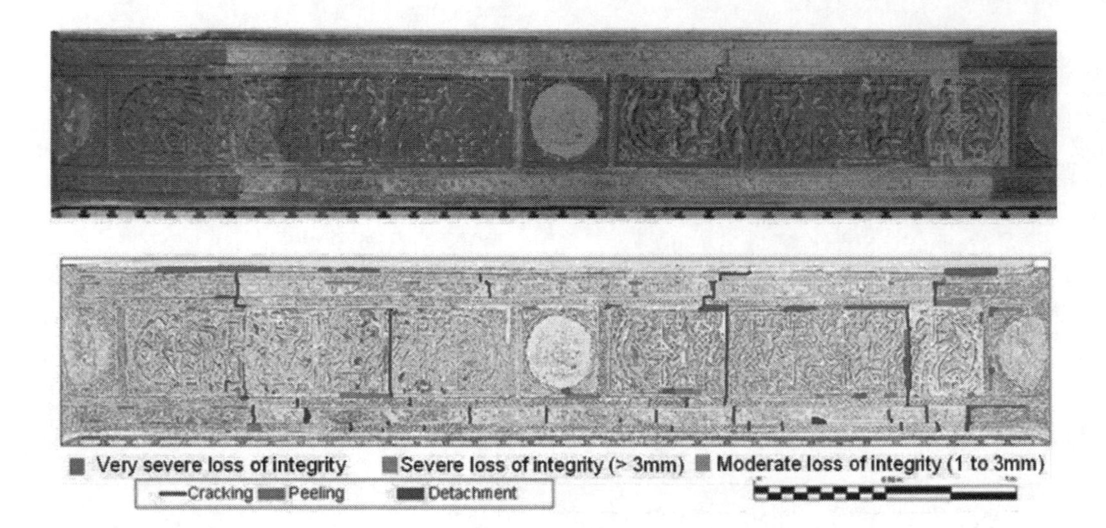

Figure 2. Assessment of the anomalies in a plasterwork of Patio de las Doncellas: mapping of the anomalies.

120	129	128	120	120	132	135	110	120	125	155	WM	125	125	90	110	110	135	145	108	WM
1,964	2,193	2,167	1,964	1,964	2,272	2,352	1,723	1,964	2,090	2,918	WM	2,090	2,090	1,283	1,723	1,723	2,352	2,628	1,677	WM
128	127	130	135	151	155	135	150	138	141	WM	WM	140	138	132	135	141	136	90	99	WM
2,167	2,141	2,219	2,352	2,800	2,918	2,352	2,771	2,433	2,516	WM	WM	2,488	2,433	2,272	2,352	2,516	2,379	1,283	1,474	WM
137	131	125	128	152	145	147	146	150	150	WM	WM	145	153	150	152	140	154	140	105	WM
2,406	2,246	2,090	2,167	2,830	2,628	2,685	2,657	2,771	2,771	WM	WM	2,628	2,859	2,771	2,830	2,488	2,889	2,488	1,608	WM

SURFACE MOISTURE IN DIGITS SURFACE MOISTURE IN PERCENTAGE OF MASS WM – Whitout Measurement

Figure 3. Assessment of the anomalies in a plasterwork of Patio de las Doncellas: humidity measurements made in situ and thermographic images.

analysis is complemented by other non-destructive diagnosis techniques, such as thermography. The presence of humidity is identified with the aid of thermography and a hygrometer.

For evaluating the pulverulence it is necessary to analyze the state of conservation of the polychromy, namely using solubility tests (Collado-Montero et al, 2016), and the surface hardness (Blasco-López & Alejandre-Sánchez, 2013). At this point, the analysis of the polychromy offers relevant information to the degradation model, being used independently as well, which allows establishing an index of severity or urgency of intervention in its conservation or restoration.

4 CONCLUSIONS

The definition of the degradation levels and the relative weighting coefficient to be applied to the different anomalies require the analysis of several plasterworks in the Real Alcazar of Seville, to determine their actual degradation conditions, and calibrate the proposed model. The use of this numerical index allows quantifying the physical degradation of the plasterworks, modelling their loss of performance over time and according to their characteristics, and the degradation agents and mechanisms. This model allows establishing a ranking of priority of interventions, according to conventional performance criteria. It also intends to reduce the risks for users, improving the conservation state and usability of the Real Alcazar of Seville.

REFERENCES

Blasco-López, F.J. & Alejandre-Sánchez, F.J. 2013. Porosity and surface hardness as indicators of the state of conservation of Mudéjar plasterwork in the Real Alcázar in Seville. *Journal of Cultural Heritage*, V. 14, pp. 169–173.

Collado-Montero, F.J., Calero-Castillo, A.I., Coba -Peña, A.C. & Medina-Florez, V.J. 2016. Colorimetric evaluation in treatment of protection and consolidation. Applications to Courtyard of the Maidens. Royal Alcazar of Seville. *Optica Pura y Aplicada*, V. 49, pp. 1–29.

Silva, A., Gaspar, P.L. & de Brito, J. 2016. Methodologies for Service Life Prediction of Buildings: With a Focus on Facade Claddings. 1st Edition, Switzerland: Springer International Publishing.

Science and Digital Technology for Cultural Heritage – Ortiz Calderón et al. (Eds)
© 2020 Taylor & Francis Group, London, ISBN 978-0-367-36368-0

Revealing Mariano Fortuny and Madrazo's technologies and materials. Identification of risk factors for conservation

E. Parra, M. Martin Gil, M.A. García, C. Ímaz, M. Bueso, E. Galiana, A. Albar, Á. Arteaga, E. García, J.A. Herráez, D. Durán, J.V. Navarro & S. Marras
Instituto del Patrimonio Cultural de España, Madrid, Spain

A. González & L. Llorente
Museo del Traje, CIPE, Madrid, Spain

G. de Osma & A. Roquero
Private consultants

ABSTRACT: As a part of an extensive research Project focused on the textile collection of Mariano Fortuny Madrazo (1871-1949) at Museo del Traje - CIPE (Madrid) some technical – scientific examinations have been achieved in textile pieces. Fortuny's patents for manufacturing, stamping and staining textiles have been consulted, and chemical analysis using IR spectroscopy, HPLC, gas chromatography/mass spectrometry, thin layer chromatography and elementary microanalysis with an electron microscope and a coupled X-ray detector have been conducted on materials from the collection. As a result some risk factors arise in order to design optimal conservation conditions for store and exhibition locations.

1 INTRODUCTION

Mariano Fortuny y Madrazo was an artist well known in its epoch, but also inventor and avant-garde designer. He was famous for its handling of silk in the design and manufacture of costumes. His dresses, clothes, in general textile designs are recognized cultural symbols of the happy twenties of the last century (de Osma 2012). He developed his own technique of stamping on fabrics with designs resembling those of classical Greece, and he patented them. Most of its core production was done at Venice, where the artist lived its last and more productive period. For instance, in his patented system for continuous stamping in fabric he uses a device with mechanically moved metal rods that makes with its pressure, the transfer of paints and dyestuffs from the model cloth (previously painted or photographically printed) in a continuous form, to a plane silk cloth[1]. There are more contributions to textile manufacturing procedures. One that demonstrates Fortuny's brilliance in handling silk is the invention of a permanent folding procedure on silk to get what is so called "*pleeting*". In the registered patent Fortuny's pleeting consists of produce a series of uniform and parallel foldings in order to resemble the aspect of the Old Greek and Roman Antiquity period robes. To achieve its famous pleeting, pressure and heat were applied to a silk cloth to get a permanent folding (see Figure 1).

These technical innovations imply necessarily new materials, or may be old materials used in different manners: dyeing procedures, stamping or pleeting. As time goes by, Fortuny's silks show different ageing patterns, as new risk factors arise and expresses themselves nowadays as deterioration. Through the data obtained from chemical analysis we intend to go deep into the knowledge

1. http://www.madrimasd.org/blogs/patentesymarcas/2018/fortuny-un-espagnol-a-venise/. Consultada el 10 de mayo de 2019

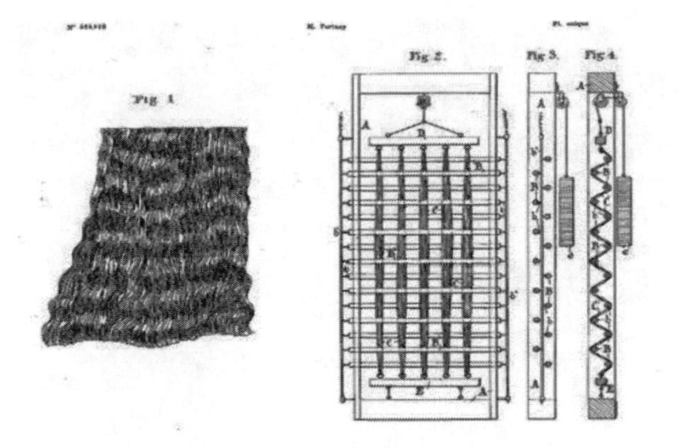

Figure 1. Ilustrations captured from French Republic public patent registry od August 26 th, 1910(code nr *FR414119*) entitled "Genre d'etoffe plissée-ondulée".

of Fortuny's materials and methods, and to identify those risk factors, materials physical – chemical properties or circumstances that may explain actual state of conservation or that may be an intrinsically deterioration cause in the near future. With the intention to guide these work lines that arise when evaluating the conservation - restoration problems, a research project was designed "Unveiling Mariano Fortuny and Madrazo's Technique and materials", actually being financed by National Research Plan in Cultural Heritage (project nr PNIC 2017/01 – IPCE - Ministry of Culture and Sports. Spanish Government), whose core objectives are the characterization of materials and research on manufacture and artistic procedures Identification of risk factors for conservation – restoration of the pieces, the control the climatic conditions of the environment of the pieces by Mariano Fortuny and to provide recommendations on conservation – restoration.

2 METHODS FOR MATERIALS' CHARACTERIZATION

Microsamples (less than 1 mg) were taken to perform micro destructive chemical analysis. So 0'5-1 cm threads were extracted from borders and hems of the clothes and dresses. Fourier Transform Infrared Spectroscopy (FT-IR) was performed using a Hyperion spectrometer (Bruker corp.) working between 400 and 4000 cm^{-1}. Gas Chromatography/Mass Spectrometry (GC/MS) was done with a Shimadzu apparatus. To do this, a fragment of the sample was hydrolyzed in a closed vial with 50 μ of HCl 6M for 24 hours at 110°C. After cooling and drying with nitrogen current, tert-buthyldimethylsilyltrifluoroacetamide (TBDMSTFA) and pyridine (1:2 volume, 50 μL) was added and heated to 65°C for 2 hrs (Parra 2005). The mixture was cooled and centrifuged. 5 μL of supernatant were injected in split less mode at 300°C (injector temperature). With 1 ml/min of He of mobile phase flow and a temperature program of: 85°C (2 min), 10°C/min to 300°C (15 min), 310°C of transfer line temperature, and a quadrupole MS detector working with 70 eV of ionization energy. Microanalysis was performed in a Hitachi electron Microscope with an Oxford instrument X- ray detector. Samples were metalized using carbon. Liquid chromatography (HPLC) analysis of dyestuffs was done with Waters 600 chromatograph with a Waters 996 diode array diode (DAD), and with an Agilent 1200 Infinity liquid chromatograph with an Agilent G1315C DAD detector and an Agilent 6530 QTOF MS/MS detector.

3 RESULTS AND DISCUSSION

The conservation state of dresses and clothes is heterogeneous. There are pieces in a perfect conservation state (apparently) and some other have some clearly recognizable damages as discolorations, loss of physical resistance of the thread leading to breakage, particularly in

Figure 2. Left, Fortuny's pleetings of Classical Greece dresses. Center, plain clothes. Left, reinforcing textile breakage.

pleated fabrics and in general in folds and hems. It is remarkable the pleated areas of the textil always show (Figure 2) breakage of the original fabric due to fatigue in the folding. Also, rests of old restoration products (adhesives, consolidants, and may be residues of old cleaning products, oxidized residues of materials coming from the transfer from bodies (lipids, sweat) and other materials derived from the use, i.e, stains and dust.

Dyestuff analysis has been yet performed on a big number of threads of Fortuny's designs. As a result of these analysis it's known that Fortuny used mainly natural dyestuffs. The most important of then found during HPLC analysis were cochinille, indigotine, palo Campeche, young fustic, and synthetic amido black II. Some of them were reported yet (Cardon 2007). Mordant and mineral salts added during dyeing process is also an important part of the color development on the fiber. Silk, cotton or wool may have iron, tin or cupper salts sometimes at high concentrations. A special case where metal salt concentration is higher than normal is the black color of so called "Eleanora's dress" whose black color is especially deep and well conserved. The reason for that so deep shade of black is the presence of hematein and hematoxylin from *palo de Campeche* dyestuff, but also the high proportions of iron, from iron sulfate, found within the fibers[2]. Acidity is a parameter not too high in the samples analyzed. Nevertheless some of the above mentioned dyes suffer from a poor photochemical behavior, mainly two hematein - hematoxylin (Palo Campeche) (Figure 3) and saffron. This last, in fact, is scarcely detected in old textile analysis, due to its instability against light (de Graaf et al. 2004).

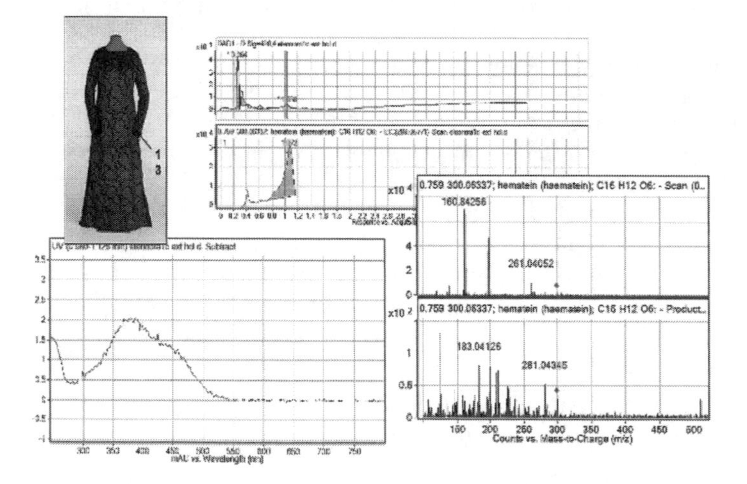

Figure 3. Analytical data for dyestuffs of Eleonora's Dress, with the detection of hematein (Museo del Traje CIPE. Madrid).

Sizing of cloths is also possible. It has been observed that pleeted cloths have fibers with some kind of organic coating spread over their surface. Using water extraction and hydrolysis and derivatization of the resulting hydrolysate an albumin – like protein was detected.

Painting is a different way used by M. Fortuny to get a color fabric. It can be achieved by stamping, the application of a paint in a fabric by transference, pressing wood relieve or another freshly painted fabric against it (as in serigraphy or any other engraving technique).

The analysis of painting materials on the surface of fibers and threads was performed directly using just stereoscopic microscope observation, gas chromatography to determine the type of binder, and SEM/EDX to describe pigments. The result was the use mica, aluminum and brass powder for metallic bright painting, and vermilion, copper arseniate and carbon black for the corresponding red, green and black colors. In the organic part, a protein was extracted with water extraction. The amino acid ratios resemble to those of albumin (Mills et al. 2012), but some interference of the soluble part of the silk fiber is expected, so that sure identification is not feasible. Finally some rests of conservation products were found in those cases where and ancient restoration treatment was performed. In this sense, polyamide and polyvinyl acetate (PVAc) where detected as adhesives use to bind reinforcing fabric.

4 INTRINSICAL RISK FACTORS

When silk was impregnated with PVAc kind of adhesives to reinforce, some acetic acid can be evolved, during the decomposition of the adhesive to polyvinyl alcohol, but unless the fabric is placed in a closed environment concentrations of the acid gas are not expected to be so high. Some pigments may act as oxidizers because the nature of their metal ions. Copper (II) or iron (III) pigments are an example. These cations are incorporated to fiber mainly as a part of a mordant or as color modifiers. Only in the case of the deep black obtained with high concentrations of Campeche dyestuff and an iron (III). In that specific case some deterioration of the fiber occurs as the microscopic image shows. In these cases where the fiber is peeled longitudinally a kind of microscopic micro fibrils joined to the main fiber appear. The deterioration leads to the loss of protein material and surely to an appreciable loss of the mechanical resistance of the yarn.

Sizes and binders as albumins are partially soluble proteins so that its presence, in cases where relative humidity is high can be colonized by microorganisms. Pleating parts of dresses and drapery are also weak parts, mainly due to fatigue of the folding, with the result of breakage of the most stressed part. This breakage is transmitted also to reinforcing cloths used to repair the dresses.

5 CONCLUSIONS

Condition state report indicates that main damages of the collection come from the use of dresses, as well as physical damage due to pleeting process and the fatigue that causes in the threads. pH and conductivity measures on wet fabric will also provide the necessary data to choose a water based system for cleaning to remove or neutralize intrinsic chemical risk factors as acidity, oxidation, and discoloration of dyestuffs or stains on the textile. The proposal of systems for tissue reintegration/consolidation, should be done, as a general rule, through physical reintegration or reinforcing, avoiding the application of chemical products. Storing and showing the collection in good and safe condition it's a priority that must conduct all future actions. It is important to take into account that mechanical stress on pleetings and hems must be relaxed by designing appropriate exhibition and store cases. Preventive conservation takes a very important role in long term conservation, and climatic conditions of stores and exhibition halls and cases, with too much attention focused on illumination is being controlled at the present moments.

REFERENCES

Cardon D. 2007 "Natural Dyes. Sources, tradition, technology and science", Archetype Books London.

H. De Graaf, J.,Roelofs & W. G., van Bommel M. 2004 "The Colourful Past: The Origins, Chemistry and Identification of Natural Dyestuffs". Archetype Books, London.

De Osma G. 2012, "Mariano Fortuny, arte, ciencia y diseño", Madrid, Ed. Ollero y Ramos.

Parra E. & García B. 2005 "Derivación con MTBSTFA de aminoácidos y ácidos grasos. Una determinación simultánea de aglutinantes proteicos y grasos en capas de pintura". Actas del II Congreso del GEIIC. Investigacion en Conservacion y Restauración. Barcelona 2005..

Pritchard F 2001. "Mariano Fortuny (1871-1949): his use of natural dyes". Dyes in History and Archaeology 16/17..

Mills J.S. & White R. 2012 "The organic chemistry of museum objects" (2 nd Ed.). Ed. Routledge. London.

Science and Digital Technology for Cultural Heritage – Ortiz Calderón et al. (Eds)
© 2020 Taylor & Francis Group, London, ISBN 978-0-367-36368-0

Chemical and morphological decay in maize stem sculptures

A.A. Ortega-Ordaz, E. Sánchez-Rodríguez, L. Rojas-Abarca & J.E. Bojórquez-Quintal
Laboratorio de Análisis y Diagnóstico del Patrimonio, El Colegio de Michoacán, Michoacán, Mexico

A. Ku-González
UBBMP-Centro de Investigación Científica de Yucatán, Mérida, Mexico

C.I. Cruz-Cárdenas
Centro Nacional de Recursos Genéticos-INIFAP, Tepatitlán de Morelos, Mexico

D.I. Quintero-Balbás
Microchemistry and Microscopy Art Diagnostic Laboratory- Universita di Bologna, Bologna, Italy

ABSTRACT: Despite the importance of maize stem lightweight sculptures, in the conservation field little is known about the maize stem ageing and its degradation by relative humidity and temperature. Therefore, were evaluated the effects of the temperature and relative humidity on maize stem samples from the lower (INF) and upper (SUP) parts of the plant. In addition, dissociated maize stems of a New Spain Sculpture (NSS) were characterized and the results were compared with those of the samples. The morphological and histochemical characterization indicate that the less lignified tissues, such as the parenchyma and metaphloem, are the most susceptible to deformations and ruptures. Besides, the structural sugars characterization suggests that the hemicellulose is the most susceptible compound to thermal damage, followed by cellulose, whose degradation is accelerated by the high relative humidity.

1 INTRODUCTION

Most artworks are susceptible to degradation by temperature and humidity (Camuffo et al. 1999) especially the objects made with organic materials rich in soluble proteins, starches and sugars (Michalski 2016), because they tend to be hygroscopic and their compounds oxidize and/or hydrolyze easily. According to Michalski (2016), inadequate temperatures can cause chemical, physical and biological damage to materials. Thermal degradation can also cause an increase in the water vapor pressure, the production of carbon dioxide and traces of organic compounds (Borrega & Kärenlampi 2009, Hunt 2012). On the other hand, Matsuo et al. (2010, 2011) suggests that treatments with high temperatures can accelerate the transformation of materials in a similar way to natural ageing.

Exposure to relative humidity causes the absorption of water and the expansion of the materials (Camuffo et al. 1999) causing the decrease of their mechanical properties (Gerhards 1982), the diffusion of water vapor (Glass & Zelinka 2008), mechanical stresses in the internal structures and the proliferation of microorganisms (Hunt 2012). Relative humidity catalyzes the reactions that cause thermal degradation, because the carboxylic acids cause the hydrolysis of the polysaccharides (Borrega & Kärenlampi 2009).

In New Spain –Mexico today–, maize stem was used in a lightweight sculpting technique that was a synthesis of the Prehispanic and European traditions. These sculptures are mostly made from the stem of the maize plant, with or without bark. Despite the importance of maize stem sculptures in Latin America and Spain, little is known about its degradation mechanisms and appropriate methods for their conservation. Therefore, this preliminary research determines the effects of different relative humidity and temperature treatments on morphology and chemical composition of the debarked maize stems. Also, dissociated and naturally aged samples from a New Spain sculpture were analyzed and compared with the treatments results.

2 MATERIALS AND METHODS

2.1 *Plant material*

Maize stems (*Zea Mays C.*) recollected during the harvest period were used in this study. To prepare the stems, they were debarked, and the nodes, buds and punctual zones with physical damages or by pathogens were removed. The stem internodes were cut into 5 cm long fragments and classified into lower (LSS) and upper stem sections (USS) according to the distance from the root plant. Also, dissociated and naturally aged samples from a New-Spain Sculpture (NSS) were analyzed and compared with the treatments results

2.2 *Temperature treatments and FTIR/ATR spectroscopy characterization*

To evaluate the isolated effect of the temperature on the maize stems and identify the riskier temperatures, the LSS and USS samples were treated with five different thermal treatments by 168 hours (one week). The parameters of the temperature treatments are observed in Table 1.

The milled and homogenized treated samples were analyzed in a Pelkin-Elmer® Frontier FT-IR/MIR spectrometer using the middle infrared region (4000 to 400cm-1) with a resolution of 4cm-1 and using 10 scans in the modality of attenuated total reflection (ATR). The results were processed in the Spectrum 10.03 software and were graphed with the Origin-Pro 2017 software.

2.3 *Relative humidity and temperature treatments*

LSS and USS samples were placed in sealed chambers and subjected to 9 different treatments of relative humidity and temperature (RH-T) for 14 days (two weeks). To keep the relative humidity conditions stable at 25% RH, 50% RH and 75% RH, different solutions of lithium chloride (LiCl) were used in each chamber. On the other hand, the temperature ranges were kept constant by incubating the chambers with the samples in ovens at 25 °C, 50° C and 75 ° C. The parameters of the RH-T treatments are observed in the Table 2.

Table 1. Temperature treatments.

T1	T2	T3	T4	T5
4°C	25°C	50°C	75°C	100°C

* The T2 corresponds to the control.

Table 2. Relative humidity and temperature treatments.

	Temperature (°C)		
Relative humidity (%RH)	25°C	50°C	75°C
25%RH	T1	T2	T3
50%RH	T4	T5	T6
75%RH	T7	T8	T9

** The T4 corresponds to the control.

2.4 *Histochemical staining*

To observe variations in the accumulation of phenolics components at the secondary cell walls (SCW), the slides of the samples were incubated in phloroglucinol-HCl (Wiesner´s staining) according to the methods Mitra & Loque (2014) and Liljebren (2014) with some modifications.

Bright field microscopy was performed with a Leica DM 4000M microscope. All images shown in this study were captured with the Leica Application Suite 4.0 software.

2.5 *Structural sugars characterization*

A modification of the methods described by Xiao et al. (2001), Rabemanolontsoa & Saka (2012) and Jung et al. (2015) were used to determine the structural sugars.

2.5.1 *Extractive free sample*

The samples were dried, milled and homogenized. The extractive-free samples were prepared in two Soxhlet extraction phases using toluene-ethanol (2:1) and ethanol, respectively.

2.5.2 *Lignin determination*

The lignin was determined according to the Klason method procedure. Acid Insoluble Lignin (AIL) content was determined by gravimetric analysis. Acid Soluble Lignin (ASL) was determined by UV-VIS method.

2.5.3 *Hollocelulose determination*

The holocellulose preparation was by delignification process. Sodium chlorite and glacial acetic acid (0.2 M) were added to the extractive-free sample and incubated in hot water bath. The procedure was repeated 5 times. The collected sample was dried and determined by gravimetric analysis.

2.5.4 *Cellulose and hemicellulose*

For cellulose determination, 17.5% sodium hydroxide were added to the holocellulose and vortexed. The samples were incubated in a water bath and filtered with a Büchner funnel. The weight of the remaining was the cellulose content. Hemicellulose weight was determined by subtracting the weight of hollocelulose from the weight of cellulose.

2.5.5 *FTIR/ATR spectroscopy*

The FTIR/ATR spectroscopy characterization was done as described in the method of the section 2.2.

2.5.6 *Statistical analysis*

Data were analyzed using a One-Way analysis of variance (ANOVA) (Sigma Plot Version 11.0). Treatment averages were compared using Tukey´s range test.

3 RESULTS

3.1 *Effects of the temperature treatments on the chemical composition of the maize stems*

By comparing the LSS and USS spectra's, is observed that the LSS show a low intensity signals at 1249 cm^{-1} and 1736 cm^{-1} (Figure 1); the bands are assigned to amide III of proteins (Wang et al. 2012) and stretching vibration in the C=O alkyl-ester bond of the hemicellulose (Alonso-Simón et al. 2012), respectively. The transmittance of the bands began to decrease from T3. However, in the USS, the signal remained constant during the different treatments. In the NSS sample the transmittance of the band 1736 cm^{-1} is missing (Figure 1).

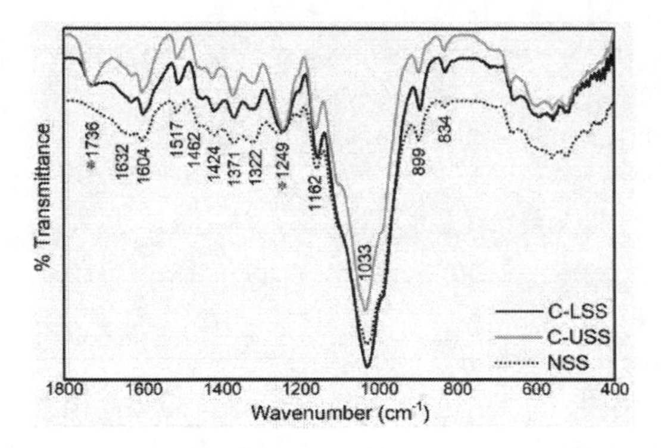

Figure 1. FTIR/ATR spectra´s of the LSS, USS and NSS. The 1736 cm^{-1} band could be a chemical a marker.

3.2 *Effects of the relative humidity and temperature treatments on the maize stems morphology*

The relative humidity and temperature treatments affected the physical characteristics of the two stems sections samples. The treatments caused darkening, decrease in the size of the samples and changes in texture. In the LSS samples, the effects of all the treatments subjected to a temperature above 25 °C were evident. Those effects were graver as the relative humidity increased. In the USS samples, the most relevant changes were observed with the treatments T8 and T9.

The Wiesner staining of the LSS and USS samples does not show relevant changes in the coloration of the SCW, this suggest that there was no loss of lignin content during the treatments. However, it is observed that the metaphloem and the parenchima surrounding the vascular bundles shows deformations and ruptures (Figure 2).

3.3 *Effects of the relative humidity and temperature treatments on the structural sugars content*

Structural sugars determination results show that the USS samples have a higher content of hemicellulose than LSS (Figure 1). The HR-T treatments had significant effects on the hemicellulose content. In the LSS samples, the graver treatments were 50°C and 75°C temperatures. The temperature effects were catalyzed with the 50%RH and 75%RH conditions.

In the USS, the isolated temperature treatments did not affect the samples. However, when the high relative humidity conditions were applied (T8 and T9) the hemicellulose percentage decreased significantly. On the other hand, in the NSS sample, the cellulose and hemicellulose contents are notably lower compared with the LSS and USS controls (T4). The lignin content did not significantly change in neither of the samples and treatments.

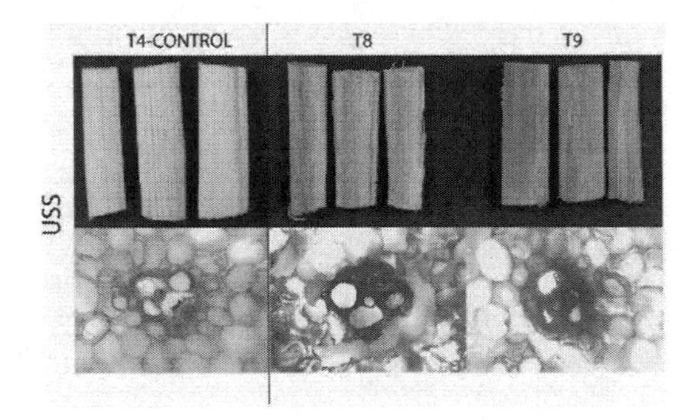

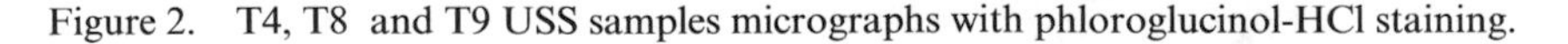

Figure 2. T4, T8 and T9 USS samples micrographs with phloroglucinol-HCl staining.

4 CONCLUSIONS

The different sections of the maize stem samples have different composition, the LSS has a lower content of hemicellulose in the cell walls than the USS. Therefore, the LSS are more susceptible to morphological and chemical degradation. In addition, the hemicellulose is the weakest structural component, due to the presence of uronic acids ramifications in its structure. Whereas, the lignin is not affected. The degradation degree in maize stems can be estimated by FTIR/ATR spectroscopy, due the 1736 cm-1band could be a chemical marker related to the hemicellulose content. Also, the Wiesner's staining is an efficient method to determine the morphological damage by evaluating the more vulnerable tissues (phloem and parenchyma). Actually, more microscopical techniques analysis and structural sugars characterization are developed.

REFERENCES

Alonso-Simón, A., P. García-Angulo, M.H. Martínez, A. Encina, J.M Álvarez & J.L Acebes. 2011. The use of FTIR spectroscopy to monitor modifications in plant cell wall architecture caused by cellulose biosynthesis inhibitors. *Plant signaling & behavior*, 6(8), 1104–1110.

Camuffo, D., G. Sturaro, A. Valentino & M. Camuffo. 1999. The conservation of artworks and hot air heating systems in churches: Are they compatible? The case of Rocca Pietore, Italian Alps. *Studies in Conservation*, 44(3), 209–216.

Gerhards, C. C. 1982. Effect of moisture content and temperature on the mechanical properties of wood: an analysis of immediate effects. *Wood and Fiber Science*, 14(1), 4–36.

Glass, S. V.& S.L. Zelinka. 2010. "Moisture relations and physical properties of wood" in Ross R.J. (ed.) *Wood handbook: wood as an engineering material: Chapter 4. General technical report FPL; GTR-190.* EUA: Department of Agriculture, Forest Service, Forest Products Laboratory, 4.1–4.19.

Hunt, D. 2012. Properties of wood in the conservation of historical wooden artifacts. *Journal of Cultural Heritage*, 13(3), S10–S15.

Jung, S. J., S.H. Kim & I.M. Chung. 2015. Comparison of lignin, cellulose, and hemicellulose contents for biofuels utilization among 4 types of lignocellulosic crops. *Biomass and Bioenergy*, 83, 322–327.

Liljegren, S. 2010. Phloroglucinol stain for lignin. *Cold Spring Harbor Protocols*, 2010 (1).

Matsuo, M., M. Yokoyama., K. Umemura., J. Gril., K.I Yano & S. Kawai. 2010. Color changes in wood during heating: kinetic analysis by applying a time-temperature superposition method. *Applied Physics A: Materials Science & Processing*, 99(1), 47–52.

Matsuo, M., M. Yokoyama., K. Umemura, J. Sugiyama, S. Kawai, J. Gril, S. Kubadera, T. Mitsutani, H. Ozaki, M. Sakamoto & M. Imamura. 2011. Aging of wood: analysis of color changes during natural aging and heat treatment. *Holzforschung*, 65(3), 361–368.

Michalski, S. [On line]. Agent of Deterioration: Incorrect Temperature. *Canadian Conservation Institute* [Consulted at May 2019] Available at: http://www.cci-icc.gc.ca

Rabemanolontsoa, H. & S. Saka. 2012. "Holocellulose determination in biomass" en Yao, T. (ed.), *Zero-Carbon Energy Kyoto 2012*. Japón: Springer, 135–140.

Wang, J., J. Zhu, R. Huang & Y. Yang. 2012. Investigation of cell wall composition related to stem lodging resistance in wheat (Triticum aestivum L.) by FTIR spectroscopy. *Plant signaling & behavior*, 7(7), 856–863.

Xiao, B., X. Sun & R. Sun. 2001. Chemical, structural, and thermal characterizations of alkali-soluble lignins and hemicelluloses, and cellulose from maize stems, rye straw, and rice straw. *Polymer Degradation and Stability*, 74(2), 307–319.

Science and Digital Technology for Cultural Heritage – Ortiz Calderón et al. (Eds)
 ISBN 978-0-367-36368-0

Interaction of heritage pigments with volatile organic compounds (VOCs). A laboratory study

O. Vilanova & B. Sanchez
FOTOAIR - Centro de Investigaciones Energéticas, Medioambientales y Tecnológicas (CIEMAT), Madrid, Spain

M.C. Canela
Centro de Ciências e Tecnologia, Universidade Estadual do Norte Fluminense Darcy Ribeiro, Campos dos Goytacazes-RJ, Brasil

ABSTRACT: In this study, different pigments were exposed to Volatile Organic Compounds (VOCs), and their physicochemical properties were studied. The pigments chosen were: hematite, malachite, white-lead, cinnabar, gypsum. Two binders were tested too: animal glue and Arabic gum. Each pigment and binder was exposed to 50μl of each of the major pollutants separately (m-xylene, 1,2,4-trimethylbenzene, 1-methoxy-2-propyl acetate, hexanal, 2,2,4-trimethylpentane, nonane, 2-butanone oxime, octamethylcyclotetrasiloxane, dodecamethylcyclohexasiloxane, methanol, acetic acid, and formaldehyde). After one month of exposure, the excess of VOCs on the pigment was volatilized at 40°C, and the samples were analyzed by XRD, FTIR and UV-Vis. The stability of these pigments against most VOCs showed that the acids and aldehydes alter the structure of the pigments based on carbonates and sulfates. Alteration of color intensity has been observed in the malachite, in the presence of hexanal and acetic acid. Malachite, gypsum, and lead white showed changes in the crystalline structure when exposed to acetic acid. Spectrometric results showed that there are interactions of VOCs with different pigments.

1 INTRODUCTION

The heritage conservation over time is the primary concern of the professionals of the museums. Many processes can cause alterations in the artwork. The most studied ones have been temperature, humidity, and radiation. Studies on the presence of contaminants in the air in close contact with the exhibition elements have paid little attention until now. In (1987 Hatchfield and Carpenter) published “Formaldehyde: How Great Is the Danger to Museum Collections?” which attracted international attention on the problem of environmental pollution in museums. The Getty Conservation Institute (GCI) began in 1985, the first research program on risk to which collections are exposed to air pollution (Preusser & Druzik, 1989). Initially, this program was focused on outdoor contamination and later expanded to indoor contamination (Grzywacz, 1989). Forty-eight Volatile Organic Compounds (VOCs) have been already detected and quantified in the atmosphere inside and outside showcases of the National Archeological Museum (MAN) in Madrid (Sanchez et al. 2015; Vilanova et al. 2018). However, there is limited information about the effect of these compounds on heritage. In this study, different pigments were chosen based on the polychromies of ancient Egyptian art and exposed to the most abundant and representative compounds identified in the museum, and their physicochemical properties were studied. These first studies are aimed to obtain rapid response to the influence of some VOCs on selected pigments at the new conditions the new pollutants represent for the works of art exhibited in a museum.

2 METHODOLOGY

The powered pigments chosen were based on used historically in works of art and were purchased commercially: hematite-Fe_2O_3, malachite-$CuCO_3Cu(OH)_2$, white-lead-$2PbCO_3$ $Pb(OH)_2$, cinnabar-HgS, gypsum-$Ca(SO_4)_2*2H_2O$, animal gum, and Arabic gum. One gram of each pigment and gum were put in the Petri plate and added 50 µl of the analytical quality standards (m-xylene, 1,2,4-trimethylbenzene, 1-methoxy-2-propyl acetate, hexanal, 2,2,4-trimethylpentane, nonane, 2-butanone oxime, octamethylcyclotetrasiloxane, dodecamethylcyclohexasiloxane (D6), methanol, acetic acid, and formaldehyde), separately (Figure 1). After one month of exposure, the plates were opened, and the excess of VOC in the sample was removed and heated surface at 40°C. The samples were analyzed by X-ray diffraction (XRD - PANanalytical X PERT), Fourier transformed Infrared (FTIR - Thermo-Nicolet 5700) and UV-Vis spectroscopy (PerkinElmer Lambda 650 UV/Vis). The experiments were performed at a controlled ambient temperature of 25 °C and 38% relative humidity.

3 RESULTS AND DISCUSSION

Malachite was the only pigment that showed visible color change after being exposed to acetic acid and hexanal. This change was based on intensity rather than color change (Figure 2).

Few changes in FTIR spectra were observed for malachite in the presence of hexanal and acetic acid (Figure 3). The bands to 820 and 877 cm-1 have been identified as the two modes of flexion (ν2) of CO_3^{2-}. The vibrational modes of symmetric tension (ν1) and asymmetric (ν3) of CO_3^{2-} have been given at 1111 cm-1 for (ν1) and at 1392 and 1486 cm-1 for (ν3). For exposure to aldehyde, a very subtle double band was observed at approximately 2900-2800 cm-1, which has been assigned with the Fermi double of the aldehydes. While in the case of the

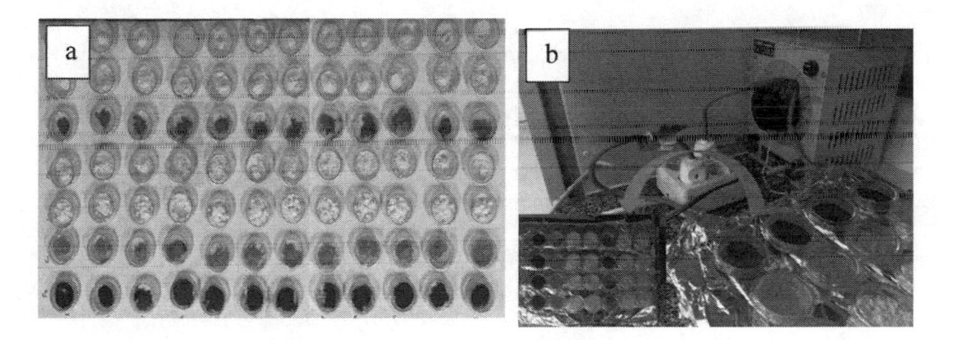

Figure 1. a) Experiment using powered pigments with VOCs; b) VOCs evaporation.

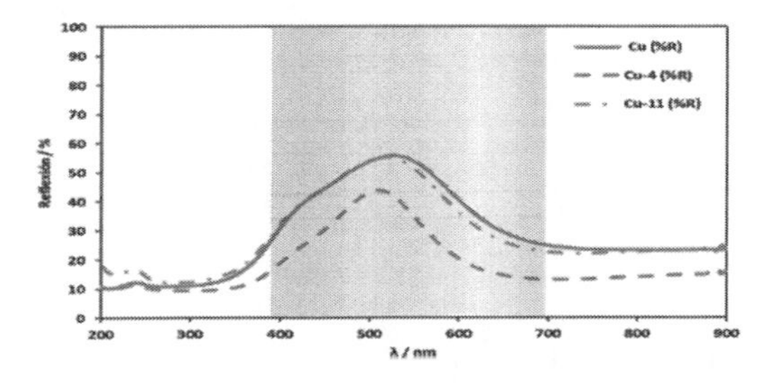

Figure 2. UV-VIS spectra of malachite after exposition. Solid line (pure malachite); Dashed line (Malachite + hexanal); and dot-dashed line (Malaquite + acetic acid).

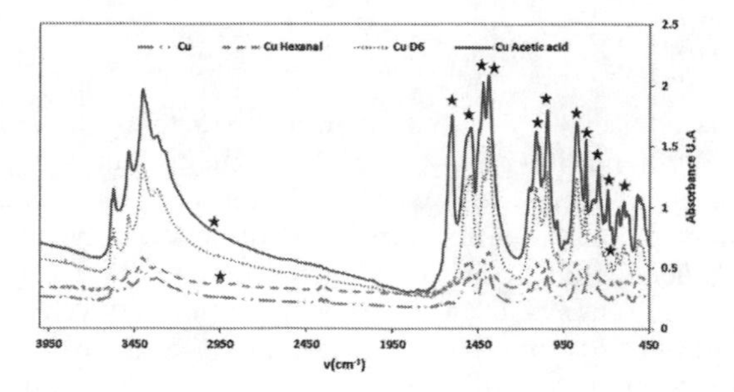

Figure 3. FTIR spectra of malachite before and after exposition at hexanal D6, and acetic acid.

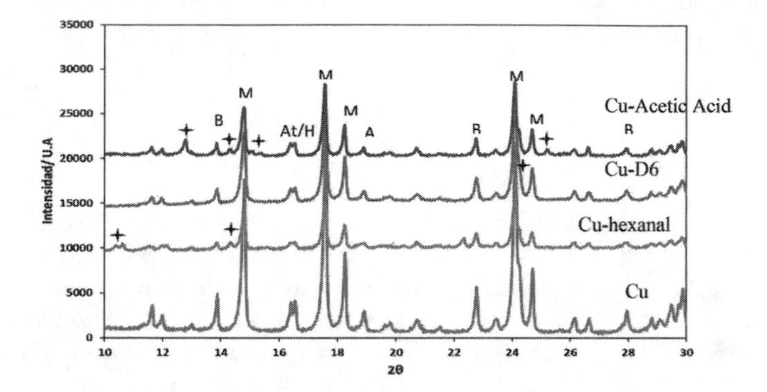

Figure 4. DRX of malachite before and after exposition at hexanal D6, and acetic acid. M- Malaquite - $CuCO_3Cu(OH)_2$ B – Brochantite - $Cu_4(SO_4)(OH)_6$ At- Atacamite - $Cu_2Cl(OH)_3$ H – Hoganite – $Cu_2(OH)_3CH_3COO$.

siloxane an increase in the Si-O vibration band has been observed at about 1250 cm-1. In this case, the acetic acid seems to have altered this pigment, since a high-intensity band is observed at approximately 1650 cm-1 corresponding to the symmetric tension of group C=O.

DRX analysis also shows alterations between contaminants (Figure 4). These alterations have been reflected in the perceptible change of color of the pigment. The shade of green has varied, reaching more bluish tones, as in the case of the hoganite after exposure to acetic acid.

Physicochemical alterations of gypsum were observed on the DRX analysis (Figure 5). The crystalline phase found when carrying out the analysis without exposure to VOCs has been gypsum, that is, the calcium sulfate dehydrate. After the exposure to the different VOCs, it is observed how the loss of these water molecules takes place, generating in all cases another crystalline phase, the one corresponding to anhydrite and in the case of exposure to acetic acid the loss of this hydration has been more severe which has led to bassanite and anhydrite.

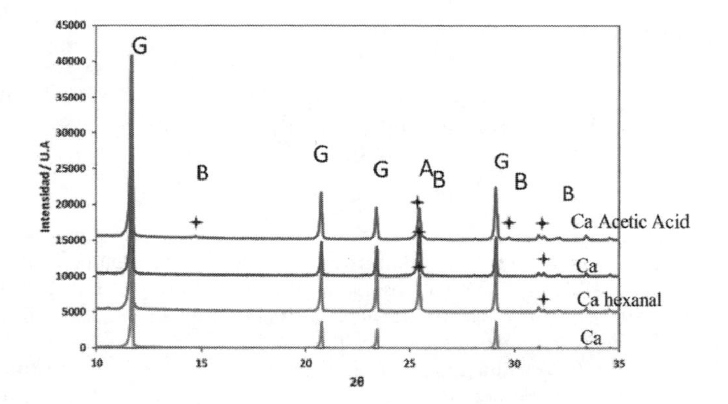

Figure 5. DRX spectra of gypsum before and after exposition at hexanal D6, and acetic acid. G-Gypsum ($Ca(SO_4)_2x2H_2O$); B- Bassanite ($Ca(SO_4)_2x0.5H_2O$); A - Anhydrite ($CaSO_4$).

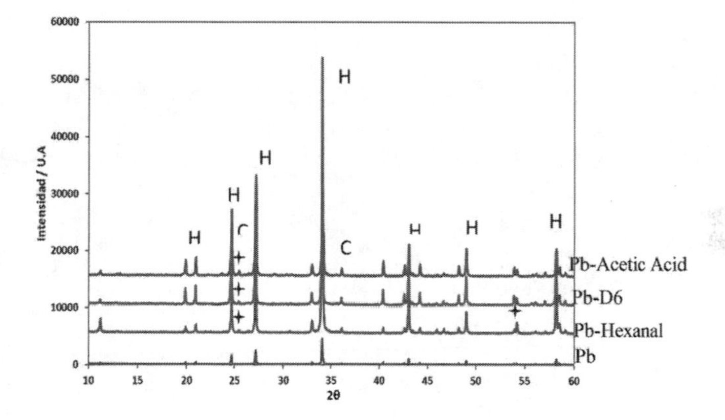

Figure 6. DRX spectra of white lead before and after exposition at hexanal D6, and acetic acid. H-Hydrocerussite $2PbCO_3Pb(OH)_2$, C- Cerussite $Pb(CO_3)$.

3.1 *White Lead*

Crystalline phases have been found in the diffractogram of lead white without exposure, which has been assigned to hydrocerussite, the basic carbonate of lead (II). When exposed to the different VOCs, another crystalline phase corresponding to cerussite ($PbCO_3$) has been generated in all cases (Figure 6). The higher acidity of the VOCs has neutralized the basic groups.

4 CONCLUSION

The experiment with acetic acid and some aldehydes with pigments generates changes in carbonates and sulfates. Hematite and red cinnabar are apparently stable for 12 VOCs, and no changes were observed during this experiment. Color and composition were modified for Malaquite and White Lead. With this contaminant, no interaction in the structure is observed. From this work, new experiments simulating paintings are being performed with colorimetric measurements.

ACKNOWLEDGEMENTS

The authors want to express their gratitude to Teresa Gómez Espinosa, Carmen Perez Die and MAN authorities for their support as well as Constanza Miliani from Perugia University.

REFERENCES

Grzywacz C.M. 1989. GCI general survey of U.S. museums: Carbonyl and organic acid pollutants in the museum environment. Final Report. Getty Conservation Institute, Marina del Rey, California.

Hatchfield P.B, Carpenter J.M. 1987 The Problem of formaldehyde in Museum Collections. International Journal of Museum Management and Curatorship 5:183–188.

Preusser F, Druzik J.R. 1989. Environmental research at the Getty Conservation Institute. Restaurator 10:160–196.

Sanchez B.C., Vilanova, O., Canela, M. C., Espinosa, T. G. 2015. Calidad del aire interior de las vitrinas en el nuevo Museo Arqueológico Nacional. Boletín del Museo Arqueológico Nacional. 33: 367–381.

Vilanova, O.; Gilaranz, J.; Parrondo, S.; Sanchez B.C.; Canela, M. C. 2018. Caracterización química y biológica del ambiente en el que se conserva la momia guanche del Museo Arqueológico Nacional. Boletín del Museo Arqueológico Nacional. 37: 119–130.

Science and Digital Technology for Cultural Heritage – Ortiz Calderón et al. (Eds)
© 2020 Taylor & Francis Group, London, ISBN 978-0-367-36368-0

Vulnerability of the Tomás Terry theater (Cienfuegos, Cuba)

D. Abreu, B. Rodríguez & A. Cepero
Faculty of Arts of the Conservation of Cultural Heritage, University of Arts, ISA, Havana, Cuba

P. Ortiz & R. Ortiz
Department of Physical, Chemical and Natural Systems, Universidad Pablo de Olavide, Seville, Spain

ABSTRACT: The Tomás Terry Theatre (Cienfuegos, Cuba) was founded in 1889. The main weathering forms in its façades are fractures, dampness and erosion, inside the building the weathering forms are mainly due to water that produces lacuna, chromatic alteration, concretions and loss of material. A matrix of vulnerability was employed in order to evaluate the state of conservation of the theatre and their main hazards. The results showed that the main agents that caused this weathering forms are dampness by capillarity and rainfall. Those agents produce total or partial losses of architectonic elements or their alteration inside and outside the building. On the other hand, the use of the building and the interventions are the main anthropogenic agents that evidence the influence of the viewers and the maintenance in the state of conservation. Further studies are being carried out in order to evaluate the preventive conservation strategies in the theatre that allow identifying, evaluating and prioritizing the maintenance and restoration for mural paintings.

1 INTRODUCTION

The Tomás Terry Theatre was founded in 1889 (Figure 1) it was declared a National Monument by Resolution No. 3, 10th, 1978 and is located in the historic center of Cienfuegos (Cuba). It was built in eclectic style with as a rectangular masonry building having galleries, porticos, five gate arcs and three entrance doors. The center roof is decorated with balustrades and vases. The ceiling placed on the living room or stall is made up of metal girders that support the structure the plaster slab is held by the meshes. There is a large mural design with 23 figures, made by Camilo Salaya on the ceiling (Sueiro, 1998).

The central hall has a horseshoe shape and it has four floors of galleries. The first or plateau caontains more than 300 lunettes and 21 boxes, the second one has seats and 10 boxes, the third is a gathering forming an amphitheater and the fourth floor called "the pot", also has three proscenium boxes or grilles on each side. The theatre façade belongs to neoclassicism style and it harmonizes with the urban context. At the highest point of the front façade (Figure 1) there are three Venetian mosaics, by Salvati factory (Batista, 2014).

Nowadays, the theatre is under restoration. For this reason, this study was carried out to evaluate the weathering forms of the building to contribute to its conservation.

2 MATERIALS AND METHODS

2.1 *Vulnerability analysis*

The expanded vulnerability index (VIe) was studied, based on a vulnerability matrix similar to that reported by Ortiz & Ortiz (2016). This matrix was adapted to the extent and the causes of the weathering forms in the monuments of Cuba.

Figure 1. The Tomás Terry Theater in 2017. General view.

The vulnerability matrix was made by inserting the hazards into the rows and the characteristics of the construction material as a degree of conservation structure and the anthropogenic factors in the columns. The pathologies were described according to CNR-ICR Normal 1/88 (1988), Fitzner (2007) and the glossary ICOMOS-ISCS (2008). The diagnosis was carried out in ten different locations, it was applied to different materials and environmental conditions.

After studying the weathering forms, the vulnerability index (VI) was calculated by dividing the total value of the deterioration patterns (Vx) for the theater by the sum of the total value of the deterioration patterns in the worst case scenario (Σvdp), when the frequency would be maximum, (Ortiz et al, 2014). Later, an expanded vulnerability index was calculated according to a DELPHI, developed by Cuban experts, a multidisciplinary group of architects, biologists, engineers and technicians with more than 10 years of experience in conservation of buildings.

$$VI = \frac{Vx}{\sum_{f=3} vdp} x100 \tag{1}$$

The expanded vulnerability index was established in accordance with this DELPHI that allows the evaluation of the influence of different characteristics on the vulnerability.

$$\text{VIe} = \sum \text{fiVi} \tag{2}$$

Where: fi=factor according to DELPHI forecasting and Vi=vulnerability associated to the variable i.

The expanded vulnerability index (VIe) was classified by degree of vulnerability using ordinal classes as was described by Galán et al. (2006).

3 RESULTS AND DISCUSSION

The weathering forms study was the first step to develop the vulnerability matrix. The main weathering forms found were loss of materials, areas of stains due to humidity and saline efflorescences (Figure 2). In the case of biological colonization, it was mainly due to the attack of xylophagous insects on the wooden elements that are part of the structure of the building and its furniture. The problem was worsened in those zones were wood was replace in previous interventions or located without a previous treatment.

When comparing the different levels in the interior of the theater, greater affectations were observed in the pictorial layer on the roof of the boxes and in the mural paintings. The mural painting of the ceiling had loss of the support and important loss of pictorial layers. Xylophages attacks where the main agent in the second and third level.

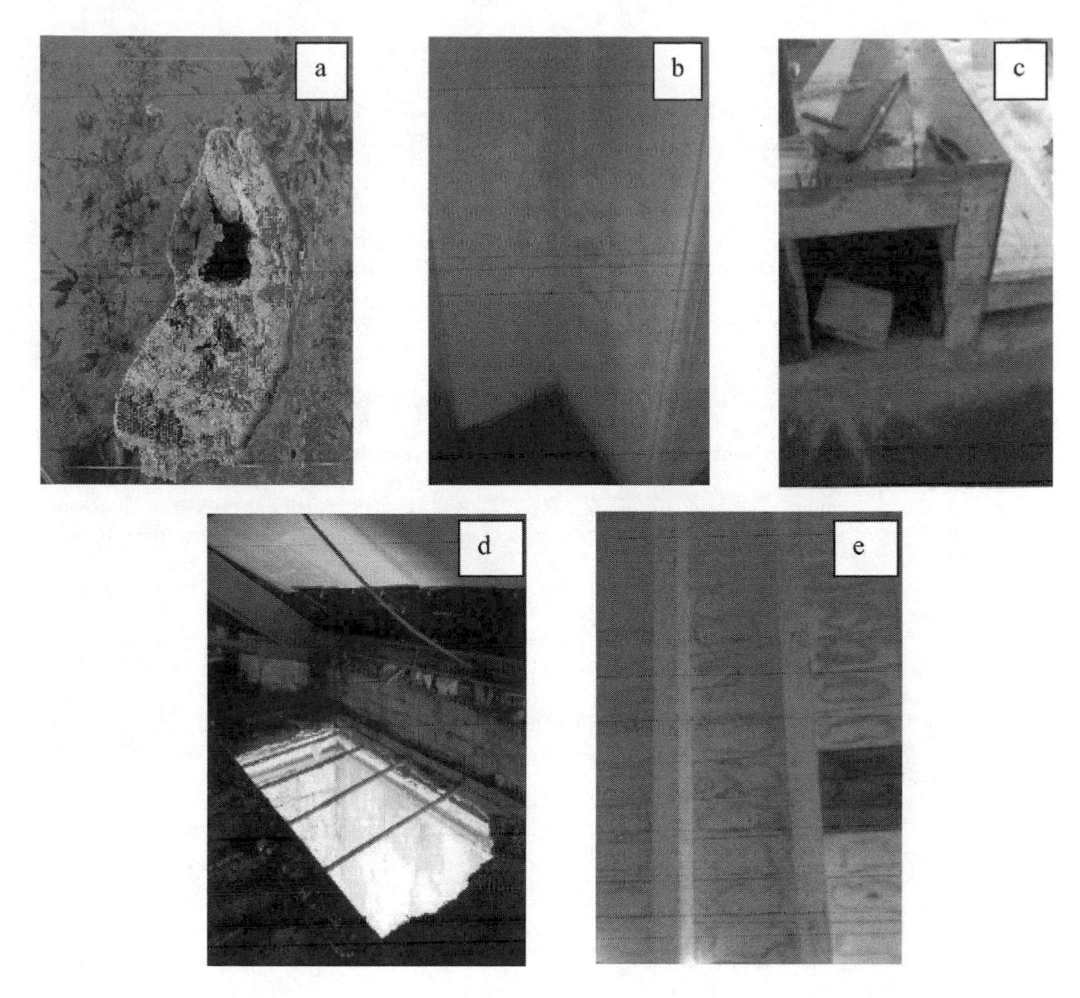

Figure 2. Main weathering forms detected inside the Terry Theater, *a, c* and *d* loss of material, *e* humidity stains and *b* saline efflorescence.

The frequency of the weathering forms allowed to calculate the expanded vulnerability index, it was equal to 37%, a value that allowed us to state that the Terry Theater had a moderate degree of vulnerability. This result reaffirmed the need to perform an intervention in a short period of time, giving high priority to building research and the mural painting of the ceiling that was very damaged and need a special restoration.

4 CONCLUSIONS

The monument presented a moderate degree of vulnerability, that recommended an intervention focused specially on the mural paintings of ceiling. These restoration works were carried out in 2018-2019. Cataloguing files and vulnerability calculation must be updated after the current intervention, and it is advisable to revise them at least every three years. Further studied must carry out to develop a new methodology to analyze wall painting, the most affected zone inside the building.

ACKNOWLEDGEMENT

This paper has been supported on the methodology developed by projects: RIV-UPH, an Excellence Project of Junta de Andalucia (code HUM-6775) and Art- Risk, a RETOS project of Ministerio de Economia y Competitividad and Fondo Europeo de Desarrollo Regional (FEDER), (code: BIA2015-64878-R (MINECO/FEDER, UE) and the project of Consejeria Fomento (code: 18.09.20.02.02). We also thank the funding of CEI CamBio and collaboration of the office of the Curator of the City of Cienfuegos and the Provincial Culture Directorate.

REFERENCES

Batista, S. 2010. Teatro Tomás Terry, un ateneo con Historia. https://cienfuegospatrimonio.wordpress.com/ 192008/02/23/teatro-tomas-terry-un-ateneo-con-historia. Consulted in february 2019.

CNR-ICR.1988. Normal 1/88. Alterazioni Macroscopiche dei Materiali Lapidei: Lessico, Istituto Centrale per il Restauro, 1–21.

Comisión Nac de Monumentos. 1978. Resolución No 3 de la Comisión Nacional de Monumentos de Cuba. http://www.planmaestro.ohc.cu/recursos/papel/documentos/resolucion3.pdf. Consulted in february 2019.

Fitzner, B. 2007. Evaluation and documentation of stone damage on monuments. *In International Symposium of Stone Conservation*, Seoul, Korea, 25–30.

Galán, E., J.B. González, and R.M. Ávila. 2006. La aplicación de la evaluación de impacto ambiental en el patrimonio monumental y el desarrollo sostenible de las ciudades. *Revista de Enseñanza Universitaria. Extraordinario* 123–140.

ICOMOS-ISCS. 2008. International Council on Monument and Sites- International Scientific Committee for Stone, Illustrated glossary on stone deterioration patterns, Champign/Marne, Ateliers 30 Impresión.

Ortiz, P., Antunez, V., Martín, J.M., Ortiz, R., Vázquez, M.A., Galán, E. 2014. Approach to environmental risk analysis for the main monuments in a historical city. *Journal of Cultural Heritage* 15: 432–440. http://dx.doi.org/10.1016/j.culher.2013.07.009.

Ortiz R, Ortiz P. 2016.Vulnerability Index: A new Approach for Preventive Conservation of Monuments. *Int J Archit Herit* 10, 1078–1100.

Sueiro, V. M 1998. Composición social y caracterización de las principales sociedades culturales y de instrucción y recreo en la región de Cienfuegos entre 1840 y 1899. Espacio, Tiempo y Forma, Serie V. H."Contemporánea", t. 11, 327–342.

Science and Digital Technology for Cultural Heritage – Ortiz Calderón et al. (Eds)
© 2020 Taylor & Francis Group, London, ISBN 978-0-367-36368-0

Vulnerability analysis of three domes and a corridor at National Schools of Arts (Cubanacan, La Habana, Cuba)

B. Rodriguez, A. Cepero & D. Abreu
Faculty of Arts of the Conservation of Cultural Heritage, University of Arts, ISA, Havana, Cuba

P. Ortiz, R. Ortiz & J. Becerra
Department of Physical, Chemical and Natural Systems, Pablo de Olavide University, Seville, Spain

ABSTRACT: National Schools of Arts (Cubanacan, La Habana, Cuba) was designed between 1961-1965 in order to create a whole space to study Art. The vulnerability of three domes and the first segment of the main corridor at Faculty of Visual Arts were studied according to the methodology developed by Ortiz and Ortiz (2016). The vulnerability indexes were calculated based on a Leopold matrix that depends on intrinsic variables and the life of the monuments. A Delphi survey was carried out by Cuban experts, to determine the influence of variables on the vulnerability. According to this survey, roofs, structure, and foundations are the variables that greatly influence on the building vulnerability. The result reproduces human reasoning to study relations between vulnerability factors, risk factors and state of conservation of the contemporary monuments. This study has allowed validating the methodology for new designs and contemporary materials as those that were used in this monument.

1 INTRODUCTION

Since the end of the last century, the concept of built heritage has been changing, not only traditional buildings and cathedrals are recognized as monuments, but also the most relevant testimonies of contemporary architecture. The recognition of the historical, artistic and cultural values of the 20th century buildings has led to an increase in the actions of conservation of this type of architecture.

The diversity of forms and structures employed in modern architecture, associated with the widespread use of different materials, sometimes of handcrafted manufacture or different qualities, makes complex the analysis of deterioration processes present in this type of works. Therefore, it becomes a challenge for professionals in this field to intervene in them, encouraging debate, which has even raised the possibility of a specific way of evaluating this architecture (Prudon, 2017)

Knowing the vulnerability of these buildings and the hazards to which they are exposed, allows us to assess the risk of deterioration for this 20th century heritage. These types of studies require the adaptation of techniques usually employed for the diagnostic of historical buildings, to the new architectural context (Macias et al., 2014 and Prieto et al., 2017). Because of that, this paper presents a new approach based on the vulnerability index for risk analysis of contemporary artworks, considering the environmental factors and the state of preservation of the building.

2 MATERIALS AND METHODS

2.1 *Case of study*

The set of buildings known as the Art Schools located in the old grounds of the Country Club of Havana constitutes one of the most outstanding examples of the Cuban architecture of the Modern Movement and a landmark of the architecture of the Revolution. This civil construction for its historical, artistic and architectural values was declared National Monument in 2010 and was included by Cuba in its indicative list to be part of the World Heritage, attending to its exceptional universal value.

The Cuban architect Ricardo Porro Hidalgo (Cuba, 1925-2014) was the general coordinator and designed the schools of Plastic Arts (where today is the Faculty of Visual Arts) and Modern Dance. Vittorio Garatti (Italy, 1927) and Roberto Gottardi (Italy, 1927-2017), were the two Italian architects who projected the schools of Ballet, Music, and the Dramatic Arts, respectively; the last is today Faculty of Theatrical Art in restoration plan in collaboration with Italian institutions.

It is a set of architecture of organic geometry, built using the Catalan vault. The traditional materials of modernist architectures such as steel and cement were replaced by brick and terra-cotta tiles. Taking into account the deterioration over the years, non-completion of some of the buildings, and the recognition of its values, the whole ensemble is the target of several initiatives for its recovery.

Ricardo Porro designed the School of Plastic Arts, inspired by the heritage of Cuban and African nations and assumed an archetypical village structure made up of series of oval – shape pavilions of various sizes, connected with curved, shaded colonnades. Three domes and one segment of corridors are chosen for this study of vulnerability index in order to develop a methodology that could be employed in the rest of the monument.

2.2 *Vulnerability Index Study*

To determine the vulnerability index of each zone, the vulnerability index (VI %) and vulnerability expanded indexes were calculated, based on a vulnerability matrix (VM) according to the methodology developed by Ortiz and Ortiz (2016), but adapted to suit the nature of this contemporary heritage in order to compare the three domes and segment of corridor mentioned. The adaptation implies to take into account the environmental conditions and the materials employed in this building and balance the influence of different deterioration agents according to the criteria of Cuban experts. Weighted factors were obtained using DELPHI process by consulting a multidisciplinary group of architects, biologists, chemists, geologist, engineer, and technician with more than 10 years of experience in conservation of buildings in Cuba. The mean value and standard deviation of the opinion of the experts regarding the influence of each variable on the vulnerability index are shown in Figure 1.

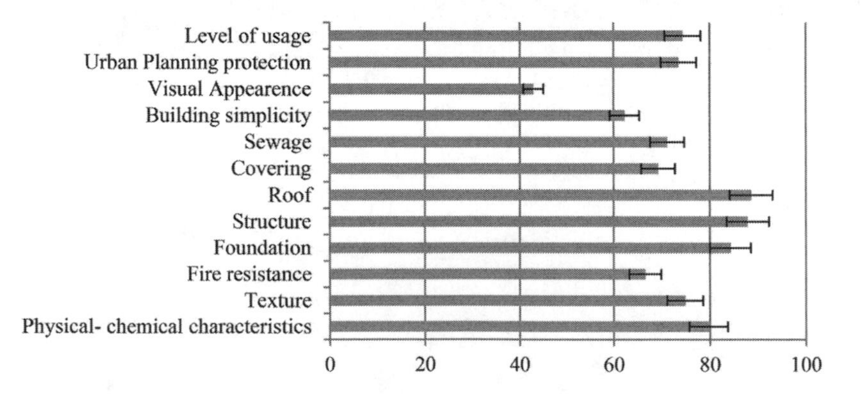

Figure 1. Factors that affect vulnerability according to average value of experts opinion.

According to the opinion of the experts, variables such as roof, structure, and foundation, with weights more than 80% have the greatest influence on the buildings vulnerability. The rest of the variables have an influence between 60 and 80% except visual appearance with a 43% influence. Each impact (matrix cell) was described with all the potential weathering forms that could be found in the domes and corridors of Faculty of Plastic Arts. Experts of the school of Art carried out the in-situ inspection of weathering forms according to CNR-ICR Normal 1/88, Fitzner (2007) and the ICOMOS-ISCS (2008) glossary. The vulnerability matrix was prepared by inserting the hazards of this areas in the rows and the building material characteristics, degree of structural conservation and anthropogenic factors in the columns of the adapted vulnerability matrix. These characteristics were included in a preliminary classification to obtain the vulnerability matrix.

According to Ortiz & Ortiz (2016), an evaluation of the frequency of weathering forms was set between 1 and 3, while the degree of weathering was classified into six relative categories adapting the categorization developed by Fitzner (2007). Frequency and damage level were combined to obtain a numerical value for the intensity of weathering forms in each monument.

After studying the weathering forms, the vulnerability index (VI) was calculated by dividing the total value of the deterioration patterns (Vx) for a monument by the sum of the total value of deterioration patterns in the worst case (vdp), when the frequency would be maximum Ortiz et al. (2014)

$$VI = \frac{Vx}{\sum_{f=3} vdp} x100 \tag{1}$$

An expanded vulnerability index was developed according to a DELPHI assessment of the influence of different characteristics in the vulnerability matrix:

$$\mathrm{VIe} = \sum \mathrm{fiVi} \tag{2}$$

where fi: is associated weighting factor according to the DELPHI forecasting; Vi: is the vulnerability associated to the variable i

Finally, the expanded vulnerability index (VIe%) was classified by degree of vulnerability using ordinal classes as described by Galán et al. (2006): very low (<10%), low (10-25%), moderate (25-50%), high (50-75%) and very high vulnerability (>75%).

3 RESULTS AND DISUSSION

The main weathering forms (frequency=3) were missing part, coloration or discoloration, scaling, moist areas, efflorescence, surface deposit, black crust and biological colonization (Figure 2). Most of this damage is associated with the environment with high humidity and salts, which imply that the external agents that cause these conditions clearly affect the conservation of the structures and buildings. Efflorescence, black crust, surface deposits and biological colonization appear abundantly, and are associated with capillarity dampness, water percolation, biological environment, the use of incompatible materials and rainfall.

Results of expanded vulnerability index are shown in table 1. Currently, the dome 1 is the most vulnerable with vulnerability index VIe average of 52% that implies a high vulnerability according to the classification of Galán et al. (2006) while the other two domes and corridor have a moderate vulnerability, with a VIe average between 37 and 42%. These results imply that the state of conservation of Dome 1 makes it more vulnerable to extrinsic factors than the other areas.

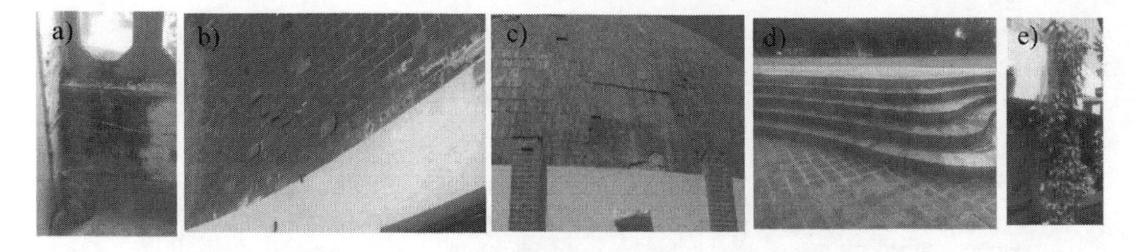

Figure 2. Weathering forms found during the in-situ studies. *a*) biological colonization, *b)* efflorescence, scaling and coloration or discoloration, *c*) missing part *d)* black crust and *e*) plant.

Table 1. Expanded vulnerability index (VIe) and priority based on level of vulnerability.

Area	VIe (%)	Priority based on level of vulnerability*
Dome 1	52%	High priority for research; short-term mitigation strategy is recommended; cost-benefit analysis of the mitigation strategy is recommended.
Dome 2	42%	Prioritize by cost-benefit analysis of mitigation strategies.
Dome 3	37%	Prioritize by cost-benefit analysis of mitigation strategies.
Corridor	36%	Prioritize by cost-benefit analysis of mitigation strategies.

**Source: based on ICCROM–CCI–ICN (2007) for risk and uncertainty*

Priorities defined by ICCROM-CCI-ICN for the conservation must consider the magnitude of risk and uncertainty. Table 1 shows the valuation of the index of vulnerability combined with the feasibility and costs of risk reduction according to Ortiz & Ortiz (2016), and include the magnitude of the vulnerability. Nevertheless, we must understand that in this analysis, three domes and one corridor of the whole monuments was compared to validate the diagnosis method to the whole due to the complexity of this 20th Century monument. Further analysis must be carried out to analyze the whole building.

4 CONCLUSIONS

In this study, the domes and corridor showed different level of conservation that correspond from moderate to highly vulnerable. The most vulnerable dome studied is Dome 1, which is highly vulnerable due to appreciable detachment of the roof, abundant missing parts and chromatic alteration due to the alteration of bricks, or efflorescence, black crusts, vegetation, and signs of previous intervention with apparently incompatible bricks, while the other two domes and corridor show moderate vulnerability.

The novelty of this approach is a multidisciplinary approach that includes the analysis of vulnerability in contemporary architectural heritage and the analysis of environmental factors around the monuments, made by in situ experts and the application of the Delphi method by experts in the Cuban-built heritage to determine the influence of each variable on the vulnerability of the buildings. Further studies are recommended to improve the methodology to study the whole monument and the complexity of the relationships between the domes, corridor and landscape.

ACKNOWLEDGEMENT

This paper has been supported and based on the Methodology developed by Projects: Art-Risk, a RETOS project of Ministerio de Economía y Competitividad and Fondo Europeo de Desarrollo Regional (FEDER), (code: BIA2015-64878-R (MINECO/FEDER, UE) and project of Consejería de Fomento, Infraestructura y Adecuación de Territorio (code: UPO-03). We also thank the funding of CEI CamBio. Pilar Ortiz thanks Salvador Madriaga funds for her research stay in the University of Oxford (2017).

REFERENCES

Fitzner B. 2007. Evaluation and documentation of stone damage on monuments, International Symposium of Stone Conservation, Seoul, Korea, pp. 25–30.

Galán E, Gonzalez JB, Ávila. RM. 2006. La aplicación de la evaluación de impacto ambiental en el patrimonio monumental y el desarrollo sostenible de las ciudades, *Rev Enseñanza Univ Extraordin*, 123–140.

ICOMOS, ISCS. 2008. Illustrated glossary on stone deterioration patterns, http://www.icomos.org/publications/monuments_and_sites/15/pdf/Monuments_and_Sites_15_ISCS_Glossary_Stone.pdf

ICCROM -CCI –CIE. 2007. Curso de evaluación de riesgos ICCROM-CCI-CIE Istituto Centrale per il Restauro, 1988. CNR-ICR Normal 1/88. Alterazioni Macroscopiche dei Materiali Lapidei: Lessico, 1–21.

Macías-Bernal JM, Calama JM, Chávez. MJ. 2014. Modelo de predicción de la vida útil de la edificación patrimonial a partir de la lógica difusa, *Inf la Constr* 66, e006.

Ortiz P, Antunez V, Martín JM, et al., 2014. Approach to environmental risk analysis for the main monuments in a historical city, *J Cult Herit* 15, 432–440.

Ortiz R, Ortiz P. 2016. Vulnerability Index: A New Approach for Preventive Conservation of Monuments, *Int J Archit Herit* 10, 1078–1100.

Prieto AJ, Macías-Bernal JM, Chávez M-J, et al. 2017. Fuzzy Modeling of the Functional Service Life of Architectural Heritage Buildings, *J Perform Constr Facil* 31, 04017041.

Prudon T.2017. Preservation, design and modern architecture: The challenges ahead, *J Archit Conserv* 23, 2017, pp. 27–35.

Science and Digital Technology for Cultural Heritage – Ortiz Calderón et al. (Eds)
 ISBN 978-0-367-36368-0

Indicators, sustainability and vulnerability in the diagnosis of historical cities

J. Benítez, R. Ortiz & P. Ortiz
Department of Physical, Chemical and Natural Systems, University Pablo de Olavide, Seville, Spain

ABSTRACT: The objective of this work is to study sustainability indicators, its relationship with vulnerability of buildings and their application in the diagnosis of historical cities. Different organizations have developed a large number of indicators for cities to assess sustainability at a social, economic and environmental level, so it seems interesting to stablish the relationship with the vulnerability of monuments in historical cities, because of culture was not considered as a main variable of sustainability until recent studies. As a result, 31 sustainability indicators were obtained, grouped according to the Pressure – State – Response model and related to each other according to how they affect the vulnerability of buildings.

1 INTRODUCTION

One of the most important moments, since the founding of the United Nations, was the 1992 World Conference on Environment, from which Agenda 21 emerged, a program to help governments make decisions related to sustainable development in cities (United Nations, 2007). Since then, different indicators, indicator systems and models have been developed to study sustainable development (Braulio-Gonzalo et al., 2015).

The Pressure – State – Response (PSR) model is the most used scheme for sustainability studies. This model was developed by the Organization for Economic Co-operation and Development in the nineties and classifies the indicators into 3 levels based on: the pressure exerted in the environment, the real situation at a given moment, and the response to changes that takes place (OECD, 2003).

Nowadays, international organizations insist on the need to take action, and develop adequate indicators that evaluate sustainability from all angles, including cultural heritage (Nocca, 2017).

2 METHODOLOGY

Sustainability has been studied taking into account social, economic and environmental aspects, because there are a large amount of sustainability indicators associated with these that influence the vulnerability of monuments (Di Turo et al., 2016).

The importance and influence of culture and historical heritage in the economy, society and the environment have been increasing over the years (Figure 1). Now we know that culture is fundamental to achieve adequate sustainable development and has been incorporated into sustainability studies due to the influence it exerts on the economy, society and the environment, considered the basic pillars of sustainability (Tweed and Sutherland, 2007).

The objective of this work is to study sustainability indicators, its relationship with vulnerability of buildings and their application in the diagnosis of historical cities. In this study we have analyzed sustainability indicators present in different programs, systems special plans, etc., applied in different countries and scales, in order to get a relationship between the information that provide indicators and the vulnerability of monuments in a historical city. The

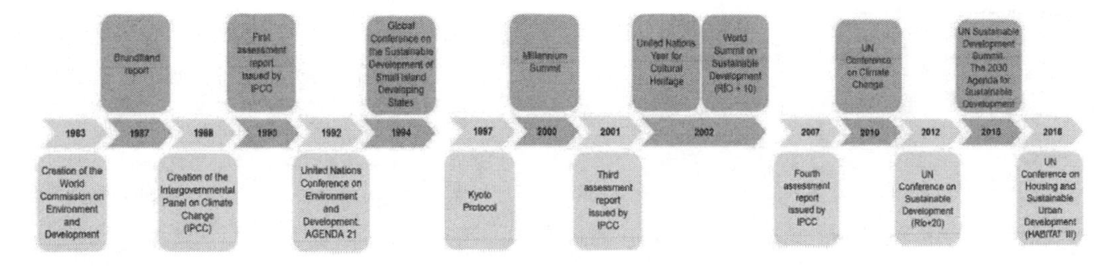

Figure 1. Main meetings of the United Nations related to sustainable development and cultural heritage.

Figure 2. Indicators grouped by topics: environment, society, urban planning, economy and cultural heritage.

final list has been grouped according to the aspects that each indicator deals (Figure 2); environment, society, urban planning, economy and cultural heritage.

The indicators founded (more than 500) have been classified according to the Pressure – State – Response model. It is expected that pressure indicators have a negative influence on vulnerability, increasing it; state indicators will increase or decrease the vulnerability depending on the situation of the city is at the time of the study; finally response indicators reduce vulnerability, because they bring changes to unfavorable situations. Relationships have been established between the 3 groups of indicators using a three-color code: red, negative; green, positive; and yellow, when the relationship can be positive or negative depending on the values of each indicator.

3 RESULTS

The Figures 3-5 show the relationships established between the different groups of indicators. The largest number of interactions are generated between pressure and state indicators. These interactions can be unidirectional, bidirectional, positive or negative, being this the most numerous.

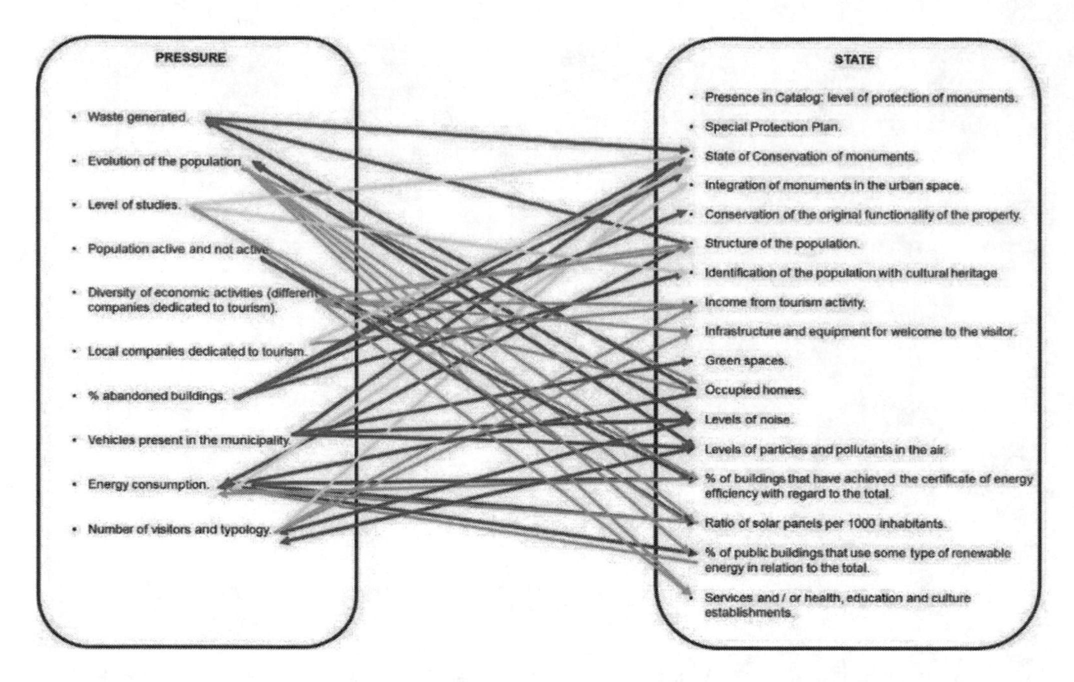

Figure 3. Relationship between Pressure and State indicators: red, negative; green, positive; and yellow, when the relationship can be positive or negative.

Among the pressure indicators (Figure 3), percentage of abandoned buildings, energy consumption and vehicles present in the municipality are those that may generate higher negative

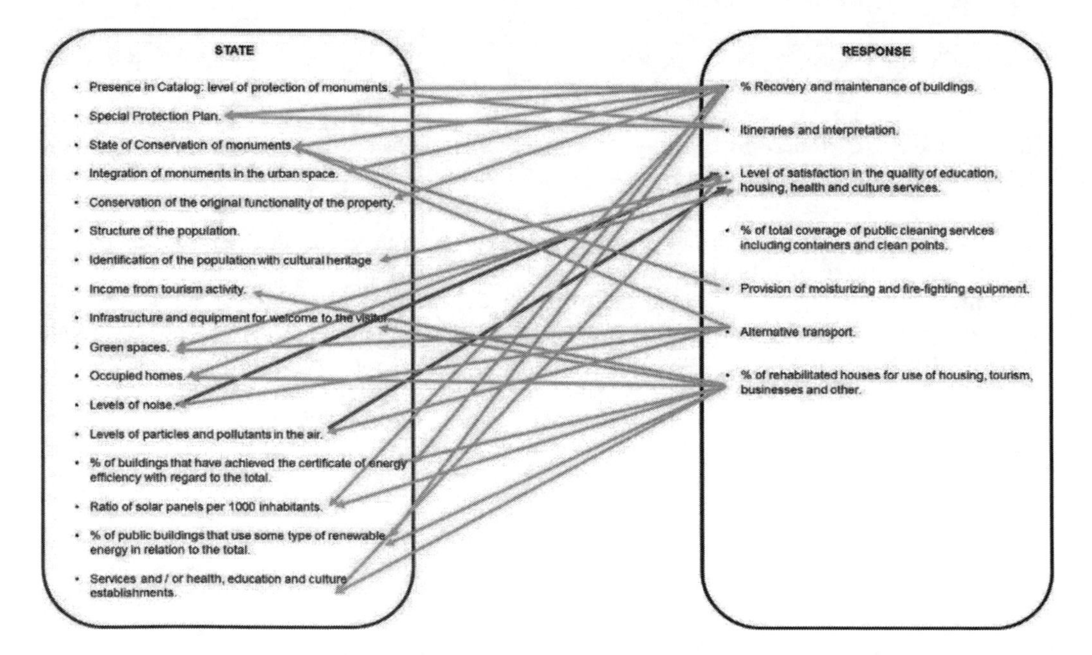

Figure 4. Relationship between State and Response indicators: red, negative and green, positive.

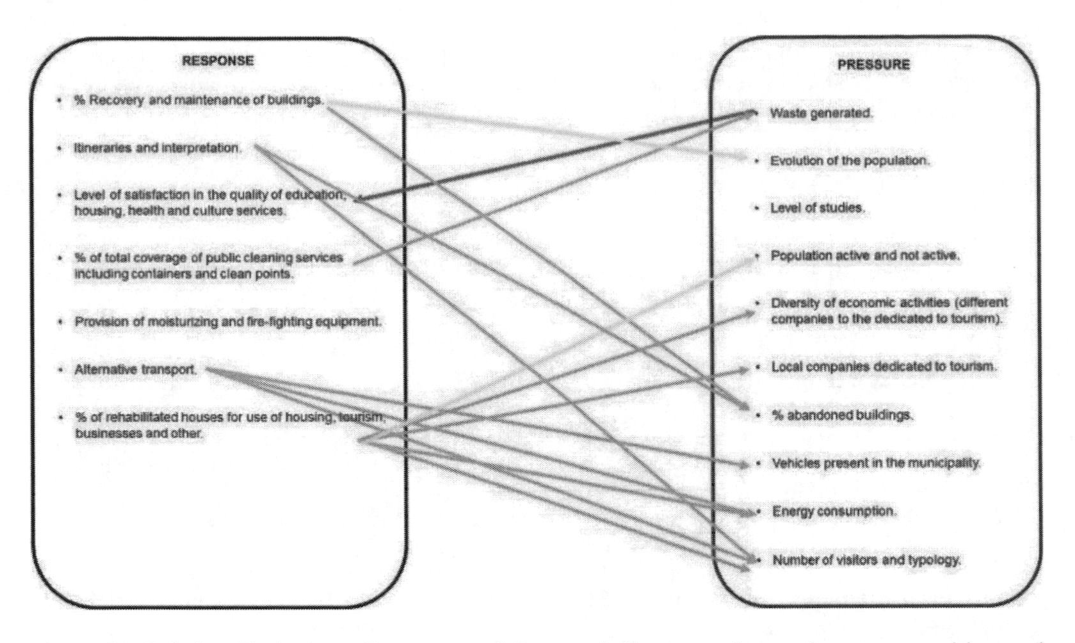

Figure 5. Relationship between Response and Pressure indicators: red, negative; green, positive; and yellow, when the relationship can be positive or negative.

impacts for vulnerability of heritage buildings in short term; while number of visitors and typology or evolution in population might influence in long term development.

In the group of state indicators (Figure 4), green areas and use of renewable energy may have positive impacts for sustainable development; while levels of noise, levels of particles and pollutants in the air and the structure of population may generate negative impacts for vulnerability.

Response indicators (Figure 4-5) generate positive interactions in most cases, reduce unfavorable situations for the city and counteract indicators that exert pressure.

For instance, the existence of alternative transport, might decrease the vehicles present in the municipality as well as levels of particles and pollutants in the air.

Between these indicators (Figure 5), recovery and maintenance of buildings, alternative transport and rehabilitated houses for the use of housing, tourism, businesses and other, are mainly positive for vulnerability and development, because of they might reduce unfavorable situations for the city.

4 CONCLUSIONS

Sustainability indicators have been studied and filtered in relationship with the vulnerability of buildings. According to the preliminary results, indicators can be a valuable source of information on the vulnerability of buildings and monuments, as they provide data on the state of these and the surrounding environment. The recovery and maintenance of buildings, alternative transport and rehabilitated housing could have a positive influence on the vulnerability of monuments, reducing the effects of pressure indicators, and helping to improve the degree of sustainable development in the cities. On the other hand, abandoned buildings, energy consumption and vehicles present in the municipality are the indicators that exert the greatest pressure and may generate a negative influence on the vulnerability of monuments.

The relationship established need further studies to assess the vulnerability of buildings and the degree of sustainable development of historical cities, for that reason indicators will be evaluated by a group of experts using the Delphi technique.

ACKNOWLEDGMENTS

This paper has been supported and based on the Methodology developed by Art-Risk, a RETOS Project of Ministerio de Economía y Competitividad and Fondo Europeo de Desarrollo Regional (FEDER), (code: BIA2015-64878-R (MINECO/FEDER, UE)).

REFERENCES

Organization for Economic Co-operation and Development (OECD). 2003. OECD environmental indicators: development, measurement and use. Paris. http://www.oecd.org/env/indicators-modelling-outlooks/24993546.pdf.

United Nations. 2007. Indicators of Sustainable Development: Guidelines and Methodologies. United Nations publication. https://sustainabledevelopment.un.org/index.php?.

Tweed, C., Sutherland, M. Built cultural heritage and sustainable urban development. 2007. Landscape and Urban Planning, vol. 83, pg. 62-69. https://www.sciencedirect.com/science/article/pii/S0169204607001442.

Braulio-Gonzalo, M., Bovea, M.D., Ruá, M.J. 2015. Sustainability on the urban scale: Proposal of a structure of indicators for the Spanish context. Environmental Impact Assessment Review, vol. 53, pg. 16-30. https://www.sciencedirect.com/science/article/abs/pii/S0195925515000311.

Di Turo, F., Proietti, C., Screpanti, A., Fornasier, M. F., Cionni, I., Favero, G., De Marco, A. 2016. Impacts of air pollution on cultural heritage corrosion at European level: What has been achieved and what are the future scenarios *. Environmental Pollution. 218. 10.1016/j.envpol.2016.07.042. https://www.sciencedirect.com/science/article/pii/S0269749116306170?via%3Dihub.

Nocca, F. 2017. The Role of Cultural Heritage in Sustainable Development: Multidimensional Indicators as Decision-Making Tool. Sustainability vol. 9, 1882. https://www.mdpi.com/2071-1050/9/10/1882.

Science and Digital Technology for Cultural Heritage – Ortiz Calderón et al. (Eds)
© 2020 Taylor & Francis Group, London, ISBN 978-0-367-36368-0

Interlaboratory experience to evaluate the vulnerability of churches in Seville (Spain)

A. Tirado-Hernández, R. Ortiz, P. Ortiz & J. Becerra
Department of Physical, Chemical and Natural Systems, Pablo de Olavide University, Seville, Spain

R. Radvan, A. Chelmus & L. Ratoiu
National Institute for Research and Development in Optoelectronics, Măgurele, Romania

ABSTRACT: Ground Penetration Radar (GPR), thermovision and thermohygrometrics records were used to assess the conservation degree of 5 Churches located in Seville (Spain). The results were compared with the vulnerability index evaluated by DELPHI methodology in order to carry out an interlaboratory diagnosis and the validation of this multidisciplinary approach. The GPR was employed to detect structural damages on the ground floors of buildings. Thermovision was carried out to assess active dampness, while sensors of humidity and temperature were located to evaluate microclimate inside the churches. The diagnosis carried out by GPR and thermovision has not detected structural damage according to the vulnerability indexes. Infrared images do not evidence important active water movement on roofs, while vulnerability methodology allows to assess the weathering forms due to dampness by capillary and percolation, and the use of buildings. The combination of vulnerability indexes and these non-destructive technologies provide very useful results to evaluate and prioritize preventive conservation strategies.

1 INTRODUCTION

In order to facilitate the comparison of the conservation degree, historical buildings with similar constructive date and constructive material have been chosen. Five churches from 13th -14th centuries located in the historical center of Seville (Spain) (Figure 1) were studied by ground penetration radar, thermovision and thermohygrometrics sensors. The conservation degrees obtained by DELPHI methodology (Ortiz & Ortiz, 2016) were compared with techniques employed in situ in order to validate the multidisciplinary approach and to make an interlaboratory exercise. The DELPHI method consists in the selection of a group of experts who are asked about issues related to future events. In this case, the experts were asked about the vulnerability of the churches.

2 METHODOLOGY

2.1 *Ground Penetration Radar (GPR)*

The GPR is a non-invasive geophysical method used for in depth investigations by emitting and studying the propagation of electromagnetic pulses to highlight the heterogeneity or discontinuities of the electrical properties of the propagation medium (Daniels, 2000). The reflected energy was recorded and it was viewed as radargrams. This technique was employed to detect structural damages on the ground floor of buildings. In all five churches the measurements were conducted using a perpendicular grid ensuring a better coverage of the surface. The different amount of records is due to the churches size and available space to do the

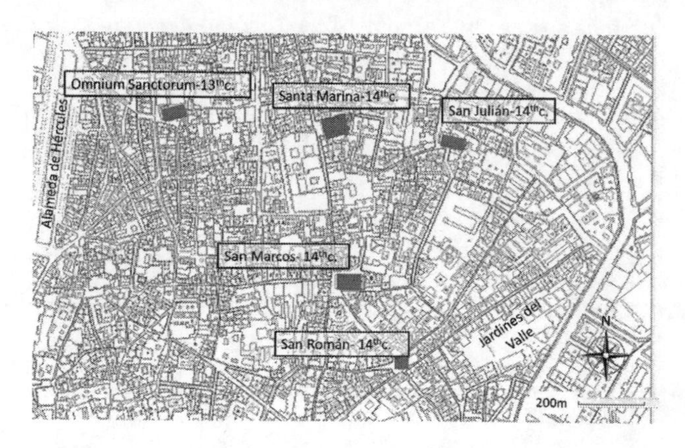

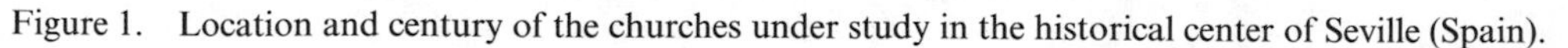

Figure 1. Location and century of the churches under study in the historical center of Seville (Spain).

studies. Mala X3M GPR system and 500 MHz and 800 MHz antennas were used for the data acquisition.

2.2 *Thermovision*

The buildings thermal anomalies are captured through the use of infrared digital sensors, aimed at defining judgments about the building envelope quality and related energy losses (Cardinale et al. 2015). Thermography reveals the distribution of surface temperature and temperature discontinuities for similar materials and conditions. Thermovision was carried out on walls at different heights and on roof to assess active dampness thanks to OPTRIS PI LightWeight Kit. The IR images were taken one day in autumn at different hours of the day.

2.3 *Microclimate study*

The thermo-hygrometric sensors, HOBO U23-002 Pro v2 logger, were located inside Omnium Sanctorum and San Marcos Churches. The data loggers have been programmed to take values of temperature, relative humidity and dew point every hour, 24 hours a day during one year. Two sensors have been placed in each church as far as possible in the horizontal plane and at different heights to analyze if there are changes in environmental conditions at different orientations and heights.

2.4 *Expanded vulnerability index (Vie%)*

The vulnerability indexes (VI %) were calculated, based on a Leopold matrix that depends on intrinsic variables and the life of the monuments (Ortiz & Ortiz, 2016). The influence of different deterioration agents was balanced with a Delphi forecast based on the experience of architects. The Vie% allows to weigh the variables so that those that affect the structure have priority over those that affect the aesthetic values or the materials.

3 RESULTS

3.1 *Ground Penetration Radar (GPR)*

Different stratigraphies were detected in each church. The presence of buried objects or structures was detected whereas structural damages were not identified in any church. The Table 1 shows the results of GPR in the churches under study.

Table 1. GPR results in the churches under study.

Churches/ No of radargrams	GPR observations
Omnium Sanctorum/82	– Inhomogeneity till the depth of 0.5-0.6m. – Some reflections due the presence of columns. – Some hyperbolic reflection indicating a buried structure or object at 1.5m approx. – In a radargram at the depth of 2.2m a reflection can be seen, which marks the change of the propagation medium.
San Julián/40	– The soil has an inhomogeneity till the depth of 0.6m. – In all the radargrams at 0.3m, change of the propagation medium which can be an old foundation or different layer.
San Marcos/58	– Inhomogeneity till the depth of 0.4m. – At the depth of 0.6m and 1.3m a buried object or structure is found. – Reflections can be seen till 1m in the columns areas which can be interpreted as foundation.
San Román/78	– Propagation medium inhomogeneous till the depth of 0.5m. – At the depth of 4.4m, presence of columns, maybe a crypt is present below the floor. – At the depth of 0.5 and 1m existence of buried structures/objects. – In the center of the church, at 1.7m, columns can be detected.
Santa Marina/76	– Propagation medium is inhomogeneous till the depth of 0.5m. – At 1.5m and 4.4m, presence of columns, maybe a crypt is present below the floor. – Reflections can be seen at 0.3-0.8 m due to the column foundation. – At 0.6m and 1.1m a buried object or structure is found.

3.2 *Thermovision*

Regarding thermovision, the temperature varies around 7 degrees maximum, between 23 and 30°C, approximately. No anomalies, structural damages, water percolation or loss of tiles were detected (Figure 2) so the results obtained by thermography showed general good conservation conditions of roofs. However, dampness by capillarity was detected in Omnium Sanctorum (Figure 2f) and Santa Marina churches.

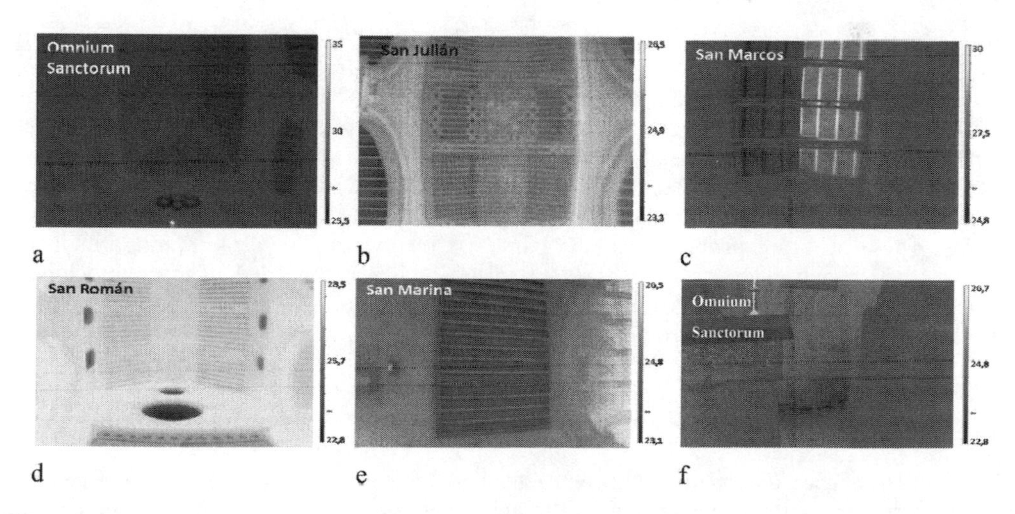

Figure 2. IR images of the roofs of Omnium Sanctorum (a), S. Julián (b), S. Marcos (c), S. Román (d) and Sta. Marina (e) churches and dampness by capillarity on walls in Omnium Sanctorum Church (f).

Table 2. Temperature, relative humidity and dew point recorded in Omnium Sanctorum and San Marcos churches.

Churches	Sensor location	Temp. average	Temp. max.	Temp. min.	RH% average	RH% max.	RH% min.	DewPt average	DewPt max.	DewPt min.
Omnium	High	20.7	32.3	11.5	59.2	87.3	29.5	12.4	21.6	2.1
Sanctorum	Low	20.1	30.5	11.6	60.5	83.1	30.6	12.2	21.0	2.0
San	High	20.6	31.9	12.0	60.4	81.8	27.6	12.5	21.5	2.6
Marcos	Low	20.3	31.7	12.7	66.6	88.9	29.0	13.8	22.6	2.3

Table 3. Expanded Vulnerability Index in the churches under study.

Churches	VIe%	Vulnerability degree
Omnium Sanctorum	36	Moderate
San Julián	29	Moderate
Santa Marina	26	Moderate
San Marcos	16	Low
San Román	11	Low

3.3 *Microclimate study*

Regarding the microclimatic study, the thermo-hygrometric measurements throughout one year period (from July 2017 to June 2018) (Table 2) showed variations of temperature (32°C-11°C) and relative humidity (89%-29%) that depends on climate and the use of the churches.

In Omnium Sanctorum and San Marcos churches the average temperature is higher in the upper part of the temples while the average relative humidity is higher in the lower part of the churches. San Marcos church is the one that presents major variations of humidity. The high relative humidity and the difference between maximum and minimum increase the vulnerability of buildings, because of efflorescence, crystallization and dissolution can be activated.

3.4 *Expanded vulnerability index*

Regarding the vulnerability, Omnium Sanctorum has the highest Expanded Vulnerability Index (Table 3), that corresponds to a moderate vulnerability degree.

In all the churches, loss of material was associated to the use of the building and its maintenance. Other weathering forms, observed to a greater or lesser extent in all the studied churches, are produced by percolation, such as concretions and moist areas; and/or by capillarity such as dampness, efflorescences, spalling or disintegration. Moisture and efflorescence are especially important in Omnium Sanctorum and Santa Marina according to thermovision records. These moisture problems can be increased by the lack of cross ventilation in all churches, which only happens when they are in use as it was recorded in the microclimate studies. San Julian vulnerability index may be increased due to structural and foundation problems apart from capillary problems.

4 CONCLUSIONS

The diagnosis carried out on ground floor, walls and roof by GPR and thermovision has not detected structural damage according to the vulnerability indexes. Roof infrared images do not evidence important active water movement on roofs, while capillarity dampness was detected on Omnium Sanctorum and Santa Marina. The vulnerability methodology allows to assess the weathering forms due to dampness by capillarity and percolation, and the use of buildings, according to thermohygrometric records. The combination of vulnerability indexes and non-destructive technologies provide very useful results to evaluate and prioritize preventive conservation strategies, highlighting the importance of ventilation and roof conservation. Further studies of interlaboratory experience must be carried out to combine a monitoring of vulnerability with non-destructive techniques for a continuous evaluation.

ACKNOWLEDGEMENTS

This paper has been supported by ART-RISK, a retos project of Ministerio de Economía y Competitividad and Fondo Europeo de Desarrollo Regional (FEDER), (code: BIA2015-64878-R (MINECO/FEDER,UE)) and the research team PAI-TEP-199 from Junta de Andalucía. Infrastructure assured by Program OPTRONICA & INOVA OPTIMA.

REFERENCES

Cardinale, T. Balestra, A. & Cardinale, N. 2015. Thermographic mapping of a complex vernacular settlement: the case of study of Casalnuovo District within the Sassi of Matera (Italy). *Energy Procedia*, 76, 40–48.

Daniels, J. 2000. Ground Penetrating Radar Fundamentals. DOI: 10.4133/1.2921864.

Ortiz, R. & Ortiz, P. 2016. Vulnerability index: a new approach for preventive conservation of monuments. *International Journal of Architectural Heritage*, Vol. 10 (8), 1078–1100. DOI: 10.1080/15583058.2016.1186758.

Science and Digital Technology for Cultural Heritage – Ortiz Calderón et al. (Eds)
© 2020 Taylor & Francis Group, London, ISBN 978-0-367-36368-0

Author index